AWS Certified Cloud Practitioner Exam Guide

Build your cloud computing knowledge and build your skills as an AWS Certified Cloud Practitioner (CLF-C01)

Rajesh Daswani

Pack<t>

BIRMINGHAM—MUMBAI

AWS Certified Cloud Practitioner Exam Guide

Copyright © 2021 Packt Publishing

Group Product Manager: Rahul Nair
Publishing Product Manager: Preet Ahuja
Senior Editor: Sangeeta Purkayastha
Content Development Editor: Nihar Kapadia
Technical Editor: Nithik Cheruvakodan
Copy Editor: Safis Editing
Project Coordinator: Neil Dmello
Proofreader: Safis Editing
Indexer: Tejal Daruwale Soni
Production Designer: Shankar Kalbhor

First published: December 2021

Production reference: 2020623

Published by Packt Publishing Ltd.
Livery Place
35 Livery Street
Birmingham
B3 2PB, UK.

ISBN 978-1-80107-593-0

www.packt.com

To my mother, Vandana, and to the memory of my father, Devkrishin,
for their sacrifices and for exemplifying the power of determination.
To my beautiful wife, Divya, for being my loving partner and anchor
always. And to my amazing daughter, Ryka, for showing me how simplicity
is the key to creativity.

– Rajesh Daswani

Contributors

About the author

Rajesh Daswani is a senior solutions architect with over 20 years' experience in IT infrastructure services and cloud computing. His work has been focused on both AWS and Microsoft 365 platforms. He has delivered cloud computing training on AWS as a corporate trainer for several clients globally.

Rajesh has helped thousands of IT professionals to appreciate real-world applications of cloud technologies and become better equipped to facilitate clients' adoption of cloud technologies.

When Rajesh is not immersed in the world of cloud computing, he can be caught watching re-runs of his favorite Star Trek shows (TNG). He also likes to be known as a food connoisseur, conjuring up his own mix of fusion dishes for his family and friends.

I would like to thank the team at Packt Publishing, especially Neil D'mello and Nihar Kapadia, for guidance and support in making this study guide a reality. I would also like to thank Renato Martins, for providing a technical review of this study guide.

About the reviewer

Renato Martins has been working in IT since 1997, having worked directly for, or consulted on behalf of, small start-ups, government agencies, and large multinationals in four countries. As a trainer, he has delivered hundreds of classes and a few dozen lectures on all topics, from development to databases, operating systems, and cloud computing in all flavors.

He has been approved in more than 200 certification exams, including AWS, Microsoft, PMI, Red Hat, Sun, and IBM, among others.

When not working, he loves to be with his family, enjoying museums and concerts, and likes to travel across the globe in whatever spare time he has.

Passionate about the fast-paced, constantly changing landscape of technology, he is always on the lookout for the next new thing, eager to learn and share it.

I'd like to thank my lovely wife, Sandra, and our incredible children, Theo and Laís, for their support, knowing I might become absent at any time as a result of diving into a big project, studying for the next exam, or delivering a class. Thanks for knowing that after, I am there to have the best moments of our lives together.

Thanks for the relaxing moments, for all the laughs and cuddles.

A big thank you also to my dog, Ted, who makes sure I am awake every day at 6 A.M. for a walk!

Table of Contents

3
Exploring AWS Accounts, Multi-Account Strategy, and AWS Organizations

Section 2: AWS Technologies

4
Identity and Access Management

5

Amazon Simple Storage Service (S3)

6

AWS Networking Services – VPCs, Route53, and CloudFront

7

AWS Compute Services

8
AWS Database Services

9

High Availability and Elasticity on AWS

10

Application Integration Services

11

Analytics on AWS

12
Automation and Deployment on AWS

13
Management and Governance on AWS

Section 3: AWS Security

14

Implementing Security in AWS

Section 4: Billing and Pricing

15
Billing and Pricing

16
Mock Tests

Answers

Other Books You May Enjoy

Index

Preface

Amazon Web Services (**AWS**) is the leader in cloud computing and has successfully maintained that position for many years now. Companies across the globe, in both the public and private sectors, continue to embrace cloud technologies at an ever-increasing rate, and the demand for IT professionals with experience of AWS far outstrips supply.

The *AWS Certified Cloud Practitioner Exam Guide* is your one-stop resource to help you prepare for one of the most popular AWS certification exams. This study guide will help you validate your understanding of the core cloud computing services offered by AWS as well as learn how to architect and build cloud solutions for your clients.

Passing this foundation certification doesn't require you to know how to build and deploy complex multi-tier application solutions. However, this book gives you the necessary skills to understand how the various services offered by AWS can be used to architect end-to-end solutions for your client. In addition, you will learn about cloud economics, security concepts, and best practices.

This study guide is designed to be used even after you pass the AWS Certified Cloud Practitioner exam, with multiple exercises that can be used as a reference to help you start building real-world solutions for your clients. Each chapter builds on the previous one, with each new exercise extending a previously configured service. This will allow you to determine how the individual services offered by AWS can be used to design end-to-end solutions.

Each chapter also includes a summary followed by a set of review questions. The book ends with two mock tests that are designed to test your knowledge and help you further prepare for the official AWS Certified Cloud Practitioner exam.

Who this book is for

This study guide is designed for anyone looking to fast-track their career in cloud computing. The *AWS Certified Cloud Practitioner* Exam Guide is designed for both IT and non-IT professionals who wish to learn the fundamentals of cloud computing and the AWS offering. Non-IT professionals can benefit from this study guide as they learn not only the theoretical concepts of a wide range of cloud services offered by AWS, but also gain valuable experience in configuring those services from a technical perspective. IT professionals who have primarily worked in on-premises environments will learn how to provision and deploy technical solutions in the cloud and understand strategies for migrating their on-premises workloads to AWS.

What this book covers

Chapter 1, What Is Cloud Computing?, discusses the fundamentals of cloud computing, and outlines the six advantages of cloud computing. We also look at the various cloud computing models and cloud deployment models.

Chapter 2, Introduction to AWS and the Global Infrastructure, introduces you to the AWS ecosystem, its global infrastructure, and how it enables you to start deploying solutions on a global scale. We discuss the excellent support services offered by AWS and the importance of choosing the right support plan.

Chapter 3, Exploring AWS Accounts, Multi-Account Strategy, and AWS Organizations, details the concept of AWS accounts, which enable you to access the vast array of AWS services securely. We also look at the use case for setting multiple AWS accounts and the best practices to follow to manage multiple accounts using the AWS Organizations service.

Chapter 4, Identity and Access Management, introduces you to one of the core fundamental security features of AWS. AWS **Identity and Access Management** (**IAM**) enables you to manage access to AWS services and resources securely. With AWS IAM, you can design policies and permissions to ensure access and authorization to services using the principle of least privilege.

Chapter 5, Amazon Simple Storage Service (S3), explains how AWS offers a wide range of different storage options, including block, object, and file storage services. Amazon S3 is AWS's object storage solution and in this chapter, you learn how to harness the power of this virtually unlimited and highly scalable storage offering from Amazon.

Chapter 6, AWS Networking Services – VPCs, Route53, and CloudFront, covers networking services in the cloud. Amazon **Virtual Private Cloud** (**VPC**) enables you to launch AWS resources in a logically isolated virtual network in the cloud. We also look at AWS's **Domain Name System** (**DNS**) offering, which enables you to register new domain names and design traffic routing services for your workloads. Finally, in this chapter, we look at Amazon CloudFront, which helps you design a content network delivery service for your digital assets and applications.

Chapter 7, AWS Compute Services, details the various compute services on offer from AWS. These include the **Elastic Compute Cloud** (**EC2**) service, which enables you to launch Linux, Windows, and macOS virtual servers in the cloud, through to containers and serverless compute offerings such as Amazon Lambda. In this chapter, you will also learn about block storage and file storage services on AWS.

Chapter 8, AWS Databases Services, examines the wide range of database solutions on offer from AWS, capable of supporting almost any use case. From traditional relational database services such as Amazon RDS through to NoSQL databases solutions such as Amazon DynamoDB, we examine their use cases and learn how to configure these databases for your applications. Additional niche database solutions are also discussed in this chapter.

Chapter 9, High Availability and Elasticity on AWS, covers one of the fundamental benefits of cloud computing and, specifically, AWS. Designing solutions that are highly available and capable of withstanding outages is of paramount importance for any organization and, in this chapter, you will learn how to use the tools to build highly available solutions. In addition, we also discuss how you can automatically scale your application, expanding your resources when demand increases and terminating them when demand drops. This enables you to manage costs much more effectively and avoid the need to guess capacity.

Chapter 10, Application Integration Services, examines various AWS services that enable you to build applications that adopt a decoupled architecture design. This enables you to move away from traditional monolithic design in favor of the more modern microservice architectures.

Chapter 11, Analytics on AWS, examines the vast array of tools and services on AWS that can help you analyze the massive amounts of data that organizations collect, much of which is collected in real time.

Chapter 12, Automation and Deployment on AWS, looks at several automation tools and processes to help you deploy infrastructure and applications that not only speed up the deployment process but also reduce configuration errors.

Chapter 13, *Management and Governance on AWS*, examines several AWS services that can be used to monitor your resources, manage them centrally, and help you follow best practices.

Chapter 14, *Implementing Security on AWS*, outlines the wide range of security tools, services, and processes offered by AWS that can help you design your application solutions, with security being at the forefront, thereby enabling you to adhere to any compliance and regulatory environments and ensure that your customers' data is always protected.

Chapter 15, *Billing and Pricing*, discusses cloud economics and examines the vast array of AWS tools to help you manage your cloud computing costs effectively. We discuss strategies for minimizing costs without compromising performance, reliability, and security.

Chapter 16, *Mock Tests*, enables you to test your knowledge acquired throughout this study guide by undertaking two complete practice exams. These mock tests will help you gauge your readiness to take the official AWS certification exams and also provide answer explanations to help you prepare for the AWS exams.

To get the most out of this book

To get the most out of this book, you must follow the chapters in the order in which they have been presented. This is because each new chapter builds on the previous one. In addition, it is highly recommended that you gain the necessary practice experience by completing all the exercises in this book.

Software/hardware covered in the book	OS requirements
OS	Windows, macOS X, or Linux (any).
AWS account	You will need to set up an AWS Free Tier account.
Tools	You will need access to the AWS CLI and PuTTY tools (for Windows users).

If you are using the digital version of this book, we advise you to type the code yourself or access the code via the GitHub repository (link available in the next section). Doing so will help you avoid any potential errors related to the copying and pasting of code.

Download the example code files

You can download the example code files for this book from GitHub at https://github.com/PacktPublishing/AWS-Certified-Cloud-Practitioner-Exam-Guide. In case there's an update to the code, it will be updated on the existing GitHub repository.

We also have other code bundles from our rich catalog of books and videos available at `https://github.com/PacktPublishing/`. Check them out!

Download the color images

We also provide a PDF file that has color images of the screenshots/diagrams used in this book. You can download it here: `https://static.packt-cdn.com/downloads/9781801075930_ColorImages.pdf`.

Conventions used

There are a number of text conventions used throughout this book.

`Code in text`: Indicates code words in the text, database table names, folder names, filenames, file extensions, pathnames, dummy URLs, user input, and Twitter handles. Here is an example: "For the key, type in `Name`, and for the value, type in `Windows-BastionSrv`."

A block of code is set as follows:

```
{
    "Id": "Policy1613735718314",
    "Version": "2012-10-17",
```

Bold: Indicates a new term, an important word, or words that you see on screen. For example, words in menus or dialog boxes appear in the text like this. Here is an example: "From the **Instances** console, select the **Launch instances** button from the top right-hand corner of the screen."

> **Tips or Important Notes**
> Appear like this.

Get in touch

Feedback from our readers is always welcome.

General feedback: If you have questions about any aspect of this book, mention the book title in the subject of your message and email us at `customercare@packtpub.com`.

Errata: Although we have taken every care to ensure the accuracy of our content, mistakes do happen. If you have found a mistake in this book, we would be grateful if you would report this to us. Please visit www.packtpub.com/support/errata, select your book, click on the Errata Submission Form link, and enter the details.

Piracy: If you come across any illegal copies of our works in any form on the internet, we would be grateful if you would provide us with the location address or website name. Please contact us at copyright@packt.com with a link to the material.

If you are interested in becoming an author: If there is a topic that you have expertise in and you are interested in either writing or contributing to a book, please visit authors.packtpub.com.

Reviews

Please leave a review. Once you have read and used this book, why not leave a review on the site that you purchased it from? Potential readers can then see and use your unbiased opinion to make purchase decisions, we at Packt can understand what you think about our products, and our authors can see your feedback on their book. Thank you!

For more information about Packt, please visit packt.com.

Share Your Thoughts

Once you've read *AWS Certified Cloud Practitioner Exam Guide*, we'd love to hear your thoughts! Scan the QR code below to go straight to the Amazon review page for this book and share your feedback.

https://packt.link/r/180107593X

Your review is important to us and the tech community and will help us make sure we're delivering excellent quality content.

Section 1: Cloud Concepts

In this section, we look at the fundamentals of cloud computing. We then look at **Amazon Web Services (AWS)** and an overview of its cloud offering as well as its value proposition. We examine AWS cloud economics and the advantages of cloud computing.

This part of the book comprises the following chapters:

- *Chapter 1, What Is Cloud Computing?*
- *Chapter 2, Introduction to AWS and the Global Infrastructure*
- *Chapter 3, Exploring AWS Accounts, Multi-Account Strategy, and AWS Organizations*

1
What Is Cloud Computing?

Cloud computing has become the default option to design, build, and implement **Information Technology** (**IT**) applications for businesses across the globe. In the old days, you would host the entire infrastructure, hire a group of developers, and design each component and process required to build your applications. This approach not only ate into the bottom line, but also often did not follow best practices. It also lacked flexibility and scope for innovation.

Understanding cloud computing has become vital for IT professionals worldwide if they are to sustain their jobs and make progress in their careers. You can no longer deliver old-school solutions to your clients—it is simply not cost-effective in today's fast-paced IT world.

In addition, architecting solutions for the cloud comes with its own challenges, such as security considerations and network connectivity. This makes it crucial to upskill so that you can gain a deep understanding of how to build resilient, scalable, and reliable solutions that can be hosted in the cloud.

In this chapter, we introduce you to the concept of cloud computing, what it includes, and the key advantages of moving to the cloud. We also discuss the various cloud computing models, as well as deployment options for the cloud. Understanding the key differences between the models and deployment options and their use cases and benefits is fundamental to formulating an effective cloud-adoption strategy for your business.

We also look at a high-level overview of virtualization—a principal ingredient that has made cloud computing possible.

This chapter covers the following topics:

- What is cloud computing?
- Exploring the basics of virtualization
- Exploring cloud computing models
- Understanding cloud deployment models

What is cloud computing?

Cloud computing is a term used to describe the on-demand access to IT services that comprise compute, network, storage, and software services from third-party suppliers, usually via the public internet or some form of direct **wide-area network** (**WAN**) access. Companies can provision necessary IT applications for their organization without having to procure and manage their own infrastructure to host those applications. Instead, they lease/rent the required IT infrastructure from such third-party providers.

Cloud computing has existed for many years in some form, since the invention of the internet. In the old days, Hotmail (first launched in 1996 and now branded as Microsoft Outlook) was a prime example of early cloud computing. You could set up email accounts for your colleagues and yourself on Hotmail and use them to communicate. An alternative would be to host your own email servers' (the infrastructure) network connectivity, as well as the email application (the email software). This would ultimately mean additional costs as well as management overheads to maintain the email servers you hosted.

Today, cloud computing has become mainstream and is, in several cases, the default option for many companies and start-ups. Currently, **Amazon Web Services** (**AWS**) is the largest provider of cloud computing services, offering a variety of cloud IT services in the form of infrastructure, platform, and software solutions. You can opt to consume these services rather than creating your own dedicated environment to host your business applications. The sheer size of AWS enables it to actually provide the necessary components to host your business applications at a fraction of the cost, while providing **high availability** (**HA**), security, and resilience.

The six advantages of cloud computing

Let's take a look at the six advantages of cloud computing, according to AWS (*AWS, Six Advantages of Cloud Computing,* `https://docs.aws.amazon.com/whitepapers/latest/aws-overview/six-advantages-of-cloud-computing.html`), as depicted in the following screenshot:

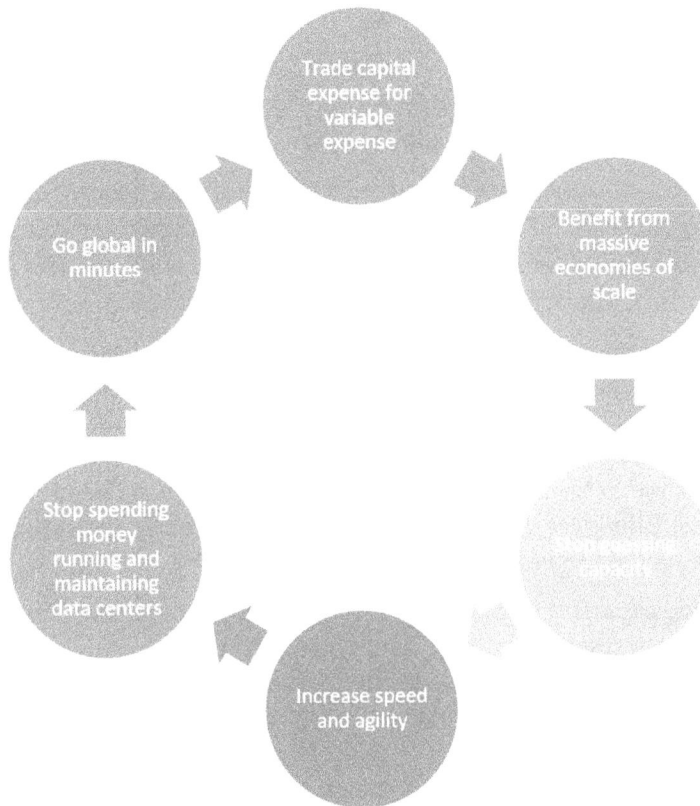

Figure 1.1 – The six advantages of cloud computing

> **Tip**
> The *AWS Certified Cloud Practitioner* examination assumes that you have these six advantages memorized when testing the *Define the AWS Cloud and its value proposition* objective.

Let's look at these advantages in detail, as follows:

- **Trade capital expense for variable expense**: One of the primary benefits of moving to cloud computing instead of hosting your own on-premises infrastructure is the method of paying for that infrastructure. Traditionally, you would have to procure expensive hardware and invest precious business capital to acquire infrastructure components necessary for building an environment to host applications.

  With cloud computing, you pay for the same infrastructure components only as and when you consume them. This on-demand, pay-as-you-go model also means that you save costs when you are not utilizing resources.

  The shift away from **capital expense (CAPEX)** for **variable expense**, also known as **operating expense (OPEX)**, means that you can direct your precious business capital to more important areas of investment, such as developing new products or improving your marketing strategy.

- **Benefit from massive economies of scale**: As an individual business, you would generally have to pay retail rates to purchase necessary IT hardware and build an environment that can be used to host your applications. Cloud providers such as AWS, however, host infrastructure for hundreds of thousands of customers, and even get involved in innovating and having components manufactured to their specifications. This gives even greater economies of scale and allows them to offer lower pay-as-you-go rates to customers.

- **Stop guessing capacity**: Traditionally, while carrying out capacity planning, you would procure necessary hardware components for future growth. Predicting future growth is extremely difficult, and this often meant that you would overprovision your environment. The result would be expensive idle resources simply going to waste. The fact that you would have made large CAPEX to acquire those components would ultimately be detrimental to the balance sheet due to the rapid loss in value arising from depreciation. On the flip side, some companies may end up underprovisioning capacity to save on costs. This can have an adverse effect on corporate image, if—for example—due to underprovisioned resources your customers are not able to complete transactions or suffer from poor performance.

With cloud computing and sophisticated management software, you can provision the necessary infrastructure when you need it most. Moreover, with monitoring and automation tools offered by cloud vendors such as AWS, you can automatically scale out your infrastructure as demand increases and scale back in when demand falls. Doing so will allow you to pay only for what you consume, when you consume it.

- **Increase speed and agility**: Cloud vendors such as AWS enable you to launch and configure new IT resources in a few mouse clicks—for example, you can provision a new fleet of servers for your developers within minutes, allowing your organization to exponentially increase its agility in building infrastructure and launching applications. If you are building test and development environments or performing experimental work as part of researching a new product/service, then once those tasks are complete you can just as quickly terminate those environments. Equally, if a particular project is being abandoned midway, you do not need to be worried about having any physical wastage—you just turn off or terminate what you no longer need. By contrast, prior to the invention of virtualization technologies (discussed later), provisioning a new server to host a database would often take weeks. This would include the time it takes to place an order with a supplier for suitable hardware, having it delivered, installing additional components such as storage and memory, and then finally implementing the manual process of installing operating systems and securing them. This process of building data centers also means that you are diverting money away from the main business—precious capital that could be spent on innovating existing products or developing new ones.

- **Stop spending money running and maintaining data centers**: Hosting your own on-premises infrastructure consumes several hidden costs. In addition to using up precious capital to purchase expensive hardware, you also need a team of engineers to efficiently configure every infrastructure component and lease necessary real estate to rack, stack, and then power up your servers. You would also be required to keep the servers cool with appropriate air-conditioning systems—and that's not all. You would also have to spend money on expensive maintenance contracts to handle the wear and tear of the hardware.

  By hosting your applications on AWS's infrastructure, you no longer need to worry about these hidden costs. Your real-estate costs and utility bills can be dramatically reduced, making your business more competitive.

- **Go global in minutes**: AWS host their data centers in various regions across the globe. Although you may be based in one country, you will have complete access to all regions. This will help you offer lower latency and a superior customer experience, regardless of where your customers are located. Hosting copies of your resources in additional regions can also help you design for **disaster recovery** (**DR**) and business continuity requirements.

By way of contrast, the cost of setting up physical data centers in other countries in which you may not have a presence may be cost-prohibitive and might prevent you from rapid global expansion. Access to multiple regions also enables you to meet any compliance or regulatory requirements related to where data is stored and how it is managed.

In this section, we learned about the basics of cloud computing and discussed its six key advantages. We understood that adopting cloud technologies helps customers manage their costs better, while also enabling them to scale their applications much faster and become more agile. In the next section, we'll discuss one of the most important underlying components of a cloud computing service—virtualization.

Exploring the basics of virtualization

Virtualization is one of the core technologies that has enabled cloud computing to go mainstream and has given birth to cloud providers such as AWS, Microsoft Azure, and **Google Cloud Platform** (**GCP**), who provide a vast array of services and applications, along with capabilities such as **high availability** (**HA**), elasticity, and the ability to provision services for their customers, usually within minutes.

Before the adoption of virtualization, if you wanted to outsource your infrastructure requirements, an IT services provider would have to provision physical infrastructure components such as a physical server for your business and grant access via the internet. Provisioning physical servers, however, often involves long lead times, from sourcing and installing all the hardware components such as the **central processing unit** (**CPU**), memory, and storage, to configuring an operating system and any necessary applications. This could mean waiting for days to have your environment configured.

The advancement of hardware technologies such as CPUs, memory, and storage has seen a substantial increase in performance and capability, to the extent that physical servers hosting a single operating system and a few applications often remain idle. Software engineering and the improvements in software design have, by way of contrast, ensured that hardware resources are efficiently consumed to power those applications. The net result has been that physical hardware resources are rarely consumed to their maximum capability by a single operating system and a small set of applications.

This relationship between hardware and software has contributed to the invention of virtualization. Virtualization technologies and **hypervisors** have made it possible to emulate the physical hardware components of a single physical server as multiple virtual components. These components are then deployed as multiple **virtual machines** (**VMs**), each running its own operating system and suite of applications.

A hypervisor is essentially a piece of software that sits between the actual physical hardware and the VMs. It is responsible for enabling the operating systems and applications running on those VMs to access the resources of the physical hardware in a manner that is controlled and that isolates the resources from each other. The hypervisor and its associated management software are used to *carve* out virtualized representations of the physical hardware components into smaller virtual components, which are then presented as VMs. Each VM can then have its own operating system installed, along with any required applications.

One of the greatest advantages of virtualization is the speed at which resources can be provisioned. With software being used to emulate *existing* physical hardware (so that the hardware is available when a customer makes a request), the lead times to provision virtual servers, storage, or network environments are drastically reduced.

In the following diagram, we can see how virtualization enables us to allocate virtual storage devices to our individual VMs from the physical storage attached to the server:

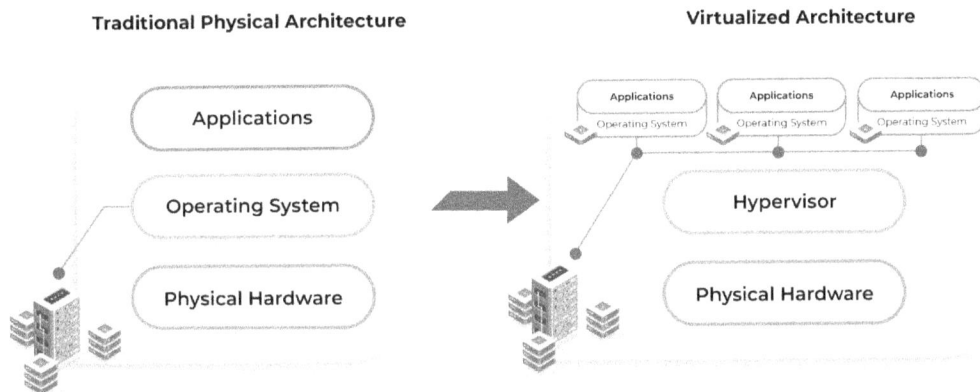

Figure 1.2 – Traditional physical architecture versus virtualized architecture

One of the greatest advantages of virtualization is the speed at which resources can be provisioned. Since software is designed to emulate *existing* physical hardware (to enable the availability of hardware when a customer makes a request), the lead times to provision virtual servers, storage, or network environments is drastically reduced.

Virtualization versus cloud computing

Virtualization, in itself, is not cloud computing. The technology, however, is responsible for making it possible to deliver cloud computing services. One of the primary characteristics of a cloud computing provider is the ability to provision virtualized infrastructure resources using a self-service management tool. AWS offers such tools in the form of its Management Console (accessible via a web browser), **command-line interface** (**CLI**), and direct access to its software **application programming interfaces** (**APIs**), to enable customers to provision their resources such as servers, network, storage, and databases. By offering well-defined APIs and enabling automation, cloud providers have made it possible for customers to provision necessary resources using a *self-service* model. Customers do not have to wait in a queue to get their resources deployed while a cloud engineer performs the necessary configuration for them. Customers can interact with the cloud services directly using API calls, and spin up their own resources in a matter of minutes.

Ultimately, cloud computing providers make use of virtualization and modern hardware technologies that are aware of virtualization, as well as software to deliver shared computing resources, **Software-as-a-Service** (**SaaS**)-based products, and other on-demand services via the internet. In addition, providers such as AWS offer solutions to enable elasticity, automation, scalability, and HA—all on a pay-as-you-go pricing model, which makes their services accessible to almost any type of client in any location.

In summary, here are the benefits of virtualization:

- Efficient use of powerful hardware by setting up multiple VMs to offer different applications

- Enables server consolidation, which translates to reduced costs

- Allows you to manage large-scale installations and deployments at a faster pace

- Improves security through infrastructure isolation and efficient management of underlying hardware resources

- Enables you to host various operating systems that serve different applications on the same hardware

In this section, we learned that virtualization technology has been a primary driving force in the evolution of cloud computing. The technology enables the provisioning of resources such as servers, networking components, and storage services in a matter of minutes. In addition, virtualization management applications enable us to build self-service platforms. Customers can simply log in to a management console and provision the necessary resources to build an architecture to host their application.

In the next section, we'll explore the cloud computing models available. Different models require varied levels of management and accordingly offer different levels of flexibility.

Exploring cloud computing models

Cloud computing today offers businesses the ability to offload the cost and complexity of hosting and managing their applications—for example, many providers offer mainstream applications as a complete service that does not require any kind of infrastructure management by the customer. Examples include **Microsoft Office 365**, which is a suite of desktop productivity applications including email, messaging, and collaboration services offered via the internet. At the same time, many organizations also need to host bespoke **line-of-business** (**LOB**) applications such as those developed in-house. Often, this means that they need access to configure the necessary infrastructure in a manner best suited to the needs of the application.

To that end, companies can enlist the services of cloud providers such as AWS, which offers different cloud models to suit the specific needs of the business. The following are three main cloud models offered by most cloud vendors such as AWS.

Infrastructure as a Service

The **Infrastructure as a Service** (**IaaS**) model offers the greatest flexibility in giving the customer access and the ability to configure the underlying network, storage, and compute services that power their LOB applications. This model is very similar to owning and managing your own physical infrastructure. However, with cloud computing, a clear difference lies in the fact that you work with virtualized infrastructure components rather than having access to the underlying physical components.

The IaaS cloud computing model is ideal if you need greater control over how your infrastructure components need to be configured (usually from the operating system layer up) to support a given application.

Platform as a Service

Platform as a Service (**PaaS**) is another cloud computing model designed to remove the burden of configuring and managing underlying infrastructure resources such as compute, storage, and network services. PaaS is designed to allow your organization to focus on developing your application code and offers you a *platform* to deploy and manage your application releases, updates, and upgrades.

As your developers deploy their application code on the PaaS environment, the provider provisions the infrastructure required to support the application. This will include the necessary network architecture, firewall rules, storage, compute services, operating system management, and runtime environments.

Depending on your vendor, the PaaS model may still offer some degree of flexibility in how the underlying infrastructure is configured. AWS, for example, gives you the option to make necessary modifications to the underlying infrastructure, offering an additional level of flexibility. Example of such services include AWS Elastic Beanstalk, AWS OpsWorks, AWS Lambda, and Amazon **Relational Database Service** (**RDS**). While the PaaS model offered by AWS removes the need to minutely configure every infrastructure component (something you would have to do with an IaaS model), it still offers the flexibility of deciding just which components are deployed to support your application.

SaaS

With a **SaaS** model, the applications are completely hosted and managed by the provider. SaaS services take away any need to set up physical infrastructure to host an application. Instead, you simply connect to those applications via the internet and consume the services offered. A majority of SaaS applications today are fully functional via a standard web browser. This also means that there is no requirement to install any client software.

While the need to set up and configure any infrastructure to host a SaaS application is solely owned and managed by the vendor, many SaaS-based applications still require some form of configuration to meet the specific requirements of your business. You will still need to either have in-house expertise to configure the application to your specification or get support from the provider/third parties. For example, Microsoft Office 365 is a SaaS-based online suite of productivity applications that combines email, file-share, and collaboration services. Although you do not need any physical hardware on premises to host the application since it is accessible as a complete product over the internet, you will have to configure the software elements to meet your business needs. This includes security configurations, configuring your domain name to be associated with the email services offered, or enabling encryption services.

Let's look at some typical examples of IaaS, PaaS, and SaaS models, as follows:

Cloud platform	Common examples
IaaS	Amazon Elastic Compute Cloud (EC2); Amazon Elastic Block Store (EBS); Amazon Elastic File System (EFS); Azure Virtual Machines (Azure VM)
PaaS	AWS Elastic Beanstalk; Azure Functions; Google App Engine
SaaS	Salesforce; GoToMeeting; Microsoft Office 365; Amazon Chime

Table 1.1 – Cloud computing models

In this section, we explored cloud computing models. We gained an understanding of the key differences between core models such as IaaS, PaaS, and SaaS. Each model comes with its own set of management overheads and with it, the flexibility to design, build, and deploy your applications.

In the next section, we examine cloud deployment models. Here, we assess the differences between hosting your own on-premises cloud (or private cloud) and using the services of a public cloud provider. We also look at how to connect your private cloud environment with the resources you might host with a public cloud provider.

Understanding cloud deployment models

When it comes to deploying cloud services for your organization, you need to consider which deployment model will suit your business. The decision will be taken based on several factors, such as the industry you are in, compliance and regulatory issues, and also cost management and flexibility of configuration.

There are three primary models of deployment, listed as follows:

- Public cloud
- Private cloud
- Hybrid cloud

These models are represented in the following diagram:

Figure 1.3 – Cloud deployment models

Let's look at each model in a little more detail.

Public cloud

A public cloud is a cloud deployment model in which a business consumes IT services from a third-party vendor, such as AWS, over the internet. This is the most popular model of cloud computing due to the vast array of services on offer. Public cloud providers such as AWS are in the business of delivering IT services across all industry verticals and for businesses of all sizes.

Public cloud services are generally paid for on a pay-as-you-go model and can help your organization move away from a CAPEX of mode of investment in IT to an OPEX mode. This frees up precious capital for more important investment opportunities. Services offered by public cloud vendors will include free services, subscription-based, or on-demand pay-as-you-go, where you are charged based on how much you consume. Providers of public cloud services are also able to offer greater scalability and agility that would otherwise have been too expensive to achieve on your own.

With a public cloud model, customers are offered a *self-service* capability and access to management consoles and command-line interfaces, as well as having API access to configure and consume the services on offer.

Private cloud

By contrast, a private cloud is a cloud deployment model in which your business procures, installs, configures, and manages all the necessary infrastructure and software components in-house. This may sound very similar to traditional *on-premises* IT. However, the cloud element of it comes from the fact that additional management software is usually deployed to allow different parts of the business to carry out *self-service* tasks in provisioning compute, storage, network, and software services from an available catalog of services.

While public cloud providers offer their services to all businesses across the globe and the services are therefore publicly available, a private cloud is designed solely for your business, where you will not be sharing underlying compute resources with anyone external to your organization.

A private cloud is highly customizable to suit the needs of your organization, giving maximum control on key areas such as designing security and infrastructure configuration options. This does not necessarily mean that a private cloud provider (for example, Red Hat OpenStack) is more secure than a public cloud provider. Public cloud providers such as AWS invest vast amounts of money to design security features for the services they offer—features that may be cost-prohibitive if an organization tried to implement them on its own.

Hybrid cloud

This is a combination of IT services deployed both on-premises (and managed solely by your business) and integrated with one or more third-party cloud providers.

Many companies that venture into the public cloud generally start with some form of hybrid model. Often, businesses will move/migrate services to the public cloud to reduce CAPEX investment as they opt for a *pay-as-you-go* model for the consumption of IT services. An example of this is where companies may need to increase the number of servers deployed for their applications, and rather than procuring more expensive physical hardware, they can set up network connectivity between on-premises infrastructure and the public cloud provider, where they would spin up those additional servers as required. Connectivity options between an on-premises environment and a cloud provider can include setting up a secure **Internet Protocol Security** (**IPsec**) **virtual private network** (**VPN**) tunnel over the public internet, or even establishing a dedicate fiber-based connection, bypassing the public internet altogether and benefiting from greater bandwidth.

A hybrid cloud is generally also used to help start off your **disaster recovery** (**DR**) projects, which often need network communication between the private cloud infrastructure and the services offered by public cloud vendors where the DR solution will be hosted. This enables replication of on-premises data and applications to the DR site, hosted with vendors such as AWS.

Hybrid cloud deployments can also help businesses to start testing out new cutting-edge technologies or adopt a phased migration approach to ensure minimum interruption to normal business functions while the migration is underway. In addition, HA solutions can also be implemented. To cite an example, if the on-premises infrastructure is experiencing downtime, consumers of those services can be redirected to replica services hosted with the public cloud provider.

Summary

In this chapter, we explored the basics of cloud computing and how it can help businesses consume necessary IT services to host their applications. We discussed six key advantages of cloud computing and the reasons it offers greater flexibility and resilience, as well as opportunities for innovation and cost reduction.

We also examined three cloud computing models, identifying their key differences and comparing the level of flexibility offered by each model. We also assessed the three cloud deployment models and identified how companies can begin their cloud journey easily by building hybrid cloud solutions.

In the next chapter, we introduce you to AWS. We will discuss its history and provide a brief overview of its services. We will also examine the AWS Global Infrastructure, which gives businesses access to globally dispersed data center facilities within which they can deploy their applications. This will enable businesses to expand their customer reach on a global scale. Then, we will look at the support plans offered by AWS, which are vital to any business looking to consume cloud services.

Questions

Here are a few questions to test your knowledge:

1. Which of the following six advantages enables small start-up companies to immediately start consuming IT services from public cloud vendors such as AWS?

    A. Trade capital expense for variable expense

    B. Go global in minutes

    C. Stop guessing capacity

    D. Increase speed and agility

2. Which feature of cloud computing enables customers to deploy their resources in a matter of minutes using a self-service model?

    A. Access to cloud provider APIs

    B. Access to cloud provider engineers to rack and stack servers

    C. Scalability features

    D. Multiple server options

3. What is a hypervisor?

    A. Software that enables you to create and managed virtualized resources running on physical hardware, such as VMs

    B. Software used to monitor the health of your Windows servers

    C. Software used to create HA websites

    D. Hardware that enables you to increase the performance of your physical servers

4. Which of the following are the primary benefits of server virtualization?
    (Select two answers.)

    A. Efficient use of physical hardware resources

    B. Ability to provision virtual servers in a matter of minutes

    C. Enhanced encryption services

    D. Ability to meet compliance requirements

5. Which of the following is a prime example of IaaS?

    A. A service that gives you access to configure underlying virtual compute, storage,
        and network resources to host your application

    B. A service that abstracts the underlying infrastructure, allowing you to focus on
        your application code deployment process

    C. A service that hosts and delivers a complete application via a public network,
        with no access to any underlying infrastructure

    D. A service that allows you to consume hardware resources for a short lease
        period and pay on a metered basis

6. Which of the following is a prime example of PaaS?

    A. A platform that hosts and delivers a complete application via a public network,
        with no access to any underlying infrastructure

    B. A service that gives you access to configure underlying virtual compute, storage,
        and network resources to host your application

    C. A service that abstracts the underlying infrastructure, allowing you to focus on
        your application code deployment process

    D. A service that allows you to build infrastructure using code for repeat
        deployments in different environments

7. Which of the following is a prime example of SaaS?

    A. A service that gives you access to configure underlying virtual compute, storage, and network resources to host your application

    B. A service that abstracts the underlying infrastructure, allowing you to focus on your application code deployment process

    C. A service that hosts and delivers a complete application via a public network, with no access to any underlying infrastructure

    D. A service that allows developers to adopt DevOps strategies for their software development life cycle

8. Which cloud deployment model enables you to connect your on-premises workloads with resources you have deployed with a public cloud provider such as AWS?

    A. Private cloud

    B. Public cloud

    C. Hybrid cloud

    D. Hyper cloud

2

Introduction to AWS and the Global Infrastructure

In this chapter, we discuss what **Amazon Web Services** (**AWS**) is, examine its brief history, and also aim to understand the AWS Global Infrastructure. The Global Infrastructure gives you access to AWS data centers across different continents, enabling you to build highly available, fault-tolerant, and scalable solutions for your customers. In addition, you can also ensure that you place workloads closer to the location of your customers and fulfill any compliance or regulatory requirements.

This key offering from AWS enables you to access and launch resources across different Regions. An in-depth understanding of this will help you meet your clients' requirements—adherence to regulatory and compliance requirements, **disaster recovery** (**DR**) solutions, and even cost savings—all leading to a better customer experience.

We also look at the support plans offered by AWS to its customers. Your clients may require different levels of support, depending on the number and complexity of the applications they need to host. Choosing the right plan will be key to ensuring that you have all the support you need while managing your costs effectively and meeting any specific requirements.

This chapter covers the following topics:

- What is AWS?

- Exploring the AWS Global Infrastructure

- Choosing the right AWS support plan for your business

- Overview of the AWS Service Health Dashboard

- The AWS **Acceptable Use Policy** (**AuP**)

What is AWS?

Amazon Web Services (**AWS**), a subsidiary of Amazon, is the largest public cloud-computing provider in the world. It offers over 175 distinct services to its clients from its data centers located across the globe. These services are accessible over the internet (with some on-premises options available as well) on a metered pay-as-you-go model. Its customer base comprises start-ups, enterprise clients, and even governmental organizations such as the **United States** (**US**) Navy.

Gartner Research creates a yearly report known as the *Magic Quadrant for Cloud Infrastructure and Platform Services*, and over the last few years has awarded AWS top position in the Leaders quadrant.

You can access a wide selection of analytical research reports at `https://aws.amazon.com/resources/analyst-reports`. In the left-hand menu, simply filter your search for `Magic Quadrants` to get a selection of Gartner reports, including the aforementioned Magic Quadrant for the *Cloud Infrastructure and Platform Services* report.

A quick history of AWS

AWS started its journey in 2002 when it began to offer a few ad hoc services to the public. The timeline shown in the following screenshot provides an overview of some of its key milestones and its journey to date:

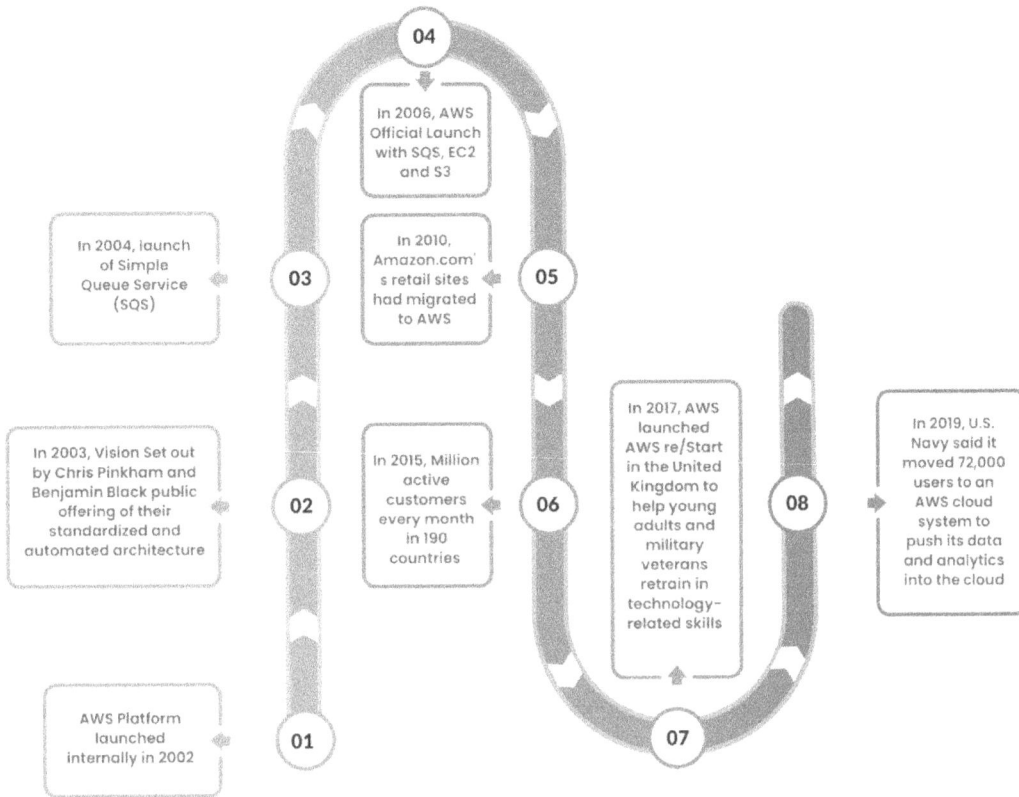

Figure 2.1 – AWS history: timeline

In this section, we looked at a brief history of AWS and how it emerged as a cloud leader in the market. In the next section, we will examine the AWS Global Infrastructure, which is critical to its functionality and in being able to offer a vast array of cloud services to its clients.

Exploring the AWS Global Infrastructure

The AWS Global Infrastructure comprises multiple data centers that house all the servers, storage devices, and networking equipment across different geographical **regions** around the globe.

As AWS continues to expand its global footprint, it builds additional data centers, which ultimately leads to an increase in the number of Regions accessible to its customers.

At the time of writing this document, the following screenshot depicts the current live Regions across the globe and includes upcoming ones too:

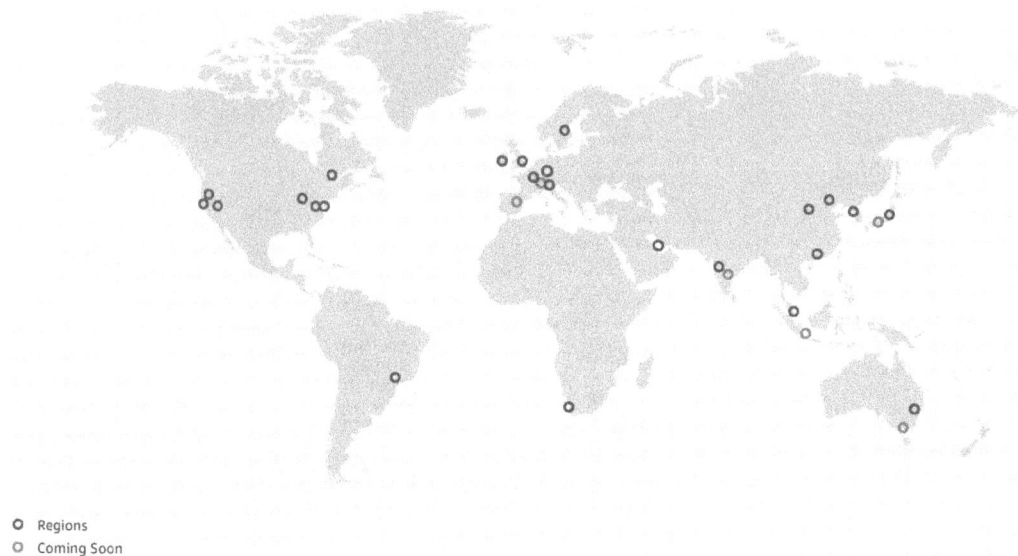

○ Regions
○ Coming Soon

Figure 2.2 – AWS Global Infrastructure. Image courtesy of AWS (`https://aws.amazon.com/about-aws/global-infrastructure/`)

An **AWS Region** is a physical location where AWS will host a cluster of **data centers**. Within a given Region, these data centers are built such that small groups of the larger cluster are logically and physically separated from each other by a distance that falls within 100 **kilometers** (**km**) (60 miles) of each other. These logically and physically separated groups of data centers form what we call **Availability Zones** (**AZs**).

AWS currently spans 77 AZs within 24 geographical regions around the world, and has announced plans for 18 more AZs and 6 more AWS Regions in Australia, India, Indonesia, Japan, Spain, and Switzerland.

Let's take a closer look at what Regions and AZs are.

Regions

A Region will consist of a minimum of two AZs, and many even consist of three or more. The North Virginia Region (N. Virginia or `us-east-1`) consists of six AZs. Usually, when AWS launches a new service, it is first deployed in the North Virginia Region.

You may be curious about the purpose of having so many Regions and how this would be useful. Simply put, having multiple Regions across the globe will allow AWS to further entice customers in those locations to sign up for its services. However, this may lead to another question. If there is nothing to stop you from accessing the services in an AWS Region on the other side of the globe, why have so many? Read on to find out why.

AWS's multi-Region strategy enables you (the customer) to derive the following benefits:

- Identify infrastructure resources closer to your end users, where you can host your application and reduce network latency, resulting in a good **user experience** (**UX**)

- Identify infrastructure within political and national borders to adhere to strict data sovereignty and compliance regulations

- Isolate groups of resources from each other to allow you to fulfill any failover or DR scenarios in case major regional outages occur

With regard to the second benefit, many services are based on Regions. This means that before you deploy a resource using that service, you will need to select the Region in which you want to deploy it. For example, if you wish to launch a new Windows-based virtual server (we call these **Elastic Compute Cloud** (**EC2**) instances on AWS), you will need to choose the Region in which you want to place it. In fact, with EC2 instances, you also need to specify the AZ within the Region where you want to launch the server. Let's look at AZs next.

AZs

AZs are the logical and physical grouping of data centers within a given Region. Each Region will have two or more AZs, as mentioned earlier. An AZ is a logical representation of a metropolitan area where AWS has deployed one or more data-center facilities. These data centers will house hardware components such as servers, storage, and network equipment, all fitted with redundant power, connectivity, cooling, and security controls.

The primary purpose of having multiple AZs is to offer customers the opportunity to build highly available, fault-tolerant, and scalable solutions. This is made possible by the fact that the AZs within a Region are connected to each other over high-bandwidth, low-latency private metro-fiber links, delivering high throughput connections between the zones.

An important aspect of this configuration is that you can achieve synchronous replication between AZs. This means that you can deploy multiple copies/replicas of your application on servers across the AZs. If there is an outage at one of the AZs, you can continue to serve your customers from the replica workloads running in the other AZs.

In *Figure 2.3*, if there was an outage in AZ **A**, users' traffic would be directed only to the servers offering the exact same services in AZ **B**. The architecture is such that users will not be made aware of the outage, and this will improve the UX substantially.

This is a viable design because we can place our replica application servers across multiple AZs and distribute end-user traffic across them using an Amazon **Elastic Load Balancer** (**ELB**). The ELB not only distributes user traffic to the application servers, but also monitors the health of those servers and sends traffic only if those servers are online and responding. We examine Amazon ELBs in detail later, in *Chapter 9*, *High Availability and Elasticity on AWS*. You can see a diagram of the aforementioned design here:

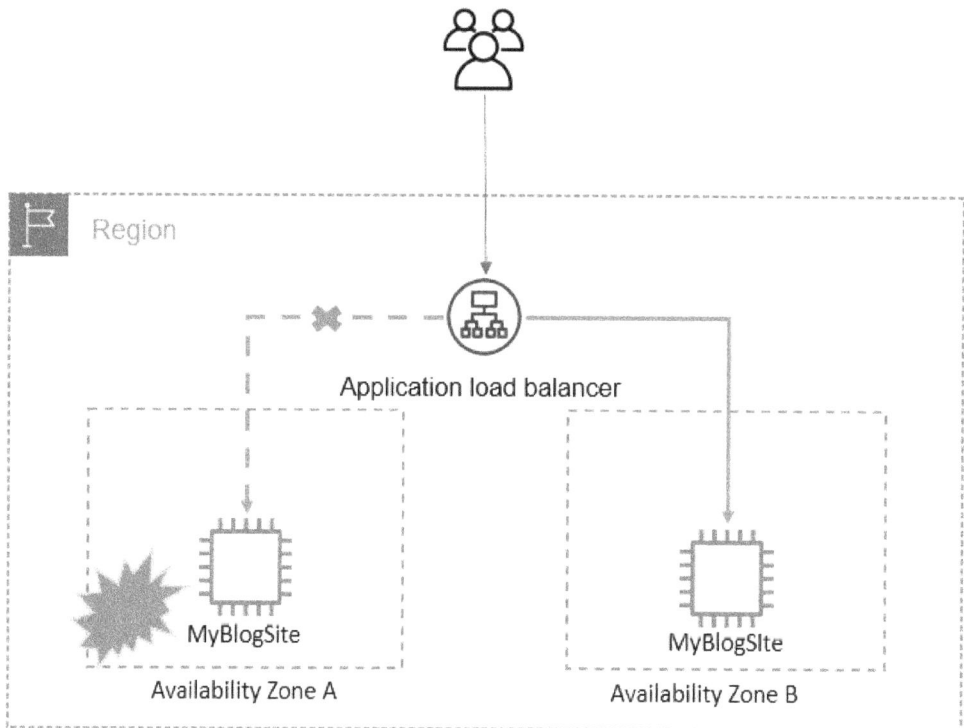

Figure 2.3 – User traffic is temporarily directed to only the server(s) in AZ B, until AZ A comes back online

So, now that we understand the differences between Regions and AZs and how AZs can be used to help build highly available solutions, we can proceed to look at another component of the AWS Global Infrastructure: Edge locations.

In this next section, we will discuss how edge locations can be used to build a **content delivery network** (**CDN**) and offer caching services.

Edge locations

In addition to Regions and AZs, AWS also offers another type of hosting service, called edge locations. These edge locations also host the physical server infrastructure, massive amounts of storage devices, and high-bandwidth networking equipment. In addition, AWS edge computing services provide infrastructure and software that enable data to be processed and analyzed closer to the end customer. This includes deploying AWS-managed hardware and software to locations outside AWS data centers, and even onto customer-owned devices themselves.

As discussed previously, most resources that you deploy on AWS are going to be Region-based. For example, an EC2 instance can be deployed in the North Virginia Region, within a specific AZ. Let's assume that the servers host media files and you want to distribute those files to your end users globally. For users based in Sydney, Australia, this would mean pulling these large media files across the public internet directly from the server located in the US, each time they make a request for those files.

With edge locations, you can cache frequently accessed files on servers located closer to those users based in Sydney. This means that the time it takes to download those frequently accessed files is drastically reduced, and this improves the UX significantly.

A prime tenant of edge location services is Amazon's CloudFront. This is a CDN service designed to help you create distribution points for your content. The distribution points are created at these edge locations in each Region, depending on your configuration. Your content is then cached at the edge location closest to the end users who attempt to access your content.

Edge locations do more than just cache content. For example, **Amazon Simple Storage Service** (**Amazon S3**) is an object-storage solution that allows you to create containers (we call them **buckets** in AWS) in each Region. You can upload any type of data to a bucket and store it. If you need to upload large files to a bucket anywhere in the world, you may experience high latency and lower throughput, as the data needs to traverse the public internet.

Amazon S3 Transfer Acceleration is a feature of Amazon S3 that allows you to upload your content to AWS buckets via these edge locations. Once your data reaches an edge location closer to you, it then traverses the AWS backbone network using high-speed links. This greatly improves the upload speed of your data to Amazon S3 buckets. You can actually test your upload transfer rates using a tool provided by AWS, which is available at `https://s3-accelerate-speedtest.s3-accelerate.amazonaws.com/en/accelerate-speed-comparsion.html`. The tool will attempt to transfer a sample file from your browser (your location) to various Amazon S3 Regions and display transfer rates comparing transfer with AWS S3 Transfer Acceleration and without.

Edge locations are connected to AWS Regions through the AWS backbone network. This comprises fully redundant, multiple 100 **Gigabit Ethernet** (**GbE**) parallel fiber connections that substantially improve throughput and offer low-latency connectivity.

The following screenshot illustrates the current edge locations across the globe. AWS continues to build and deploy more edge locations based on consumer demand and business requirements:

Figure 2.4 – Edge locations and regional edge caches across the globe. Image courtesy of AWS
(`https://aws.amazon.com/cloudfront/features/`)

Edge locations do have some limitations in that their cache size will fill up over time, and to cater for new content, they will evict content not recently accessed. This would mean that the next time someone tries to access that particular piece of content, it will need to be retrieved from the original source. To help reduce the number of times this happens, AWS also offers *regional edge caches*, which we discuss next.

Regional edge caches

In addition to edge locations, we also have regional edge caches. Regional edge caches have more storage and offer large cache sizes. As you can imagine, edge location storage devices will contain massive amounts of cached content—these could be images, videos, documents, and so on. Their primary purpose is to store frequently accessed content. As content becomes stale, it is evicted from the cache to make room for newer content. This means that when some content is infrequently accessed, end users will again have to pull those files from the original server to access them.

Regional edge caches are highly useful in such situations. There are far fewer regional edge caches than edge locations deployed across the globe, but they are strategically placed. They offer additional storage and cache that will continue to hold the data not accessed frequently for a longer period of time than with standard edge locations.

Ultimately, regional edge caches will also evict data that is not being accessed, but in the interim period, if another end user on the other side of the globe makes a request for that same data, then the process repeats, whereby that data is then cached at an edge location closer to that specific user.

Regional services

We discussed earlier that most services on AWS are Region-based, and you need to first select a Region and subsequently create resources of that service in the Region. For example, if you wish to launch a **Relational Database Service** (**RDS**) database, you must first select the Region you want to deploy your database in and then launch your chosen database engine in that Region.

This is because the underlying physical infrastructure that will host that service (servers, storage, or databases) will reside in the Region you are working in. This is why most services on AWS are Region-based.

In the following screenshot, you will need to select a Region before you can launch a new Windows server. Similarly, if you need to access a given server that you have previously deployed, you must know the Region in which it is located:

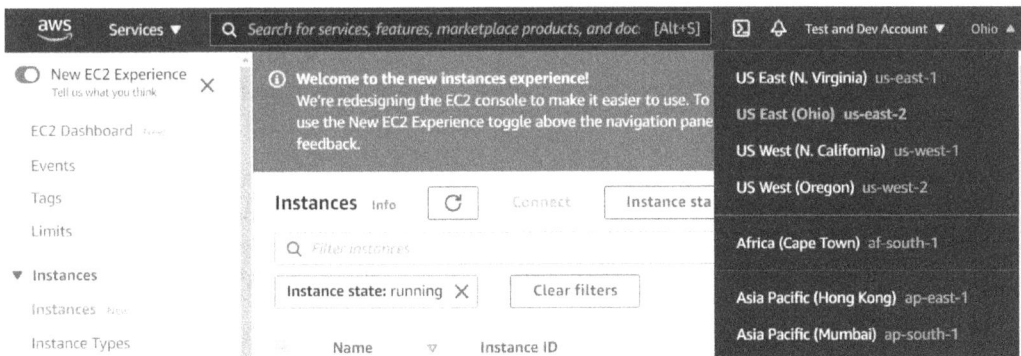

Figure 2.5 – Selecting a Region to access the EC2 instances (servers) deployed in that Region

Now that we have understood the core components of the AWS Global Infrastructure, we can move on to discuss how various types of services are offered to clients. Some services require you to specify the Region in which you wish to consume it, and others are accessible globally, without the need to specify a specific Region. Let's discuss this in the following section.

Global services

Although most services are Region-based, as we have mentioned already, there are some services that fall under the category of global services. This may seem contradictory to the logic given previously that a given service needs to obviously run from physical infrastructure located in each Region. While this is true, AWS will, nevertheless, present the service as a global offering. In other words, regardless of which Region you happen to be working in, you can directly access those global services from any location.

The reason for these few services being presented as global services is that you want the resources created in those services to be accessible globally and, in many cases, to be unique across all Regions within your AWS account.

Consider, for example, individual user accounts (**Identity and Access Management (IAM)** users) in your AWS account. It makes sense to have unique accounts for each member of staff in your organization who needs to work with services and resources in your AWS account. Imagine having to create multiple accounts for a given developer in each Region. That would only mean a huge management overhead.

Although not an exhaustive list, the following services are presented as global services on AWS:

- **AWS IAM**—A service offered by AWS to enable you to grant access to services and resources in your AWS Account. AWS IAM allows you to create IAM users for your staff who need access to those services, define permissions, configure groups, and set up roles. We discuss AWS IAM in *Chapter 4, Identity and Access Management.*

- **Amazon CloudFront**—A CDN service that allows you to create distribution points for your content for a specific origin server. The Amazon CloudFront services will cache content locally at edge locations closest to those users who request access to your content.

- **Amazon Route 53**—A highly available, scalable, and fully managed cloud **Domain Name System** (**DNS**). You can use Amazon Route 53 to register new domain names, configure domain records, and design global routing policies for various use cases, such as an active/passive solution for building a highly available solution.

- **Amazon S3**—Although Amazon S3 buckets need to be created in each Region and those buckets are therefore Region-specific, the service itself is presented as a global service. When you access the Amazon S3 console, you do not need to select a given egion. Instead, you are presented with a list of buckets across all Regions that you have created, as illustrated in the following screenshot:

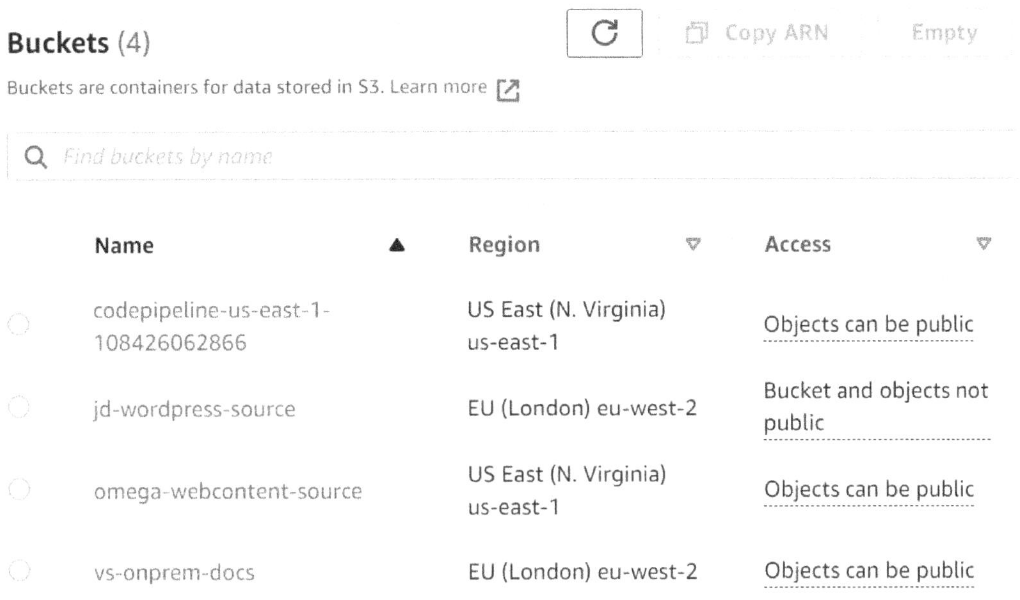

Buckets (4)   Copy ARN   Empty

Buckets are containers for data stored in S3. Learn more

Find buckets by name

Name ▲	Region ▽	Access ▽
codepipeline-us-east-1-108426062866	US East (N. Virginia) us-east-1	Objects can be public
jd-wordpress-source	EU (London) eu-west-2	Bucket and objects not public
omega-webcontent-source	US East (N. Virginia) us-east-1	Objects can be public
vs-onprem-docs	EU (London) eu-west-2	Objects can be public

Figure 2.6 – Amazon S3, displaying all buckets in a specific AWS account across all Regions where they have been deployed

In addition to global and regional services, AWS also offers certain services that are meant to be consumed on premises. These include services that can be used to build hybrid cloud models or assist in the migration of your on-premises workloads to AWS. We examine some of these in the next section.

On-premises services

Although AWS is a public cloud vendor, they also provide certain services that are hosted at their clients' data centers on-premises. This enables clients to bring the management capabilities of AWS services for certain services to their local data centers.

Some on-premises services are intended to also facilitate a hybrid cloud deployment solution, whereas other on-premises services can be used to help migrate data from local data centers to the AWS cloud via offline routes where bandwidth constraints may be an issue. Another reason for such on-premises offerings is that it helps address the needs of certain clients who have strict data-residency laws.

Some services that are designed to be hosted or consumed on-premises include the following:

- **Amazon Snow Family**—These are physical enclosure units that contain **solid-state drives** (**SSDs**), compute hardware, and networking components that are shipped to client sites. The AWS Snow Family comprises Snowball Edge Devices, Snowcone, and Snowmobile. They can be used for copying **terabytes** (**TB**) of data to **petabytes** (**PB**) of data onto the devices, which can be returned to AWS so that the data can be copied into Amazon S3. They also offer compute capability so that data can be processed and analyzed as required while it is being copied to the devices.

- **Amazon Storage Gateway**—This enables users to connect their on-premises storage with Amazon S3, offering different gateway options designed to enable offloading of their storage data to Amazon S3. They will continue to have seamless connectivity to that data from their on-premises servers. Depending on the configuration option chosen, Amazon Storage Gateway can maintain a small subnet of frequently accessed data locally, with the bulk of that data in Amazon S3, reducing the total storage hardware needed on premises, which leads to lower capital costs. Alternatively, if on-premises applications are extremely sensitive to network latency, then the Amazon Storage Gateway service can provide data backup capabilities, with the ability to send snapshots of locally stored data to Amazon S3.

- **Amazon Outposts**—This is a 42U rack that can scale from 1 rack to 96 racks to create pools of compute and storage capacity, hosted at your local data center. The U refers to rack units or "U-spaces" and is equal to 1.75 inches in height. A standard height is 48U (a 7-foot rack). Amazon Outposts offers all the management software and capabilities of AWS to manage your EC2 instances, as well as storage and RDS database services locally on-premises.

  This enables low-latency connectivity, local data processing, and adherence to local data-residency requirements. Other services that can be run locally on Amazon Outposts are **Elastic Container Service** (**ECS**), **Elastic Kubernetes Service** (**EKS**), and **Elastic MapReduce** (**EMR**) clusters. Yet another service that you can run locally in your data center is host-object storage services using Amazon S3 on Outposts. This can help you meet your local data-processing and data-residency needs.

In this section, we examined the AWS Global Infrastructure. This is the underlying core set of infrastructure components hosted across a vast array of AWS data centers around various strategically placed geographical regions. AWS gives you access to all its Regions, enabling you to host your applications to meet a wide range of use cases—from ensuring that your applications are located as close as possible to end users, to enabling you to meet any compliance or data-sovereignty laws.

We also reviewed the key difference between Regions, AZs, and edge locations, and learned how you can use these to build highly available, scalable, and fault-tolerant architectures at a very high level. We also discussed how edge locations can be used to offer low-latency access to your content and digital assets.

Next, we moved on to discuss the difference between regional-based services and global services. We also examined some services that are offered to run on-premises at clients' data centers.

In the next section, we will review AWS support plans. As you start to build solutions on the AWS platform, choosing the right support plan is extremely important to ensure that you have the correct level of support for your use case, while ensuring you manage your costs effectively.

Choosing the right AWS support plan for your business

AWS is strongly focused on delivering excellent customer support. Different businesses, however, require varied levels of support, and at AWS, each customer is catered for. In addition to standard customer support services such as resolving issues with account setups or billing, AWS offers an extensive range of technical support to help clients adopt cloud technologies faster and more efficiently.

Even if you have vast technical experience within your on-premises data centers, moving to the cloud requires learning a new set of skills, understanding how to architect solutions for the cloud, and being cost-effective at the same time. AWS hires and trains some of the best engineers, and their support team is highly proficient with the technologies they offer.

There are *four support plans* on offer from AWS, discussed in the following sections.

Basic support plan

Regardless of which support plan you ultimately decide to choose for your business, every account is subscribed to the Basic support plan. Now, before this piques your interest, I should point out that this plan does not come with *any real technical support*. The Basic support plan is completely free and offers customer support for any account-related issues such as bill payment or if you have issues logging in to your account. You also get access to publicly available documentation, whitepapers, and support forums. You can access the Basic support services via email, chat, and phone 24/7, and the phone support involves getting Amazon to call you on your landline or mobile—so, they pay the call charges.

In addition, you also get access to seven basic checks on the Trusted Advisor tool, which helps you to identify best practices for increasing performance and improving security. We will take a look at the Trusted Advisor tool in *Chapter 13, Management and Governance on AWS*. Finally, you also get alerts regarding interruptions to any AWS services that may impact your deployed resources via the **Personal Health Dashboard** (**PHD**). The PHD is discussed in detail in the *Chapter 13, Management and Governance on AWS*, as well.

Developer support plan

AWS recommends subscribing to the Developer support plan if *you are experimenting or testing in AWS*; refer to `https://aws.amazon.com/premiumsupport/plans/` for more information. The Developer support plan is a cost-effective solution for getting support with non-production workloads. These are websites or applications that are still at the development stage and do not yet support any critical business requirement. If you are satisfied with periods of extended downtime for such workloads, the Developer support plan is offered at a very reasonable price.

While you get technical support with the Developer support plan, it is limited to generic support primarily around technical configurations with AWS use cases, and the support team will not be able to discuss specific application-layer problems that you might be having. Support is also only available via email (no phone support is offered) during business hours, with access to Cloud Support associates. While you can raise an unlimited number of cases, the case severity and response times are *within 24 hours* for general guidance and *within 12 hours* for system-impaired issues.

As with the Basic support plan, you only get access to the seven core checks on the AWS Trusted Advisor tool under the Developer support plan.

Business support plan

The Business support plan is recommended for production environments and enables companies to get technical support with their workloads on AWS. Examples range from resolving an issue with an RDS database that may have failed, where you need to perform a restore from backup, to more complex troubleshooting issues and problem resolutions.

The Business support plan offers full 24/7 support via email, chat, and telephone. Depending on the severity of the issue, different response times are offered. For example, if you have a production system that is down, you can expect support from a Cloud Support engineer within 1 hour. Also, unlike the Developer support plan, which offers more generic support covering typical AWS use cases, the Business support plan includes helping you troubleshoot interoperability issues between AWS resources and third-party software. The level of support offered is therefore contextual to your use case.

For an additional cost, you also get access to AWS **Infrastructure Event Management** (**IEM**). This service offers guidance and operational support to help you with your project launch events or migration tasks. This includes evaluating your AWS environment readiness, identifying any potential risks and ways to mitigate them, followed by ongoing support during your actual launch, and a post-event review.

Finally, the Business support plan also entitles you to receive a full set of AWS Trusted Advisor checks that allow you to cross-reference your workloads against best practices and receive recommendations across five categories: **Cost Optimization**, **Security**, **Fault Tolerance**, **Performance**, and **Service Limits**. We will discuss the key features of the AWS Trusted Advisor tool in the *Chapter 13*, *Management and Governance on AWS*.

Enterprise support plan

The Enterprise support plan is the *crème de la crème* of all the support plans on offer from AWS. You would expect the best level of service under this plan given its price tag, which starts at **US Dollars** (**USD**) $15,000 per month. The Enterprise support plan is naturally appropriate for very large organizations, such as multinational companies or companies that have extensively large workloads spread globally.

Examples of such companies include Netflix, Amazon Prime, and Dropbox. The Enterprise Support plan stands out because of its different VIP-style offerings such as a designated **Technical Account Manager** (**TAM**). Your TAM will actively monitor your environment and work closely with you to actively guide your team through planning, design, and implementation of your cloud projects.

Your TAM will assist with optimization tasks and suggest various best-practice methodologies, and also provide access to the best experts within AWS. Another key offering is access to Well-Architected reviews. This allows you to get access to a senior AWS solutions architect who can conduct an audit of your solutions deployed on AWS. AWS will provide guidance and best practices to help you design reliable, scalable, fault-tolerant, and cost-effective solutions.

In terms of **service-level agreements** (**SLAs**), you get full 24/7 email, chat, and phone support, with access to senior cloud engineers and with a 15-minute response time for business-critical technical issues.

Here is a table highlighting some key benefits of the different AWS support plans:

Basic	Developer	Business	Enterprise
Offers only non-technical customer support	Technical support offered during business hours alone (8:00 a.m. to 6:00 p.m.—customer local time zone)	24/7 phone, email, and chat access to Cloud Support engineers	24/7 phone, email, and chat access to Cloud Support engineers
	General guidance offered within 24 business hours	General guidance offered within 24 hours	General guidance offered within 24 hours
	System impaired troubleshooting offered within 12 business hours	System impaired troubleshooting offered within 12 hours	System impaired-troubleshooting offered within 12 hours
		Production system-impaired support within 4 hours	Production system-impaired support within 4 hours
		Production system-down support within 1 hour	Production system-down support within 1 hour
			Business-critical-system down support within 15 minutes

You can find a full breakdown of each plan at `https://aws.amazon.com/premiumsupport/plans/`.

In this section, we reviewed the four AWS support plans, ranging from a free offering of the Basic support plan to the all-inclusive VIP-style treatment of the Enterprise support plan. A key difference between the plans is to know which plans offer 24/7 technical support via phone and chat and which do not. Additionally, not all plans give you full access to all Trusted Advisor reports that can help you ascertain whether your workloads and solutions follow best practices and security guidelines.

Make sure you know the key difference in the response times for production and business-critical issues and which additional features are offered with the Enterprise plan, such as access to a designated TAM.

In the next section, we will look at the AWS Service Health Dashboard, which enables you to get a visual on the current status of each service across every Region. Any outages of a given service are highlighted on the service dashboard.

Overview of the AWS Service Health Dashboard

AWS publishes service health status across all data centers located in its various Regions. This is the first place you should consider investigating if a service appears to be non-responsive. AWS offers SLAs for its various service offerings and you can get a full list of these at `https://aws.amazon.com/legal/service-level-agreements/`.

You can see an example overview of the AWS Service Health Dashboard here:

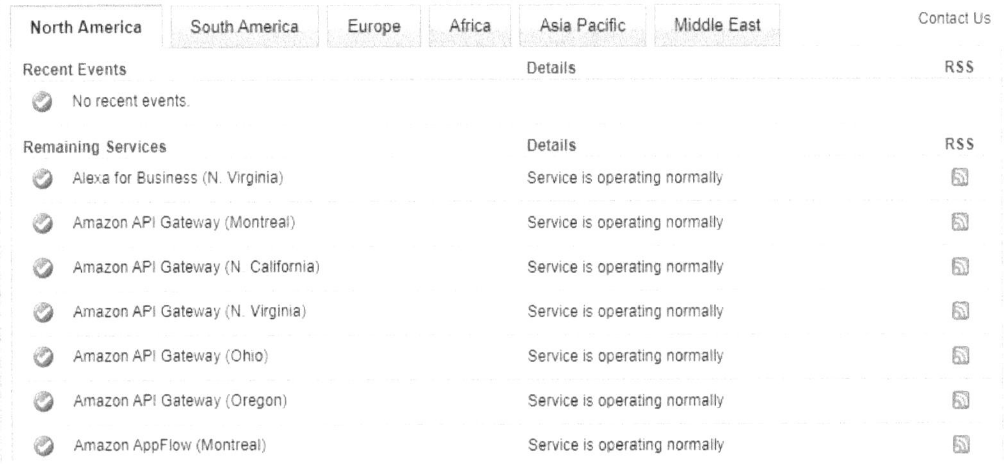

Figure 2.7 – AWS Service Health Dashboard. Image courtesy of AWS
(`https://status.aws.amazon.com/`)

In the next section, we move on to look at the AWS PHD. Similar to the previously discussed Service Health Dashboard, the PHD provides information on services that affect your resources and workloads. You can access the PHD once logged in to the AWS Management Console of your AWS Account. We look at how to create your AWS Account in *Chapter 3*, *Exploring AWS Accounts, Multi-Account Strategy, and AWS Organizations*, as well as understanding the AWS PHD.

While the AWS Service Health Dashboard provides general information on all of AWS services and their availability, if there are any outages, the PHD provides more tailored information, wherein it reports on issues that may affect your applications and resources, as illustrated in the following screenshot:

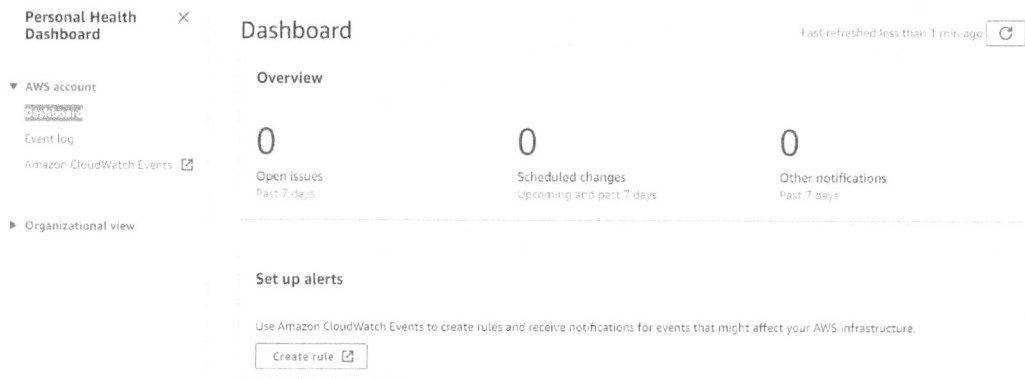

Figure 2.8 – AWS PHD

Some of the key benefits of the PHD include the following:

- **Personalized view of service health**—This includes information that will allow you to quickly identify any issues with any AWS service that may impact your resources or applications.

- **Proactive notifications**—You also get access to notifications such as those for upcoming schedule changes that may impact you. This will enable you to plan for those scheduled changes accordingly. You can also set up alerts to receive timely and relevant information using a service called Amazon CloudWatch Events.

- **Detailed troubleshooting guidance**—Alerts that you receive will also include necessary remediation details and any action that you may need to take.

- **Integration and automation**—The PHD integrates with CloudWatch events, which can be used to trigger automated tasks such as Lambda functions.

- **Aggregate health events across AWS Organizations**—AWS Organizations is a service that helps you manage multiple AWS accounts centrally. You can use the PHD to aggregate all notifications for every account in your organization. This gives you a holistic view of real-time events, operational issues, or any scheduled maintenance across all accounts.

So far, we have covered some basic components of the AWS Global Infrastructure, compared regional services with global services, and examined the AWS Service Health Dashboard and the PHD. You should note that you must agree to certain terms and conditions before using any service on the AWS Platform. Specifically, you must follow the AWS AuP, which we take a look at next.

The AWS AuP

Although it's obvious, you need to remember that when you sign up for an AWS account for your personal or business use, you must comply with the AWS AuP. This policy provides information on acceptable use of the services offered by AWS and describes prohibited uses. It is imperative that you adhere to the guidelines in the policy or you risk your account being suspended or terminated, which may well affect the workloads you deploy for your business.

You can review the full policy at `https://aws.amazon.com/aup/`. By signing up for the services offered by AWS, you automatically agree to the latest version of this policy.

Summary

In this chapter, we examined the AWS Global Infrastructure in detail. We understood that, as the customer, you can choose which AWS Region to deploy your resources and applications in, and looked at the key factors that should be taken into consideration when choosing a specific Region.

We also looked at the concept of AZs, which enable you to build your solutions that offer high availability, fault-tolerance, and scalability options. Understanding which services are classified as global or regional is also critically important to your design architecture.

We then looked at edge locations and their importance in the overall design of your cloud-application solutions. Edge locations and regional edge caches enable you to make your web applications' content and digital assets available to customers on a global scale via low-latency and high-speed network links. This is achieved by caching frequently accessed content locally at edge locations closer to the end users, substantially improving the UX in accessing that content.

Next, we examined the four different AWS support plans available and discussed the key differences between them. Choosing the right support plan is vital to ensure that you get technical assistance when you need it most, depending on your specific use case. Furthermore, understanding the offering of each plan will allow you to decide how much support is going to cost your organization, enabling you to budget accordingly.

Finally, we also examined the benefits offered by the PHD service—a vital tool to receive notifications and alerts of any AWS service outage or planned maintenance task that is likely to affect your workloads and applications. This will enable you to make proactive decisions as and when required.

In the next chapter, we will learn how to access the vast array of AWS services using an AWS account. An AWS account is required in order to consume its various services in a secure and isolated manner and it ensures that any workloads to deploy on AWS are not accessible to any other customers or entities—unless, of course, you grant such access.

We also discuss multi-account strategies and why you might need to have more than one AWS account, as well as how to centrally manage multiple AWS accounts.

Finally, we walk you through a step-by-step guide on creating your very first free AWS account.

In the next section, we present a series of review questions to help you test your knowledge gained so far.

Questions

Here are a few questions to test your knowledge:

1. Which of the following AWS support plans gives you access to all AWS Trusted Advisor reports? (Select two answers)

    A. Basic support plan

    B. Developer support plan

    C. Business support plan

    D. Enterprise support plan

    E. Global support plan

2. You have spent months developing a new application for your customers. You are now ready to go live and want to ensure that you have access to AWS technical support engineers if there are any issues with your application servers or backend database. Your organization is comfortable with 1-hour response times for production-system down issues. Which support plan is the most cost-effective option for you?

    A. Basic support plan

    B. Developer support plan

    C. Business support plan

    D. Enterprise support plan

3. Which AWS support plan gives you access to a technical account manager who will monitor your environment and provide guidance to optimize your workloads on the AWS platform?

    A. Basic support plan

    B. Developer support plan

    C. Business support plan

    D. Enterprise support plan

4. You are planning to build a test and development environment on AWS as a precursor to ultimately migrating your workloads to the platform. In the interim period, your developers require some basic technical support as they are new to cloud computing. Which AWS support plan offers cost-effective access to Cloud Support associates during business hours?

    A. Basic support plan

    B. Developer support plan

    C. Business support plan

    D. Enterprise support plan

5. Which of the following services is provided across all AWS support plans and allows support access 24/7 via telephone, chat, and email?

    A. Access to technical support via telephone and chat

    B. Access to customer support services to resolve any billing or account login issues

    C. Access to a technical account manager to help you manage your account

    D. Access to a full range of reports from the AWS Trusted Advisor

6. Which feature of the AWS Global Infrastructure enables you to launch applications and store data in a manner that is compliant with regulatory requirements?

    A. Regions

    B. AZs

    C. Edge location

    D. CloudFront

7. Which component of the AWS Global Infrastructure enables you to distribute your content to users across the globe such that cached versions of your digital assets are available locally to those users?

    A. Regions

    B. AZs

    C. Edge locations

    D. AWS RDS

8. Which component of the AWS Global Infrastructure enables you to architect your application solution to offer high-availability capabilities within a specific Region?

    A. Regions

    B. AZs

    C. Edge locations

    D. Regional edge caches

9. Which of the following services are considered global services on the AWS platform? (Select two answers)

    A. AWS IAM

    B. Amazon **Virtual Private Cloud** (**VPC**)

    C. Amazon Snowball

    D. AWS EC2

    E. Amazon CloudFront

10. Which of the following services are designed to be set up, configured, and consumed on premises? (Select two answers)

    A. AWS Outposts

    B. Amazon Storage Gateway

    C. Amazon DynamoDB

    D. AWS **Simple Notification Service** (**SNS**)

    E. AWS PHD

11. As part of the signup process, you are required to adhere to policy guidelines that describe prohibited activities. Which policy does this fall under?

    A. Compliance policy

    B. Password policy

    C. AuP

    D. Vulnerability testing guidelines

12. Which AWS service publishes up-to-the-minute information regarding any outages or issues with any service across all Regions of the AWS ecosystem?

    A. PHD

    B. Outage and issues dashboard

    C. Service Health Dashboard

    D. Amazon CloudWatch

3
Exploring AWS Accounts, Multi-Account Strategy, and AWS Organizations

To access services on the AWS platform, you need to have an AWS account. AWS offers hundreds of different services, which you, as a customer, can consume to build cloud IT solutions for your business and clients.

AWS offers public cloud services that are accessible to anyone on the internet. An AWS account provides a means of accessing these public AWS services in an isolated boundary separate from other customers. This means that users outside your account cannot access your resources unless, of course, you grant them access. An AWS account thus offers security, access isolation, and billing boundaries for the services that you consume and the resources you deploy. In addition, the cost of consuming any AWS service will be allocated to your AWS account.

In this chapter, we explore the benefits of having multiple AWS accounts and we also discuss how to manage those accounts using a service called **AWS Organizations**. We also demonstrate how to set up your first AWS account, which you will be using throughout the various exercises in this training guide.

This chapter covers the following topics:

- Why have a multi-account AWS environment?
- Introduction to AWS Landing Zone
- Automating landing zone creation with AWS Control Tower
- Exploring AWS Organizations
- Exercise 3.1: Setting up your first AWS Free Tier account
- Exercise 3.2: Setting up a billing alarm

Why have a multi-account AWS environment?

While you can host all your business resources in a single AWS account, this can very quickly become too complex to manage. Imagine hosting multiple resources for your various non-production applications under development, **User Acceptance Testing (UAT)**, and production workloads, all within the same AWS account. This can rapidly become a huge management overhead. The complexity is further compounded because you would have to ensure that many of these applications are isolated from each other for compliance or security reasons. This would require you to define highly complex policies and permissions to ensure proper segregation of different workload types and effective management of resources.

Above all, having a single AWS account prevents you from limiting the blast radius of any major disasters. Separating your workloads using an appropriate strategy will help limit the blast radius of catastrophic disasters. So, for example, you can have a separate account for all your experimentation work such as developing new applications (what we call a sandbox environment) and other accounts for your actual development, testing, and production environments. In the event of any disaster (such as accidental exposure of root account credentials or a misconfiguration of an autoscaling group, causing a repeat loop of EC2 instance deployments) occurring in the sandbox environment, it will not affect the other accounts if those workloads are separated and isolated from one another.

In the diagram shown in *Figure 3.1*, we can see that a number of accounts have been created for a given company. In this example, we have development, UAT testing, and production accounts as well as a sandbox account for experimental work:

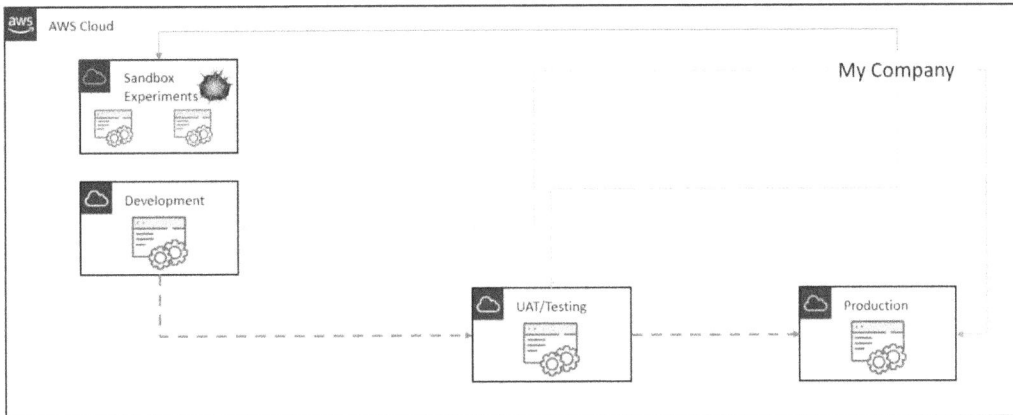

Figure 3.1 – Multiple accounts help to limit the blast radius of your workloads

Some key benefits of a multi-account architecture include the following:

- **Administrative isolation between workloads**: With a multi-account architecture, you can offer different business units varied levels of administrative controls, depending on several factors including development life cycles. For example, you may *not* want to grant your developers full access to the production account, where you deploy applications under public release.

- **Limited visibility and discoverability of workloads**: AWS accounts offer a natural boundary and enable you to isolate your workloads from any identity external to the account. Applications and resources deployed in an AWS account are not accessible to identities (users and applications) from any other AWS accounts unless permission is explicitly granted.

- **Isolation of security and identity management**: Instead of creating multiple user identity accounts for your team (for example, developers, server administrators, UAT testers) in each AWS account that they need to work in, you can host all your users in a *separate AWS identity management account*. This enables you to avoid duplicate accounts for your staff and reduces management overhead. Ultimately, your users can be granted access to any of your other AWS accounts using **cross-account access**, where policies and permissions limit the tasks they can perform following the **principle of least privilege**.

- **Isolation of recovery or audit accounts**: Many organizations will need to have a disaster recovery and business continuity strategy so that in the event of failure, the business can easily continue their operations using duplicate workloads and quickly recover from any major loss. Such workloads should be placed in separate accounts as best practice.

In this section, we briefly introduced you to the concept of AWS accounts and how they grant you isolated access to AWS services, ensuring that no other customer can gain visibility of your workloads and applications unless you explicitly grant them that level of access.

In the next section, we look at services offered by AWS to help you quickly design and architect your multi-account strategy. We discuss AWS Landing Zone, which was a service offered by AWS solutions architects to create a customized baseline architecture for your multi-account deployment. We also take a look at AWS Control Tower, a service that automates the build of landing zones in accordance with industry best practices.

AWS Landing Zone

Building a multi-account environment can become very complex and time-consuming. AWS offers its customers a set of best practice methodologies to follow when designing a multi-account ecosystem. Previously, AWS offered a solution called *AWS Landing Zone*, which has now been deprecated in favor of the new AWS Control Tower.

The previous AWS Landing Zone service offered customers a baseline blueprint to design and architect a multi-account environment, which offered identity and access management, governance, data security, and logging features.

> **Important note**
> Although AWS Landing Zone is currently in long-term support and will not receive any additional features, it is still likely to show up in the exam.

AWS Control Tower

Customers who are now looking to set up a landing zone in accordance with the updated architectural best practices should use the new AWS Control Tower. This service automates the setup of a new landing zone using the latest blueprints. Some AWS accounts created as part of this landing zone include the following:

- Creation of an AWS Organizations and multi-account setup
- Identity and access management with AWS **Single Sign-On** (**SSO**) default directory services
- Account federation using SSO
- Centralized logging using AWS CloudTrail and AWS Config

The landing zone deployed by AWS Control Tower comes configured with recommended security policies called guardrails and customers can choose how their accounts are configured to comply with their overall organizational policies.

In this section, we looked at two services that can be used easily to architect your multi-account architecture. If this was to be carried out manually, it would be time-consuming and complex. We also examined AWS Landing Zone and the new AWS Control Tower.

In the next section, we will look at how you centrally manage your various AWS accounts using a service called AWS Organizations. As a cloud practitioner, understanding the requirements for multiple accounts and knowing which tools can be used to manage them is vital to help your clients build a robust and secure cloud architecture for their businesses.

Managing multiple accounts – AWS Organizations

It is all very well creating n number of AWS accounts to help you separate out different workloads or different application life cycles. But you also need to think about how you effectively manage them – who has permissions to what and which set of services can be launched and configured in which AWS account.

Introducing AWS Organizations

The AWS Organizations service enables you to centrally manage all of your AWS accounts. This service is **offered free**; the resources launched within are, however, chargeable as they would be even if you were not using the AWS Organizations service. AWS Organizations allows you to create one **management account** (previously termed **master account**) and then invite or create additional AWS accounts that will become member accounts of the organization. AWS Organizations offers tight integration with a number of AWS services and allows you to scope out which services can be consumed in your individual member accounts.

You may also have multiple member accounts in each AWS Organizations instance, where some accounts share similar types of workloads or functions. You can, therefore, collate common accounts under what are known as **Organization Units** (**OUs**). An OU is a logical group of one or more AWS accounts in your AWS Organizations that allows you to organize your numerous AWS accounts into hierarchies. This makes it easier to manage your numerous AWS accounts.

You can then apply **Service Control Policies (SCPs)** to your OUs (or directly to the AWS account, although this is not considered best practice) to apply guardrails to what services can be deployed and configured in each account. In the following diagram, we can see how service control policies are being applied to OUs, which determine permission boundaries for the AWS accounts contained within those OUs:

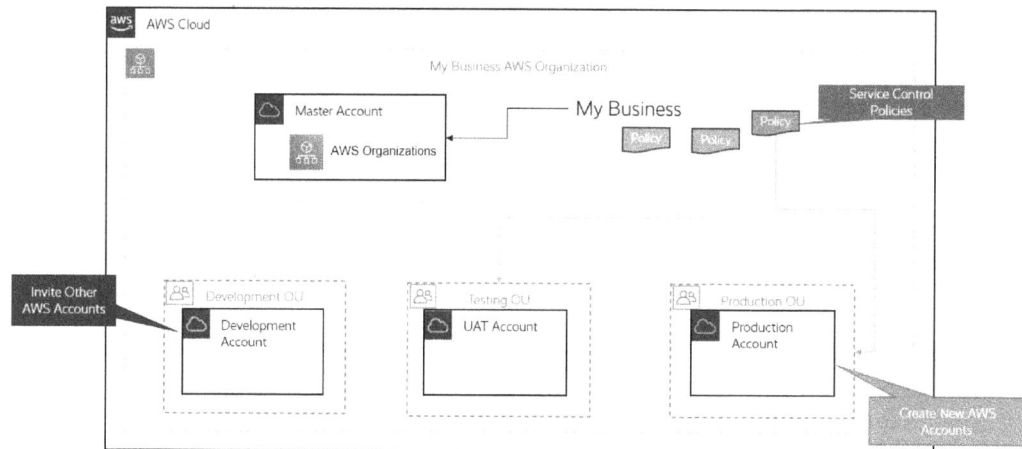

Figure 3.2 – AWS Organizations with multiple accounts

Consolidated billing

Another key feature of AWS Organizations is **consolidated billing**. AWS Organizations can be deployed using one of two options – **all features** or just the **consolidated billing feature**:

- **All features**: This allows you to take advantage of AWS Organizations' security and management capabilities for member accounts with the **Service Control Policies (SCPs)** and tag policies. The **All features** option also comes with the consolidated billing feature, which allows you to have one centralized bill for all your member accounts. The **Management Account** is ultimately responsible for the bill and charges incurred by the other member accounts.

- **Consolidated billing feature**: If you enable just the consolidated billing feature, you get basic management tools and the ability to get a centralized bill for all your member accounts.

One of the prime benefits of AWS Organizations in your multiple account strategy is the consolidated billing feature. The management account is responsible for all charges incurred by its member accounts and you are provided with a combined view of the total AWS charges incurred by the member accounts. You can also view itemized details of charges incurred by each individual member account.

Some key benefits are listed as follows:

- **Single bill**: You get a single bill, which shows charges incurred by each member account.

- **Easy tracking**: You can track individual account charges.

- **Volume discounts**: With multiple accounts combined in AWS Organizations, you benefit by combining the costs across your accounts and this enables you to receive discounts when the total charges cross certain volume discount thresholds.

- **Free service**: You can make use of the consolidated billing feature at no additional cost.

AWS Organizations offers centralized management of all your AWS accounts, enabling you to enforce security boundaries and ensure cost management. Determining how many AWS accounts you need for your organization will largely depend on your business requirements and whether you have the necessary infrastructure as per best practices.

How many AWS accounts do you need?

Deciding on the number of AWS accounts you need will involve careful consideration of your business functional requirements, the complexity of workloads, security, and compliance requirements (for example, HIPAA compliance, which requires businesses in the healthcare industry to follow various guidelines and enforce technical safeguards to protect the integrity of an individual's health information).

In general, you would expect to host a minimum number of accounts to facilitate the isolation of your different development and production life cycles and offer redundancy and resilience against failures. You should also encourage a design strategy that limits administrative overhead.

The following discussion illustrates one such example of a set of AWS accounts you might need, but remember that your architecture might be different based on your specific requirements. As much as possible, it is advisable to create accounts that meet your functional requirements and fulfill your security controls, rather than simply creating accounts based on some corporate hierarchical nature.

Depending on your architecture, you should start by defining what OUs you need to create in AWS Organizations. Before configuring AWS Organizations, you need to create your first AWS account, or at least a new AWS account that will be the management account of the organization.

In the next section, we discuss best practices for configuring AWS Organizations, specifically the importance of correctly designing your OUs.

Core AWS OUs

At a basic level, you should create an infrastructure OU and a security OU, which will contain a shared infrastructure services account and a security account:

- **Infrastructure services account**: These will contain services that can be shared across all accounts for common services (for example, directory services, shared networking environments, and other common shared IT services, such as central repositories for your **Amazon Machine Images (AMIs)**).

- **Security services**: These will contain a centralized Identity and Access Management account to host individual user accounts, groups, and roles. These identities will then be granted access to other accounts via cross-account policies, as well as other services for logging and auditing.

Both the foundational core OUs will contain non-production and production AWS accounts. Separating non-production from production ensures the isolation of workloads under development from those that are released for business consumption and limits the blast radius of any disasters.

Additional OUs

Depending on your business use case, you can have any number of additional OUs within which to contain appropriate accounts. As depicted in the following diagram, your AWS Organizations will comprise the core OUs as well as any number of additional OUs that will meet your individual business requirements:

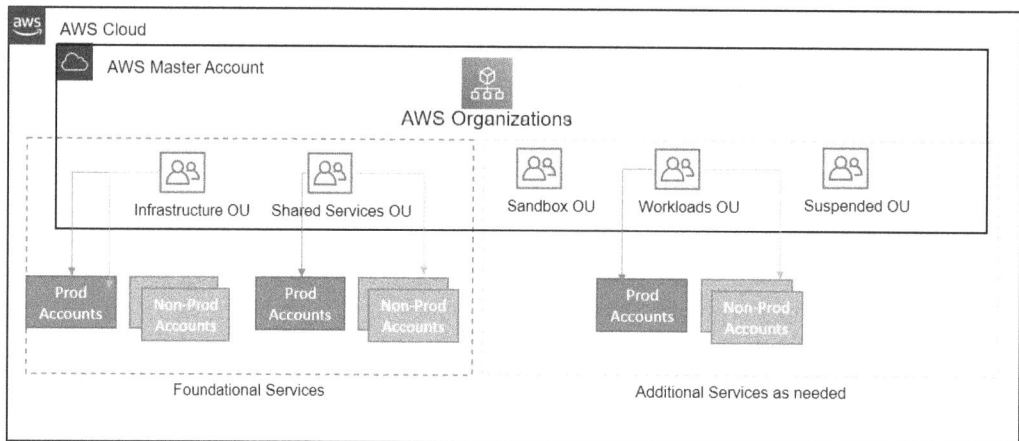

Figure 3.3 – AWS configured with core infrastructure as well as security and AWS Organizations additional OUs

To illustrate an example, you may wish to configure the following additional OUs for your organization:

- **Sandbox OUs**: As your business develops new applications or perhaps conducts experiments to improve existing workloads, you want to make sure that the blast radius is restricted to an environment you can afford to have failures in. A sandbox environment should be an account that can be detached from internal networks and you must establish limits to cap expenses and prevent overuse.

- **Workloads OUs**: This OU will contain AWS accounts, where you host customer-facing applications. Ideally, you will have several non-production environments such as Dev, Test, and Pre-Prod. You should also consider multiple production accounts, such as Prod 1 and Prod 2, for even greater resilience.

- **Suspended OUs**: Any account no longer being used may still need to be kept for auditing and compliance purposes. Apply necessary SCPs to ensure that only specific admins can access these accounts.

The preceding list is by no means exhaustive, and your OU architecture is going to be influenced by your functional and technical needs. For additional guidance on the type of OU structure you could build, refer to the following web page on recommended OUs: `https://docs.aws.amazon.com/whitepapers/latest/organizing-your-aws-environment/recommended-ous.html`.

In this section, we learned about the AWS Organizations service and how it can be used to design and architect a multi-account strategy for your business. We examine the core features of AWS Organizations, which include the ability to apply **Service Control Policies** (**SCPs**) and effectively manage your billing and costs with the consolidated billing feature.

In the next section, we will look at AWS Free Tier accounts and showcase how to create one, for which you will need to complete the various hands-on labs in this training guide.

AWS Free Tier accounts

An AWS Free Tier account is a normal standard account that can be used for any purpose or workload type. AWS offers a generous Free Tier for the first 12 months of opening any new account. The Free Tier offers access to more than 85 AWS technologies and services (at the time of writing this training guide), wherein if you consume these services up to specified thresholds, you will not be charged. For example, under the Free Tier, you can do the following:

- Consume up to 5 GB of Amazon S3 storage for up to 12 months, free of charge.

- Launch a t.2micro **Elastic Compute Cloud** (EC2) instance running either a specific distribution of the Linux OS or a base Windows OS for up to 750 hours a month. In fact, with this offering, you could potentially run one low-powered website for an entire year without incurring the cost of compute for those 12 months. The Free Tier offering for EC2 is based on the number of hours per month, so there is nothing to stop you running two or even four t2.micro instances for a few hours a month, as long as the total duration does not exceed 750 compute hours a month. This may be useful if you want to perform some proof of concept work or testing.

- Run a lightweight Amazon Relational Database Service (RDS) instance for up to 750 hours a month.

In the next section, we will look at some additional features available to you under the AWS Free Tier account offering.

Free tools

In addition to the standard 12-month Free Tier offering, certain services are offered completely free, without any time limit. These may include tools that can be used to deploy certain resources. The tools themselves are free to use, but the resources deployed will be chargeable based on their price list. Some examples of these tools are as follows:

- **AWS CloudFormation**: This is an AWS service that enables you to define templates (using code) to launch various infrastructure components. Amazon CloudFormation enables developers and architects to create a collection of related AWS and third-party resources, and provision and manage them in an orderly and predictable fashion. These templates can be used for repeat deployments. While the service itself is free to use, resources and infrastructure components deployed will be chargeable as per Amazon's price list, such as EC2 instances, RDS databases, and networking components.

- **Amazon Elastic Beanstalk**: This is an orchestration service that provisions necessary infrastructure components to support and power your applications. Such infrastructure components include S3 buckets, EC2 instances, and load balancers. Amazon Elastic Beanstalk enables developers to simply upload their code ready for deployment and AWS provisions the infrastructure required to support that application. The orchestration service is again offered free of cost, but the resources are chargeable, as indicated previously.

In addition to the preceding free tools, AWS offers certain limited resources for some services throughout the lifetime of your AWS account. In the next section, we will look at some examples.

Always free services (limited offering)

Some AWS services are offered completely free on a permanent basis. These include services where resources can be created up to specified threshold limits without incurring any charges. Examples include the following:

- **Amazon CloudWatch**: Used for monitoring your cloud resources and applications. You get 10 custom metrics and up to 10 alarms, with up to 1,000,000 API requests completely free.

- **Amazon Lambda**: A serverless compute service that allows you to run code in response to events and can help you build a serverless architecture. With AWS Lambda, you get up to 1,000,000 free requests per month and up to 3.2 million seconds of compute time per month.

- **AWS Organizations**: Centrally manage and control access to your AWS accounts and benefit from consolidated billing with volume discounts. This service is offered entirely free of charge on a permanent basis.

AWS also offers various product trials from time to time. In the next section, we will look at this in some detail.

Free trials

Some services are offered with free trials, for example a 30-day trial to test out a service. Examples include the following:

- **Amazon Workspaces**: These are virtual desktops that can run Linux or Windows OSes. As a trial, you can get up to two standard edition workspaces that come with an 80 GB root volume and a 50 GB user volume. You can use both workspaces for up to 40 hours combined usage per month, for 2 months, as part of the free trial.

- **Amazon Detective**: This service enables you to analyze and visualize security data and identify the root cause of potential security issues. You get a 30-day free trial of the service.

- **Redshift**: This is an enterprise-grade data warehousing solution through which you can query and combine exabytes of structured and semi-structured data and conduct insights and analysis of your data. AWS offers a 2-month free trial with 750 hours of a DC2.Large node per month.

In this section, we looked at details of the AWS Free Tier account. Amazon offers a wide range of services with specific limits free for the first 12 months of your account opening. This allows you to experiment with those services, create sandbox environments, and start architecting your solutions. The AWS Free Tier is of great help to students wanting to learn about the various AWS services in preparation for their exams. At times, however, you may have to go over the free tier thresholds to really learn about more complex configuration options.

In the next section, we will complete a lab exercise, where you will be provided with a step-by-step guide on creating your first AWS account.

Exercise 3.1 – Setting up your first AWS Free Tier account

The following step-by-step process will show you how to set up your first AWS account. You will need the following to complete the setup process:

- Your personal details, name, physical address, and an email address.

- A mobile phone.

- A credit card. As far as possible, the labs in this training guide will fall under the free tier and your credit card will not be charged for those resources you deploy. A couple of labs, however, may go over the free tier threshold and if you choose to do those labs, there may be a small minimum charge. We will discuss this in more detail shortly.

Now that we know about the requirements, let's get started with creating our account:

1. In your favorite browser, search for the term AWS Free Tier and you should find a link to the Amazon Web Services Free Tier offering. Click on the second link, as shown in the following screenshot:

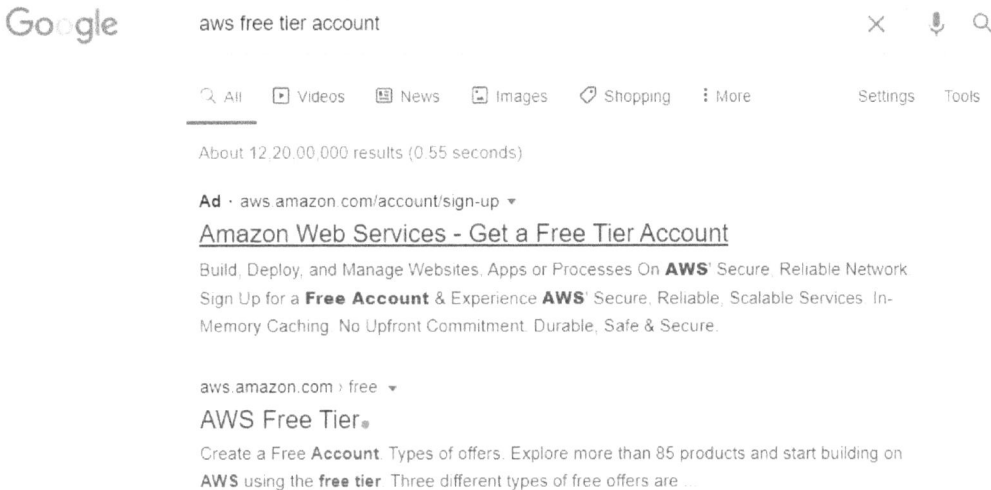

Figure 3.4 – AWS Free Tier link via Google Search

2. You will be taken to the AWS Free Tier home page. Next, click on **Create a Free Account**:

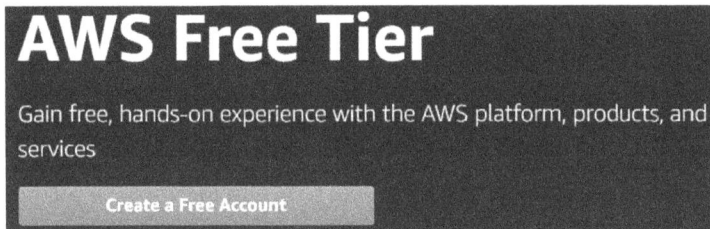

Figure 3.5 – Free account setup

3. At the AWS **Sign in** page, click, **Create a new AWS account**:

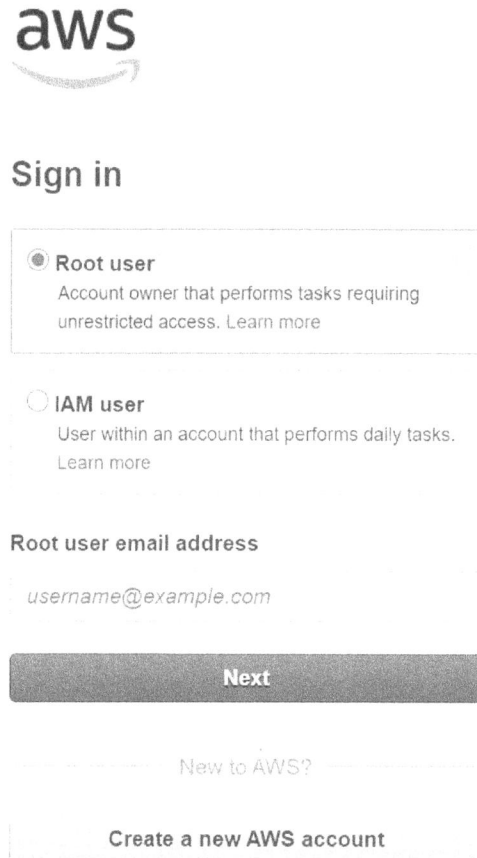

Figure 3.6 – Free account setup – creating a new AWS account

4. Next, provide an email address, choose a password, and choose an account name. The account name can be any name you like to use to identify the purpose of the account, for example, Dev or Prod:

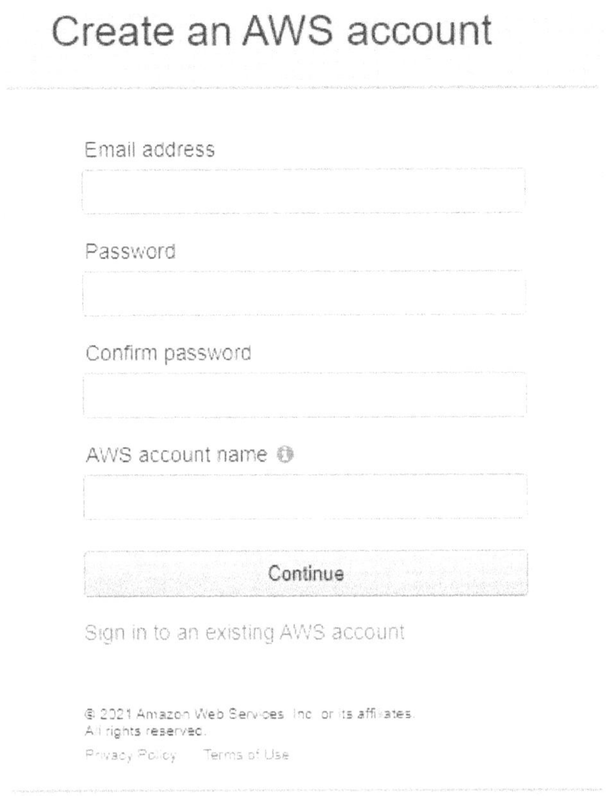

Create an AWS account

Email address

Password

Confirm password

AWS account name ⓘ

Continue

Sign in to an existing AWS account

© 2021 Amazon Web Services, Inc. or its affiliates.
All rights reserved.
Privacy Policy   Terms of Use

Figure 3.7 – Free account setup – providing an email address and an account name

5. Next, you may be asked to fill up a captcha form for security. Type the letters and numbers in the image into the textbox provided and click **Continue**:

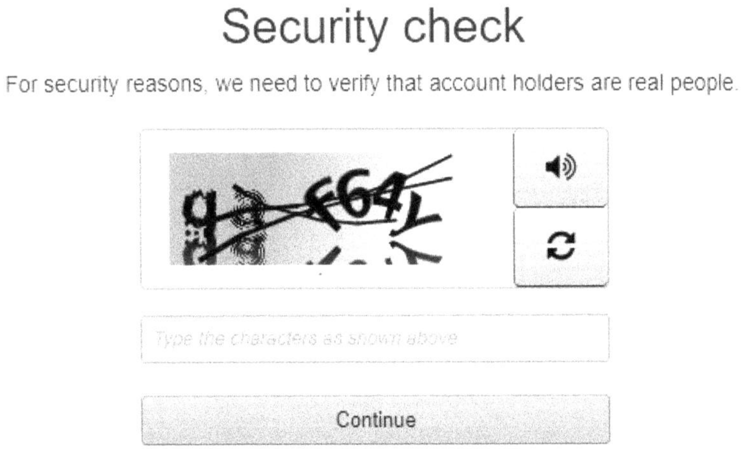

Security check

For security reasons, we need to verify that account holders are real people.

Type the characters as shown above

Continue

Figure 3.8 – Free account setup – security check screen

6. Next, you need to provide your full contact details and choose the type of account you want to create – Personal or Professional. Professional accounts allow you to get a full tax invoice and, in some countries, claim back certain types of tax, such as VAT or GST. Once you have completed the form, click **Create Account and Continue**:

Account type ⓘ
◉ Professional   ◯ Personal

Full name

Sandbox

Company name

Phone number

Country/Region

United States   ⌄

Address

Street, P.O. Box, Company Name, c/o

Apartment, suite, unit, building, floor, etc.

City

State / Province or region

Postal code

☐ Check here to indicate that you have read
    and agree to the terms of the AWS
    Customer Agreement

Create Account and Continue

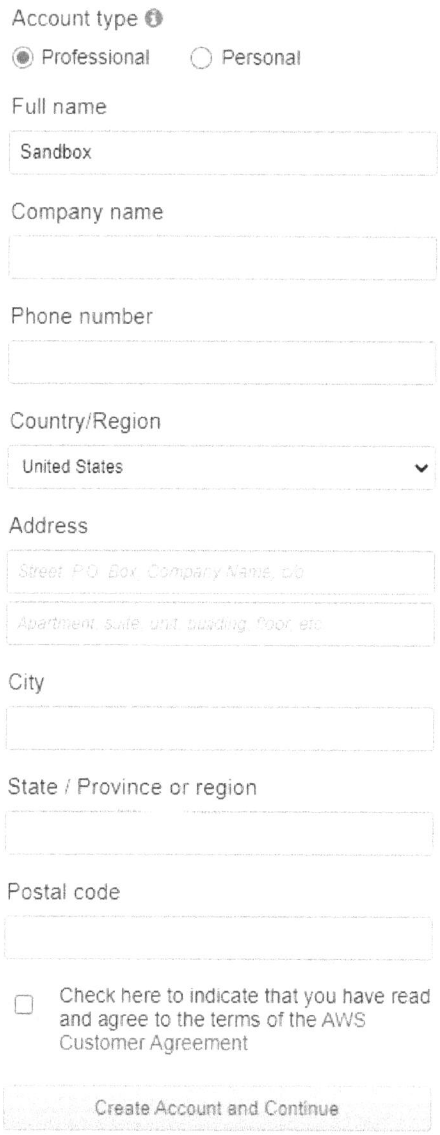

Figure 3.9 – Free account setup – providing contact details

7. Next, you need to provide your debit or credit card details, confirm your address, and click **Verify and Add**:

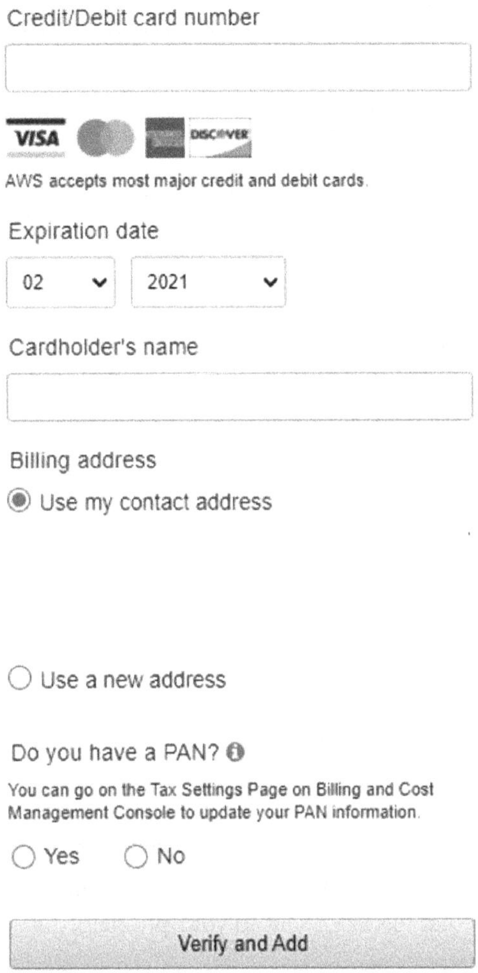

Credit/Debit card number

VISA   DISC●VER

AWS accepts most major credit and debit cards.

Expiration date

02   2021

Cardholder's name

Billing address

⦿ Use my contact address

○ Use a new address

Do you have a PAN? ⓘ

You can go on the Tax Settings Page on Billing and Cost
Management Console to update your PAN information.

○ Yes   ○ No

Verify and Add

Figure 3.10 – Free account setup – providing credit card details

8. You may need to provide a one-time PIN that will be sent to your phone to verify the card. In some countries, you may be charged a very small amount, perhaps a few cents, to verify the card, but this amount will be refunded through the banking system.

9. After this, you will receive a **Confirm your identity** dialog box that requires you to provide a phone number using which you will be sent a verification code, either as an SMS text message or voice call, depending on your preference. You will also need to complete the security checkbox and then click on, for instance, the **Send SMS** option, as shown in the following screenshot:

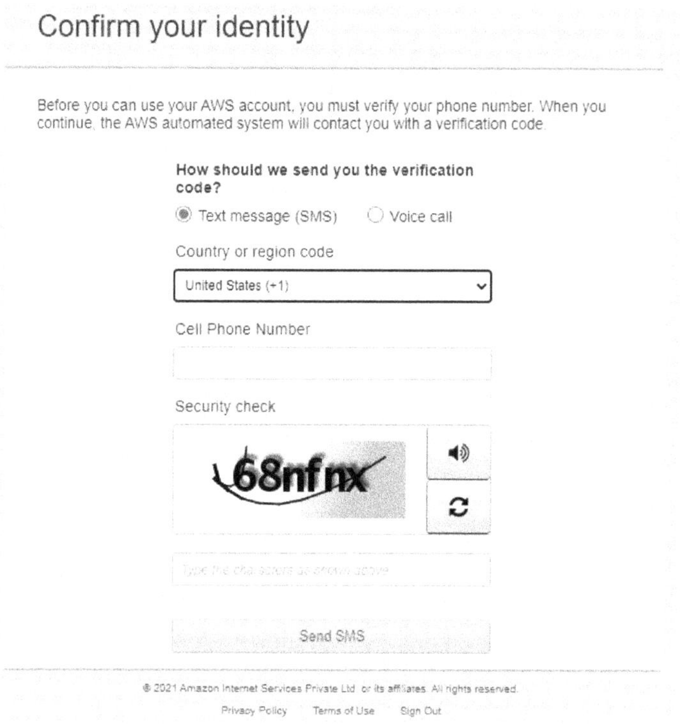

Confirm your identity

Before you can use your AWS account, you must verify your phone number. When you continue, the AWS automated system will contact you with a verification code.

How should we send you the verification code?

◉ Text message (SMS)   ○ Voice call

Country or region code

United States (+1)   ⌄

Cell Phone Number

Security check

68nfnx

🔊

⟳

Type the characters as shown above

Send SMS

® 2021 Amazon Internet Services Private Ltd. or its affiliates. All rights reserved.

Privacy Policy   Terms of Use   Sign Out

Figure 3.11 – Free account setup – providing a phone number for verification

10. You will then be taken to the following dialog box, which requires you to provide the verification code that has been sent to your phone:

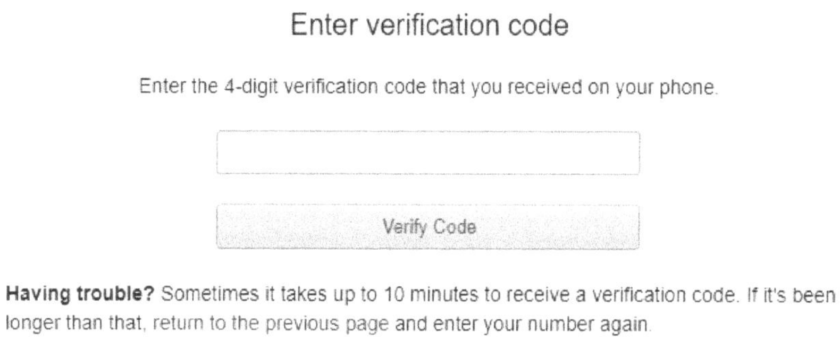

Enter verification code

Enter the 4-digit verification code that you received on your phone.

Verify Code

Having trouble? Sometimes it takes up to 10 minutes to receive a verification code. If it's been longer than that, return to the previous page and enter your number again.

Figure 3.12 – Free account setup – providing a verification code received on your mobile phone

11. Once you enter the code, your account will be verified, and you should get the following dialog box. Click **Continue**:

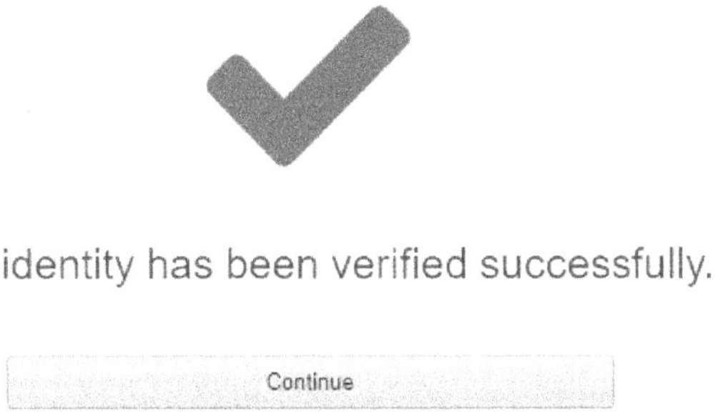

Your identity has been verified successfully.

Continue

Figure 3.13 – Free account setup – completing the verification process

12. You will be taken to the **Select a Support Plan** screen, where you will have the option to select your support plan for this account. For the purpose of this account, go ahead and select **Basic Plan**:

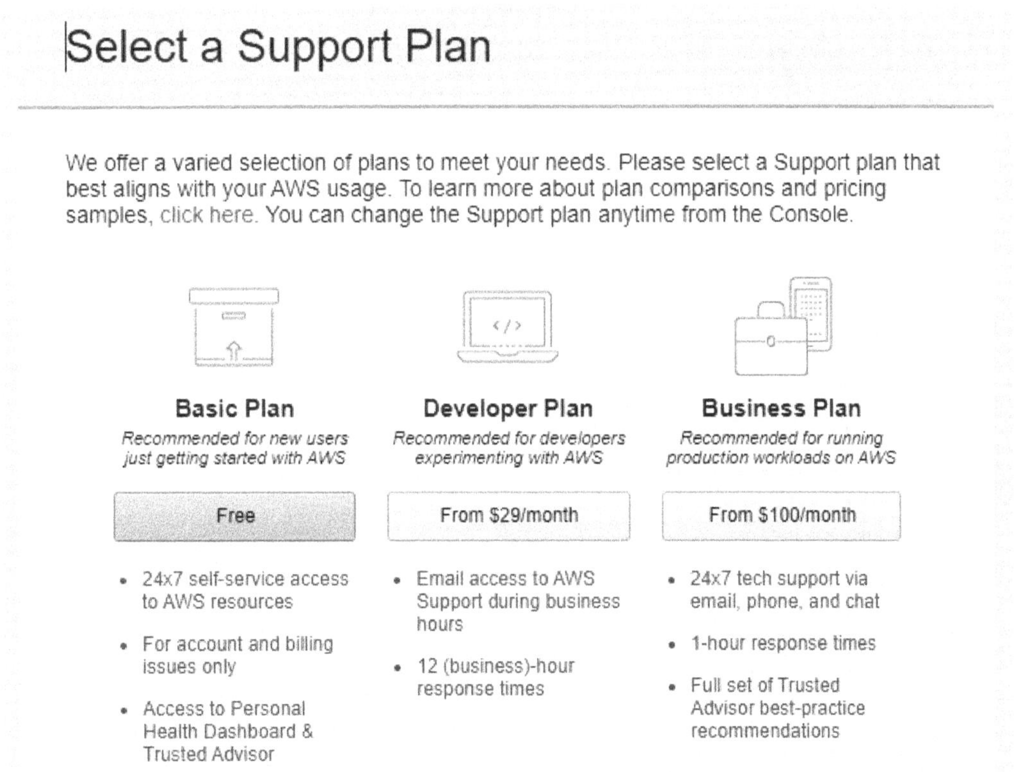

Select a Support Plan

We offer a varied selection of plans to meet your needs. Please select a Support plan that best aligns with your AWS usage. To learn more about plan comparisons and pricing samples, click here. You can change the Support plan anytime from the Console.

Basic Plan	**Developer Plan**	**Business Plan**
Recommended for new users just getting started with AWS	Recommended for developers experimenting with AWS	Recommended for running production workloads on AWS
Free	From $29/month	From $100/month
• 24x7 self-service access to AWS resources	• Email access to AWS Support during business hours	• 24x7 tech support via email, phone, and chat
• For account and billing issues only	• 12 (business)-hour response times	• 1-hour response times
• Access to Personal Health Dashboard & Trusted Advisor		• Full set of Trusted Advisor best-practice recommendations

Figure 3.14 – Free account setup – selecting the basic support plan

13. You will now be taken to the **Welcome to Amazon Web Services** screen. Click the **Sign In to the Console** button:

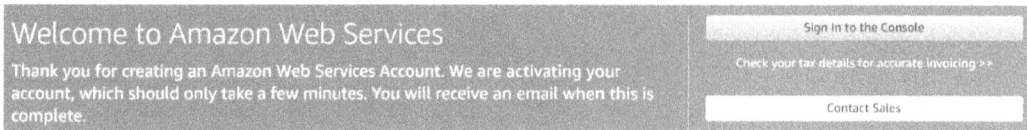

Figure 3.15 – Sign In to the Console

Ensure that you select **Root user** and provide the email address and password you used to create your new Free Tier account. This email address and password combination is also known as the **root user** of your AWS account and has complete control over your account:

Figure 3.16 – Free account setup – signing in as the root user to test account setup

14. Once logged in, you will be presented with the **AWS Management Console** page:

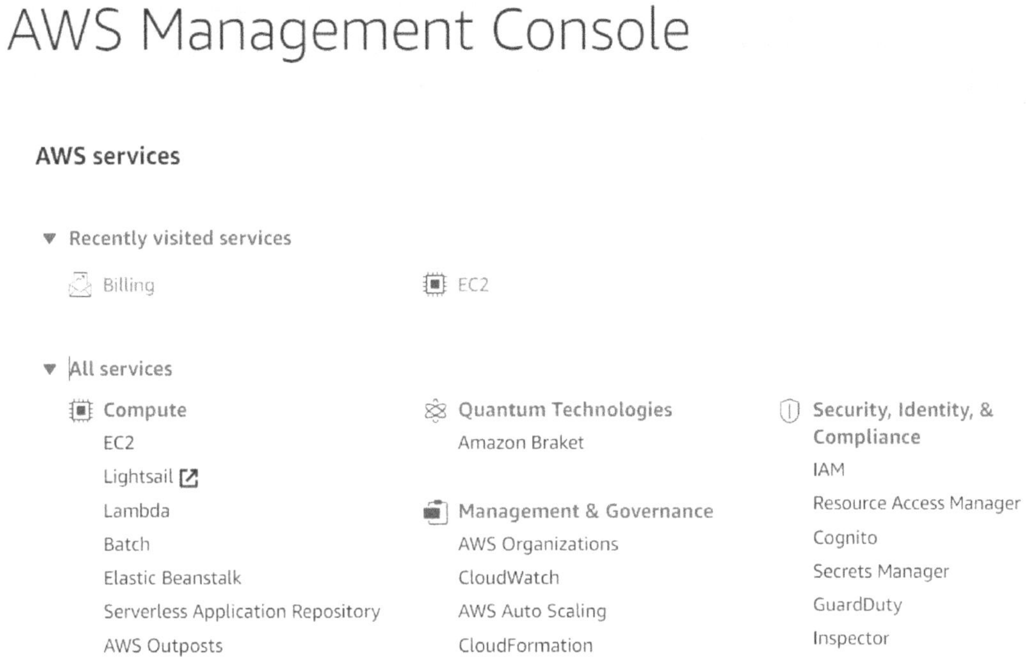

AWS Management Console

AWS services

▼ Recently visited services

    ⊠ Billing   ▣ EC2

▼ All services

▣ Compute	⚛ Quantum Technologies	ⓘ Security, Identity, & Compliance
EC2	Amazon Braket	
Lightsail ↗		IAM
Lambda	▤ Management & Governance	Resource Access Manager
Batch	AWS Organizations	Cognito
Elastic Beanstalk	CloudWatch	Secrets Manager
Serverless Application Repository	AWS Auto Scaling	GuardDuty
AWS Outposts	CloudFormation	Inspector

Figure 3.17 – Free account setup – accessing the AWS Management Console

In this section, you created your first AWS account, which will enable you to access all AWS services, across its global infrastructure. Using your AWS account, you can now set up and deploy a wide range of resources to support and host any application workload. In the next exercise, we look at how to set up your billing alarm.

To maximize your benefit from this training, you are encouraged to complete all of the exercises. This will help you to gain the necessary hands-on experience and the confidence to start building real-world solutions on AWS. While we ensure that most of the exercises fall within the Free Tier thresholds, there are a few that will incur some costs. You will be notified of this as part of the exercise and you may choose to complete them if you wish.

In addition, you may deploy certain resources and forget to terminate them, resulting in crossing some of those Free Tier thresholds. By having a billing alarm, you can set yourself a small budget sufficient to complete all of the labs. With a billing alarm in place, you will be alerted if your total charges cross the budgeted thresholds.

We discuss how to set up your billing alarm in the next section.

Exercise 3.2 – Setting up a billing alarm

When you configure a billing alarm, you define a dollar amount as a threshold value as your maximum budget. If the total charges on your AWS account cross this value, you are alerted with a notification and can take remedial action.

As previously discussed, this training guide offers several hands-on labs and exercises to enable you to gain real-world hands-on experience in configuring various services to host your workloads in the cloud. Most of the labs will fall within the free tier, except for a few that may incur very minimal charges. We indicate the labs that may incur such charges. It is also important to terminate any labs you complete to ensure you do not forget about them.

To complete all exercises in this training guide, we recommend you set a billing alarm of USD 10, although you can choose any value you are comfortable with. Should you exceed this dollar amount, you will be alerted with a notification via email to take immediate action. You can then terminate any labs you no longer need.

So, let's now proceed with configuring your AWS billing alarm. We start by enabling the option to receive billing alerts. This is a prerequisite step and must be completed before configuring billing alarms:

1. Log in to your AWS Management Console using your root account credentials. This is the email address and password you will have configured at the time of account sign-up.

2. Access the **Billing and Cost Management** dashboard at `https://console.aws.amazon.com/billing/`.

3. In the left-hand navigation pane, click **Billing Preferences**.

4. Click on **Receive Billing Alerts** and then click **Save Preferences**.

The following screenshot illustrates enabling the billing alerts option:

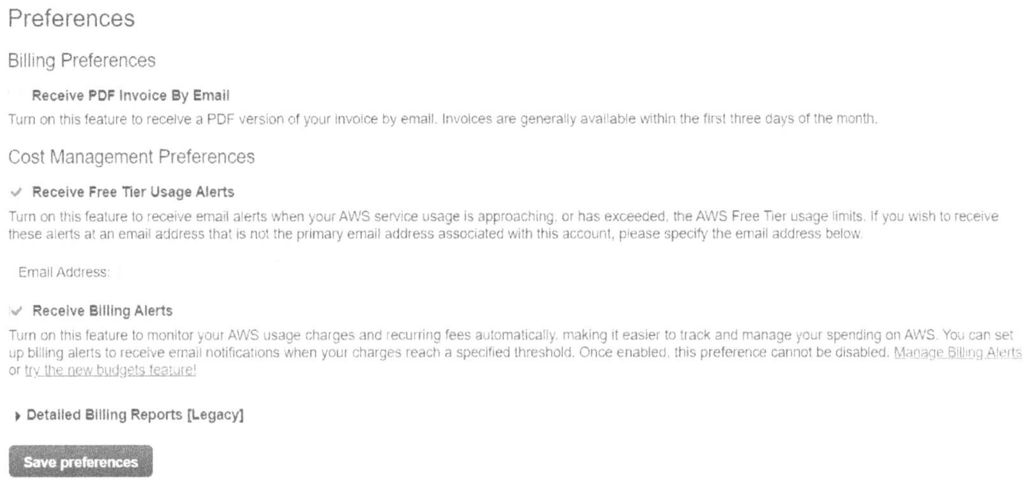

Preferences

Billing Preferences

Receive PDF Invoice By Email

Turn on this feature to receive a PDF version of your invoice by email. Invoices are generally available within the first three days of the month.

Cost Management Preferences

✓ **Receive Free Tier Usage Alerts**

Turn on this feature to receive email alerts when your AWS service usage is approaching, or has exceeded, the AWS Free Tier usage limits. If you wish to receive these alerts at an email address that is not the primary email address associated with this account, please specify the email address below.

Email Address:

✓ **Receive Billing Alerts**

Turn on this feature to monitor your AWS usage charges and recurring fees automatically, making it easier to track and manage your spending on AWS. You can set up billing alerts to receive email notifications when your charges reach a specified threshold. Once enabled, this preference cannot be disabled. Manage Billing Alerts or try the new budgets feature!

▸ **Detailed Billing Reports [Legacy]**

`Save preferences`

Figure 3.18 – Setting up billing alarms – enabling the option to receive billing alerts

Now that you have enabled the option to receive billing alerts, you can proceed with setting up your billing alarm. Note that once this setting has been enabled, it cannot be disabled.

In the following step-by-step process, you create an alarm that sends an email message when your estimated charges for your AWS account exceed a specified threshold:

1. Access the CloudWatch console at `https://console.aws.amazon.com/ cloudwatch/`. Note that billing metric data is stored in the US-East-1 Region. From the top right-hand menu, ensure that you are in the N. Virginia (us-east-1) Region as per the following screenshot:

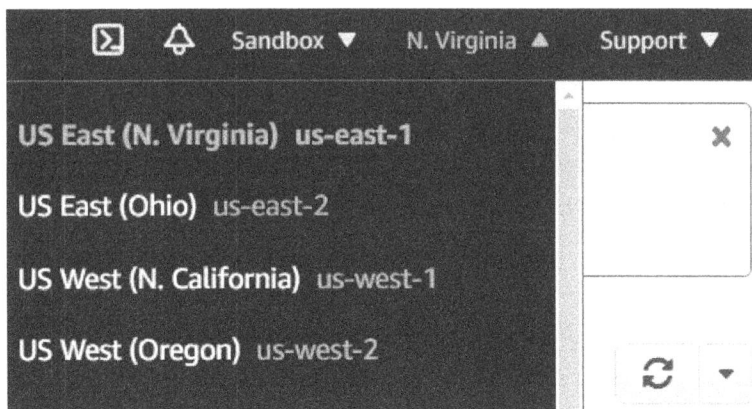

Figure 3.19 – Setting up billing alarms – navigating to the N. Virginia Region

2. In the left-hand navigation pane, select **Alarms** and then click **Create Alarm** in the far-right hand corner of the screen. You will be presented with the four-step **Create alarm** wizard:

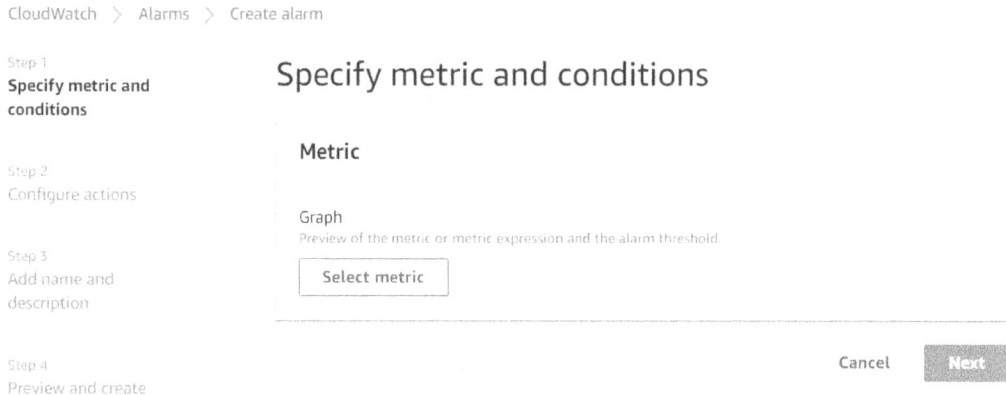

CloudWatch > Alarms > Create alarm

Step 1
Specify metric and conditions

Step 2
Configure actions

Step 3
Add name and description

Step 4
Preview and create

Specify metric and conditions

Metric

Graph
Preview of the metric or metric expression and the alarm threshold

Select metric

Cancel   Next

Figure 3.20 – Setting up billing alarms – setting up billing alarms in CloudWatch

3. Under **Step 1**, click **Select metric**.

4. Under **All metrics**, click on **Billing**.

5. Next, click on **Total Estimated Charge**:

All metrics   Graphed metrics   Graph options   Source

All > Billing   Q Search for any metric, dimension or resource id

2 Metrics

By Service
1 Metric

Total Estimated Charge
1 Metric

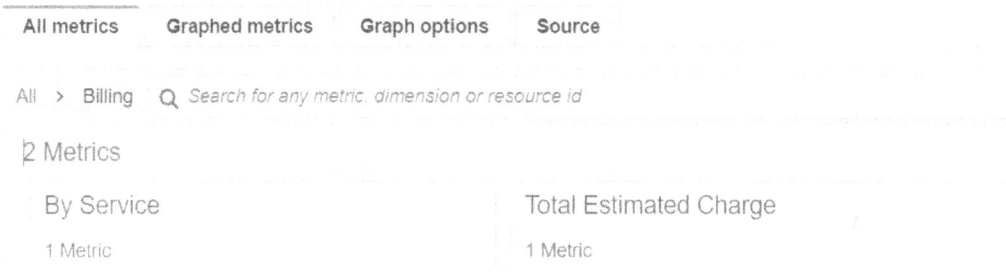

Figure 3.21 – Setting up billing alarms – defining metrics for Total Estimated Charge

6. Click the checkbox next to the **USD** currency for the metric name **EstimatedCharges**.

7. Click **Select metric**.

8. In the dialog box entitled **Specify metric and conditions**, scroll down and select **Static** as the threshold type, under **Conditions**.

9. Select **Greater > threshold**, under the heading **Whenever EstimatedCharges is....**

10. Finally, set the dollar amount to **10** for USD under the **Define the threshold value** sub-heading, and then click **Next**, as per the following screenshot:

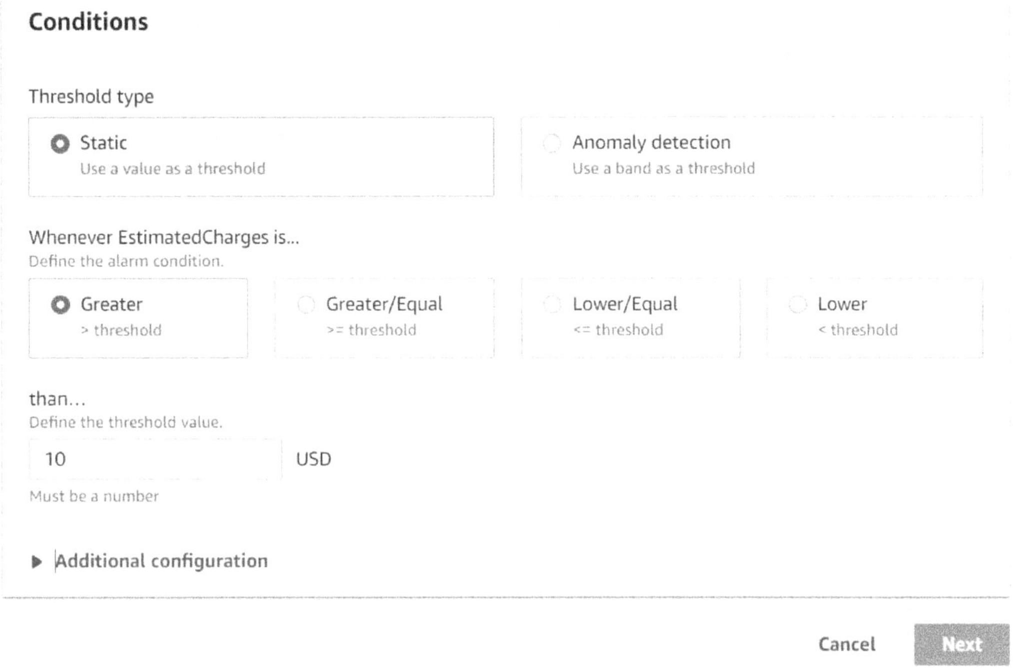

Conditions

Threshold type

○ Static
  Use a value as a threshold

○ Anomaly detection
  Use a band as a threshold

Whenever EstimatedCharges is...
Define the alarm condition.

○ Greater
  > threshold

○ Greater/Equal
  >= threshold

○ Lower/Equal
  <= threshold

○ Lower
  < threshold

than...
Define the threshold value.

| 10 | USD |

Must be a number

▶ Additional configuration

Cancel   **Next**

Figure 3.22 – Setting up billing alarms – setting up a billing alarm threshold value

11. In *Step 2*, you can now configure actions when the alarm breaches. For this, click the **Add notification** button:

Configure actions

Notification

Add notification

Figure 3.23 – Setting up billing alarms – setup notification

12. Under the alarm state trigger, make sure that the **In alarm** state is selected.

13. Next, under **Select an SNS topic**, select the option for **Create new topic. SNS** stands for **Simple Notification Service** and is a push-based messaging service. You can configure an SNS topic to send you email alerts when the alarm is in the **In alarm** state. We discuss Amazon SNS in detail in *Chapter 10, Application Integration Services*.

14. Under the heading **Create a new topic…**, provide a topic name, for example, `MyBillingAlerts`.

15. Next, under the heading **Email endpoints that will receive the notification…**, provide an email address that you have access to where the alerts will be sent. You can use the same email address that you created your AWS account with.

16. Next, click **Create topic**:

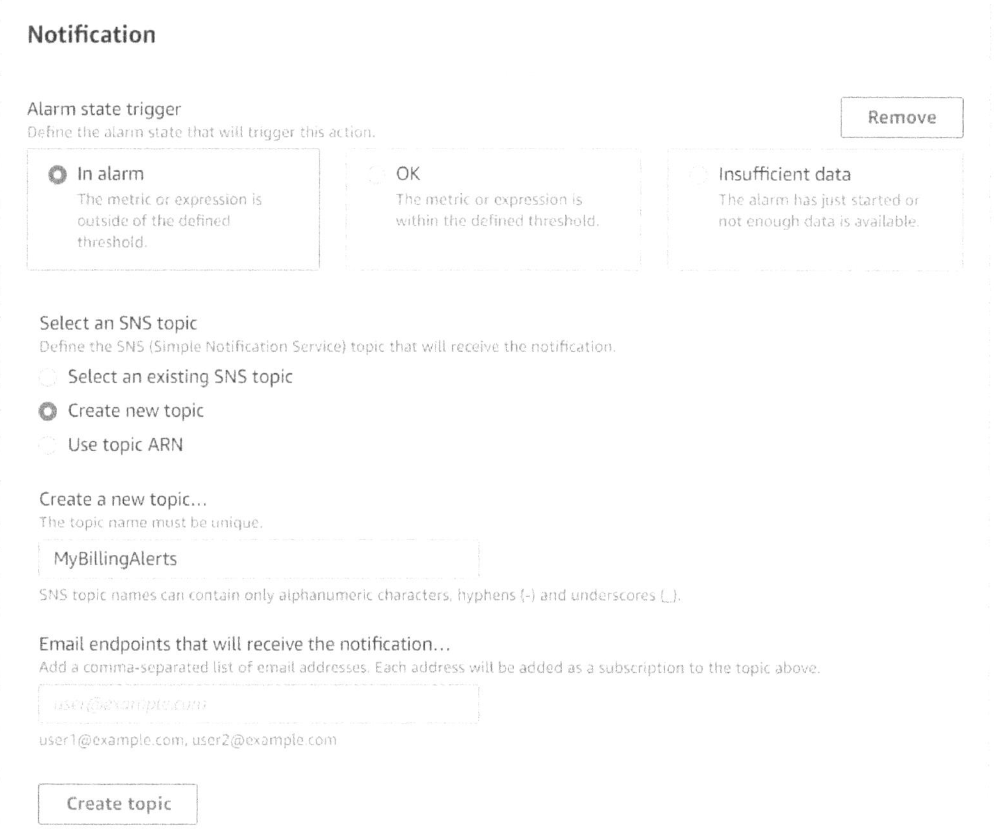

Notification

Alarm state trigger
Define the alarm state that will trigger this action.

		Remove
In alarm The metric or expression is outside of the defined threshold.	**OK** The metric or expression is within the defined threshold.	**Insufficient data** The alarm has just started or not enough data is available.

Select an SNS topic
Define the SNS (Simple Notification Service) topic that will receive the notification.

- Select an existing SNS topic
- Create new topic
- Use topic ARN

Create a new topic...
The topic name must be unique.

MyBillingAlerts

SNS topic names can contain only alphanumeric characters, hyphens (-) and underscores (_).

Email endpoints that will receive the notification...
Add a comma-separated list of email addresses. Each address will be added as a subscription to the topic above.

user@example.com

user1@example.com, user2@example.com

Create topic

Figure 3.24 – Setting up billing alarms – defining SNS topics for notifications

17. Your SNS topic with the name you chose will be created and you will see the email address that notifications will be sent to.

18. Scroll toward the bottom of the screen and click on **Next**.

19. In *Step 3*, provide an appropriate name and description for the alarm and click **Next**:

Name and description

Alarm name

Alarm name

Alarm description - *optional*

Alarm description

Up to 1024 characters (0/1024)

Cancel   Previous   Next

Figure 3.25 – Setting up billing alarms – alarm setup

20. In *Step 4*, you can review your settings and confirm by clicking on the **Create alarm** button.

21. You are then presented with the alarm configuration status in Amazon CloudWatch:

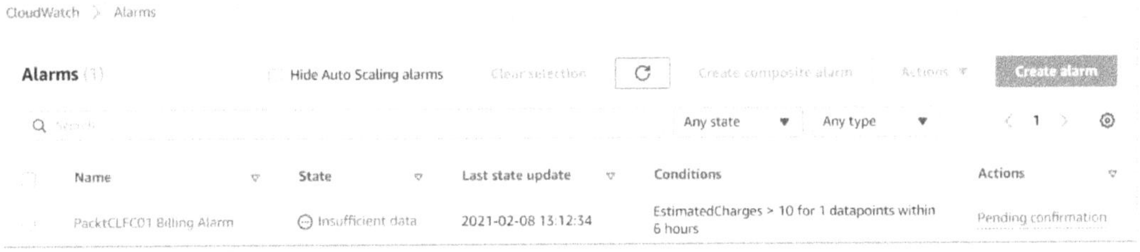

Figure 3.26 – Setting up billing alarms – verifying the Actions status

22. You will notice from the preceding screenshot that you have a **Pending confirmation** link highlighted in red. When you configure an SNS notification to send alerts to an email, you are, in effect, subscribing to the SNS topic you created earlier. For security purposes and to avoid rogue messages, you are required to log in to your email account and confirm the subscription.

23. Log in to your email account and you should find an email from Amazon asking you to confirm your subscription, as in the following screenshot:

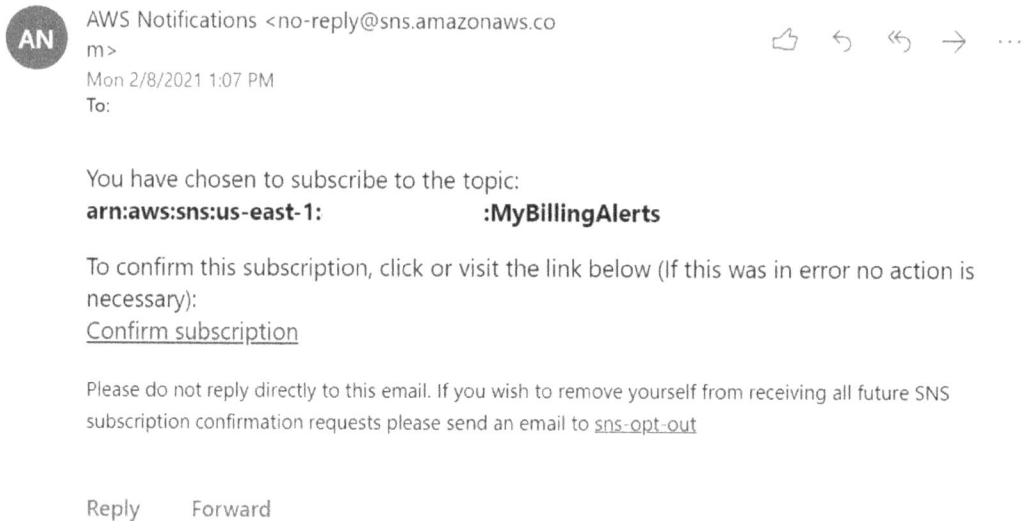

AN AWS Notifications <no-reply@sns.amazonaws.com>

Mon 2/8/2021 1:07 PM
To:

You have chosen to subscribe to the topic:
arn:aws:sns:us-east-1:   :MyBillingAlerts

To confirm this subscription, click or visit the link below (If this was in error no action is necessary):
Confirm subscription

Please do not reply directly to this email. If you wish to remove yourself from receiving all future SNS subscription confirmation requests please send an email to sns-opt-out

Reply   Forward

Figure 3.27 – Setting up billing alarms – confirming a notification subscription

24. Click on the **Confirm subscription** link so that your subscription can be activated. Should your monthly charges exceed the USD 10 threshold, an email notification will be sent to you and you can terminate any unwanted or forgotten labs.

25. If you now refresh your alarms by clicking on the circular arrow, you will see that the pending confirmation message disappears. Furthermore, as this is a brand-new account, you will not yet have incurred any charges and you will note that the alarm is in the **OK** state:

Figure 3.28 – Setting up billing alarms – verifying alarm status

Now that you have set up and configured your AWS billing alarms, you can be rest assured that you will be alerted if you exceed the threshold you selected previously. Alarms can be set for a wide range of services, enabling you to effectively monitor and maintain your workloads on AWS. Alarms are configured in Amazon CloudWatch, which we will discuss in detail in *Chapter 13, Management and Governance on AWS*.

Summary

In this chapter, we discussed the fact that to access any AWS services to deploy and configure resources, workloads, and applications, you will first need to set up an AWS account. We also discussed the importance of having multiple AWS accounts, which will help manage your cloud ecosystem better, offer greater levels of security, implement necessary workload isolation, and limit your blast radius in the event of catastrophic disasters.

By understanding the thought process involved in building a multi-account strategy, you have already embarked on your learning of certain architectural best practices related to building your cloud solutions. Specifically, you have learned that a multi-account setup enables you to offer greater levels of security and scalability.

We have also examined the AWS Organizations service and how it enables us to centrally manage all our AWS accounts by placing common accounts into OUs and applying appropriate guardrails using SCPs. An additional benefit of AWS Organizations is the consolidated billing feature, which enables you to benefit from volume discounts by combining the charges across all your accounts.

We discussed the AWS Free Tier offering and how you can start experimenting with a 12-month free AWS account. A full step-by-step guide has also been provided to easily set up your own AWS account, which you will need throughout this training guide for the various labs on offer. Finally, you learned how to set up and configure your own billing alarms, which can be used to alert you if you exceed a given budgeted threshold for your total AWS expenses.

In the next chapter, we look at AWS **Identity and Access Management (IAM)**. AWS IAM is integral to designing a highly secure environment with authentication and authorization services of your AWS account. With IAM, you can create additional identities that can have access to your AWS account. For example, if you have a team of developers who will be building an application on AWS, you would not want to share the root account credentials with them. Each of your developers should have their own individual IAM user accounts. This enables better auditing and allows you to enforce the principle of least privilege when granting access to services in your AWS account.

Questions

Here are a few questions to test your knowledge:

1. Before setting up your billing alarms, which preference setting needs to be enabled first?

   A. Enable billing alerts

   B. Enable alarms

   C. Set up AWS Organizations

   D. Configure MFA

2. Which AWS service enables you to centrally manage multiple AWS accounts with SCPs to establish permission guardrails using which services can be enabled in those accounts?

   A. AWS Organizations

   B. AWS IAM

   C. AWS VPC

   D. AWS GuardDuty

3. Which of the following services are offered completely free by AWS? (Select two answers.)

   A. AWS **Identity and Access Management (IAM)**

   B. AWS Elastic Beanstalk

   C. **Amazon Simple Storage Service (Amazon S3)**

   D. **Amazon Relational Database Service (Amazon RDS)**

   E. AWS **Simple Notification Service (SNS)**

4. Which feature of AWS Organizations enables you to combine the costs of each member account to take advantage of any volume discounts on offer?

   A. Consolidated billing

   B. AWS EC2 savings plan

   C. AWS Control Tower

   D. AWS IAM

5. Which of the following is required when creating an AWS Free Tier account?

    A. A credit card

    B. A bank statement

    C. A passport or driving license

    D. An invitation letter from Amazon

6. Which AWS service enables you to automatically set up a new landing zone in accordance with best practices?

    A. AWS Landing Zone

    B. AWS Control Tower

    C. AWS Organizations

    D. AWS Free Tier Account

7. Which feature of the AWS Organizations service enables you to combine AWS accounts in a container that has common workloads and then apply a common set of policies to those accounts?

    A. AWS Control Tower

    B. AWS Landing Zone

    C. **Organization Units (OUs)**

    D. **Service Control Policies (SCPs)**

Section 2: AWS Technologies

In this section, we examine various AWS technology solutions that can be used to architect end-to-end application solutions on the AWS platform. We examine the core AWS services comprising compute, network, storage, and database services, as well as services such as analytics, management and governance, and security.

This part of the book comprises the following chapters:

- *Chapter 4, Identity and Access Management*
- *Chapter 5, Amazon Simple Storage Service (S3)*
- *Chapter 6, AWS Networking Services – VPCs, Route53, and CloudFront*
- *Chapter 7, AWS Compute Services*
- *Chapter 8, AWS Databases Services*
- *Chapter 9, High Availability and Elasticity on AWS*
- *Chapter 10, Application Integration Services*

4

Identity and Access Management

So far, you have learned about the basics of cloud computing and its advantages. You have also been introduced to AWS and had a quick overview of its services. We have discussed the AWS Global Infrastructure and its support plans for customers looking to use the services offered. We have also emphasized the importance of AWS accounts and how they help you gain isolated and secure access to the wide range of AWS services within which you can build your cloud solutions. We also identified the architectural reasons behind having multiple AWS accounts and using AWS Organizations to centrally manage all your accounts.

You have also learned how to set up your first AWS Free Tier account, which will enable you to start configuring resources on the platform.

In this chapter, we will look at the AWS **Identity and Access Management** (**IAM**) service. This is a critical foundational service designed to secure access to your AWS account. AWS IAM is an authentication and authorization service that enables you to decide who or what can access the AWS services in your account (known as **authentication**), and what these entities are permitted to do in your account (known as **authorization**).

This chapter covers the following topics:

- Introduction to the AWS IAM service
- The root user account and implementing **Multi-Factor Authentication** (**MFA**)
- The importance of defining IAM password policies
- Key differences between IAM users and IAM groups
- Defining permissions with IAM policies
- Reviewing credential reports
- Exercise 4.1 – creating an IAM group
- Exercise 4.2 – creating an IAM user and attaching it to the developers group
- Exercise 4.3 – logging in to your AWS account as an IAM user
- Accessing AWS via the CLI
- Exercise 4.4 – accessing the AWS platform using the AWS CLI on a Windows computer

Introduction to the AWS IAM service

At this point, you should have a single AWS Free Tier account that you can log in to using your chosen email address and password combination. This email address that you have used to create the account is also known as the **root user**. The root user is the person who created the account and holds the keys to the kingdom. This privileged user must be guarded well, and knowledge of its credentials (passwords and so on) should not be given to anyone who is not authorized to use it. The root user should also not be used for day-to-day operations. You can create additional user accounts (what we call IAM users) to perform daily tasks. We'll look at IAM users shortly in this chapter.

The AWS IAM console

To access any AWS service, including the IAM service, you can either use the web-based management console, the **command-line interface** (**CLI**), or AWS SDKs if you are writing code. We will start with the AWS Management Console. Log in to your AWS account using your root user credentials and you will be presented with the following splash screen:

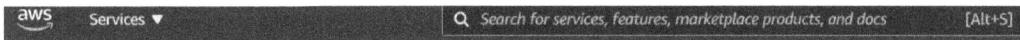

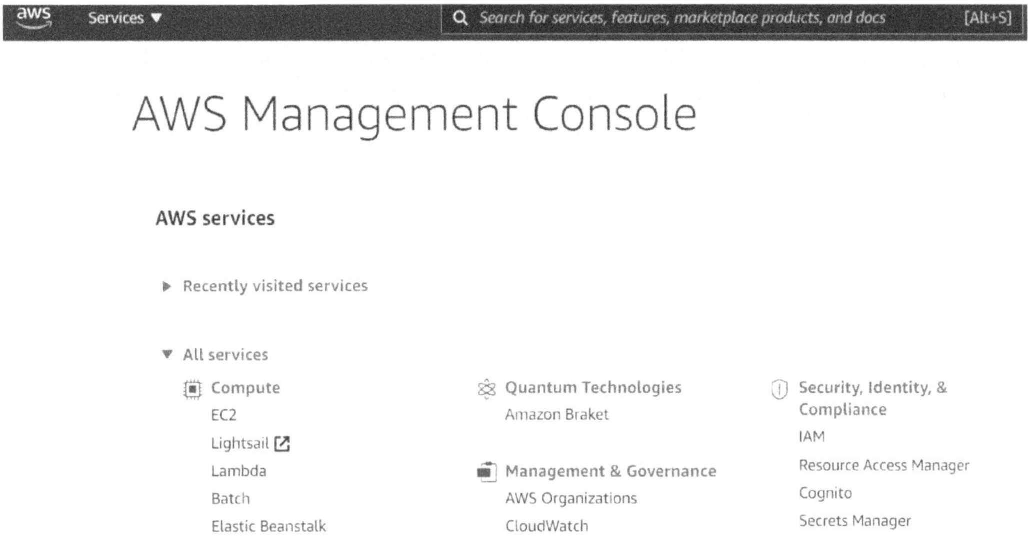

Figure 4.1 – AWS Management Console

As you can see from the screenshot, the wide range of services is displayed under category headings, depending on the type of service. So, for example, we have categories such as **Compute**, **Network**, and **Storage** and under these categories, relevant services are listed.

AWS IAM falls under the **Security, Identity, & Compliance** category, but you can also search for the service using the search bar at the top of the page.

The AWS IAM services

The first time you navigate to the IAM console, you will note some security alerts and a list of best practices to follow.

You will also note a *sign-in URL* for IAM users in this account. This is a special URL that your IAM users can browse to access your account. This URL is customizable and you can replace the series of digits shown after the `https://` portion with a custom name that is easier to remember. The series of digits you see is, in fact, your AWS account ID.

Simply click the **Customize** link after the URL, as shown in the following screenshot, and choose an appropriate name for your AWS account, for example, PacktDevAccount. Names chosen must be unique and you may need to associate a common name with your company name to create a name that is unique:

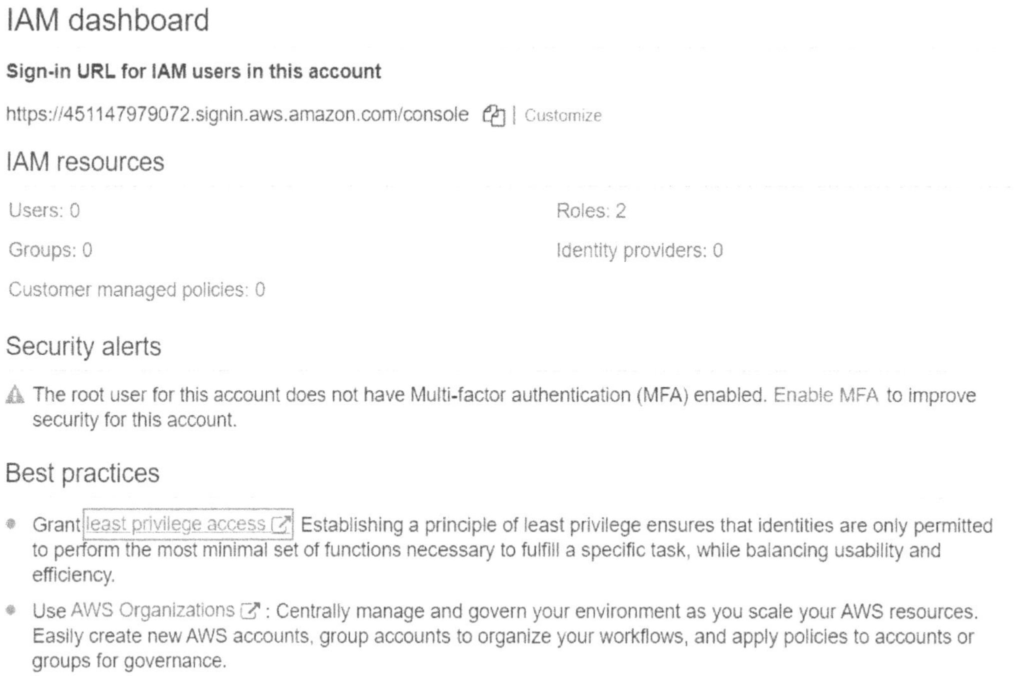

IAM dashboard

Sign-in URL for IAM users in this account

https://451147979072.signin.aws.amazon.com/console 🗐 | Customize

IAM resources

Users: 0	Roles: 2
Groups: 0	Identity providers: 0
Customer managed policies: 0	

Security alerts

⚠ The root user for this account does not have Multi-factor authentication (MFA) enabled. Enable MFA to improve security for this account.

Best practices

- Grant least privilege access ☑ Establishing a principle of least privilege ensures that identities are only permitted to perform the most minimal set of functions necessary to fulfill a specific task, while balancing usability and efficiency.

- Use AWS Organizations ☑ : Centrally manage and govern your environment as you scale your AWS resources. Easily create new AWS accounts, group accounts to organize your workflows, and apply policies to accounts or groups for governance.

Figure 4.2 – IAM dashboard

Once you have customized your IAM sign-in URL, you can provide this link to other IAM users, who can use it to sign in to your account. This can only happen if your users have an IAM user account to sign in with. We discuss how to create IAM users later in this chapter.

In the next section, we look at the root user account and discuss the implementation of an additional security measure, using **Multi-Factor Authentication** (**MFA**).

The root user account and implementing Multi-Factor Authentication (MFA)

One of the first things you want to do is configure MFA for your root user account. Normally, when you log in to an AWS account, you simply provide a username and password. You are probably aware that you must choose a highly complex password – one that has lowercase letters, uppercase letters, numbers, and symbols, and must be randomly generated rather than dictionary words that can be guessed easily.

However, a username and password combination alone is not sufficient in this age of malware attacks, hacking, and brute force attacks. MFA is a mechanism where you are prompted to verify your identity using more than one set of credentials. Instead of just having two passwords, however, MFA uses two separate secrets to verify your identity – *something you know* and *something you have*. So, for example, something you know would be your username and password, and something you have would be a one-time password pin that is generated on a device that you possess. An example of such a device could be an RSA token, a **Universal Second Factor** (U2F) device, or an authentication app that you can install and use on your smartphone.

We strongly recommend that you set up MFA for your root user account.

Setting up MFA

Let's quickly set up MFA for our root user account. To complete the step-by step guide, you will need access to a smartphone, either an Android or iOS device. There are several smartphone-based authenticators that are supported by AWS, including Google Authenticator and Microsoft Authenticator. You can review the supported apps at `https://aws.amazon.com/iam/features/mfa/`. For this exercise, go ahead and install the **Google Authenticator app** on your phone, which is available free of charge from your Google or iPhone play store. If you do not currently have the app installed, make sure you do this first before proceeding.

Ensure that you have navigated to the IAM management console, then take the following steps:

1. On the IAM console, click on **Enable MFA**, under **Security alerts**:

Security alerts

⚠ The root user for this account does not have Multi-factor authentication (MFA) enabled. Enable MFA to improve security for this account.

Figure 4.3 – Security alerts

2. You are then presented with the **Security Credentials** page. Click on **Activate MFA.**

3. A pop-up dialog box is presented, and you need to select **Virtual MFA device**:

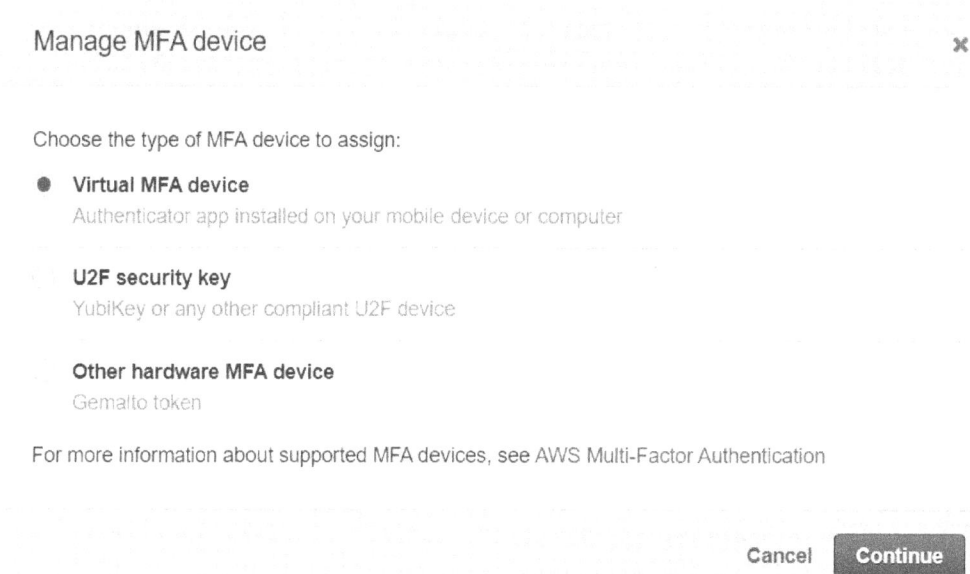

Manage MFA device   ✖

Choose the type of MFA device to assign:

● **Virtual MFA device**
   Authenticator app installed on your mobile device or computer

   U2F security key
   YubiKey or any other compliant U2F device

   Other hardware MFA device
   Gemalto token

For more information about supported MFA devices, see AWS Multi-Factor Authentication

                            Cancel   Continue

Figure 4.4 – Setting up a virtual MFA device

4. Click **Continue**.

5. You are then presented with the **Set up virtual MFA device** dialog box. You will have the option to scan a QR code to link your Google Authenticator app on your mobile phone with your AWS account. To get started, launch the Google Authenticator app on your phone and select the **Scan a QR code** option (you may have to tap the plus sign (+) first). This will activate your phone camera.

6. On the AWS console, in the **Set up virtual MFA device** dialog box, click **Show QR code** under list item number **2**:

Set up virtual MFA device   ✖

1. **Install a compatible app on your mobile device or computer**
   See a list of compatible applications

2. **Use your virtual MFA app and your device's camera to scan the QR code**

   Show QR code|

   Alternatively, you can type the secret key. Show secret key

3. **Type two consecutive MFA codes below**

   MFA code 1

   MFA code 2

                                        Cancel   Previous   Assign MFA

Figure 4.5 – Setting up a virtual MFA device

7. The QR code will be displayed. At this point, you need to position your phone camera so that it captures the QR code while in the Google Authenticator app.

8. Once the QR code is captured, you will be presented with an MFA code, which lasts for a few seconds. You will then need to type in the code in the textbox next to **MFA code 1**. Wait for the next code to be displayed and type that code into the textbox next to **MFA code 2**.

9. Finally, click **Assign MFA**.

Your root user has now been configured with MFA authentication. The next time you log out and then log back in, you will be prompted to enter the MFA code shown on your Google Authenticator app after you provide your root user's email address and password. Remember that the MFA one-time pin expires after a few seconds, so you need to promptly provide the code shown in your app or wait for the next code.

Once you provide the MFA pin, you will then be logged in and redirected to the AWS Management Console.

In this section, we introduced you to the basic AWS Management Console, and AWS IAM. We also looked at how to configure MFA for your root user account.

In the next section, we will discuss password policies, which will allow you to enforce strong and complex passwords for any IAM users that you create in your AWS account.

The importance of defining IAM password policies

Now that you have secured your root user account, you should start creating additional accounts for users in your organization. Remember that the root user is the most privileged account, and you should not use the root user account for daily operations. Each member of your organization that needs to access the AWS services in your AWS account must be provided with an IAM user account. Never share your root credentials with other team members, even with other administrators, as they should be using their own IAM accounts with the appropriate administrative permissions.

We discuss IAM users in the next section, but for now, it becomes obvious that an IAM user account will be configured with a password. And if you have a hundred different IAM user accounts, you want to enforce some sort of password policy so that those accounts do not have weak passwords that are easy to crack.

AWS password policies enable you to define rules to enforce password complexity. This means that users will be forced to configure a password that is in compliance with your complexity rules and this will ensure strong passwords across all IAM user accounts.

The password policies can be configured from the **Account Settings** section of the IAM dashboard.

In the next section, we discuss the importance of creating additional IAM users and IAM groups. IAM users are additional identities that you can create in addition to your root user account. An IAM user can represent a physical person who needs access to your AWS account, such as a developer in your development team or a server administrator in your shared IT services team.

Key differences between IAM users and IAM groups

In this section, we look at the importance of setting up additional identities that need to access your AWS account. We also look at best practices in managing what those identities can or cannot do in your AWS account, using IAM groups.

IAM users

As discussed in the preceding section, in addition to the **root user**, you can create additional users known as **IAM users**. IAM users can be used to represent physical people in your organization, such as members of your development team or server administrators. These users can then use their IAM user accounts to log in to your AWS account and perform tasks based on permissions you grant them.

IAM user accounts can also be used by applications and other services that need to authenticate themselves against a given AWS service. For example, if an application needs to update a backend Amazon RDS database, you want to make sure that the application is authorized to do so. The application can be assigned with an IAM user account that it can use to authenticate itself against the database and depending on the permissions you grant that IAM user account, the application will be able to modify the database as required. While this method of allowing an application to authenticate against AWS services is possible, it is not considered best practice for most use cases. This is partly since the credentials are usually stored in some configuration file in plain text. Furthermore, you would need to implement a process of regularly rotating the credentials to improve security somewhat and this can be a management overhead. AWS offers another type of identity specifically for this purpose known as IAM roles, which we discuss later in this chapter.

As previously discussed, to access any AWS service, you can use the web-based management console, the CLI, or AWS SDKs. To access your AWS account using the AWS Management Console as an IAM user, you will create a username and password combination that will enable the user to authenticate against your AWS account. If you wish to access your AWS account using the CLI for programmatic access, you will need to configure a set of access keys. Access keys are similar to usernames and passwords and consist of an **access key ID** (such as a username) and a **secret access key** (such as a password). Access keys are for programmatic access, whereas the username and password combinations are used for web-based console access.

In this section, we introduced you to IAM users, which can be used to represent physical people such as your colleagues who may need access to your AWS account. IAM user accounts can also be set up for specific applications to authenticate themselves against your account. These are known as **service accounts**.

In the next section, we talk about IAM groups, which allow you to manage a group of IAM users who may share a common job role.

IAM groups

When you create an IAM user for a particular colleague who needs access to your AWS account, you need to also configure a set of permissions for that user to ensure that they are permitted to perform tasks as required by their role in your organization. These permissions are assigned via IAM policies, which we will discuss shortly.

Although you can create separate sets of IAM policies for each user in your account, an even better way of managing your users and the permissions you assign them is by combining those users who share a common job role into an AWS IAM group. By doing so, you can centrally manage the users assigned to that group by simply applying policies at the group level, which get filtered down to the users within the group.

For example, if you have a team of 10 developers, all of whom need the ability to create and manage Amazon S3 storage buckets, then instead of assigning the same permission to each user individually at the IAM user level, you can assign a single permission to the developers group. Your developers, who are then made members of the group, will inherit that permission to access Amazon S3.

In this section, we looked at the importance of setting up IAM users and IAM groups. This allows you to ensure that you can grant varying levels of access to different identities in your organization and use IAM groups to manage those identities.

In the next section, we will take a look at IAM policies in detail. IAM policies enable us to define permissions for those additional identities and control what they are permitted to do in our AWS account.

Defining permissions with IAM policies

IAM policies are objects attached to a given IAM identity, such as an IAM user, groups of IAM users, or an IAM role. These policies define what the identity can or cannot do within the AWS account and are written as **JSON documents**. In the following diagram, user **Bob** can access and read the content of an S3 bucket. The policy is attached to the group that **Bob** is a member of and therefore inherits the ability from the group.

When a principal (IAM user or IAM role) tries to access an AWS service, AWS will evaluate the policy document attached to it and determine what action the principal can or cannot perform:

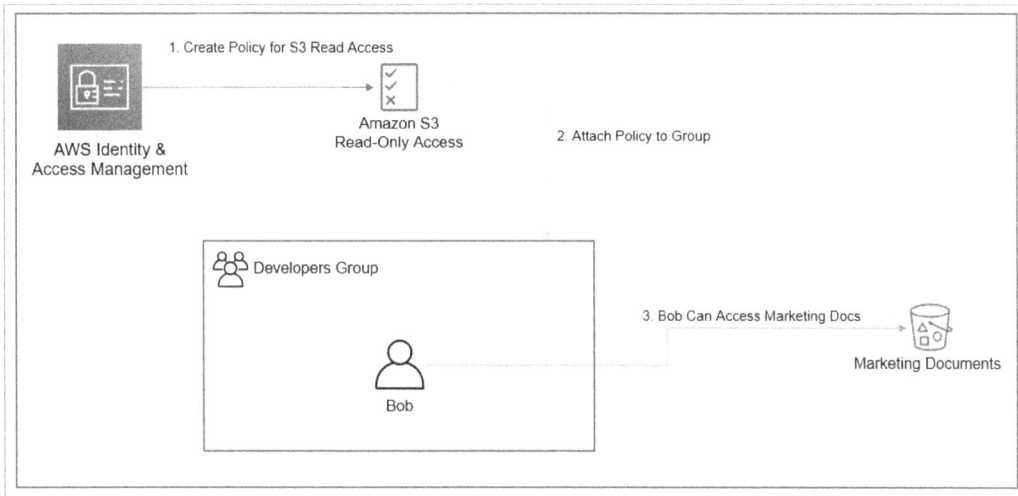

Figure 4.6 – Policy document attached to the developers group, granting Bob read access to the Marketing Documents bucket

IAM policies enable you to follow the principle of least privileges, which means that identities can be configured to only have access to services and configuration options necessary to fulfill their roles, and nothing more. This greatly enhances the security of your AWS account and follows best practices.

AWS offers six types of policies that you can create:

- **Identity-based policies**: These are policies attached to your IAM identities that specify what those identities can or cannot do in your AWS account. IAM identities are your IAM users, groups of IAM users, or IAM roles within your AWS account. Note that you cannot attach an IAM identity policy to an identity in another AWS account. However, identities in other accounts can be configured to assume IAM roles in your AWS account, which can provide your external identities with access to certain services and resources in your account. We discuss IAM roles later in this chapter.

- **Resource-based policies**: These are policies that are attached to the resource in each AWS account and are inline policies. Examples include the Amazon S3 bucket policies. The policy specifies the principal that is granted access to the resource to which the policy is attached. This enables you to grant access to principals in the same account as the resource or in external AWS accounts too. In addition, the policy requires you to define one or more principals that can be granted access to. You can also specify the principal as a wildcard (*), which means that you can grant anonymous access to some of your resources. Understandably, you should be extremely careful when granting anonymous access.

- **Permission boundaries**: You can define a policy as a permission boundary for an IAM entity (user or role), which defines the maximum set of permissions that can be granted by an identity-based policy.

- **Organization Service Control Policies** (**SCPs**): As discussed in the last chapter, organization SCPs enable you to define the maximum permissions for account members of an organization. SCPs enable you to restrict what permissions you can define in an identity- or resource-based policy in the member account. However, they are by themselves not able to grant permissions.

- **Access Control Lists** (**ACLs**): These are permissions you use to manage access to certain resources such as Amazon S3 buckets and objects. You can use ACLs to grant basic read/write permissions to other AWS accounts; you cannot grant permissions to users in your account. They are used in certain cases where S3 bucket policies cannot be used, such as S3 server access logging. They are like resource-based policies, but the permission sets that can be configured are limited in granularity. In addition, they do not use a JSON structure to construct the policy.

- **Session policies**: These allow you to pass session policies when you access services programmatically (for example, the CLI) to assume a role or federate a user. Session policies allow you to limit permissions for a specific session.

Next, let's further study the types of identity-based policies in detail.

Types of identity-based policies

As previously mentioned, identity-based policies are written in JSON format and these policies are attached to IAM identities, defining what those identities can or cannot do within your AWS account.

Identity-based policies can be categorized as follows:

- **Managed AWS policies**: These are standard policies pre-configured with specific permissions that AWS provides. A *managed* policy is one that is created independent of any IAM identity and can be attached to one or many identities. You can use these typical policies for most of the permission sets you need to define, for example, granting *read-only* access to Amazon S3. AWS managed policies cannot be edited by the customer and only AWS can update and manage them.

- **Customer-managed policies**: Customers can create their own managed policies and then attach them to any IAM identity as required. Creating your own managed policies allows you to generate more granular sets of permissions. In addition, IAM creates up to five versions of your managed policies whenever you update them and this allows you to revert to older versions if necessary.

- **Inline policies**: These policies are created and attached directly to the IAM identity, for example, an IAM user. This also allows you to maintain a strict one-to-one relationship between a given policy and an identity. However, too many inline policies can be more difficult to manage and must be used sparingly. Ultimately, inline policies are tied to the life cycle of the entity it is associated with.

Let's understand IAM policies with the help of an example.

Example of an IAM policy

In *Figure 4.6*, we saw an example of Bob, who is a member of the developers group, being able to access the Marketing Documents bucket because of the IAM policy attached to the group.

Here is an example of what a JSON policy document looks like for the preceding scenario:

Show Policy   ✖

```
{
  "Version": "2012-10-17",
  "Statement": [
    {
      "Effect": "Allow",
      "Action": [
        "s3:Get*",
        "s3:List*"
      ],
      "Resource": "*"
    }
  ]
}
```

Figure 4.7 – AWS managed policy: AmazonS3ReadOnlyAccess

The preceding screenshot shows a managed AWS policy that grants the identity it is attached to the ability to read contents from any S3 buckets in your AWS account.

Identity policies do not need to specify a principal within the policy, because by its very nature, it will be attached to the identity that requires the level of access specified in the policy.

Let's look at some of the components of the policy document:

- `Version`: This is the JSON document version currently supported by AWS and should be listed as `2012-10-17`.

- `Statement`: This is the beginning of the policy statement. You can have multiple statement blocks within a single policy, allowing you to grant various levels of access across different services.

- `Effect`: This specifies whether the statement block is going to allow some level of access or deny access.

- `Action`: This is the actual permission that is either permitted or denied based on the preceding `Effect` statement. The `Action` component of your JSON document will refer to the service of concern and the actual permission that is being granted. So, for example, in the preceding JSON statement, we have two sets of actions being permitted, namely the `Get` action and the `List` action of the Amazon S3 service. This means that the identity associated with this policy will be able to *list* your Amazon S3 buckets and perform various other list operations and be able to read the contents of the buckets along with other `Get` operations. These action statements actually have more granular-level operations that can be performed. So, for example, the `Get` operation includes a number of operations such as `GetObject` (which allows you to access an object) and `GetBucketVersion` (which returns the versioning state of a bucket) among others. You will notice the asterisks (*) after the `Get` and `List` actions in the preceding screenshot. You use the asterisks (*) when you want to allow all operations within a given `Action` statement. So, in the preceding example, `Get*` would include both the `GetObject` and `GetBucketVersion` operations.

- `Resource`: This component describes the actual resource against which this policy permits or denies access. A wildcard (*) denotes all resources of that service. So, in this case, the resources allow the `Get` action and the `List` action against all S3 buckets in your AWS account. To restrict access to a specific resource, you will need to specify the actual **Amazon Resource Name (ARN)** of that resource. The ARN is a unique identifier of the resource in your AWS account. For the preceding example, if we wanted to restrict the permission sets to just the marketing bucket, we would need to specify the ARN of the bucket. In our example, it looks something like this: `arn:aws:s3:::packt-marketing-docs`. So, our JSON policy will then need to be amended to look like this:

```
1  {
2      "Version": "2012-10-17",
3      "Statement": [
4          {
5              "Effect": "Allow",
6              "Action": [
7                  "s3:Get*",
8                  "s3:List*"
9              ],
10             "Resource": "arn:aws:s3:::packt-marketing-docs"
11         }
12     ]
13 }
```

Figure 4.8 – Customer-managed policy restricting access to a single Amazon S3 bucket

ARNs are composed of segments and each segment is delimited with a colon (`:`). The following are the formats in which ARNs can be constructed with their individual segments:

a) `arn:partition:service:region:account-id:resource-id`

b) `arn:partition:service:region:account-id:resource-type/` `resource-id`

c) `arn:partition:service:region:account-id:resource-` `type:resource-id`

Let's look at the individual segments:

a) **partition**: This represents the partition of AWS that the resource resides in. AWS has several partitions, representing different customer groups. For example, the standard partition accessible to retail customers in most countries is `aws`. A separate partition also exists for China and this is named `aws-cn`, and so on.

b) **service**: The service namespace that identities the AWS product. For example, the Amazon S3 service is represented by the name `s3`.

c) **region**: The Region that the resource resides in. Note for some services, such as Amazon S3, the ARN does not require you to list out the region segment. This is because S3 bucket names are globally unique, and you do not need to specify a Region as part of the ARN. For these resources, the Region name segment is simply replaced with an additional colon (`:`).

d) **accountID**: This is the ID that owns the resource. Note that the account ID is a series of 12-digit numbers with the hyphens omitted. Here, again, for certain resources, you do not need the account ID for the ARN of the resource to be listed. So, with Amazon S3 buckets, you omit the region and account ID from the ARN, replacing them with just colons (`:`). For example, here is a valid ARN for our marketing docs bucket:: `arn:aws:s3:::packt-marketing-docs`.

e) **resource** or **resourcetype**: This part of the ARN varies by service. A resource identifier can be a name or ID of the resource, for example, `user/John` or `instance/i-1234567890xydcdeg0`.

- `Condition`: In addition to the preceding list of components defined in a policy document, you can also create conditional elements for your statement so that you can further restrict the application of the policy based on a predetermined condition. For example, you might want to restrict access only if the source of that access is from within your corporate network IP address space.

In the next section, we'll study the IAM policy simulator.

IAM policy simulator

If you need to test out your policies or troubleshoot any access issues, you can use the IAM policy simulator. This tool, located at `https://policysim.aws.amazon.com/`, can help you troubleshoot identity-based policies, IAM permissions boundaries, organizations' SCPs, and resource-based policies. Note that the simulator only simulates attempted access and whether that access will be granted or denied. It does not actually make calls to the service APIs.

In the screenshot shown next, you will note that we have currently selected the **Database Team (DB Team)** IAM group. In this example, you would have thought that **DB Team** can access the AWS **Relational Database Service (RDS)** to create database instances. However, you will see that when we ran the simulation, for the `CreateDBInstance` action, we got a **denied** result. This can be cross-referenced with the list of policies attached to the group on the left-hand pane. The **DB Team** group seems to have only the `AllowS3FullAccess` policy attached. While this example is fairly simple, you can image an IAM user, group, or role having multiple complex policies attached to it and running the simulator can enable you to test out specific levels of access:

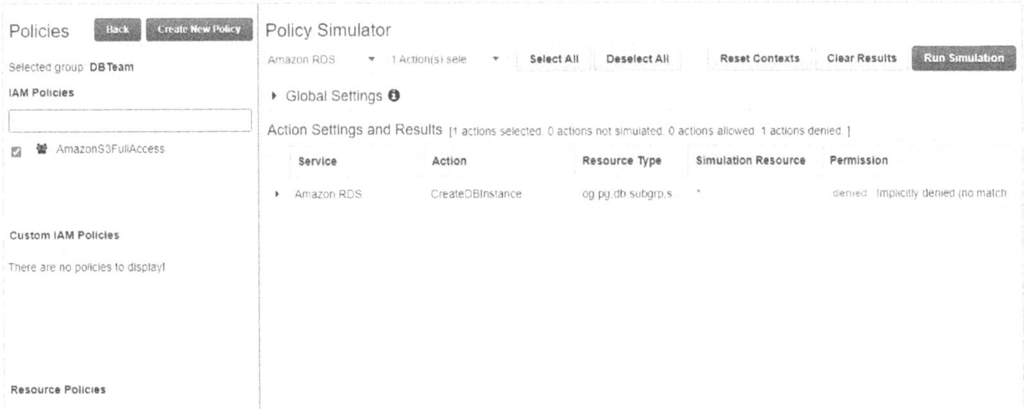

Figure 4.9 – Amazon policy simulator

In this section, we examined how IAM policies can be used to grant access based on the principle of least privileges. We looked at the different types of policies and the different categories of IAM policies. We also examined the JSON structure of a typical IAM policy.

In the next section, we will discuss IAM roles, which are another type of identity that can be used to grant access to services and resources in your AWS account.

Assigning temporary credentials with IAM roles

While an IAM user account either represents a physical person or can be used as a "service account" for an application that requires authentication, IAM roles are, in fact, independent identities that can be assumed by other entities to gain access to AWS services and resources. In other words, IAM roles are not attached to a specific user. IAM roles also have IAM policy documents attached to them to determine what services and actions can be granted or denied.

IAM roles are generally used to grant access for the following use cases:

- An AWS service that needs access to another service in your own AWS account, for example, an application running on an EC2 instance that needs access to a database to update customer records.

- An IAM user in another account that needs access to services in your account via cross-account access.

- A federated user from another web **Identity Provider** (**Idp**) such as Google, Facebook, or Amazon that needs access to resources in your AWS account. IAM roles can be used to grant those external users with only the specific rights to specific services and resources in your account, without the need to create yet another IAM user account for them.

- A federated corporate user using an identity service such as Microsoft Active Directory, who needs access to a service in your AWS account.

As mentioned in the preceding list, IAM roles can be used to grant access to federated users. Identity federation is a process where you trust an external Idp to verify a given user's identity and then grant the user access on that basis. After authenticating the user, the Idp sends an assertion, which contains the user's login name and any attributes that AWS needs to establish a session with the user. The policy attached to the IAM role is then used to determine the level of access that can be granted to the user.

Identity federation thus allows you to grant external identities access to your AWS environment, whether via the Management Console or APIs. It also allows access to resources without the need to create an IAM user account for each external user.

Temporary credentials

A key benefit of using IAM roles is that they make use of temporary credentials that are rotated on a regular basis by AWS. For example, with IAM roles, you can grant a third-party mobile app access to resources in your AWS account without storing long-term credentials on the mobile phone of the user. This greatly enhances security and reduces management overhead. Imagine having to create an IAM user account for every individual that uses the app and then finding a secure method to distribute those IAM user credentials. IAM roles come to the rescue, because by themselves, they do not contain any credentials.

IAM roles make use of a service called **Security Token Service (STS)**. The **STS** service assigns temporary credentials to the identity that assumes the role. These temporary credentials will include an **access key ID**, a **secret access key**, and a **security token**. This security token is valid for only a short term and becomes useless after expiration. The STS service will renew the temporary credentials before expiry for the identity if the identity is still permitted access, and this happens in the backend and is managed by AWS. Ultimately, temporary credentials grant users temporary access to resources in your AWS account and are much more secure than using long-term access credentials.

When you create an IAM role, you also define a **trust policy**. Within the trust policy, you specify the entities that will be trusted to assume the role. These entities can be AWS services or identities external to your organization that need access to your AWS resources. Furthermore, the entities themselves will need to have permissions to be able to assume that role.

Where possible, using IAM roles instead of creating an IAM user account is the recommended approach, primarily for entities external to your organization and because of the benefits of using temporary credentials. So, using the example of the mobile app that needs to update a database in your AWS account, it is safe to say that using an IAM role will ensure that long-term credentials are not stored locally on the phone. This also greatly reduces the chances of malicious attacks using stolen credentials.

In this section, we looked at IAM roles, which allow you to grant secure access to your AWS services and resources using temporary security credentials.

In the next section, we'll look at credential reports, which allow you to audit your IAM identities.

Reviewing credential reports

AWS enables you to download a **comma-separated values** (CSV) file, updated every 4 hours, which allows you to audit your IAM user security state and review important information. The information could be a list of all your IAM users in your AWS account and the status of their credentials (such as if they have been configured with passwords and access keys). The report also highlights if your user accounts have been configured with MFA.

Monitoring your credentials report will also help you pick up on identities that may not have accessed resources in your AWS accounts recently. You can then work out whether those users still need access and delete unwanted users from your AWS accounts.

In this section, we looked at credential reports, which allow you to generate details of your IAM users and their current access status. In the next sections, we provide a number of exercises to help you build hands-on experience of using the IAM service to secure access to your AWS accounts.

Exercise 4.1 – creating an IAM group

In this exercise, you will create an IAM group for a development team that is going to require full access to Amazon S3:

1. Log in to your AWS account.

2. Click on the **IAM** link under the **Security, Identity, & Compliance** category on your **AWS Services** home page.

3. From the left-hand menu, click **Groups**.

4. Next, click on the **Create New Group** button.

5. You will then be presented with a step-by-step wizard. Provide a group name for your new group. For this exercise, type in `Developers`.

6. Click the **Next Step** button in the bottom right-hand corner of the screen.

7. You now need to attach a policy. You can create your own customer-managed policies but for the purposes of this exercise, type `S3` in the **Policy Type** filter search box. This will narrow down the available policies that relate to Amazon S3.

8. Tick the checkbox next to the **AmazonS3FullAccess** policy.

9. Click the **Next Step** button in the bottom right-hand corner of the screen.

10. Finally, click the **Create Group** button in the bottom right-hand corner of the screen.

You will now see that your group has been created and listed under **Group Name**. Now that you have created a group, you can proceed to create an IAM user and add it to the group. This allows you to manage multiple users more effectively. In the next exercise, we will create an IAM user and add it to the developers group.

Exercise 4.2 – creating an IAM user

Now that you have created a developers group, you can add your developers to this group. To illustrate this, we will create a new user, John. John is one of our senior developers at Packt and we would like to ensure that he is a member of the developers group, which will give him full access to Amazon S3:

1. In the IAM dashboard, click on **Users** from the left-hand menu.

2. Click the **Add user** button.

3. In the **User name** textbox, type in john (all lowercase).

4. Next, you need to select the type of access you want to grant John. John is a developer and will require both console access and programmatic access. This means that sometimes, John will use the web-based console to configure resources in Amazon S3, and at other times, he may use the CLI. For this exercise, tick both boxes – **Programmatic Access** and **AWS Management Console access**.

5. To access the AWS account via the console, you need to create a password for the user. For john, you can have AWS automatically generate a password for you or you can create a custom one yourself. Select **Custom password** and choose a complex password of your choice. Type that password in the textbox provided.

6. An additional setting, **Require password reset**, enables you to force your IAM users to change their password at the next login. That way, you will not know what their password is when they change it and it is best practice to follow. For the purposes of this lab, disable this checkbox for now.

7. Click the **Next:Permissions** button in the bottom left-hand corner of the screen.

8. You now have the option to set permissions for the user. As we have already created a group with the right set of permissions attached to it, we simply need to make this user a member of the group. For this exercise, under **Add user to group**, tick the box next to the **Developers** group and click the **Next:Tags** button in the bottom right-hand corner of the screen.

9. Tags are key-value pairs that you can attach to any resource. You can attach up to 50 tags to each resource and they enable you to classify your resources better. You can then use tags to understand cost allocation and to identify and manage your resources. Tags can include user information, such as an email address, or can be descriptive, such as a job title. For this exercise, set a single tag, with the key set to `Name` and the value set to `John`.

10. Click the **Next:Review** button in the bottom right-hand corner of the screen.

11. You can now review all your settings for the user and once satisfied, go ahead and click the **Create user** button in the bottom right-hand corner of the screen.

12. You are now presented with a **Success** screen, which confirms that the user has been created. You are also informed of the option to download your access keys. Access keys are like usernames and passwords and comprise an access key ID (similar to the username) and the secret access key (similar to the password). Access keys are used to grant programmatic access via the AWS CLI or using the AWS SDKs. It is important that you download these keys now and keep them safe on your computer. If you move away from this screen, the secret access key (such as the password) is no longer visible, and you would need to recreate the keys. For this exercise, ensure that you download the keys by clicking on the **Download .csv** button. Store the `.csv` file.

13. You will also note that you have been provided with a special link to log in to your AWS Free Tier account using an IAM user account. The AWS account sign-in page for IAM users is slightly different from that of the root user. This is because when you sign in as an IAM user, you need to specify the AWS account ID you are trying to sign into. In our example, we have a link like this:

    ```
    Users with AWS Management Console access can sign-in
    at: https://111222333444.signin.aws.amazon.com/console
    ```

    The series of numbers just before `.signin` represents your AWS account ID.

14. Click the **Close** button in the bottom right-hand corner of the screen to exit from the user setup wizard.

We have now created a user who represents a developer in our organization and we have added the user to the developers group. In the next exercise, you will learn how to access your AWS account as the IAM user you created previously.

Exercise 4.3 – logging in to your AWS account as an IAM user

In this exercise, we log out of the AWS account as the root user and re-login as the IAM user you just created. If you did not make a note of the special sign-in link, you will need to know what your AWS account ID is. You can easily discover this by clicking on your account name in the top right-hand corner of your web-based management console and noting the account ID:

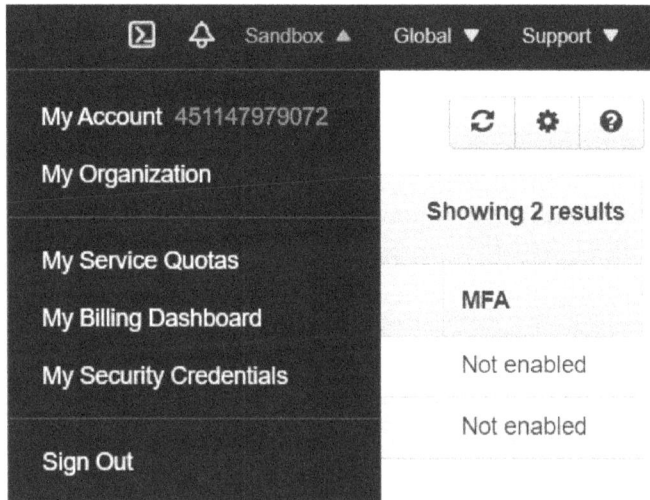

Figure 4.10 – Drop-down box when clicking on the account name to discover the account ID, which is shown after My Account

Now that you have the account ID (or you can make a note of the sign-in URL previously discussed), we can proceed to log in as the IAM user you created earlier:

1. Sign out of the AWS account (remember, you are currently signed in as the root user).

2. You will be taken to the AWS console home page where you can click the **Log back in** button or the **Sign in to the Console** button.

3. Clicking on either of these will take you to the AWS sign-in page.

4. Click on **IAM user** and provide the account ID in the available textbox:

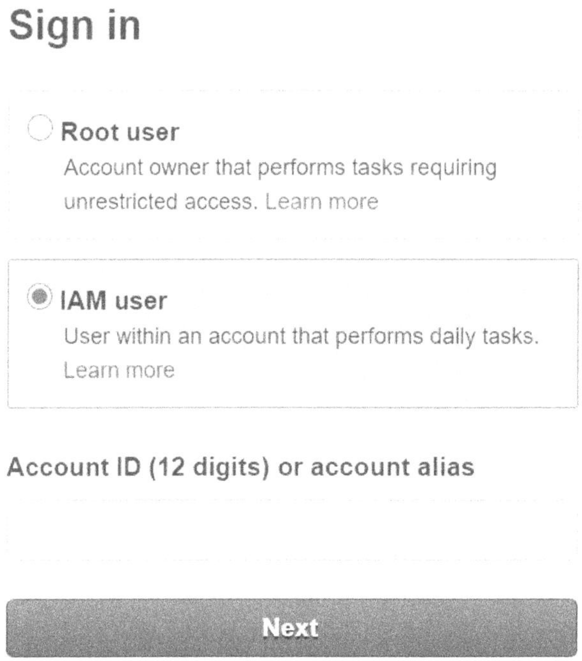

Sign in

○ **Root user**
  Account owner that performs tasks requiring
  unrestricted access. Learn more

◉ **IAM user**
  User within an account that performs daily tasks.
  Learn more

Account ID (12 digits) or account alias

[   ]

[**Next**   ]

Figure 4.11 – AWS signing page

5. Once you have provided the account ID, click **Next**.

6. Now, provide your username, in this case, john, and the password you chose when you created the user.

7. You will now be logged in to the AWS Management Console as our developer, John. You will note that your username is displayed in the top right-hand corner of the screen:

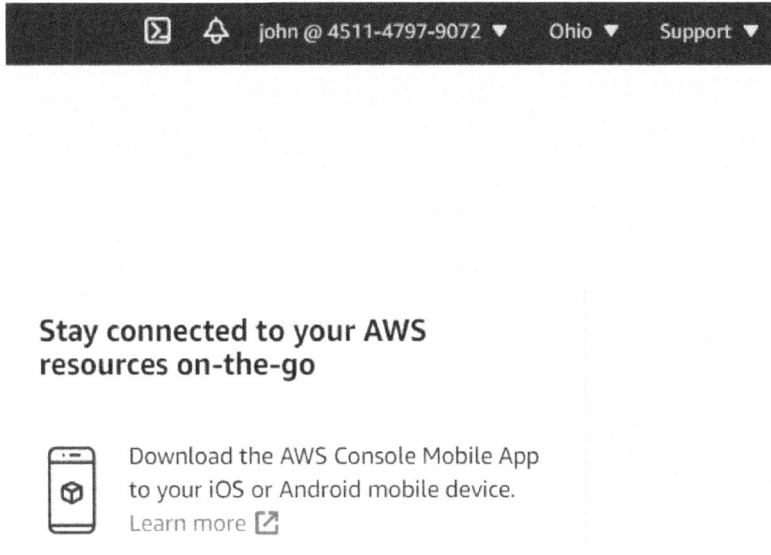

Stay connected to your AWS
resources on-the-go

Download the AWS Console Mobile App
to your iOS or Android mobile device.
Learn more ↗

Figure 4.12 – IAM user John has successfully logged in to this AWS account

In this section, we completed a series of exercises using the AWS IAM service. We demonstrated how to create IAM groups and users and how to log in to the AWS Management Console as an IAM user.

In the next section, we look at how you can access your AWS account using the AWS CLI.

Accessing the AWS platform using the CLI

As previously discussed, you can access the vast array of AWS services using the web-based management console or the CLI. The AWS CLI is a unified tool to manage your AWS services programmatically. The AWS CLI gives you access to the **application programming interface** (**API**) of each AWS service. This enables you to remotely access your AWS account and run commands from your Terminal application on Linux and Mac computers or use Command Prompt on Windows computers.

The AWS CLI is therefore ideal for running simple commands to complete repetitive tasks and because a single command string can contain necessary attributes of your request, you will find it a lot faster than using mouse clicks within the web console. The CLI is something you will need to get accustomed to—for one, you will need to know how to construct commands, but AWS offers complete reference documentation for this.

You can run your CLI commands line by line individually, or you can even create scripts to run a series of commands for a set of tasks. For example, you can create a script that launches a new EC2 instance, configures it as a web server, and installs any third-party applications.

Accessing your account via the CLI

In this section, we will walk you through the steps of configuring the CLI on your computer and accessing your AWS platform using John's account.

> **Important note**
>
> It is highly recommended that you avoid creating and using access keys for the root user account unless you have certain specific requirements. For example, to enable MFA-delete on Amazon S3 buckets (which we look at in *Chapter 5, Amazon Simple Storage Service (S3)*), you need to use the CLI, for which you will need a set of access keys. The problem is when you use the AWS CLI to access your AWS account, your credentials (access key ID and secret access keys) get stored locally on the computer that you are using. This means that you should consider rotating your keys on a regular basis and ensure you secure your computer, for example, by enabling local disk encryption. As a best practice, if you need to create a set of access keys for the root user account, you should follow a process of creating the keys, using them for the specific task, and then disabling or deleting the keys once the task is complete.

Accessing your AWS account using the CLI from your local computer as the root user therefore means that the root user's credentials are also stored on the local machine. This is considered a major security risk because you really should not be using the root user account for day-to-day operations. Even if you need full administrative privileges, you should ideally create an IAM user and assign it with the necessary admin rights. You can then configure your AWS CLI with the IAM administrator user account credentials instead.

The root user is the ultimate owner of your account and can even close your account. This is the reason why extreme caution is advised in how you manage the root user's credentials.

Downloading the CLI tools

To use the AWS CLI, you first need to download and install it on your local computer. You can access the AWS CLI tools here: `https://aws.amazon.com/cli/`.

Depending on your operating system, you will need to download and install the appropriate tool. The AWS CLI tool is currently available in version 2 and comes with new and improved installers, new configuration options, such as AWS **Single Sign-On (SSO)**, and various interactive features. You can download the installers for your specific operating system, whether it is Windows, Mac, or Linux. Previously, you had to have Python installed to use the AWS CLI, but this is no longer the case.

Download and install the appropriate installer for your operating system from the preceding link:

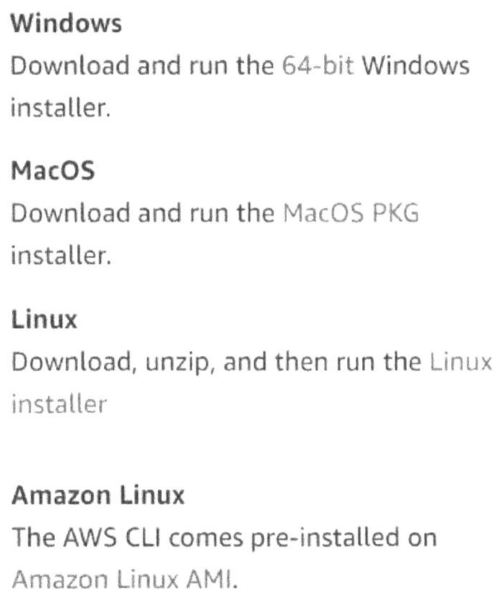

Windows

Download and run the 64-bit Windows installer.

MacOS

Download and run the MacOS PKG installer.

Linux

Download, unzip, and then run the Linux installer

Amazon Linux

The AWS CLI comes pre-installed on Amazon Linux AMI.

Figure 4.13 – Links to download the appropriate AWS CLI installer for your operating system

Once you have installed the AWS CLI, the next step is to configure it with a set of access keys to log in to your AWS account programmatically. If you are using Windows, you can access the AWS CLI from Command Prompt. If you are using Linux or Mac, you can use the Terminal application. The installer would have already set up any environmental variables, so you can access the tool from anywhere in your Command Prompt.

In the next exercise, you will learn how to access the AWS platform using the Amazon CLI on a Windows computer.

Exercise 4.4 – accessing the AWS platform using the AWS CLI on a Windows computer

Once you have installed the AWS CLI tools on your Windows machine, the next step is to configure it to access your AWS account as the IAM user you created earlier:

1. Open the credentials file, which you downloaded earlier when you created the IAM user john. This file has the access keys for your user, which comprises the **access key ID** and the **secret access key**.

2. On your Windows desktop, click on the **Start** button and search for Command Prompt by typing in CMD:

Figure 4.14 – Command Prompt on a Windows computer

3. At the prompt, type in AWS configure to start the configuration process.

4. You will then be prompted to enter the **AWS access key ID**, followed by the **AWS secret access key**. These keys are in your credentials document that you downloaded earlier.

5. For **Default region name**, type in us-east-1.

6. For **Default output format**, leave this blank and press *Enter*:

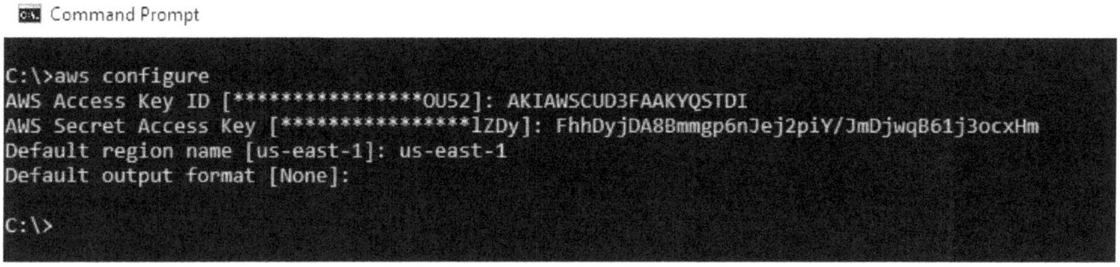

Figure 4.15 – Configuring the AWS CLI with access key ID and secret access keys

7. Your AWS CLI tool has now been configured with John's credentials.

8. You can try running a command such as aws s3 ls. This command lists out any Amazon S3 bucket you have in your account. If you have not created any buckets yet, you will just have the prompt return. In my case, I have already got one bucket in my account as you can see in the following screenshot:

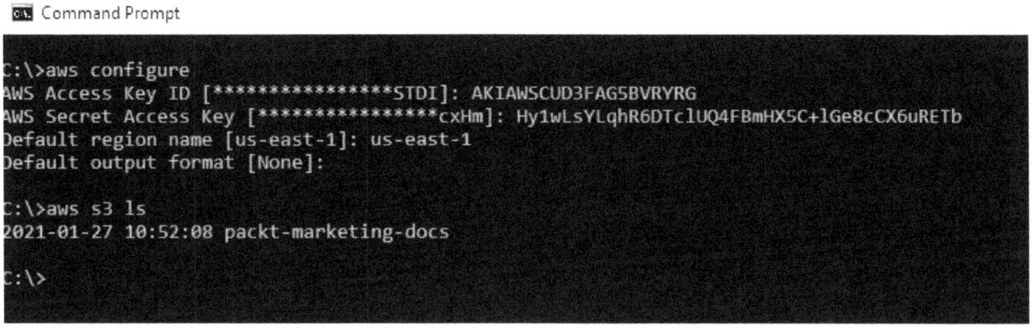

Figure 4.16 – Configuring the AWS CLI with access key ID and secret access keys

In this section, we discussed the necessary steps required to configure your AWS CLI tool, so that you can access your AWS account as an IAM user using Command Prompt on a Windows machine. You can also use Terminal on a Mac or Linux computer to complete the same tasks. You would need to install the appropriate tool for your operating system.

Using the CLI can be very efficient, especially if you are trying to perform repeat tasks as you can also create scripts to automate the whole process.

You should try and avoid using the root account to perform any day-to-day operations in your AWS account. Instead, you must log in with an IAM user account that has only the necessary privileges to carry out the task at hand. This is known as following the principle of least privilege.

In the next exercise, you will create another IAM user account that you will use for all the exercises in the upcoming chapters.

Exercise 4.5 – creating an IAM user with administrative privileges

In this exercise, you will create another IAM user account that you will use to log in to your AWS account. This IAM user will be provided with full administrative access to help you work through the upcoming exercises easily, although in the real world, you would want to restrict permissions to only the job function of the users in question. By getting used to logging in as an IAM user, you will build a habit of avoiding the use of the root user credentials for your day-to-day tasks:

1. Ensure that you are logged in to your AWS account as the root user (the email address and password combination you used to create your AWS account).

2. Navigate to the IAM dashboard.

3. From the left-hand menu, click on **Users**. Next, from the right-hand pane, click **Add user**.

    You will be redirected to the **Add user** wizard page.

4. For the username, type in `Alice`. Throughout the rest of this guide, you will be logging in as `Alice` to carry out all upcoming exercises.

5. Under **Select AWS access type**, select both **Programmatic access** and **AWS Management Console access**.

6. For **Console password**, select the **Custom password** option and provide a complex password of your choice.

7. Uncheck the box next to **Require password reset** and click the **Next: Permissions** button.

8. In step 2, under **Set permissions**, select the **Attach existing policies directly** option. This will allow you to attach an inline policy to Alice's account alone.

9. From the list of policies provided, select the checkbox next to **AdministratorAccess** as per the following screenshot:

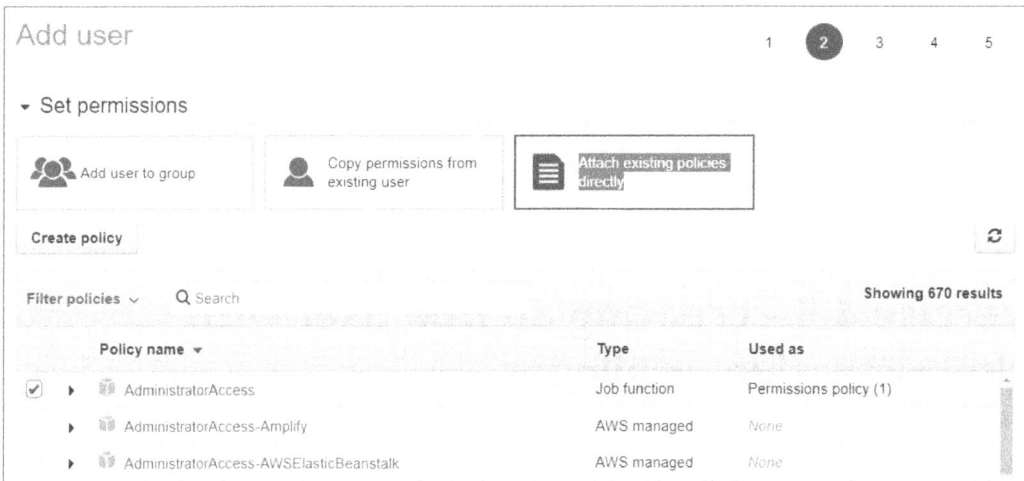

Figure 4.17 – IAM user (Alice) with administrator access permissions

10. Click the **Next: Tags** button at the bottom of the screen.

11. In step 3, under **Add tags (optional)**, provide a key-value pair where the key is set to `Name` and the value is set to `Alice`.

12. Click the **Next: Review** button.

13. Next, click the **Create user** button.

14. You will then be prompted to download the .csv file containing Alice's access keys and secret access keys. Download the file and ensure you keep it in a folder on your computer that you can easily access. Once downloaded, go ahead and click the **Close** button.

15. Your IAM user Alice is now ready to log in. Going forward, you will need to log in as the IAM user Alice for all upcoming exercises. Ensure you also make a note of either the special sign-in link for your AWS account or the AWS account ID, which you will need to log in as an IAM user.

In this exercise, you created an IAM user account that you will use to log in to your AWS account and perform all the upcoming exercises. In the next section, we'll review a summary of this chapter.

Summary

In this chapter, we discussed the AWS IAM service and how it acts as the gatekeeper to your AWS account. We discussed the root user of your account, which is the most senior administrative account for your AWS account. The root user has complete access and protecting this account with MFA is a recommended best practice.

With AWS IAM, you can create and manage identities that are granted or denied access to the various AWS services in your account. These identities can include IAM users, groups of IAM users, or IAM roles. You can also further enhance the security of your IAM users by configuring them with MFA.

We discussed best practices when configuring your IAM users and IAM groups. We emphasized that, as best practice, you should create IAM groups and subsequently place any necessary IAM users within groups that share a common task. For example, if you have a developers group, you can place all your developers in that group. In addition, you can create IAM policies, which can be attached to your IAM identities to determine what they can or cannot do within your AWS account. AWS IAM thus offers authentication and authorization services to your account.

We also examined IAM roles, which allow you to create identities not specifically attached to any physical user. IAM roles enable you to grant access to those entities that may not necessarily have access. IAM roles can be used to grant temporary credentials to entities when they need them and, depending on the use case, are more secure than using IAM user accounts.

Next, we discussed credential reports, which allow you to audit your existing user base within your AWS account and audit usage patterns. Finally, we looked at how you can access your AWS account using the AWS CLI.

At this point, you have now learned the importance of creating IAM users, groups, and roles and have understood their application. You have learned that the root user account must not be used for day-to-day operations and you must set up MFA for your root user account as well as your IAM users. You have also gained knowledge on how IAM roles can be used to grant cross-account access and federated access and enable one AWS service to access another AWS service. You have also learned how IAM policies can be used to enforce the principle of least privileges, thereby improving the security of your AWS account.

In the next chapter, we look at another core AWS service, called **Amazon Simple Storage Service** (**Amazon S3**). Amazon S3 is an object storage solution that allows you to store unlimited amounts of data in the cloud. Amazon S3 can be used for a wide range of use cases, including storage of digital assets for your web applications, such as documents, images, and video, as well as for archive storage.

Questions

Here are a few questions to test your knowledge:

1. You wish to deploy a dev and test environment on AWS. You want to ensure that your developers can access your AWS account using a highly secure authentication process and follow best practices. Which of the following two configuration options will help ensure enhanced security? (Choose two answers)

    A. Configure your IAM accounts with MFA.

    B. Configure your IAM password policy with complexity rules.

    C. Ensure you encrypt your EBS volumes.

    D. Create RDS databases with Multi-AZ.

    E. Provide the root account credential details to your developers.

2. Your developer is working from home this weekend and needs to access your AWS account using the CLI to configure your RDS database from their local computer. Which type of IAM credentials would they need to configure the AWS CLI tool on their machine?

    A. IAM username and password

    B. Access key IDs and secret access keys

C. Access keys and secret ID

D. HTTPS

3. Which AWS service enables you to troubleshoot your IAM policies and identify the sets of permissions that may be denying access to a given AWS service?

A. IAM policy simulator

B. CloudWatch

C. CloudTrail

D. IAM policy manager

4. Which of the following AWS services is a better option to securely grant your application running on an EC2 instance access to a backend database running on Amazon RDS?

A. Access keys

B. IAM role

C. IAM group

D. Security group

5. Which format are IAM policy documents written in?

A. JSON

B. YAML

C. XML

D. JAVA

6. What best practice strategy should you follow when assigning permissions to IAM users and groups?

A. Follow the principle of least privilege.

B. Follow the principle of most privilege.

C. Follow the ITIL principles.

D. Follow the GDPR principle.

7. Which IAM service enables you to effectively manage users by creating a collection of them based on their job function and assigning them permissions according to their roles to the entire collective?

    A. IAM groups

    B. IAM policies

    C. IAM collection

    D. IAM roles

8. Which feature of IAM enables you to use your existing corporate Active Directory user credentials to log in to the AWS Management Console and therefore offer an SSO service?

    A. Identity federation

    B. IAM user database

    C. Active Directory users and computers

    D. MFA

9. Which AWS service enables you to generate and download a report that lists your IAM users and the state of their various credentials, including passwords, access keys, and MFA devices?

    A. AWS policies

    B. AWS Explorer

    C. Credentials report

    D. User report

10. Which AWS service is responsible for assigning and managing temporary credentials to entities that assume an IAM role?

    A. AWS Password Manager

    B. AWS Security Token Service

    C. AWS Credentials Manager

    D. AWS Credentials Report

5
Amazon Simple Storage Service (S3)

In this chapter, we look at one of the available storage services on **Amazon Web Services** (**AWS**). Many clients who are just starting on their cloud journey often consider storage services in the cloud as a stepping stone to going cloud-native in the long run. While storage options have become cheaper over the years, the fact remains that we continue to consume more and more storage with the passage of time. That said, it is vital that organizations also have a smart life cycle policy for their storage needs. Companies may be required to keep data for many years, and for as long as 7 to 10 years for compliance and regulatory purposes. However, at some point, a substantial amount of data is no longer required, and purging this data from the network not only makes management easier but also saves on cost.

Access to AWS storage services is extremely easy, and rather than procuring new storage hardware to host on-premises, it is much easier and more cost-effective to use the services offered by a cloud vendor such as AWS. Understandably, there will some types of data that need to be stored on-premises primarily because of latency issues, but from a security standpoint, AWS offers numerous options to ensure that your data is accessible only to you.

AWS offers different storage options, and in this chapter, we look at one of its flagship products: **Amazon Simple Storage Service** (**Amazon S3**). Amazon S3 is an object storage solution and offers very high levels of availability, durability, and scalability. AWS also offers other types of storage options, which we look at in subsequent chapters.

By the end of this chapter, you will understand the fundamentals of object storage on AWS and how Amazon S3 can help fulfill core storage requirements for your business in the cloud. You will also learn how about various features that can be used to manage your cloud storage, address regulatory and compliance concerns, and design cost-effective solutions. Finally, you will learn how to access your cloud storage from on-premises locations and how to migrate large datasets to the cloud.

The topics in this chapter include the following:

- Introduction to storage options on AWS

- Introduction to Amazon S3

- Learning about archiving solutions with Amazon S3 Glacier

- Connecting your on-premises storage to AWS with Amazon Storage Gateway

- Migrating large datasets to AWS with the Amazon Snow Family

- Exercise 5.1—Setting up an Amazon S3 bucket

- Exercise 5.2—Configuring public access to S3 buckets

- Exercise 5.3—Enabling versioning on your bucket

- Exercise 5.4—Setting up static website hosting

Technical requirements

To complete the exercises in this chapter, you will need to have an AWS account via the AWS Management Console. You will need to be logged in as the IAM user, **Alice**, that you created in the last chapter.

Introduction to storage options on AWS

A storage service provides the necessary infrastructure to enable you to store and access data. However, different use cases require varied storage architectures to ensure performance, reliability, durability, and the right type of access to the data. There are three primary storage options available, and AWS offers services to cater to each of these.

Block storage

Block storage is an architectural design that enables the storage of data onto media such as a hard disk, in fixed-sized chunks. Data is broken up into small blocks and placed on the media in these chunks, with a unique address assigned that forms part of its metadata. Block storage makes use of a management software (which can be part of the operating system) to organize the blocks of data. When a user tries to retrieve a file, the management software identifies the blocks to retrieve, reassembles the data, and presents the whole file to the user.

On AWS, block storage options are available as **Elastic Block Store** (**EBS**). These can be configured as volumes attached to your **Elastic Compute Cloud** (**EC2**) instances and offer ultra-low latency required for high-performance workloads. One advantage of EBS volumes is that they are not directly attached to the EC2 instance you deploy, but instead are connected via high-speed network links. This allows you to detach an EBS volume from one EC2 instance and attach it to another if, for example, the first EC2 instance experiences some sort of failure.

Typical use cases include running and managing system files such as those used by your operating system, large databases, or for applications such as **enterprise resource planning** (**ERP**) solutions. These types of applications require very low-latency access to the data and generally make use of **direct-attached storage** (**DAS**) or **storage area networks** (**SANs**).

File storage

Another storage architectural design is **file storage**. The architecture offers a centralized location for your corporate data, and files are stored in folders and sub-folders. File storage offers a hierarchical structure to store your data, and this means you can imitate your real-life counterpart—the filing cabinet—to organize your data.

Retrieval of the data requires you to know the file and folder structure and provide this information. For example, if I need last August's balance sheet Excel document, I will need to look in the 2020 folder and the August sub-folder, and this would enable me to retrieve that specific data.

Due to the nature of file storage, metadata information can be limited, and a key limitation to be aware of is that you cannot have unlimited folders and sub-folders due to your operating system restrictions. Your hierarchical structure can also be the cause of some performance issues, and therefore you need to decide on this structure carefully.

File storage lends itself well to being used for file sharing within a corporate organization. Because of the folder/sub-folder architecture, it becomes very easy to organize your data to fit in well with your organizational structure.

Amazon offers three different file storage systems, outlined as follows:

- **Elastic File System** (**EFS**)—This is a managed elastic filesystem designed to let you share file data without provisioning or managing storage as you would with EBS. Your filesystem will grow and shrink as you add and remove data, and mount points can only be created on Linux EC2 instances.

- **FSx for Lustre**—A high-performance filesystem designed for applications that require fast storage and can scale to hundreds of **gigabytes** (**GB**) of throughput and millions of **input/output operations per second** (**IOPS**). FSx for Lustre is also designed for a wide range of Linux-based EC2 instances.

- **FSx for Windows File Server**—Designed for Microsoft Windows EC2 instances and offers a fully managed file-share solution, natively supporting the Windows file system such as the industry-standard **Server Message Block** (**SMB**) protocol. Typical use cases include file-sharing services, local archiving, application data sharing, and data protection.

Object storage

By contrast, **object storage** involves storing complete files as individual objects. Object storage presents a flat file structure—you create some form of container and place your objects within this container without using any folder or file-level hierarchy. This is also known as unstructured data. Object storage metadata (information about the object—such as its name, and so on), along with other attributes, is then used to create a unique identifier to easily locate that data in your storage pool. Due to the nature of object storage, the metadata can contain a vast array of information, enabling you to use object storage for data analytics far more easily than a file-based storage solution.

An additional benefit is that object storage lends itself well to offering higher levels of performance, durability, and scalability. In a file-level storage solution, the depth of the folder and file structure will often have a limit based on the operating system you are using. With object storage, however, you can potentially scale to having limitless amounts of data.

The flat file structure is another advantage point as this enables you to retrieve data much faster due to the extended categorization feature, as opposed to retrieving data from a file storage service.

On AWS, object storage options are available with the Amazon S3 service. Amazon S3 lets you create containers called **buckets**, within which you place your data (objects) in a flat file structure (unstructured manner). You can store and retrieve any amount of data—anytime, anywhere.

Typical use cases for object storage include storing digital assets for your websites and applications (documents, images, video), the ability to perform analytics on your objects, and offering storage solutions to cutting-edge technologies such as **Internet of Things** (**IoT**).

In this section, we looked at the three main types of storage options available, their key features, and typical use cases. We also highlighted some examples of AWS services that offer storage solutions.

In the next section, we will introduce you to the Amazon S3 service, which is an object storage solution for the cloud.

Introduction to Amazon S3

Amazon S3 is one of Amazon's flagship products, and offers a robust, scalable, durable, and cost-effective **object storage** solution in the cloud. Customers can use Amazon S3 to store any amount of data for a wide range of use cases, including digital media content for websites, data lakes, mobile applications, IoT device data, and big data analytics.

Amazon S3 can offer up to 99.999999999% durability and fulfills the storage requirements for a majority of clients and their individual business needs.

What does eleven 9s of durability mean? According to AWS, if you store 10,000,000 objects on Amazon S3, then on average you can expect to incur a loss of a single object once every 10,000 years. You can review all **frequently asked questions** (**FAQs**) for Amazon S3 here: `https://aws.amazon.com/s3/faqs/`.

Buckets and objects

Before you can upload any data to Amazon S3, you need to create a container called a bucket. Buckets need to have a unique global namespace as their contents can be made accessible over the public internet. This means that your bucket names must be unique across the AWS ecosystem. For example, a document named `blueberry-muffin.txt` stored in a bucket named `just-desserts` can be accessible via the internet in two primary ways, outlined as follows:

- **Virtual hosted-style endpoints**—Here, the bucket name forms part of the **Domain Name System** (**DNS**) subdomain name so that the preceding `blueberry-muffin.txt` object is accessible via the S3 **Uniform Resource Locator** (**URL**), such as `https://just-desserts.s3.amazonaws.com/blueberry-muffin.txt`.

- **Website endpoint**—Here, the bucket is configured with a static website hosting service and the website is available at the AWS Region-specific website endpoint in one of the following two formats:

  a) `s3-website` dash (`-`) Region—`http://bucket-name.s3-website-Region.amazonaws.com`. For example, our recipe will be available at `http://just-desserts.s3-website-us-east-1.amazonaws.com/blueberry-muffin.txt`.

  b) `s3-website` dot (`.`) Region—`http://bucket-name.s3-website.Region.amazonaws.com`. For example, our recipe will be available at `http://just-desserts.s3-website.us-east-1.amazonaws.com/blueberry-muffic.txt`. We look at static website hosting on Amazon S3 later in this chapter.

When creating buckets, you may therefore find that common names are not available. For example, if you wanted to create a bucket name called `marketing`, then you will most likely not be able to use this name as it may have been used already. You could instead create a marketing bucket with a unique prefix or suffix to get an appropriate name. Most companies will try to associate their bucket names with their organization name or project name—for example, your marketing bucket name could be `my-company-marketing`. That said, you still cannot solely depend on any specific naming convention you define for your buckets, because another customer may have chosen the exact same name that you might have wanted, and bucket names are defined on a first-come first-serve basis.

Some key attributes of Amazon S3 buckets are provided here:

- Bucket names must be between 3 and 63 characters long.

- Bucket names must always be in lowercase letters. They can, however, contain numbers, hyphens (`-`), and dots (`.`) only.

- Bucket names must also begin and end with either a letter or number and should not include spaces between the names.

- Bucket names must not be formatted as an **Internet Protocol** (**IP**) address (`192.168.1.1`).

- Buckets used with **Amazon S3 Transfer Acceleration** (**S3TA**) cannot have dots (`.`) in their names. We discuss Amazon S3TA later in this chapter.

- You cannot have nested buckets—for example, a bucket within another bucket.

> **Important note**
>
> Except for website-style endpoints, you should avoid using dots (.) as part of your bucket names as they cannot be used with virtual hosted-style endpoints using **Secure Sockets Layer** (**SSL**) and the **HyperText Transfer Protocol Secure** (**HTTPS**) protocol. Note that the only reason they work with website-style endpoints is because static website hosting is only available over HTTP. You can get around this problem if you need to serve your content over HTTPS by incorporating a Amazon CloudFront distribution point. We discuss CloudFront in *Chapter 6, AWS Networking Services – VPCs, Route53, and CloudFront*.

Any data stored in an Amazon S3 bucket is represented as an **object**. An object is stored in its entirety within a given bucket rather than with block storage, where a file may be broken up into chunks and stored on a given media such as a hard disk. As discussed previously, objects are stored in an unstructured manner as a **flat filesystem**. This means that accessing an object requires you to know its unique ID, which is generally part of the object's metadata. You can store an unlimited number of objects within a given bucket, and the maximum size of an object on Amazon S3 is 5 **terabytes** (**TB**).

The filename of an object is called a **key**, and the data contained as part of that *object* is known as its **value**. Keys can be up to 1,024 bytes long and your objects can consist of letters and numbers, as well as characters such as, !, -, _, ., *, and ( ) .

Managing your objects in a bucket

As mentioned previously, objects are stored in a **flat filesystem**, and this can sometimes make it difficult to manage your objects. As humans, we find it easier to categorize objects into folders and sub-folders, and this helps us organize our data. While object storage does not provide a folder structure, it offers the ability to use **prefix** and **delimiter** (/) parameters to help you manage and browse your objects in a hierarchical fashion.

At first glance, the usage of prefixes and delimiters (/) appears as a typical folder hierarchy, but the prefixes and delimiters themselves also form part of the object's **key**, as we can see in the following screenshot:

Amazon S3 > packt-marketing > campaign/ > cloud-practitioner/

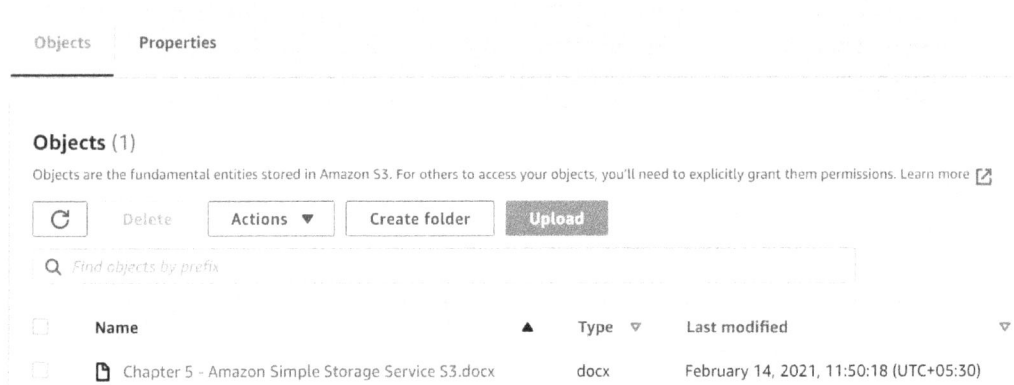

cloud-practitioner/

Objects | **Properties**

Objects (1)

Objects are the fundamental entities stored in Amazon S3. For others to access your objects, you'll need to explicitly grant them permissions. Learn more ↗

| C | Delete | Actions ▼ | Create folder | **Upload** |

Q *Find objects by prefix*

	Name	▲	Type	▽	Last modified	▽
	🗋 Chapter 5 - Amazon Simple Storage Service S3.docx		docx		February 14, 2021, 11:50:18 (UTC+05:30)	

Figure 5.1 – Amazon S3 prefixes and delimiter example

As you can see in the preceding screenshot, a file (object) called `Chapter 5 – Amazon Simple Storage Service S3.docx` appears to be stored in a `cloud-practitioner` sub-folder, under another sub-folder named `campaign`, in a bucket named `packt-marketing`.

This architecture allows you to better manage your objects, but the fact remains that the keys of the object itself comprise those prefixes and delimiters. This means that the `development/sourcecodes.php` and `production/sourcecodes.php` keys can reside in the same bucket and are considered unique objects because of the different prefixes.

Prefixes are also used to help you limit search results to only those keys that begin with a specific prefix name. In addition, delimiters enable you to perform list operations such that all the keys that share a common prefix can be retrieved as a single summary list result.

Prefixes further enhance performance when it comes to accessing your objects in Amazon S3. For example, you can achieve 3,500 `PUT/COPY/POST/DELETE` or 5,500 `GET/HEAD` operations per second per **prefix** in a bucket. This means that if you have five prefixes to categorize five different collections of objects, you can achieve your read performance of 27,500 read requests per second for your `GET` operations.

Regional hosting – global availability

Your buckets and the objects contained within them are globally accessible if you define the necessary permissions. However, it is important to realize that a given bucket and any objects it holds are stored in one specific Region alone. When you create an Amazon S3 bucket in your AWS account using either the web console, **command-line interface** (**CLI**), or via **application programming interface** (**API**) access, you must specify the **Region** in which you wish to create it.

Your choice of the Region to create a given bucket is going to depend on several factors, including the following ones:

- Optimizing latency by creating buckets closer to end users who need access to them
- Minimizing costs
- Addressing any regulatory requirements such as data-sovereignty laws

Amazon will *never* replicate your data outside of the Region in which you create it. This offers the assurance that you can meet and adhere to any data-residency laws you may be required to follow as a business.

You can, however, replicate the contents of one bucket in a given Region to another bucket in a different Region for several use cases, including **disaster recovery** (**DR**) or sharing content with your colleagues, such that the data is closer to them to reduce overall latency.

Access permissions

To create and manage your Amazon S3 buckets and upload and download objects to the bucket, you need to define the necessary permissions. By default, only the resource owner, which is the AWS account that creates the resource, can access a bucket.

However, you can also grant access to other users in your account, as well as define permissions for specific **Identity and Access Management** (**IAM**) roles to have access to those resources. In addition, you can grant access to users in other AWS accounts, and even grant access to members of the public by configuring anonymous access. Anonymous access is ideal when you wish to host publicly accessible digital assets such as documents, images, and videos for your websites.

Amazon S3 offers two primary methods of granting access to resources such as buckets and objects. You can attach a resource-based policy known as a bucket policy, or you can attach **access-control lists** (**ACLs**). Let's examine both of these methods for granting access next.

Bucket policies

A bucket policy is applied directly to an entire bucket and can be used to grant access to both the bucket itself and the objects stored within it. Bucket policies can be used to specify different levels of access for different types of objects within the same policy document. A bucket policy document is also written in **JavaScript Object Notation Format (JSON)** format, just like IAM policies are.

With bucket policies, you can also grant anonymous access to object in your buckets, such as a web page, image, or video, which means that anyone with the S3 object URL can access it.

Bucket policies are very flexible and allow you to grant cross-account access too. This means that if users in other AWS accounts are permitted, you can grant them access to your buckets by specifying their account ID and their IAM user **Amazon Resource Name (ARN)**.

Here is an example of a typical bucket policy granting anonymous access to the objects it contains:

```
{
  "Id": "Policy1613373871314",
  "Version": "2012-10-17",
  "Statement": [
    {
      "Sid": "Stmt1613373870082",
      "Action": [
        "s3:GetObject"
      ],
      "Effect": "Allow",
      "Resource": "arn:aws:s3:::packt-marketing/*",
      "Principal": "*"
    }
  ]
}
```

Figure 5.2 – Bucket policy example granting anonymous access to the
contents of the 'packt-marketing' S3 bucket

As seen in the preceding screenshot, the policy allows anonymous access by specifying the Principal attribute as a wildcard (*), which indicates everyone. The action allowed on the packt-marketing bucket is s3:GetObject, which restricts access to only being able to read/download the object(s).

Another element that can be added to a bucket policy is a **condition**. This allows you to restrict the application of a policy based on a predefined condition. You can even set the `Effect` attribute of a policy to Deny unless the condition is met. For example, you can restrict the ability for users to access and download contents from an S3 bucket, unless the request originates from a range of IP addresses specified in the condition, as illustrated in the following screenshot:

```
{
    "Version": "2012-10-17",
    "Id": "S3PolicyId1",
    "Statement": [
      {
        "Sid": "IPAllow",
        "Effect": "Deny",
        "Principal": "*",
        "Action": "s3:*",
        "Resource": [
                  "arn:aws:s3:::packt-marketing",
            "arn:aws:s3:::packt-marketing/*"
        ],
        "Condition": {
            "NotIpAddress": {"aws:SourceIp": "w.x.y.z/c"}
        }
      }
    ]
}
```

Figure 5.3 – Bucket policy defined with a conditional statement

In the preceding screenshot, you will note that the `Effect` attribute of the policy is to deny access unless the `NotIPAddress` condition is met. You will need to replace the dummy IP address range shown (`w.x.y.z/c`) with a real IP address range for your use case.

Bucket and object ACLs

ACLs are now considered legacy control systems because bucket policies tend to be flexible and offer more granular levels of access. You can mostly use ACLs to grant anonymous access to objects and buckets, or to other AWS accounts. Since ACLs do not allow you to specify an individual IAM user as the grantee of those permissions, their use case is limited and preference is given to bucket policies instead.

However, certain features require you to configure ACLs instead of bucket policies. **Server access logging**, for example, is a feature you can enable to provide detailed records for the requests that are made to an Amazon S3 bucket.

IAM policies

As previously discussed in *Chapter 4, Identity and Access Management*, you can create IAM policies that are assigned to IAM identities (such as IAM users, groups of users, or IAM roles) and define what they can or cannot do with any specific service and/or resource in your AWS account.

You can, therefore, grant your IAM principals access to S3 buckets and objects contained within those buckets. However, IAM policies cannot be attached directly to the resource. Furthermore, you cannot attach an IAM policy directly to an IAM user in another AWS account. You would first need to create an IAM role with the necessary permissions and then enable a trust policy for the IAM user in the other AWS account to be able to assume the role.

Finally, IAM policies cannot be used to grant anonymous access because of the simple principle that IAM policies can only be attached to an IAM identity.

> **Note**
>
> You can use a combination of bucket policies, ACLs, and IAM policies to grant access. You must remember, however, that any conflicting `Deny` permission will always override an `Allow` permission. However, these policy options are not mutually exclusive.

Choosing the right S3 storage class

Amazon S3 allows you store an unlimited amount of data in the cloud. However, not all data needs to be treated the same. For example, you may have some data that you require instant access to, but also other types of data that may be rarely accessed as it represents old archives stored for compliance purposes.

You may also have some data that you can afford to lose because recreating it would be easy, whereas other types of data may be simply irreplaceable. Depending on the data, its importance, and access patterns, AWS offers different storage classes for different use cases. So, for example, if you need to store old archives, you have the Amazon Glacier storage class, and if you need instant rapid access to your data, you have the Amazon S3 Standard storage class. All storage classes offer 99.999999999% **durability** for your data. Durability refers to long-term data protection, and AWS offers the necessary infrastructure and processes to manage your data, replicate copies, ensure data redundancy, and safeguard against degradation or other corruption.

Depending on the storage class that you choose to store your data in, AWS also offers different levels of **availability**, which determines the percentage of time an object is available for retrieval, based on the underlying storage system being operational (or the Region and **Availability Zones** (**AZs**) being online and accessible). Critical data such as digital assets, medical records, or financial statements would be ideal candidates for those storage classes that offer higher levels of availability. However, secondary copies of data may not need to be as available as the primary copies of the same data.

Amazon S3 charges are based around six cost components. These comprise the storage itself (comprising the amount of storage, duration, and storage class your objects are placed in), requests and data retrievals, data transfers, use of transfer acceleration, data management, and analytics, and the use of an Amazon S3 Object Lambda (which is the ability to modify and process data as it is returned to an application using Lambda functions). Note—data transferred in from the internet to an S3 bucket is free. One way of minimizing costs is to identify data that may not require instance access or that may be replaceable and store it in classes that are cheaper.

> **Important note**
> A key point to note here is that you can host different objects under different storage classes within the same bucket. You do not need to create separate buckets for each storage class.

Let's look at the different storage classes next.

Frequent access

Amazon S3 Standard—This is the default storage class when you upload objects to a bucket, unless you specify otherwise. Amazon S3 Standard offers the full eleven 9s (99.9999999%) of durability and four 9s (99.99%) of availability. With Amazon S3 Standard, your objects are always replicated across a minimum of three AZs in the Region you place them in.

Infrequent access

Amazon S3 offers two types of infrequent-access storage classes. These can be used to store objects that you are not going to frequently access, but at the same time, you have instant access to the data when you need it.

AWS offers these classes at lower costs on the condition that you do not access your data frequently, as you would with the Standard storage class. To enforce the conditions, AWS will charge additional retrieval fees. Furthermore, there is a minimum object size of 128 **kilobytes** (**KB**). You can still store objects under this minimum size, but those objects will be billed as though it is a minimum of 128 KB in size.

Amazon S3 Standard-Infrequent Access (**S3 Standard-IA**)—S3 Standard-IA is designed for data that is just as critical as with the Standard storage class but is infrequently accessed and is therefore ideal for long-term storage, such as for backups, and to act as a data store for DR purposes.

Amazon S3 One Zone-Infrequent Access (**S3 One Zone-IA**)—Data stored in this storage class is restricted to one AZ only within the Region you upload it to. This reduces your overall availability of the data to 99.5% but is also much cheaper than the Standard or the Standard-IA storage classes. This also means that if there is an outage of the AZ in which your data is stored, you would have to wait for the AZ to come back online before you can access any data. In the unlikely event of the destruction of an AZ, you may also lose that data.

Amazon recommends this class for data that can act as a secondary backup or that can be recreated.

Archive storage

Often, data needs to be retained for archival purposes so that it is available when needed for auditing or reference. More often, regulatory and compliance requirements state that certain types of data should retained for *n* number of years. These could be financial information or past medical records, for example.

Amazon offers very low-cost storage for such requirements through its archival solution, **Amazon Glacier**.

Amazon Glacier—This storage class is designed for long-term archiving of data that may need to be accessed infrequently and within a few hours.

Retrieving data from Amazon Glacier works differently, however, as it is not immediately accessible and requires you to initiate a restore request for the data. This restore process can take some time (between a few minutes to 12 hours) before the data is available to download, and this depends on the retrieval option you choose.

Amazon Glacier Deep Archive—This is the lowest-cost storage class whereby customers can store very old historical data to meet compliance and regulatory requirements. Such data may be required to be kept for 7 to 10 years. Retrieval times can take 12 hours or more, depending on the retrieval option chosen.

We discuss the Amazon S3 Glacier retrieval options in more detail later in this chapter.

Unpredictable access patterns

Normally, your business will have a predictable access pattern for most data—for example, newly created data may need to be accessed frequently, such as daily. As data gets older it is accessed less frequently, and sometimes very rarely. Your access pattern, while predictable in this case, changes over time. AWS offers a feature known as lifecycle management that allows you to move data from one storage class to another, depending on changes to your access patterns. We look at lifecyle management and lifecycle rules later in this chapter.

Sometimes, however, it is difficult to categorize data as frequently accessed or infrequently accessed, simply because of the nature of that data. You might be frequently accessing a set of objects for an initial period of a few weeks, and those objects may later become infrequently accessed. However, a few months down the line, you may need to access those objects again for analysis or some form of investigation. This data may now need to be frequently accessed over a period once again.

In these scenarios, AWS offers another storage class called the Intelligent-Tiering storage class. For this privilege, you are charged a small monitoring fee for every object to ensure it is automatically transitioned into the right tier, depending on access patterns.

Intelligent-Tiering—This storage class offers automated tiering of data depending on your access pattern. Objects are automatically transitioned across four different tiers, two of which are latency access tiers designed to move objects between frequently accessed and infrequently accessed tiers, and the other two being optional archive access tiers:

- Frequent and infrequent tiers—Objects that are frequently accessed (within 30 days) are placed automatically in the frequent access tier (Standard storage class). Any objects not accessed for 30 days are then moved into the infrequent access tier (Standard-IA), thereby incurring lower costs. Remember—the minimum object size for Standard -IA is set to 128 KB, and objects less than this size are treated and charged as if they are a minimum of 128 KB. Any object in the infrequent tier that later gets accessed is then automatically moved back into the frequent tier and charged accordingly.

- Optional archive access tiers—You can optionally choose to activate the archive access tiers. Once activated, this results in the S3 Intelligent-Tiering service transition when any object is not accessed for 90 days is moved into the Amazon Glacier **Archive Access tier**. If the object is not accessed for 180 days, it will be moved into the Amazon **Deep Archive Access tier**.

Intelligent-Tiering does not charge a retrieval fee but if objects are archived, retrieval can take some time, depending on the retrieval option chosen. The following table illustrates the retrieval options available:

Storage class or tier	Expedited	Standard	Bulk
S3 Intelligent-Tiering Frequent and Infrequent Access tier	1–5 minutes	3–5 hours	5–12 hours
Intelligent-Tiering Archive and Deep Archive Access tier	Not available	Within 12 hours	Within 48 hours

Table 5.1 – Retrieval times for S3 Glacier, Deep Archive, and S3 Intelligent-Tiering archive classes

As you can see, you have several retrieval options, and the times will vary depending on which archive storage option you select.

S3 on Outposts

Amazon Outposts is a fully managed on-premises service that comes with a 42U rack that can host the same AWS infrastructure and services at your data center. The U refers to rack units or "U-spaces" and is equal to 1.75 inches in height. A standard height is 48U (a 7-foot rack). The service allows you to create a pool of compute, storage, networking, and database services locally on-premises and is ideal if you have workloads running that are very sensitive to low-latency access. Amazon Outposts can also be used as a precursor to migrate your entire data center to the cloud at a later stage.

With Amazon Outposts already widely available, AWS offers yet another storage class called **S3 Outpost**. The service offers durability and redundancy by storing data across multiple devices and servers hosted on your outposts and is ideal for low-latency access, while also enabling you to meet strict data-residency requirements. Amazon S3 on Outpost allows you to host 48 TB or 96 TB as part of the S3 storage capacity and provides the option to create a maximum of 100 S3 buckets on each outpost. Have a look at the Amazon S3 performance chart shown in the following screenshot:

	S3 Standard	S3 Intelligent-Tiering*	S3 Standard-IA	S3 One Zone-IA	S3 Glacier	S3 Glacier Deep Archive
Designed for durability	99.999999999% (11 9's)	99.999999999% (11 9's)	99.999999999% (11 9's)	99.999999999% (11 9's)	99.999999999% (11 9's)	99.999999999% (11 9's)
Designed for availability	99.99%	99.9%	99.9%	99.5%	99.99%	99.99%
Availability SLA	99.9%	99%	99%	99%	99.9%	99.9%

	S3 Standard	S3 Intelligent-Tiering	S3 Standard-IA	S3 One Zone-IA	S3 Glacier	S3 Glacier Deep Archive
Availability Zones	≥3	≥3	≥3	1	≥3	≥3
Minimum capacity charge per object	N/A	N/A	128KB	128KB	40KB	40KB
Minimum storage duration charge	N/A	30 days	30 days	30 days	90 days	180 days
Retrieval fee	N/A	N/A	per GB retrieved	per GB retrieved	per GB retrieved	per GB retrieved
First byte latency	milliseconds	Milliseconds	milliseconds	milliseconds	select minutes or hours	select hours

Figure 5.4 – S3 storage class performance and key attributes

As shown in the preceding screenshot, you can choose which storage class to store your objects in depending on your use case. When making this decision, you need to consider durability and availability, as well as the minimum size of your objects, and ascertain whether you require instant access to those objects.

Versioning

To protect against accidental deletions or overwriting, AWS also offers a feature called S3 Versioning. By default, when you create a bucket versioning is disabled, which means that if you were to upload an object with the same name (which, as mentioned earlier, is called a **key** on AWS) as an existing object in an S3 bucket, then the original object in the bucket will get overwritten. Sometimes this may be exactly what you want, but in most cases, you might want to preserve the original version. Often, objects are overwritten simply because the name of the object was not changed before performing the upload, and this results in an accidental overwrite.

Amazon S3 offers a feature where you can enable versioning on a bucket. The setting is applied to the entire bucket and will therefore affect all objects. Once versioning is enabled, any object that is uploaded with the same name as an existing object will be tagged with a new version ID. Accessing the object will yield the latest/current version, but a toggle switch in the console will allow you to also see all previous versions in case you need to access an earlier version of the object.

If you try to delete an object (without specifying the version ID) in a bucket that has had versioning enabled, then the object is not deleted. AWS adds a delete marker to the object and hides it from the S3 management console view. Subsequently if you need to *restore* the object again to make it visible in the S3 management console, you will simply have to delete the delete marker itself.

You should note that buckets can be in one of three states, outlined as follows:

- Unversioned (default)
- Versioning-enabled
- Versioning-suspended

Once you enable versioning on a bucket, you can never return to the Unversioned state, but you can suspend versioning if you do not want new versions of objects being created.

Cross-Region and same-Region replication

As previously discussed, to help customers comply with compliance and data-residency laws, AWS will never replicate your objects outside of the Region in which you create them. However, there is nothing to stop you from replicating your data outside of the Region in which you uploaded it if there are no regulatory requirements that prevent you from doing so.

Amazon Cross-Region Replication (CRR) is used to asynchronously copy objects across AWS buckets in different AWS Regions. You can use CRR to do the following:

- **Reduce latency**—By copying objects closer to where end users are based, you can minimize latency in accessing those objects.

- **Increase operational efficiency**—If you run applications across multiple Regions that need access to the same set of data, maintaining multiple copies in those Regions increases efficiency.

- **Meet regulatory and compliance requirements**—Your organization compliance and regulatory requirements may require you to store copies of data thousands of kilometers away for DR purposes.

In addition to CRR, Amazon S3 also enables you to configure replication services between buckets in the same Region. This is known as **Same-Region Replication** (**SRR**), which can help you achieve the following:

- **Log Data Aggregation**—You may be collecting log data from several sources and applications. You can collate these datasets in a single log management bucket via replication.

- **Replicating between development and production accounts**—If you use separate development and production accounts and need to use datasets in each, you can use replication to move objects from your development accounts to production accounts.

- **Compliance requirements**—You may be required to maintain multiple copies of your data to adhere to data-residency laws. SRR can help you copy data between multiple buckets to ensure you have more than one copy for compliance purposes.

It is important to note that both buckets must be configured with versioning enabled in order to set up CRR or SRR. You can also replicate objects across Regions or within the same Region, in either the same AWS account or across multiple AWS accounts. Finally, you can replicate objects into different storage classes from its original storage class, which means that your replicated objects can reside in a cheaper storage class, if perhaps they are being used simply as a backup copy—for example, your original objects can reside in the Standard storage class, but the replicated objects can be placed into the Standard-IA storage class. This will reduce your overall costs of storage.

Another feature of the replication service is support for multiple destination buckets. You can configure S3 Replication (multi-destination), which enables you to replicate data from one source bucket to multiple destination buckets, either within the same Region, across Regions, or a combination of both.

Amazon S3 Lifecycle Management Amazon S3 offers unlimited amount of storage. This means that it is very easy to upload any amount of data you create and simply forget about it. At the same time, let's not forget that Amazon S3 charges you for the total amount of storage consumed, and the cost also depends on the storage class in which you place your objects.

Often, the bulk of your data is going to be infrequently accessed, especially after the initial period in which the data was created. This makes it essential to have some mechanism for moving objects you no longer need frequent access to into a cheaper storage class, to manage your storage costs effectively. In addition, you may also host a lot of archive data that you no longer require after a period, even for compliance and auditing purposes. For example, some regulations state that certain types of data need only be kept for a maximum of 7 years.

Manually trying to manage vast quantities of data can be a tiresome affair, often involving the creation of scripts and tools to review the data stored in the cloud. Instead, you can use a reliable solution by Amazon S3, known as Amazon S3 Lifecycle Management.

Amazon S3 Lifecycle Management actions can be applied to your Amazon S3 buckets. These can be applied to the entire bucket or a subset of data by defining a prefix. These actions fall into two main categories, outlined as follows:

- **Transition actions**—These allow you to move objects from one storage class to another after a certain period of time has passed. For example, if you know that you are going to be infrequently accessing a particular dataset after 60 days, you can set a rule to move that data from the Standard storage class to the Standard-IA storage class 60 days after creation.

- **Expiration actions**—These allow you to delete objects from the S3 storage system after a set number of days. For example, if you do not require old log files after 365 days, you can set a rule to automatically expire those objects after 365 days, which will purge them from the storage platform.

You can use a combination of transition actions and expiration actions as well. For example, you may have log data that you frequently access for the first 30 days. After that, you may still need to revisit that data for a period of 180 days, post which you no longer require it. You can set a combination rule to transition the log files after 30 days of creation from the Standard class to the Standard-IA class, then create another expiration action to purge the data after 180 days.

You can also apply different lifecycle actions to versioned data—for example, you can have one set of rules and actions against the current version of your objects and another set for previous versions. This further allows you to manage your objects more effectively.

S3 encryption

All data uploaded to Amazon S3 is encrypted in transit using the HTTPS protocol. However, data stored on S3 is not modified in any way, which means that if you are uploading sensitive data in plaintext, the data is stored unencrypted by default.

To add an additional layer of security, you can encrypt the data before storing it in S3. This is known as encryption at rest. AWS offers two options for encrypting your data at rest, outlined as follows:

- **Server-side encryption**—When you upload (create) an object, Amazon S3 encrypts it before saving it to disk, and when you download/request an object, it is automatically decrypted by the S3 service. You have three mutually exclusive options when deciding to encrypt your objects using server-side encryption, outlined as follows:

    a) **Server-side encryption** with **Amazon S3-managed keys (SSE-S3)**— Amazon encrypts your data with a 256-bit **Advanced Encryption Standard** (**AES-256**). Each object is encrypted with a unique key, and the key itself is further encrypted with a master key that AWS rotates and manages for you.

    b) **Server-side encryption** with **customer master keys (CMKs)** stored in AWS **Key Management Service (SSE-KMS)**—This is similar to SSE-S3 but with added features, including the ability to create and manage your own CMKs, as well as an auditing feature that shows when your CMK was used and by whom.

    c) **Server-side encryption** with **customer provided keys (CPKs)** (**SSE-C**)— Encryption is performed by Amazon S3, but with CPKs. This is ideal if you need to follow regulatory requirements of creating and managing your own keys.

- **Client-side encryption**—This is where data is encrypted on the client side and the encrypted data is then uploaded to Amazon S3. The full encryption process is therefore managed by the customer.

Static website hosting

In addition to storing data, Amazon S3 also offers a service for hosting complete websites for your company. The only limitation is that the web hosting service is designed to host static websites only, as opposed to dynamic websites.

The primary difference is that while the content stored on Amazon S3 to deliver the complete website can be changed and updated, it remains constant and *static*, and all users accessing the site will see the same content.

Dynamic websites use server-side scripting to deliver dynamic content that changes in-flight depending on various parameters, and generally connect to a backend database to fetch content.

Nevertheless, static websites can also provide complete end-to-end solutions and serve several use cases at a fraction of the cost. In addition to hosting and delivering digital assets such as **HyperText Transfer Markup (HTML)** files, **Cascading Style Sheets (CSS)**, **Portable Document Format (PDF)** documents, images, and videos, you can also host client-side scripts that run in the client browser to offer additional features that can include interactive elements—for example, you can run a client-side script on a S3 static website to collect email addresses for potential customers and store them with a third-party email-campaign service provider. You can also host scripts that access additional AWS services. Having lambda functions and, potentially, **Elastic Container Service (ECS)** or **Elastic Kubernetes Service (EKS)** as backends allows you to run anything starting from a static website. Ultimately, you have a wide range of use cases for hosting static websites on Amazon S3. Owing to its highly available and scalable nature to handle large volumes of traffic, Amazon S3 can prove to be a better solution than hosting static sites across a fleet of EC2 instances, depending on your use case.

Some examples of considering the Amazon S3 static website hosting service include the following:

- **Developing a product-prelaunch website**—Often, when you need to advertise a pre-launch campaign of a new product range, you may not be able to gauge the amount of traffic you might generate. Hosting the same solution on a fleet of EC2 instances may be more costly since you may need to scale out fast and utilize a large server farm if your marketing campaigns have been particularly successful. By way of contrast, the scalable nature of S3 will ensure that demand is met automatically with the inflow of large amounts of traffic.

- **Offloading**—With Amazon S3, you get a highly scalable, reliable, and low-latency data storage solution. Even if you host dynamic websites that run on expensive EC2 instances and EBS volumes, you are likely to have a large volume of static content (such as documents, images, and videos, for example). You could offload such static content to an S3 bucket and reference it via API calls from all sites hosted on EC2 instances. The benefits include low storage costs for your digital assets and, because your servers are kept lean, you also get better performance.

- **"Lite" version of website or "Under Maintenance" banners**—Sometimes, you need to host an alternative version of your site when you are performing upgrades or rolling out major updates. By hosting a *lite* static version of your site on an S3 bucket, you can easily redirect request to the S3 buckets during periods of maintenance or major updates.

In this section, we looked at the fundamentals of Amazon S3 on AWS and learned about its various features. In the next section, we look at some additional services, covering transfer of data and archiving.

Amazon S3TA

If you host an S3 bucket in a specific Region but require users across the globe to upload objects to it, your users may experience longer and unexpected variable speeds for uploads and downloads over the internet, depending on where they are based. S3TA reduces this speed variability that is often experienced due to the architecture of the public internet. S3TA routes your uploads via Amazon CloudFront's globally distributed edge locations and AWS backbone networks. This, in turn, gives faster speeds and consistently low latency for your data transfers.

Here is a screenshot of a S3TA speed checker that uploads a sample file from your browser to various S3 Regions and compares the speed results between standard internet upload versus uploads via S3TA. You can try out the speed test at `https://s3-accelerate-speedtest.s3-accelerate.amazonaws.com/en/accelerate-speed-comparsion.html`:

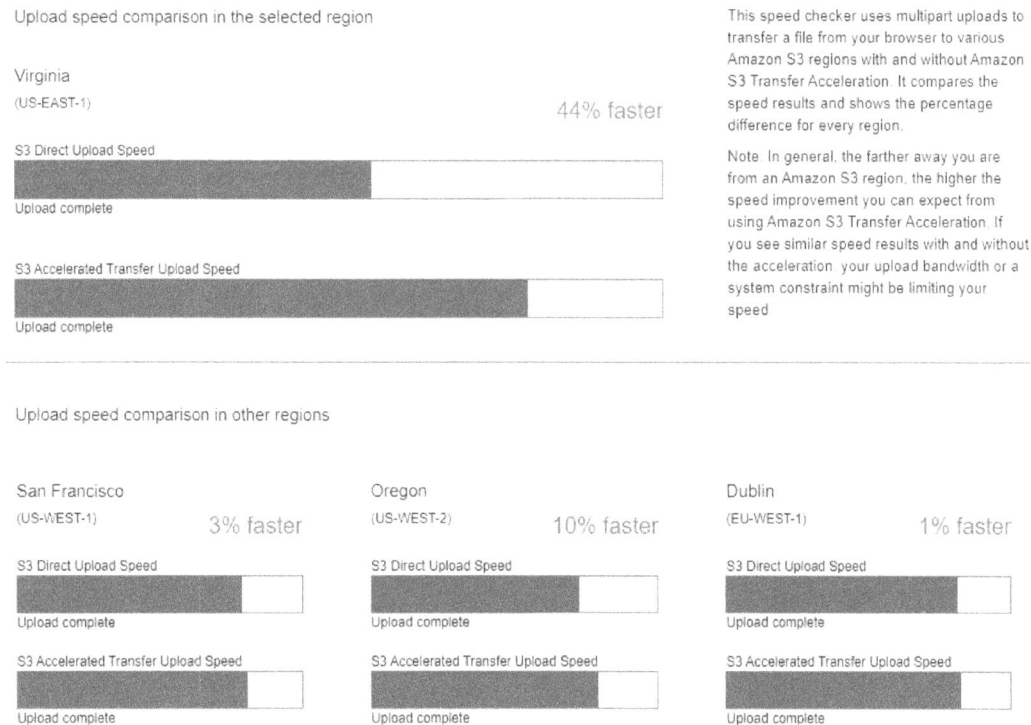

Figure 5.5 – Amazon S3TA speed test

In this section, we introduced you to the Amazon S3 service and discussed its various feature sets. Amazon S3 is an object storage solution designed to help customers store any amount of data in the cloud. With features such as versioning, CRR/SRR, encryption, and static website hosting, you can use Amazon S3 for a wide range of use cases at affordable storage costs.

In the next section, we look at some additional features of the Amazon Glacier services, which offer an archival storage solution.

Learning about archiving solutions with Amazon S3 Glacier

Earlier in this chapter, we introduced you to the Amazon S3 Glacier and Glacier Deep Archive storage classes. Amazon Glacier offers long-term storage at very a low cost and is intended to be used for archival storage. The architecture offers the same 99.999999999% (eleven 9s) of durability so that in the event of a major disaster, you can rest assured that your old archives will be available to recover if the need arises. The technology works differently from standard S3 storage. The archives need to be requested before you can access/download them, which involves a two-step process of first creating a retrieval job and then downloading your data once the job is complete.

This retrieval process can take some time and depends on your chosen retrieval options, as previously discussed. However, the upside to this delay in being able to access your data is that you get some of the cheapest storage options on the Amazon platform.

Archives and vaults

As with other Amazon S3 storage classes, you can store any amount of data in the Glacier class. However, your objects are stored as archives, and an archive can contain either a single file or multiple files clubbed together in a `.zip` or `.tar` format. The size of your archive can between 1 byte and 40 terabytes. On Amazon S3, a single object can only be a maximum of 5 TB.

Furthermore, archives can be grouped and stored in vaults. When you create a vault, you need to specify the Region in which it will be created. Vaults also let you define access and notification policies, and you can have up to 1,000 vaults per Region. A vault policy enables you to define who can access it and which actions can be performed on it. You can also define vault lock policies, such as **Write Once Read Many** (**WORM**), or time-based record-retention policies for regulatory archives.

Retrieval options

As previously discussed, access to your archives in Amazon Glacier is not instantaneous. Depending on the Glacier storage class you choose (Glacier versus Deep Archive), you have the following different retrieval options available:

a) **Amazon S3 Glacier Retrieval Options**

- **Standard**—This is the default retrieval option and typically takes between 3 and 5 hours to complete, before your data is made available to download.

- **Expedited**—If you need urgent access to just a subset of your archives, you can opt for the Expedited retrieval option. Naturally, the cost of retrieval is higher than the other options. Furthermore, Expedited retrievals are made available within 1 to 5 minutes for archives of up to 250 **megabytes** (**MB**). In addition, two types of expedited retrieval options are available: On-Demand and Provisioned. With On-Demand, your retrieval requests are generally fulfilled within a 5-minute period, although during periods of high demand, this may take longer. Optionally, you can purchase Provisioned capacity, which ensures available retrieval capacity when you need it the most.

- **Bulk** —This is designed to help you retrieve large amounts of data at the lowest-cost retrieval option and typically takes between 5 to 12 hours to complete.

b) **Amazon S3 Glacier Deep Archive Retrieval Options**

- **Standard**—Retrieval of your deep archives can be achieved within 12 hours.

- **Bulk**—Retrieval of **petabytes** (**PB**) of data within 48 hours can be achieved, and it is also the lowest-cost option available.

In this section, we looked at archiving solutions with the Amazon S3 Glacier service and how you can store data for many years to fulfill compliance and regulatory requirements. In the next section, we look at how you can connect your on-premises storage services to Amazon S3.

Connecting your on-premises storage to AWS with Amazon Storage Gateway

Amazon Storage Gateway is an on-premises solution that enables you to connect your on-premises servers and storage systems to the Amazon S3 cloud environment. The service involves installing the Storage Gateway **virtual machines** (**VMs**) on-premises and connecting your servers to them. The gateway uses industry-standard protocols to then transfer data between your servers and the Amazon S3 platform. The VM can be deployed on either VMware ESXi or a Microsoft Hyper-V hypervisor. Optionally, you can also order a hardware appliance, which is a physical server that comes pre-installed and configured with the Storage Gateway software. This reduces the administration time involved in setting up your own VMs and integrates with your existing storage systems over protocols such as **Network File System** (**NFS**), SMB, and **Internet Small Computer Systems Interface** (**iSCSI**).

Amazon Storage Gateway enables your on-premises application to connect to the AWS storage systems *transparently* over standard protocols such as NFS/SMB, **Virtual Transport Layer** (**VTL**), and iSCSI. Connectivity between the Storage Gateway VMs or hardware appliance to the AWS platform can be established over the internet, through secure IPsec **virtual private network** (**VPN**) tunnels or via AWS Direct Connect.

Amazon S3 Storage Gateway supports different use cases with the following deployment options:

- **File Gateway**—Enables you to use standard NFS SMB protocols to store data in Amazon S3. Data is also cached locally, enabling low-latency access. Here, there are two options available, as follows:

  a) **Amazon S3 File Gateway**—This service enables you to present a file-server solution to your on-premises servers and access Amazon S3, where you can store and retrieve objects in Amazon S3 using industry-standard file protocols such as NFS and SMB. Furthermore, because the data is ultimately stored in S3, you benefit from all its features such as versioning, bucket policies, CRR, and so on.

  b) **Amazon FSx File Gateway**—This service allows you to connect your on-premises Windows servers or Windows-based applications (as well as Linux and macOS systems) to the cloud-hosted Amazon FSx for Windows File Server with low latency, and the ability to set up and access a virtually unlimited number of Windows file shares in the cloud. Amazon FSx File Gateway offers full support for SMB protocol support, as well as integration with **Active Directory** (**AD**) and the ability to configure access controls using ACLs.

- **Volume Gateway**—Enables you to present block storage volumes to your on-premises servers over the iSCSI protocol. Volume Gateway can be used to asynchronously back up your data to Amazon S3 and comes in two different modes, outlined as follows:

- **Cache mode**—The bulk of your data is stored in Amazon S3, with only the most frequently accessed data stored locally in the cachefor low-latency connectivity. This means that you do not need very large amounts of local storage, which helps reduce your capital expenditure.

- **Stored mode**—Your data is stored locally and available for low-latency access on premises. This is particularly useful if your application is sensitive to latency for data access. The data is then asynchronously backed up to Amazon S3 and can be used for DR purposes.

- **Tape Gateway**—Many organizations use backup software solutions for their on-premises backup needs (for example, Veritas Backup Exec and NetBackup). Often, these applications back up data to physical tapes, which most companies store off-site. However, the tape drives, tapes, and off-site storage facilities can become very costly. The Tape Gateway solution comes to the rescue by enabling you to present virtual tapes to your backup software applications over iSCSI. Tape Gateway stores these virtual tapes in a **virtual tape library** (**VTL**), which is backed up by Amazon S3. Data is written to these virtual tapes, which results in the Tape Gateway solution asynchronously uploading the data to Amazon S3. When you need to restore the data, it is downloaded to the local cache, and the backup application can restore it to a location you specify on premises.

To manage long-term storage of old data, you can transition virtual tapes between Amazon S3 and Amazon S3 Glacier or Amazon S3 Glacier Deep Archive. If you later need to access the data on an **archived virtual tape**, you need to retrieve the tape and present it to your Tape Gateway solution.

Note that retrieval of an archived tape from Glacier will take between 3 and 5 hours and from Deep Archive can take up to 12 hours.

In this section, we looked at how you connect your on-premises applications to the Amazon S3 storage service using the Storage Gateway solution. In the next section, we look at how you can migrate large volumes of data to the cloud using alternative offline methods, which is particularly useful when you have limited internet bandwidth.

Migrating large datasets to AWS with the AWS Snow Family

Many companies looking to move to the cloud generally host vast amounts of data on premises. While it is possible to transfer data over the public internet into Amazon S3, customers with vast amounts of data may need to consider offline methods of transfer due to bandwidth limitations.

AWS offers rugged devices that can be delivered to your on-premises location. These include the **Snowcone**, **Snowball**, and **Snowmobile** devices.

AWS Snowball

At its very basic offering, you simply copy large amounts of data to the device and ship it back to AWS to have the data imported into Amazon S3. These devices are known as AWS Snowball devices and are part of the Snow Family of devices.

These edge devices come with compute and storage capabilities contained in highly rugged, tamper-proof devices. The devices feature a **Trusted Platform Module** (**TPM**) chip that detects unauthorized modifications to hardware, software, or firmware. These devices can be used for storage and data processing at your on-premises locations. Often, customers will use Snowball edge devices for migrations, data collections, and processing with or without internet connectivity.

Amazon Snowball comes in *two flavors*, as follows:

- **Snowball Edge Compute Optimized**—These devices offer both storage and computing resources and can be used for **machine learning** (**ML**), analytics, and any local computing tasks. The devices come with 52 **virtual central processing units** (vCPUs), 208 GB of memory, and an optional NVIDIA Tesla V100 **graphics processing unit** (**GPU**). In terms of storage, the device offers 42 TB of **hard-disk drive** (**HDD**) capacity and 7.68 TB of **solid-state drive** (**SSD**) capacity.

- **Snowball Edge Storage Optimized**—These devices offer larger storage capacity and are ideal for data migration tasks. With 80 TB of HDD and 1 TB of **serial advanced technology attachment** (**SATA**) SDD volumes, you can start moving large volumes of data to the cloud. The device also comes with 40 vCPUs and 80 GB of memory.

Amazon Snowcone

The smallest member of the AWS Snow Family, these devices are the smallest ever and weigh just 4.5 **pounds** (**lb**) (2.1 **kilograms** (**kg**)). Snowcone devices come with 8 TB of usable storage and are designed for outside use in areas of low network connectivity. Examples include IoT, vehicular, and drone use cases.

The device also offers compute capabilities with two vCPUs and 4 GB of memory, as well as USB-C power using a cord and optional battery.

Amazon Snowmobile

If you need to transfer exabyte-scale data to the cloud, then you are going need an extremely large 45-foot-long *rugged* shipping container. Amazon Snowmobile can transfer up to 100 PB of data to the cloud and assist in all your data center migration efforts.

The shipping container is delivered on-site and an AWS team member will work with your team to connect a high-speed switch from Snowmobile to your local network.

With 24/7 video surveillance, optional escort security, and 256-bit encryption, you have access to the most secure way of transferring sensitive data to Amazon S3.

In this section, we reviewed offline methods to transfer large amounts of data to the cloud and assist in your data migration efforts. The Amazon Snow Family offers more than just storage containers—these devices come with high levels of compute capabilities to perform data processing, analytics, and ML tasks as you copy data to them as well.

In the next section, we review some of the key points highlighted in this chapter.

Exercise 5.1 – Setting up an Amazon S3 bucket

In this exercise, we will create an Amazon S3 bucket and upload an object to it. More specifically, we will upload a single web page document and test access to it after the upload. Since we plan to later *use* this bucket to host a static website and make content accessible to anyone on the internet, you will need to disable the **Block Public Access** setting, as discussed in the access permissions settings earlier in this chapter. Proceed as follows:

1. On your computer, create a new file using a standard text editor of your choice (Notepad on Windows or TextEdit on Mac).

2. Add the following lines of code to the document:

```
<html>
<title>Blueberry Muffin Recipe</title>
```

```
<Body>
<h1>Blueberry Muffin Recipe</h1><p>
Bake the ultimate blueberry muffins for your guests and
loved ones. This recipe shows you how to create cafe-
style blueberry muffins in your own kitchen.</p>
<p><strong>Ingredients:</strong></p>
<ul><li>100g fresh <span style="color: rgb(85, 57,
130);"><strong>blueberries</strong></span></li>
<li>300g flour</li>
<li>150g granulated sugar</li>
<li>1 tsp. vanilla</li>
<li>60 ml vegetable oil</li>
<li>50g of butter</li>
</ul><p><br></p>
</body>
</html>
```

The preceding code is also available in our GitHub repository for this book
`https://github.com/PacktPublishing/AWS-Certified-Cloud-`
`Practitioner-Exam-Guide`, and you can simply download the `index.html`
file to your desktop as well.

3. Next, save it with a filename of `index` with a `.html` extension—so, the filename
    with the extension should be `index.html`. This will create a simple web page
    object for you. You may need to set the **Save as type** option to **All Files**, as
    illustrated in the following screenshot:

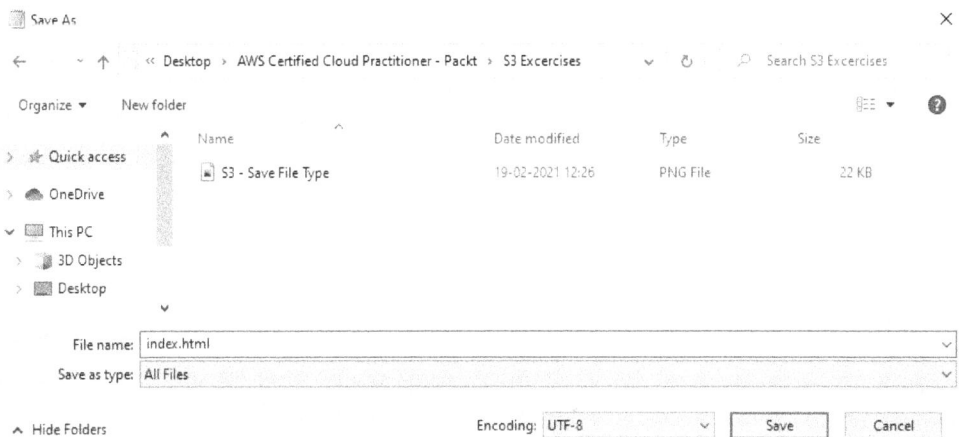

Figure 5.6 – Saving a file with a .html extension to create a web page

4. Next, log in to your AWS account as the IAM user, **Alice**.

5. Navigate to the Amazon S3 console.

6. Click **Create bucket**, as illustrated in the following screenshot:

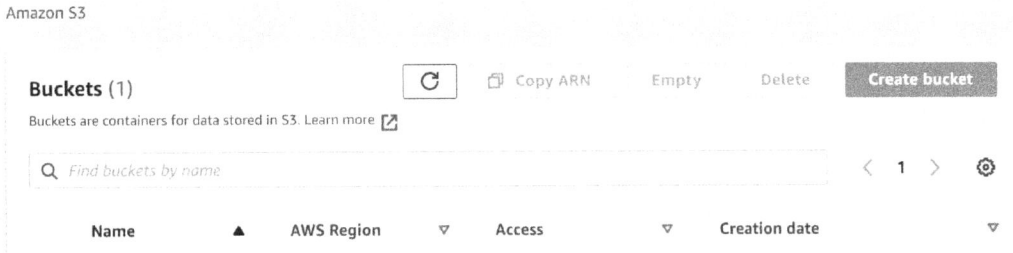

Figure 5.7 – List of buckets

7. For the name of the bucket, type in your name followed by a hyphen (-) and the word webpage. Make sure there are no spaces in the name and that all lowercase letters are used. Assuming that the name you have chosen has not already been taken by another customer of AWS, you should be able to use this bucket name. If you get an error when you create the bucket, stating that the name is not available, you will simply need to choose a different name.

8. For **AWS Region**, select us-east-1.

9. Next, under the **Block Public Access settings for bucket** sub-heading, uncheck he box for **Block all public access**. Note that for general use cases, you do not want to unblock public access unless your use case demands it, such as when trying to configure static website hosting, which we will look at later in *Exercise 5.4*. If you do not need anonymous access such as that from end users on the public internet, you must always correctly configure your permissions using bucket policies, ACLs, or access point policies, to ensure you leverage the **principal of least privilege (PoLP)**.

10. Next, check the box to state that you acknowledge that the preceding settings could make the bucket and its objects publicly accessible, as illustrated in the following screenshot:

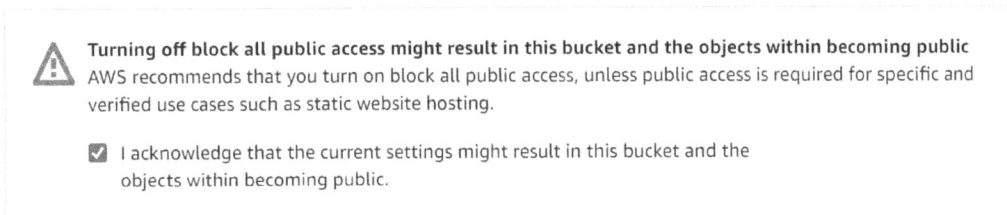

Figure 5.8 – Turning off block all public access on your bucket

11. Leave all other settings as default and click on the **Create bucket** button at the bottom of the screen. Your Amazon S3 bucket has now been created.

12. Next, in your list of buckets in the main S3 console, select the bucket you just created. This will take you to the current list of objects in the bucket. You will note that there will be none at present.

13. You will notice an **Upload** button. Click on this button and you will have the option to add files and folders, as per the following screenshot:

Upload

Add the files and folders you want to upload to S3. To upload a file larger than 160GB, use the AWS CLI, AWS SDK or Amazon S3 REST API. Learn more [↗]

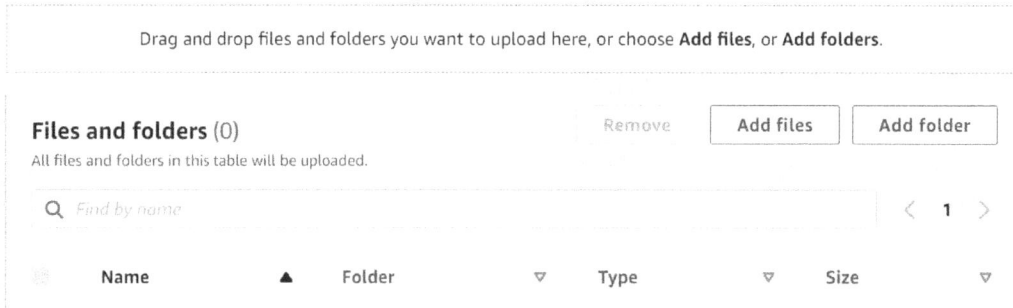

Drag and drop files and folders you want to upload here, or choose **Add files**, or **Add folders**.

Files and folders (0)   Remove   Add files   Add folder
All files and folders in this table will be uploaded.

Q Find by name   < 1 >

Name   ▲   Folder   ▽   Type   ▽   Size   ▽

Figure 5.9 – Uploading object to your bucket

14. Add the `index.html` file you created/downloaded earlier.

15. Scroll toward the bottom of the screen and click the **Upload** button.

16. Your file will be uploaded.

17. You will note a green banner at the top of the screen to say that the upload has been successful. Now, go ahead and click **Exit**.

18. You are then presented with the contents of the bucket you just created.

19. You can now click on the `index.html` file in the **Objects** list, which will take you to the Amazon S3 properties page of the file itself.

20. Note that each object has its own URL accessible from the internet (as long as the permissions are correctly set), as we can see in the following screenshot:

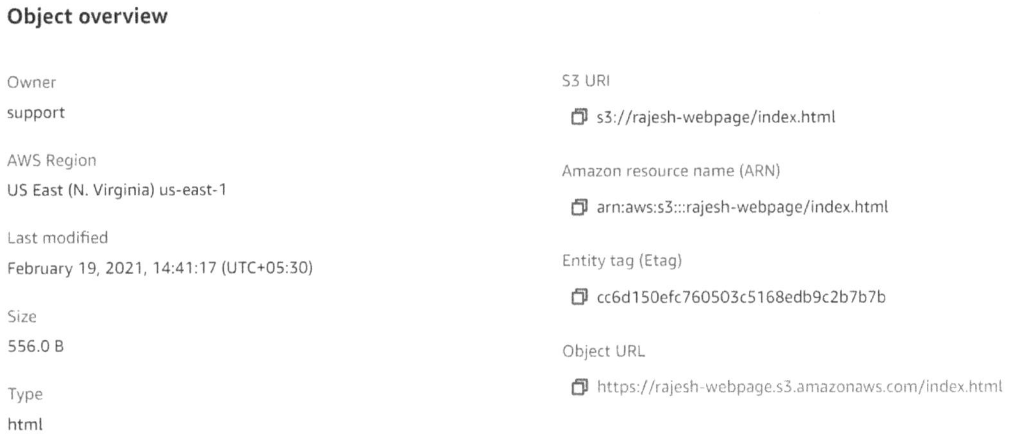

Object overview

Owner
support

AWS Region
US East (N. Virginia) us-east-1

Last modified
February 19, 2021, 14:41:17 (UTC+05:30)

Size
556.0 B

Type
html

S3 URI
s3://rajesh-webpage/index.html

Amazon resource name (ARN)
arn:aws:s3:::rajesh-webpage/index.html

Entity tag (Etag)
cc6d150efc760503c5168edb9c2b7b7b

Object URL
https://rajesh-webpage.s3.amazonaws.com/index.html

Figure 5.10 – Uploading the index.html object to your bucket

21. If you try to click on this object URL to open it up in another browser window, you will find that you cannot access it. Instead, you get an **Access Denied** error message. This is because access to an object via its URL has the same effect as trying to anonymously read the object over the public internet.

Although we disabled the **Block all public access** setting earlier, your buckets and objects still need an explicit `Allow` rule to grant access to them. You could click on the **Permissions** tab of the object itself and set up an ACL to enable public access for this object. However, as discussed previously, using bucket policies is a better option as these offer more features and granular control.

In the next exercise, we will set up a bucket policy to see how we can allow public access to this file.

Exercise 5.2 – Configuring public access to S3 buckets

In this exercise, we will configure the Amazon S3 bucket with a **bucket policy** (resource policy) that will allow users on the public internet to be able to access and read the `index.html` web page you created earlier.

Remember that you could choose to restrict access to only a set of known users—for example, if you wanted only IAM users in your AWS account to have access to the objects. You can also configure cross-account access, in which you define principals that belong to another AWS account and grant them specific levels of access.

In this exercise, we want to grant anonymous access to the `index.html` page because ultimately, we will be building out a static website hosting service using this bucket in later exercises. Proceed as follows:

1. Navigate back to the S3 console and click on the bucket you just created, as illustrated in the following screenshot:

rajesh-webpage

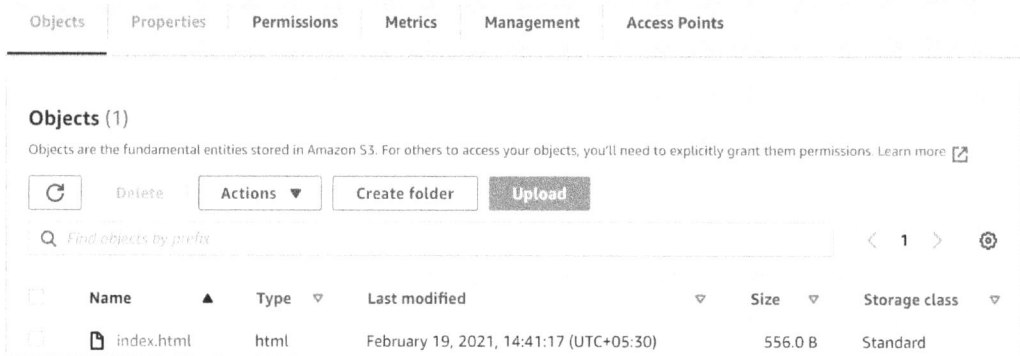

Objects	Properties	Permissions	Metrics	Management	Access Points

Objects (1)

Objects are the fundamental entities stored in Amazon S3. For others to access your objects, you'll need to explicitly grant them permissions. Learn more [↗]

C	Delete	Actions ▼	Create folder	Upload

Q Find objects by prefix   < 1 >   ⚙

Name ▲	Type ▽	Last modified	▽	Size ▽	Storage class ▽
🗋 index.html	html	February 19, 2021, 14:41:17 (UTC+05:30)		556.0 B	Standard

Figure 5.11 – Successful upload

2. Click on the **Permissions** tab.

3. You will note that the **Block public access** has been disabled and is in an **Off** state.

4. Scroll further down until you get to **Bucket Policies**, and then click **Edit**.

5. Add the following policy, replacing the values in the placeholder `Your-Bucket-Name` with the name of your S3 bucket:

```
{
    "Id": "Policy1613735718314",
    "Version": "2012-10-17",
    "Statement": [
        {
            "Sid": "Stmt1613735715412",
            "Action": [
                "s3:GetObject"
            ],
            "Effect": "Allow",
            "Resource": "arn:aws:s3:::Your-Bucket-Name/*",
            "Principal": "*"
```

```
        }
    ]
}
```

6. Click **Save Changes**. If you copied the policy correctly, the policy validator will not throw up any errors.

7. You should then get a confirmation that the policy has been saved and, more importantly, you will note that the bucket's contents are now publicly accessible, as illustrated in the following screenshot:

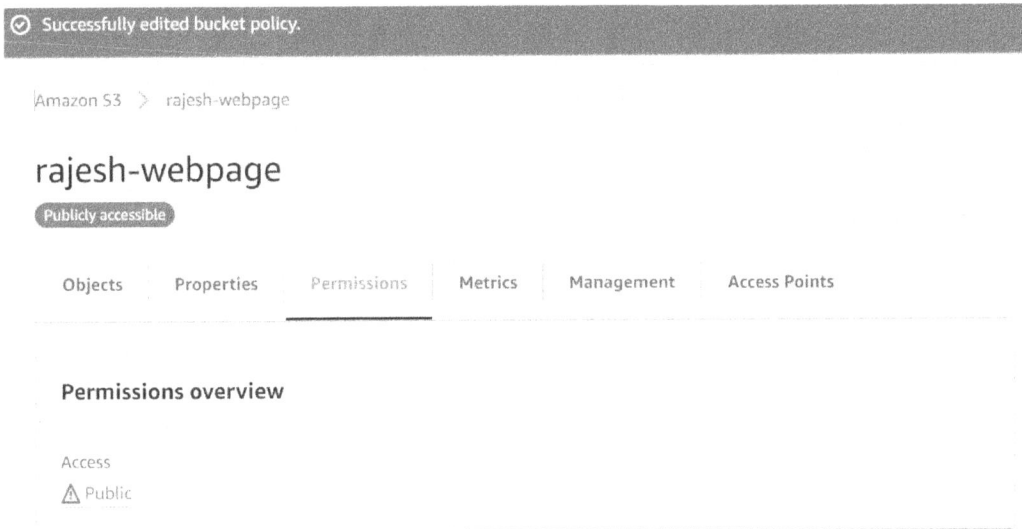

⊘ Successfully edited bucket policy.

Amazon S3 > rajesh-webpage

rajesh-webpage

Publicly accessible

| Objects | Properties | Permissions | Metrics | Management | Access Points |

Permissions overview

Access

⚠ Public

Figure 5.12 – S3 Bucket permissions tab

8. Next, click on the **Objects** tab again.

9. Click on the index.html file to open its **S3 Properties** pane.

10. Right-mouse click on the object URL and open it in a new browser tab.

11. You should find that the web page is now accessible from your browser, as illustrated in the following screenshot:

rajesh-webpage.s3.amazonaws.com/index.html

Blueberry Muffin Recipe

Bake the ultimate blueberry muffins for your guests and loved ones. This recipe shows you how

Ingredients:

- 100g fresh **blueberries**
- 300g flour
- 150g granulated sugar
- 1 tsp. vanilla
- 60 ml vegetable oil
- 50g of butter

Figure 5.13 – Your index.html page in a browser window

So far, we have disabled block public access on this bucket as we will eventually be configuring it for static website hosting. In this exercise, we also uploaded the object, index.html, which is a recipe document written in HTML code.

In the next exercise, you will learn how to configure versioning on your bucket. Versioning will help you create previous copies of an object so that new uploads of updated content for the same object are stored as new versions of the object. This will enable you to prevent against accidental changes to your objects by being able to restore a previous version, as we will see in the upcoming exercises.

Exercise 5.3 – Enabling versioning on your bucket

In this exercise, we will enable versioning on the Amazon S3 bucket. As you update existing objects with newer versions, you can rest assured that if you need to revert to an older version, those versions will still exist in your bucket. Obviously, if you try to delete a specific version of the object itself, then it will be purged from the S3 platform. However, enabling versioning can help prevent against accidental deletions and overwrites. Proceed as follows:

1. Navigate back to the S3 console.

2. Click on the bucket you created earlier in *Exercise 5.1*.

3. Click on the **Properties** tab.

4. You will see an **Edit the Bucket Versioning** option to edit the state. At present, the versioning will be set to **disabled**. Note that once again you can suspend versioning actions, but you will not be able to disable them.

5. Click **Edit** in the **Bucket Versioning** section.

6. Select **Enable**.

7. Click **Save Changes**.

Let's try to test the versioning feature next, as follows:

1. Navigate to the location where you saved the `index.html` web page on your computer. Open the file with your text editor using Notepad or TextEdit (for Mac).

2. Replace the word `Blueberry` in the existing `<H1>` tag within the document to `Chocolate`.

3. Save the file without changing the format or extension.

4. Navigate back to your Amazon S3 console in your AWS account and click on the bucket you created earlier.

5. Click on **Objects**, as illustrated in the following screenshot:

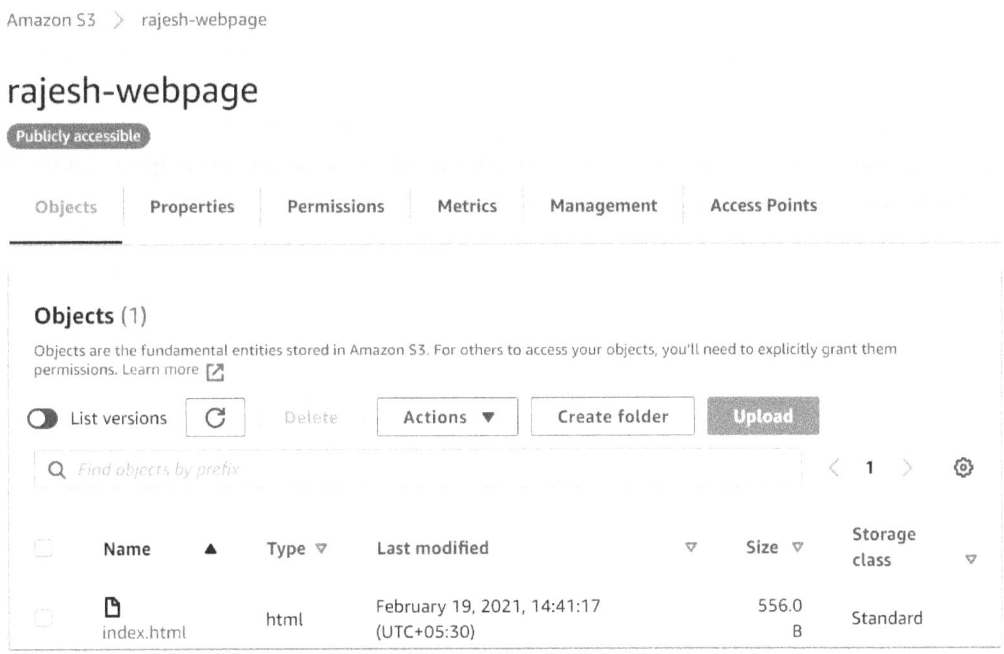

Amazon S3 > rajesh-webpage

rajesh-webpage

`Publicly accessible`

| Objects | Properties | Permissions | Metrics | Management | Access Points |

Objects (1)

Objects are the fundamental entities stored in Amazon S3. For others to access your objects, you'll need to explicitly grant them permissions. Learn more

List versions | Delete | Actions ▼ | Create folder | Upload

Q Find objects by prefix   < 1 >

	Name ▲	Type ▽	Last modified	▽	Size ▽	Storage class ▽
	index.html	html	February 19, 2021, 14:41:17 (UTC+05:30)		556.0 B	Standard

Figure 5.14 – List of objects in your Amazon S3 bucket

6. Click **Upload**.

7. Click **Add Files**.

8. Select the same `index.html` file you updated moments ago and click **Upload**.

9. Click **Exit**.

10. Click on the `index.html` fileagain to open up its **S3 Properties** window.

11. Under **Object URL**, right-mouse click on the URL and open in a new browser window.

12. You should find that the web page has been updated with the word **Chocolate**, as illustrated in the following screenshot:

← → C 🔒 rajesh-webpage.s3.amazonaws.com/index.html

Chocolate Muffin Recipe

Bake the ultimate blueberry muffins for your guests and loved ones.

Ingredients:

- 100g fresh **blueberries**
- 300g flour
- 150g granulated sugar
- 1 tsp. vanilla
- 60 ml vegetable oil
- 50g of butter

Figure 5.15 – Your index.html page showcasing the recipe

13. In the Amazon S3 management console, click on the **Versions** tab, as illustrated in the following screenshot:

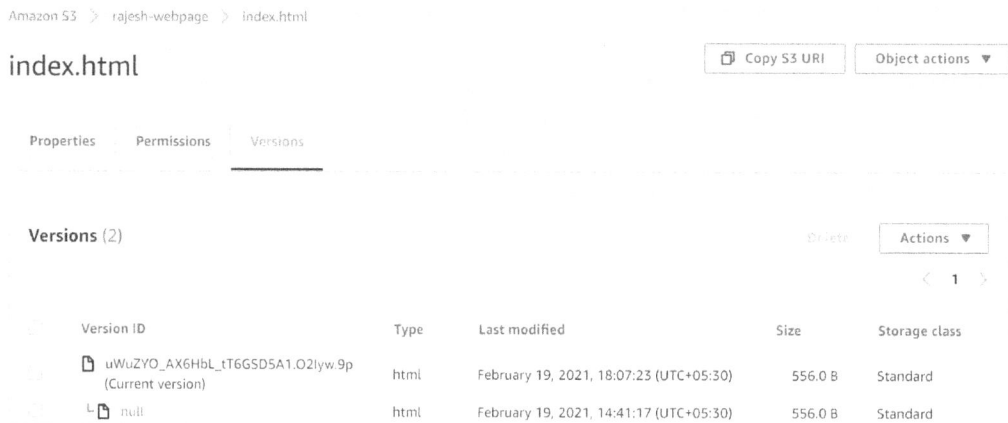

Amazon S3 > rajesh-webpage > index.html

index.html

Copy S3 URI   Object actions ▼

Properties   Permissions   Versions

Versions (2)

Delete   Actions ▼

‹ 1 ›

	Version ID	Type	Last modified	Size	Storage class
	uWuZYO_AX6HbL_tT6GSD5A1.O2Iyw.9p (Current version)	html	February 19, 2021, 18:07:23 (UTC+05:30)	556.0 B	Standard
	└ null	html	February 19, 2021, 14:41:17 (UTC+05:30)	556.0 B	Standard

Figure 5.16 – Bucket version tab

14. Note that there are two versions—the original version, which has a version ID of `null`, and a newer version with a **string of letters and numbers**, called **Current version**. The reason why the first version has a version ID of `null` is because it was created/uploaded to the bucket before we enabled versioning. Going forward, all new updates to this file will be assigned a new version ID, allowing you to preserve older versions if you ever need to access them again.

In this exercise, you learned how to configure versioning on your bucket. You were able to upload and manage multiple versions of the same object, and you discovered how unversioned objects have a version ID of `null`, whereas versioned objects have a version ID comprised of a series of characters unique to that version. You also discovered how to display a list of available versions of your objects in a version-enabled bucket.

Exercise 5.4 – Setting up static website hosting

In this exercise, we will configure the bucket to host a static website. When configured with a static website hosting service, the bucket will be configured with a website endpoint that you can distribute to your users, who can then access all the pages (assuming they are linked) using the standard HTML protocol.

To configure your bucket for static website hosting, you need a minimum of two files— an `index.html` file and an `error.html` file. An error file is simply a file that the S3 static website hosting service will redirect to if there is a problem with the `index.html` file—for example, if it cannot find the `index.html` page. You could use the `error.html` file to broadcast the fact that perhaps the site is under maintenance. Proceed as follows:

1. Create a new HTML file using your text editor as before (either Notepad on Windows or TextEdit on a Mac). However, in this file, simply add a line of text along the lines of `This site is under maintenance.`

2. Save the file as `error.html`, making sure to set the file types to **All Types** if you are using a Windows machine.

3. Navigate back to the S3 console and click on your S3 bucket.

4. Click on **Properties** and then scroll toward the bottom of the page, until you find the **Static website hosting** section heading.

5. Click **Edit** and select the **Enable** option.

6. For **Hosting type**, select **Host a static website**.

7. Under the **Index document** sub-heading, type the name of your index file—in this case, `index.html`.

8. Under the **Error document** sub-heading, type the name of your new error file—in this case, `error.html`.

9. Leave all the remaining settings at their defaults and click **Save changes** at the bottom of the page.

10. Next, click on the **Objects** tab.

11. Click **Upload** and click **Add files**.

12. Select the `error.html` file and click **Upload**. You should then see a screen like this:

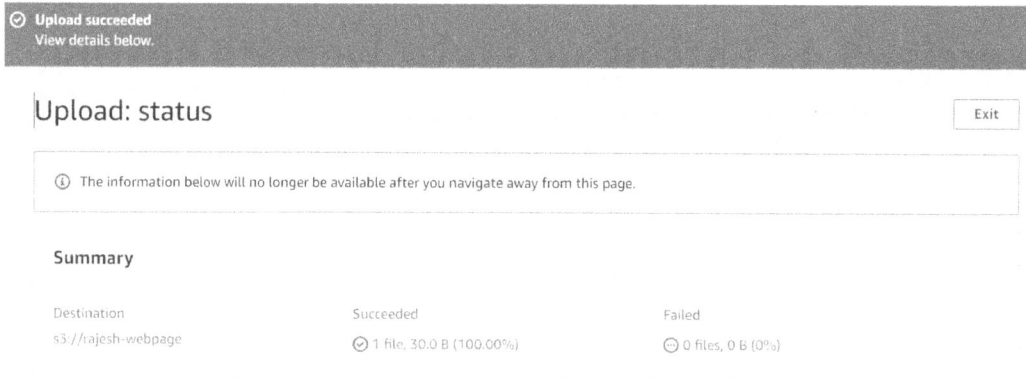

⊘ **Upload succeeded**
View details below.

Upload: status | Exit |

ⓘ The information below will no longer be available after you navigate away from this page.

Summary

Destination	Succeeded	Failed
s3://rajesh-webpage	⊘ 1 file, 30.0 B (100.00%)	⊖ 0 files, 0 B (0%)

Figure 5.17 – Upload of updated index page to your bucket

13. Click **Exit**.

14. At this stage, your bucket has been configured for static website hosting. To test it, you need to access your website via the S3 website URL endpoint.

15. In the S3 console, while still viewing the contents of the buckets (under **Objects**), click on the **Properties** tab again.

16. Scroll down till you reach the **Static website hosting** section, and you will note the URL is provided under the **Bucket website endpoint** heading, as illustrated in the following screenshot:

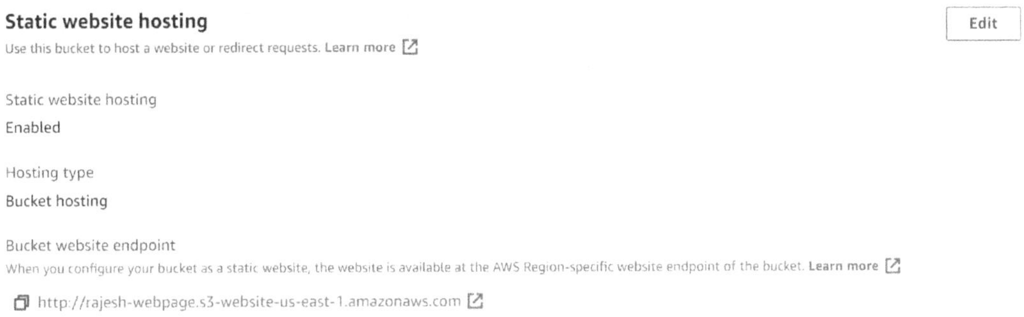

Static website hosting

Use this bucket to host a website or redirect requests. Learn more ↗

Edit

Static website hosting

Enabled

Hosting type

Bucket hosting

Bucket website endpoint

When you configure your bucket as a static website, the website is available at the AWS Region-specific website endpoint of the bucket. Learn more ↗

http://rajesh-webpage.s3-website-us-east-1.amazonaws.com ↗

Figure 5.18 – Enabling static website hosting on your bucket

17. Navigate to the provided URL in a new browser window, and you should find that the website opens with the recipe web page, as illustrated in the following screenshot:

← → C ⚠ Not secure | rajesh-webpage.s3-website-us-east-1.amazonaws.com

Chocolate Muffin Recipe

Bake the ultimate blueberry muffins for your guests and loved ones.

Ingredients:

- 100g fresh **blueberries**
- 300g flour
- 150g granulated sugar
- 1 tsp. vanilla
- 60 ml vegetable oil
- 50g of butter

Figure 5.19 – Updated index.html page with the wrong heading (Chocolate)

18. As you will note, the recipe is for a blueberry muffin, but the heading has changed to **Chocolate**. Assuming that this was an error in the update, we can easily revert to the previous version of this web page because we have already configured the bucket for versioning.

19. Navigate back to the Amazon S3 bucket so that you are looking at the actual contents of the bucket under the **Objects** tab, as illustrated in the following screenshot:

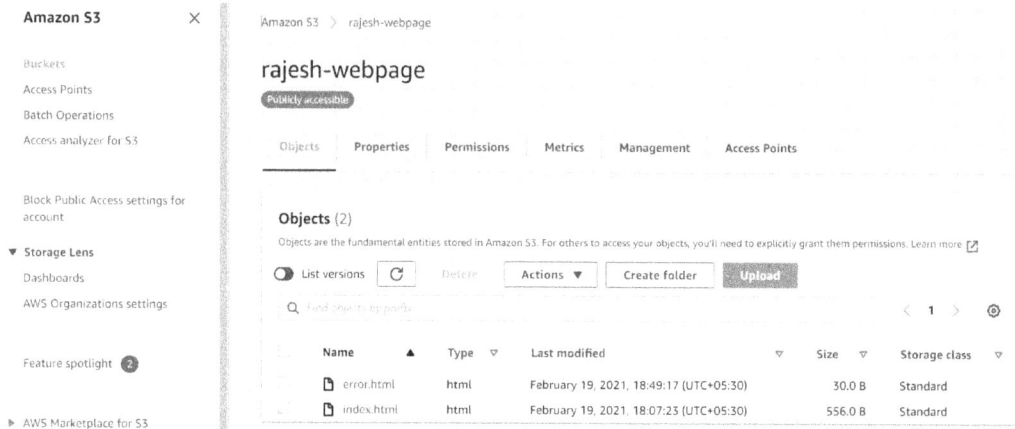

Figure 5.20 – List of updated objects in your bucket

20. Notice the **List versions** toggle just below the **Objects** sub-heading.

21. Click this toggle switch to list out all versions of all objects in your bucket, as illustrated in the following screenshot:

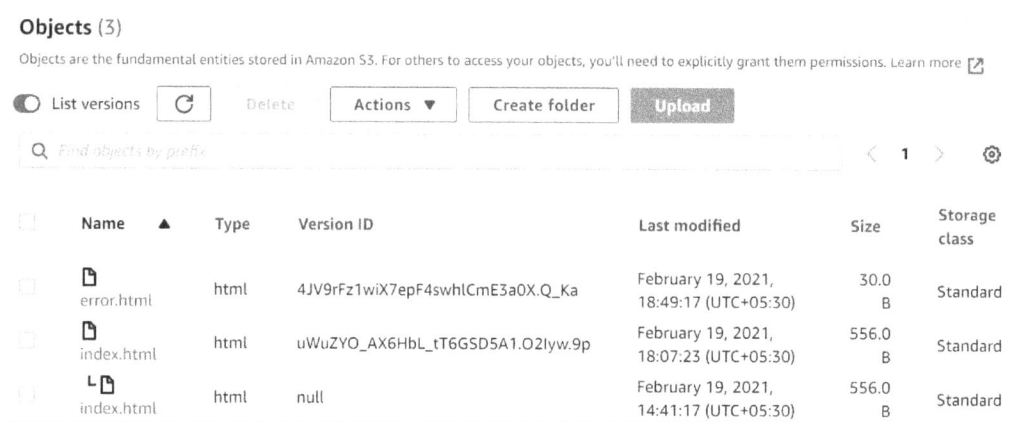

Figure 5.21 – List of objects and their individual versions.

22. As you will note, there are two versions of the `index.html` page. The latest version has got a version ID and contains an incorrect recipe heading.

23. Click on the checkbox to select this version and then click the **Delete** button.

24. You are then prompted to confirm your delete request by typing in the phrase `permanently delete` in the provided textbox, as illustrated in the following screenshot:

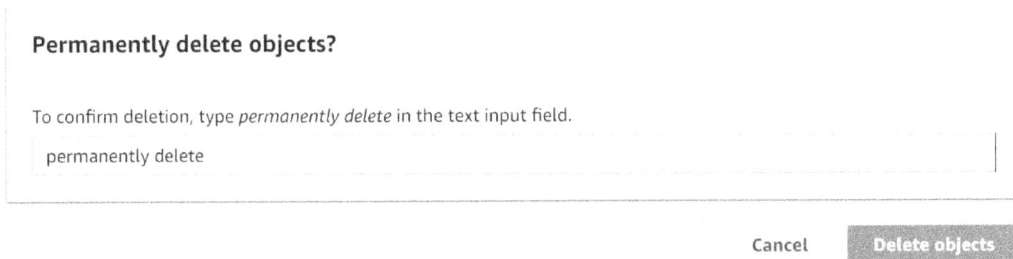

Permanently delete objects?

To confirm deletion, type *permanently delete* in the text input field.

permanently delete

Cancel   Delete objects

Figure 5.22 – Deleting incorrect version of the index.html page

25. Next, click **Delete objects**.

26. Click **Exit**, and you will note that the version has now been deleted.

27. Click on **Properties** again and then scroll down to the **Static website hosting** section. Open up the website URL in a new browser tab and you should see that the older, correct version of the web page is now displayed, as illustrated in the following screenshot:

← → C   ⚠ Not secure | rajesh-webpage.s3-website-us-east-1.amazonaws.com

Blueberry Muffin Recipe

Bake the ultimate blueberry muffins for your guests and loved ones.

Ingredients:

- 100g fresh **blueberries**
- 300g flour
- 150g granulated sugar
- 1 tsp. vanilla
- 60 ml vegetable oil
- 50g of butter

Figure 5.23 – Previous recipe page with the correct heading (Blueberry)

In this exercise, you learned how to configure your bucket with static website hosting. You learned how, in this particular lab exercise, we made an error in the title of our web page and we were able to revert to an older versioning of the same document, thanks to having versioning enabled earlier.

Summary

Amazon S3 is one of AWS's flagship storage products and comes with unlimited amounts of storage capacity that is highly scalable and durable.

In this chapter, you learned about the core feature of Amazon S3, including versioning, lifecycle management, and replication services, and how Amazon S3 meets a wide range of use cases. You also learned how you can build and deploy static website hosting on an Amazon S3 bucket and its various applications in the real world.

We also discussed how Amazon S3 comes with a wide range of security tools such as the ability to create granular access permissions via bucket policies and ACLs, as well as encryption of data in transit and at rest. You have also learned how you can connect your on-premises workloads to the Amazon S3 platform using Amazon Storage Gateway via the internet, a VPN, or AWS Direct Connect services.

If you are looking to migrate large amounts of data to the cloud, you can use the Amazon Snowball service to help you transfer large volumes of data, using rugged and tamper-resistant devices that are shipped to your on-premises location. Once you get the data copied, you simply ship it back to AWS to have your data imported into your S3 environment.

In the next chapter, we discuss networking services, focusing on the **Amazon Virtual Private Cloud** (**VPC**) service, among others. We also look at the Amazon Direct Connect service and at how you can connect your on-premises network with your AWS cloud services using VPN technologies.

In the next section, we look at some review questions for this chapter to test your knwoledge.

Questions

Here are a few questions to test your knowledge:

1. Which of the following is true regarding Amazon S3? (Select 2 answers)

    A. Amazon S3 is object-based storage.

    B. Amazon S3 is an example of file storage.

    C. Amazon S3 is an example of block storage.

    D. The Amazon S3 One Zone-IA storage class offers 99.5% of availability. Amazon S3 can be configured as shared mount volumes for Linux-based EC2 instances.

2. You wish to enforce a policy on an S3 bucket that grants anonymous access to its content if users connect to the data from the corporate and branch offices as part of your security strategy. Which S3 configuration feature will enable you to define the IP ranges from where you will allow access to the data?

   A. Security groups

   B. Bucket policy

   C. NTFS permissions

   D. **Network ACLs (NACLs)**

3. Which AWS service is the most cost-effective if you need to host static website content for an upcoming product launch?

   A. Amazon EC2

   B. Amazon EFS

   C. Amazon S3

   D. Azure ExpressRoute

4. Which Amazon S3 storage class enables you to optimize costs by automatically moving data to the most cost-effective access tier, while ensuring that frequently accessed data is made available immediately?

   A. Amazon S3 Standard

   B. Amazon S3 One-Zone IA

   C. Amazon Snowball

   D. Amazon S3 Intelligent-Tiering

5. Which Amazon S3 service can be configured to automatically migrate data from one storage class to another after a set number of days as a means of reducing your costs, especially where frequent instant access may not be required to that subset of data?

   A. Static website hosting

   B. Lifecycle management

   C. Storage transition

   D. S3 migration

6. When retrieving data from Amazon Glacier, what is the typical time taken by a Standard retrieval option to make the archive available for download?

   A. 20 minutes

   B. 24 hours

   C. 3 to 5 hours

   D. 90 seconds

7. Which feature of the Amazon S3 platform enables you to upload content to a centralized bucket from across any location via Amazon edge locations, ensuring faster transfer speeds and avoidance of public internet congestion?

   A. Amazon S3TA

   B. AWS S3 Storage Gateway

   C. Amazon VPC

   D. CloudFront

8. Your on-premises applications require access to a centrally managed cloud storage service. The application running on your servers need to be able to store and retrieve files as durable objects on Amazon S3 over standard NFS-based access with local caching. Which AWS service can help you deliver a solution to meet the aforementioned requirements?

   A. AWS Storage Gateway— Amazon S3File Gateway

   B. AWS EFS

   C. Amazon Redshift

   D. EBS volumes

9. You are looking to migrate your on-premises data to the cloud. As part of a one-time data migration effort, you need to transfer over 900 TB of data to Amazon S3 in a couple of weeks. Which is the most cost-effective strategy to transfer this amount of data to the cloud?

   A. Use the Amazon RDS service

   B. Use the Amazon Snowball service

   C. Use the Amazon VPN connection between your on-premises network and AWS

   D. Use AWS Rain

6
AWS Networking Services – VPCs, Route53, and CloudFront

Networking is a fundamental component of any IT infrastructure, whether on-premises on in the cloud. Without networking, it would not have been possible to architect the complex world of communications that we live in today. In the absence of networking, there would be no internet in the modern world.

Almost every business today needs to have some form of network connectivity if it is to collaborate with partners and end customers. In this chapter, we look at some of the core building blocks of designing a network. We will also look at network services offered on AWS and, specifically, how **Virtual Private Cloud** (**VPC**) enables customers to build multiple isolated and secure networks within their AWS accounts, allowing them to isolate workloads and applications.

We examine AWS Route53, which is Amazon's **Domain Name System** (**DNS**) that enables the routing of network traffic across the AWS ecosystem and the wider internet. Finally, we look at Amazon CloudFront, which is a **Content Delivery Network** (**CDN**), designed to help customers distribute their content to their customers much more effectively and with low latency.

The following are the key topics we discuss in this chapter:

- Introduction to on-premises networks
- Fundamentals of IP addressing, port numbers, subnet masks, and CIDRs
- Introduction to Amazon **VPC**
- Basics of DNS and global routing with Amazon Route53
- Implementation of a robust CDN with Amazon CloudFront
- Introduction to API Gateway

Technical requirements

To complete this chapter, you will need access to your AWS account. You will be creating a simple VPC in your AWS account for which you will need to be logged into your AWS account using the IAM user account `Alice` that we created in *Chapter 4*, *Identity and Access Management*. Remember, to log in using an IAM user account, you will need to provide your AWS account ID or account alias.

Introduction to on-premises networks

Almost every business will have some form of an on-premises network. Even if you are a self-employed "one-man band," you are likely to have a home office that also boasts a private network environment. Your home office may look something like the following diagram:

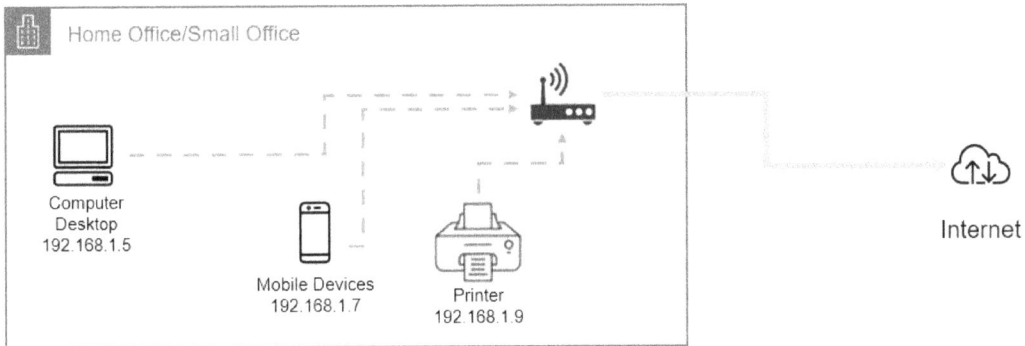

Figure 6.1 – Home network components

In the preceding diagram, the devices on the home network communicate with each other over the Wi-Fi connection. If you send a print request from the computer desktop to the printer, then your document will get printed. This communication is made possible via the connectivity established over the Wi-Fi network. For devices to communicate with each other, they each require a unique IP address. If the IP addressing element is correctly defined, each device on the network will be able to see the other. The Wi-Fi service also connects the devices to the internet via a router/modem device that has established a connection with an internet service provider. This connection is usually via some form of physical cabling or cellular network connectivity through your telecom provider.

Basic corporate networks

Companies will also have their own corporate networks, and you have probably worked in an office setting that has an IT network infrastructure in place. Corporate networks require more careful planning since many of them will also allow access to applications from outside the network. For example, your business may be publishing a website from within the corporate network to showcase its products and services to potential customers on the internet. The use of firewall routers that allow only specific types of traffic to enter the network from the internet and be directed to the appropriate server is all part of the secure planning for connectivity with the outside world.

A corporate network would usually be divided into multiple smaller networks—each being used for a specific purpose. At its very basic level, a corporate network would consist of two subnetworks: one for internal backend purposes and another to place services that are accessible from the internet.

In the following diagram, a corporate network has been divided into two separate networks: one called the internal network, and another called the **demilitarized zone (DMZ)**. The DMZ is an area where services are deployed that can be exposed on the internet, for example, a web server. Traffic to services deployed in this zone is restricted with strict inbound rules to ensure high levels of security.

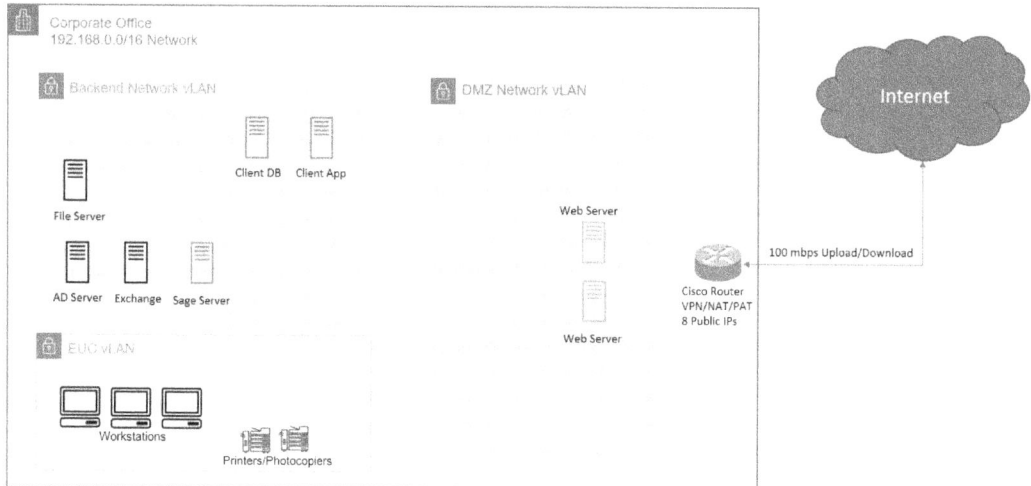

Figure 6.2 – Basic office network

As shown in the preceding diagram, the corporate network is divided into three separate subnetworks. This ensures that we can configure rules that define the type of traffic that can enter each subnetwork and from which source. For example, we could configure inbound rules from the internet to grant access to our web servers over HTTP/HTTPS traffic. This will allow members of the public to access our corporate website and review our service offerings. On the other hand, we would not expect to allow direct inbound traffic from the internet into the **End User Computing (EUC) Virtual LAN (VLAN)**, as there is no requirement for such inbound connection, and it ensures our corporate network is secure. Traffic from the workstations to the internet would, however, be permitted to allow members of staff to access online services and tools.

Similarly, when building solutions on AWS, we need to configure virtual networks in the cloud that would allow us to host our applications in a manner that offers security, isolation, and inbound access only where it is needed. However, before examining the details of how we build cloud networks, it is important to understand how devices on a network communicate with each other and how we create networks and subnetworks.

In the next section, we review some fundamentals of IP addressing, which will enable us to understand how we build such networks.

Fundamentals of IP addressing and CIDRs

For devices on your network to communicate with each other, an **Internet Protocol address (IP address)** is required. Each network device, whether it is a computer, laptop, mobile phone, printer, or network router, will need to be assigned an IP address that is routable in each network.

Furthermore, each device's IP address must be unique – you cannot have more than one device with the same IP address. This is just how telephones work. Each telephone has a unique number assigned to it. To call someone on the phone, you need to first know their telephone number and then, dial that number, which results in your call getting connected. In *Figure 6.1*, you would have noticed that each of the internal devices in the home network had an IP address.

There are two types of IP address: IPv4 and IPv6. We will discuss the key differences between them.

IP address version 4 – IPv4

IPv4 was the first version of the IP addressing system that was widely deployed and that ultimately formed the backbone of the internet.

The standard IPv4 address format that you are familiar with and as depicted in the previous diagrams follows the structure of four decimal numbers separated by dots. An example of this is 192.168.1.6. Each decimal notation in an IP address is called an **octet** and can be any number between 0 and 255, base 10 (decimal). As you are probably aware, computers use binary numbers rather than decimal numbers. Each decimal number, when converted into binary, comprises 8 bits, ones and zeros. In both cases, whether decimal or binary, an IPv4 address is 32 bits in length.

Let's take an example of an IP address of 192.168.1.6. In binary, each octet would be between 8 zeros and 8 ones (00000000 to 11111111). The individual decimal numbers in the IP address can be converted to its equivalent binary representation, which would comprise a combination of ones and zeros in each 8-bit octet. In the following diagram, the IP address 192.168.1.6 in decimal is the same as 11000000.10101000.00000 001.00000110 in binary:

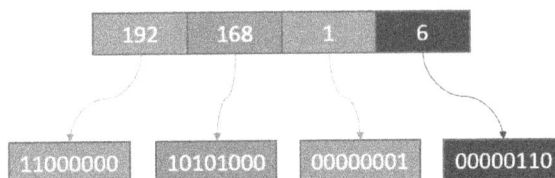

Figure 6.3 – IP address to binary conversion

While understanding how this conversion takes place is not a requirement to pass the Cloud Practitioner exam, I wanted to give you a quick overview here to further help you build your networking knowledge.

Conversion of decimal notation into binary requires remembering the place value of the individual bits in each 8-bit octet. You can also calculate the place values if you do not want to simply remember them by heart.

Let's take the example of the last octet in the IP address 192.168.1.6, which in this case is the decimal number 6.

In binary, you have 8 bits of zeros and ones to represent this decimal number. In the following table, we can see each of those bits and their place values:

X represents a '0' or '1' in binary

X	X	X	X	X	X	X	X
$X*2^7$ or	$X*2^6$ or	$X*2^5$ or	$X*2^4$ or	$X*2^3$ or	$X*2^2$ or	$X*2^1$ or	$X*2^0$ or
X * 128	X * 64	X * 32	X * 16	X * 8	X * 4	X * 2	X * 1

Place value:

Figure 6.4 – IP address place values

For each octet, starting from right to left, the following applies:

- The first value of the first bit is always equal to 1. It is 2 to the power of 0 (2^0), that is, 1.

- The second bit is double the first bit and equals 2 (*2 to the power of 1 – that is, 2*).

- The third bit is double the second bit and equals 4 (*2 to the power of 2 – that is, 4*).

- The fourth bit is double the third bit and equals 8 (*2 to the power of 3 – that is, 8*) and so on.

Calculating the decimal value of a binary representation is done as follows:

Decimal = *(X*128) + (X*64) + (X*32) + (X*16) + (X*8) + (X*4) + (X*2) + (X*1)*

Next, to convert the decimal number *6* from the IP address 192.168.1.6 into binary requires identifying which of the "Xs" should be converted to zeros and which to ones. You want to convert the minimum number of bits to ones to get your decimal representation, and furthermore, you should continue to convert only those bits to ones whose value is less than your decimal number. So, for example, as shown in *Figure 6.5, we have the following*:

- You would not convert the bit on the far left to a *1* because 128 is greater than the last octet, *6* in our IP address. Similarly, you would not convert the next bit from the far left to a *1* in binary because 64 is also greater than the last octet, and so on.

- You would convert the third bit from the far right to a *1* because its placeholder value is equal to *4*, which is less than *6* in our IP address.

- You would also convert the second bit from the far right to a *1* because its placeholder value is equal to *2*, which is also less than *6* in our IP address.

- Remember that you want to convert the fewest bits to ones to get your decimal representation. So, in this case, *4* plus *2* equals the last octet, *6* in our IP address. As such, we should not convert the last bit to a *1* as the total would then add up to *7*, which is more than *6*.

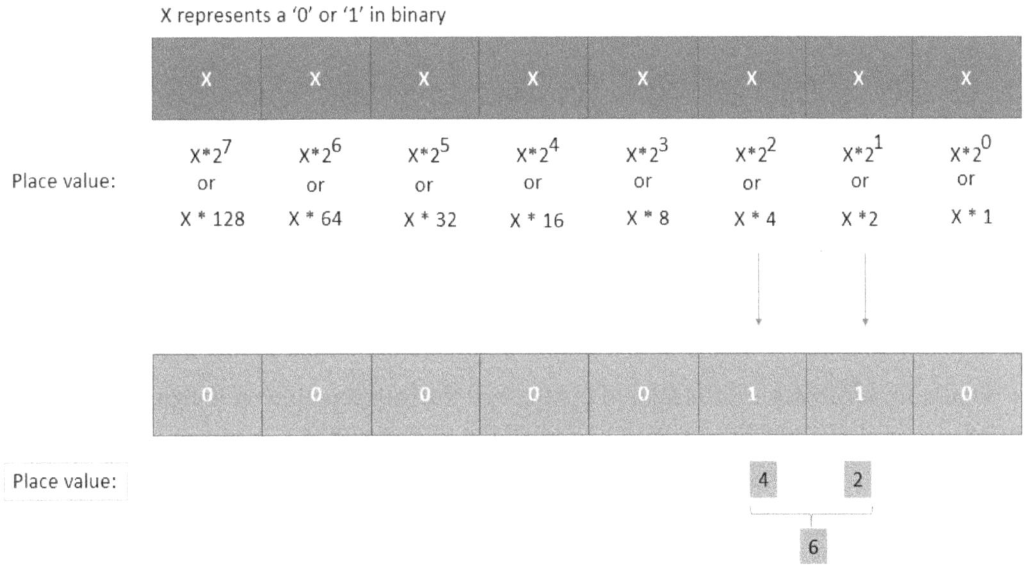

Figure 6.5 – Converting an IP address to its equivalent binary representation

Hence, the binary representation of *6* is 00000110. Similarly, converting the IP address 192.168.1.6 to binary would give you 11000000.10101000.00000001.0 0000110.

Limitations of IPv4 addresses

One of the primary limitations of an IPv4 address is that it is only 32 bits in length. This means that the maximum number of addresses you can have in an IPv4 addressing scheme is *2^32*, which is 4,294,967,294 addresses in total. Four-billion-odd addresses might seem like a large number, but the fact is that we have exhausted this range simply because of the vast number of devices that now need an IP address to participate on any given network.

Currently, the largest network on the planet is the internet. Every device that needs to communicate on the internet also requires an IP address. Furthermore, every device on a given network must have a unique IP address. You cannot have two devices in the same network using the same IP address as this would result in a conflict. Given that the four-billion-odd addresses are not sufficient to handle the huge volumes of devices, the **Internet Assigned Numbers Authority** (**IANA**) devised a brilliant plan to allocate a range of IP address for private use only. These address ranges are not routable on the internet, which means that businesses (and homes) can configure their internal private networks using these addresses without any possibility of them conflicting with other businesses' networks, particularly if those businesses do not plan to connect their networks together.

The following IP address ranges are designed for private use:

- `10.0.0.0/8` IP addresses: `10.0.0.0 - 10.255.255.255`
- `172.16.0.0/12` IP addresses: `172.16.0.0 - 172.31.255.255`
- `192.168.0.0/16` IP addresses: `192.168.0.0 - 192.168.255.255`

> **Additional Note**
>
> Another range of private IP addresses is `169.254.0.0` to `169.254.255.255`, but those addresses are for **Automatic Private IP Addressing** (**APIPA**) use only, designed for internal Microsoft networks.

The remaining addresses are considered public, and thus are routable on the global internet. To illustrate how this helps, let's look at the next diagram:

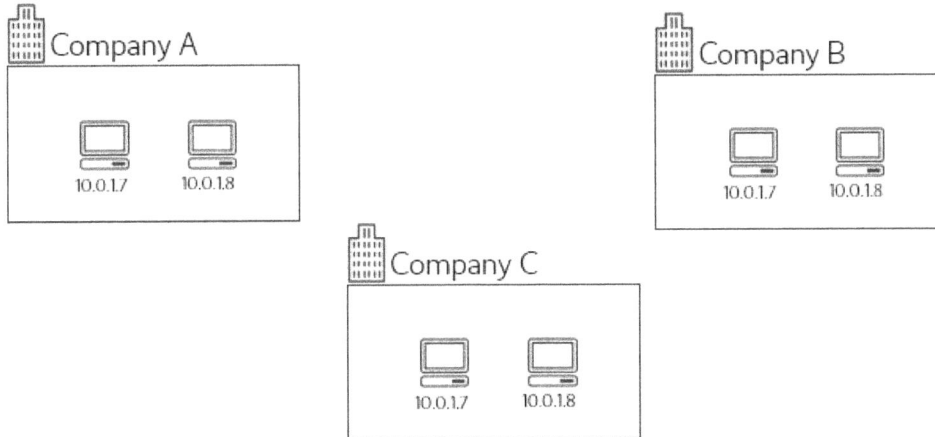

Figure 6.6 – Private IP address ranges used by businesses

In the preceding diagram, you will note that the three companies are able to use the same IP addresses for their internal devices. Since these businesses are not connected to each other, there is no possibility of an IP address conflict. Private IP addresses such as the ones designated by the IANA have helped businesses build internal networks without the need to procure any of the public addresses. The private IP address space also enhances internal network security because these addresses are not routable on the internet. We can also allow more devices to be networked, as the address ranges can be repeated among companies that do not need to be connected to each other over the same network.

Businesses need internet access

In the preceding illustration, we see that businesses can define internal network IP address ranges that are not routable over the internet. These businesses will still require access to the internet, whether to send and receive emails from their customers or host e-commerce applications that their clients would need access to from the internet. To facilitate internet connectivity, public IP addresses are required. However, having to assign each device on the internet with a public IP address would defeat the purpose of private IP ranges and pose a security risk. Instead, the internal network can be configured to access the internet via a service called **Network Address Translation** (**NAT**).

In the following diagram, we can see that businesses are now able to access the internet via a NAT service configured on their external router. The NAT service requires a minimum of one single public IP address and relays requests from the internal devices to the internet, acting as a proxy in between. Replies to those requests are also handled by the NAT service, ensuring that they are correctly redirected to the internal device that made the original request.

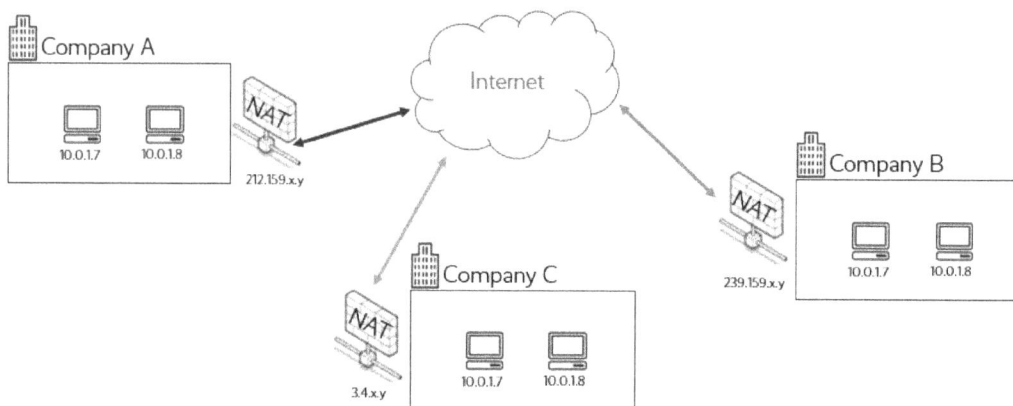

Figure 6.7 – Private IP address ranges used by businesses with internet via NAT services

Given the limitation of IPv4, IPv6 was developed by the **Internet Engineering Task Force (IETF)** in the 1990s. We take a look at IPv6 next and discuss how it overcomes the 32-bit address limitation of IPv4.

What about IPv6?

To address the limitations of IPv4, IPv6 was developed by the IETF. IPv6 uses a 128-bit address, which would give us 2^{128} addresses. IPv6 is also denoted in hexadecimal format rather than the standard decimal format. With IPv6, technically, each device could have its own public IP address. In fact, **Amazon Web Services (AWS)** offers IPv6 as an option to set up cloud networks. Even if you need to protect devices from the public internet, you can still use an IPv6 address for a virtual server in the cloud and allow it to send traffic over the internet using an egress-only internet gateway.

However, many companies continue to use IPv4, partly due to the capabilities of NAT services and partly to ensure interoperability with legacy devices that may not be IPv6-aware.

Network sizes and classes

Originally, the IETF designed different classes of IPv4 addresses to help define different network sizes and use cases.

Classes A to C represent generic unicast IP addresses (with a few exceptions) that members of the public can use to build networks of different sizes. Class D comprises multicast addresses, and class E has been reserved for experimental use.

The way these classes help define network sizes is by splitting the IP address into a network portion and a host portion. Let's look at this individually by class:

- **Class A** – The first 8 bits of a class A address define the network portion, and the remaining 24 bits are used to denote the host portion. Network bits are denoted by 1 (a one in binary) and host bits are denoted by 0 (zeroes). Also, the far-left bit of a class A address is set to 0.

- **Class B** – The first 16 bits of a class B address define the network portion, and the remaining 16 bits are used to denote the host portion. In a class B network, the two far-left bits are set to 10.

- **Class C** – The first 24 bits of a class C address define the network portion, and the remaining 8 bits are used to denote the host portion. Also, in a class C network, the two far-left bits are set to 11.

To better illustrate how these three classes of networks actually look, let's look at the next diagram:

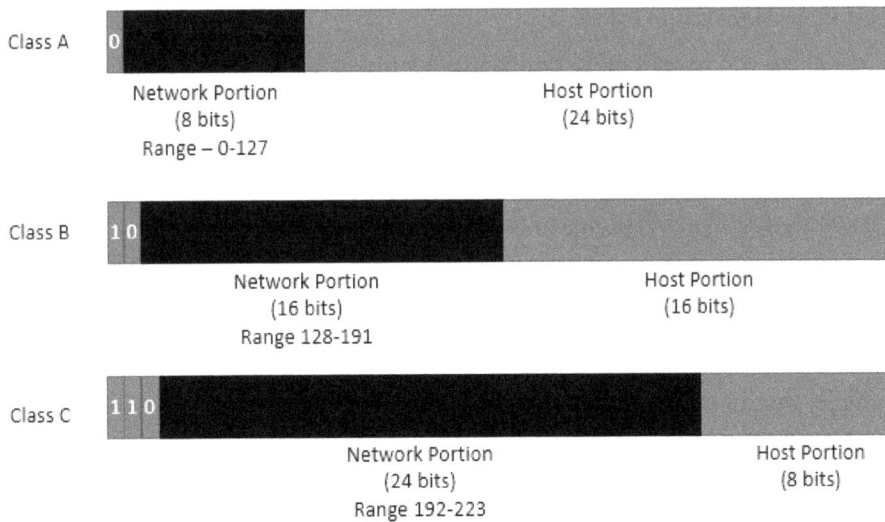

Figure 6.8 – IP address classes

In the preceding diagram, you can identify which class a particular IP address belongs to and instantly identify the potential number of host IP addresses that IP block would have. So, for example, if we take the IP address 192.168.1.6, we can confirm that it is a class C address.

What this means is that the network portion of the IP address is 192.168.1.x. In this example, x can be any number between 1 and 254. That gives you a total of 254 IP addresses in the host portion of the IP block. Although the total number of IP addresses you can have in any one octet is 256 (2^8, which equals 256), it is important to remember that the first and last IP addresses are unusable. The first IP address is always known as the network ID, which in this case is 192.168.1.0. Here, the last octet would be represented by all zeroes (or, in binary, 00000000). The last IP address is 192.168.1.255, which is known as the broadcast address. Here, the last octet would be represented by all ones (or, in binary, 11111111).

A simple formula to work out the number of usable IP addresses in an IP block is as follows:

Number of Usable IP Addresses = 2^Number of Host Bits - 2

In the preceding example, we have an IP address of 192.168.1.6, which belongs to an IP block that can only contain 254 usable IP addresses.

So far, you have taken our word for it that the IP address 192.168.1.6 belongs to a class C network and there are 254 IP addresses in the block of 192.168.1.x. Let's look at how this works next, when we discuss subnet masks.

What are subnet masks?

Subnet masks allow you to split an IP address block into a network portion and a host portion. Host devices on the same network portion can easily talk to one another and will need some form of routing to talk to hosts in other networks.

In the preceding class C network, the first three octets belong to the network portion and only the last octet (last 8 bits) belongs to the host portion.

A subnet is a 32-bit number created by setting host bits to all zeroes and setting network bits to all ones. A logical AND operation is then performed with a corresponding IP address block to define the number of host IP addresses that can be available in each block.

So, for example, we can check that the IP address 192.168.1.6 belongs to a class C address because a class C masks the first three octets (first 24 bits in binary) as the network portion and leaves the last octet (last 8 bits) for the host bits.

In binary, the IP address 192.168.1.6 and its associated subnet mask are represented as follows:

IP Address 192.168.1.6	11000000	10101000	00000001	00000110
Subnet Mask 255.255.255.0	11111111	11111111	11111111	00000000
Logical AND	11000000	10101000	00000001	00000000

Network Portion (24 Bits)   Host Portion (8 Bits)

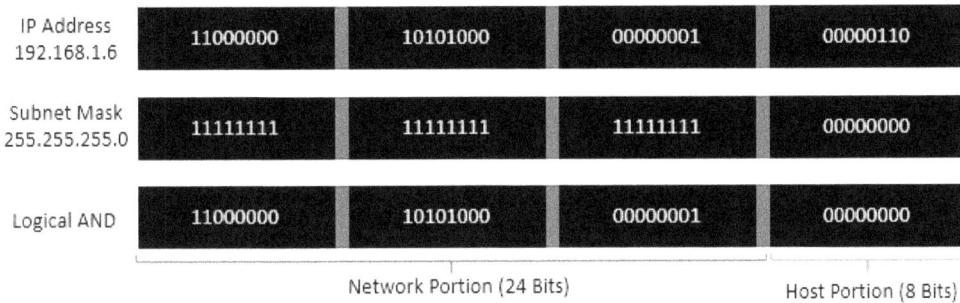

Figure 6.9 – IP address and subnet mask conversion for 192.168.1.6

A logical AND operation is calculated using the following process:

- A 0 and 0 equals 0.
- A 0 and 1 equals 0.
- A 1 and 0 equals 0.
- A 1 and 1 equals 1.

Based on the result of the preceding logical AND operation, you will note if any of the first three octets of the IP address were to change, for example, if instead of 192 we used 193, then that would yield a different result for the corresponding logical AND operation block. Since the first three octets represent the network portion of an IP address, the resulting change would effectively denote a different network.

Furthermore, because the host portion in the subnet mask is set to all zeroes and is only represented by the last octet (last 8 bits), any variation in this octet will keep the IP address within the same network.

For example, let's look at the IP address 192.168.1.12. Depicting this in the following diagram while performing a logical AND operation with the subnet mask would yield the following:

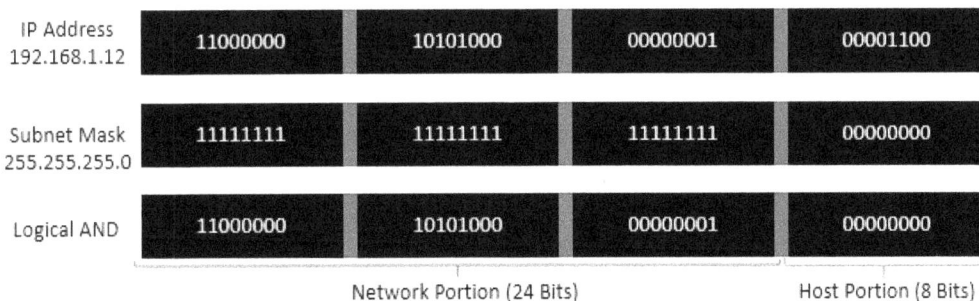

IP Address 192.168.1.12	11000000	10101000	00000001	00001100
Subnet Mask 255.255.255.0	11111111	11111111	11111111	00000000
Logical AND	11000000	10101000	00000001	00000000

Network Portion (24 Bits)   Host Portion (8 Bits)

Figure 6.10 – IP address and subnet mask conversion for 192.168.1.12

What you will note from the preceding two diagrams is that the logical AND operation yields the same pattern of binary numbers, indicating the following:

- Both IP addresses belong to the same network.
- The network portion in both cases is the same, derived by the subnet mask of 255.255.255.0 (or 11111111 11111111 11111111 00000000).

Now that we know how subnet masks are used to *mask* portions of an IP address as a network portion and host portion, let's examine the concept of subnetting.

What is subnetting?

In the preceding discussion on subnet masks, we defined three classes of networks. The networks are derived by using subnet masks with fixed lengths to create these classes. So, for example, in a class A network, the first octet of an IP block represents the network portion and the remaining three octets the host portion.

This is made possible by using what we call subnet masks. Specifically, we use a subnet mask of 255.0.0.0 to carve off the first octet for the network portion. Any change in the first octet of an IP address with a subnet mask of 255.0.0.0 would yield a different network. For example, 10.0.0.0 is on a different network to 31.0.0.0.

Subnetting is the process by which you can create subnetworks within a larger network. In each network, you may need to create smaller, isolated portions of the network, such as one portion for hosting all backend servers and another for hosting frontend web servers. Subnetting allows us to break up a large network into these smaller subnets.

The process of creating subnets involves *borrowing* additional bits from the host portion of an IP address range. These borrowed bits are used to create smaller subnetworks within a larger primary network.

For example, if we have a business requirement for eight subnetworks, each capable of hosting 30 IP addresses for 30 devices, then using a standard class C address, we could use the following IP blocks:

- Network 1 – 192.168.1.0 (subnet mask: 255.255.255.0)
- Network 2 – 192.168.2.0 (subnet mask: 255.255.255.0)
- Network 3 – 192.168.3.0 (subnet mask: 255.255.255.0)
- Network 4 – 192.168.4.0 (subnet mask: 255.255.255.0)
- Network 5 – 192.168.5.0 (subnet mask: 255.255.255.0)

- Network 6 – `192.168.6.0` (subnet mask: `255.255.255.0`)

- Network 7 – `192.168.7.0` (subnet mask: `255.255.255.0`)

- Network 8 – `192.168.8.0` (subnet mask: `255.255.255.0`)

While this network design works perfectly, we have a lot of wastage in terms of the number of IP addresses available versus the number of IP addresses required. Each of these eight networks contain 254 usable IP addresses, but we only need 30 addresses per network as per the requirement.

When using private IP ranges, this may not matter so much, but when we consider applying the same approach to the limited public IP address range, it becomes impossible to achieve. Also, note that you generally must pay for public IPs.

Instead of this approach, we can use subnetting and subnet masks to break up a single network into smaller networks, conserving and efficiently using the available IP address space within a given network.

So, for example, we can take the class C network of `192.168.1.0` with a subnet mask of `255.255.255.0` and break this up into smaller subnets. To do this, we borrow additional bits from the host portion of the IP address block to represent our subnets in the network.

Let's take a look at the next diagram to illustrate this concept:

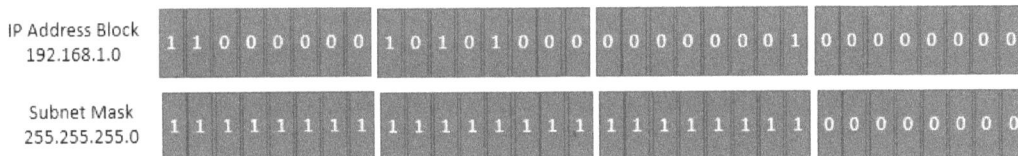

Figure 6.11 – IP class C network

In a standard class C network, the preceding IP address block of `192.168.1.0` with a subnet mask of `255.255.255.0` would yield a single network with 254 IP addresses.

This is because we have used the first three octets (24 bits) to represent the network portion and the last octet (8 bits) to represent the host portion.

Because we have 8 bits to represent host bits, we can use the formula 2^8, which equals 256. However, remember the first and last IP addresses are not usable, so that is 256-2, which gives us 254 usable IP addresses.

Now if we borrow the first 3 bits from the host portion of the IP address block to build subnetworks, we will have the following representation:

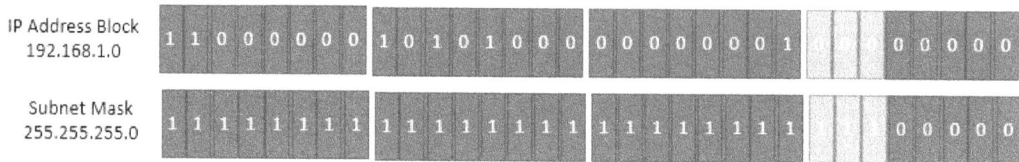

Figure 6.12 – Creating subnets by borrowing bits from the host portion of the IP address

By borrowing 3 bits from the far left of the fourth octet (the host portion), we can effectively create eight subnet networks. This is derived by the following formula:

Number of subnets = 2^number of host bits borrowed

In this case, *2^3 = 8* and so, this gives us 8 subnetworks. Furthermore, because we now only have 5 bits remaining to represent the host portion, we can work out the number of IP addresses we have per subnetwork. The formula is *Number of Hosts = 2^Number of Host Bits Remaining*. In this case, it is *2^5* (because there are 5 remaining host bits) = 32 IP addresses. In addition, as previously discussed, the first and the last IP addresses are not usable and as a result, we subtract 2 from the number of IP addresses (32-2), which gives us our 30 IP addresses, as required.

If we look at the subnet mask representation of the borrowed 3 bits, we will have a new subnet mask as follows:

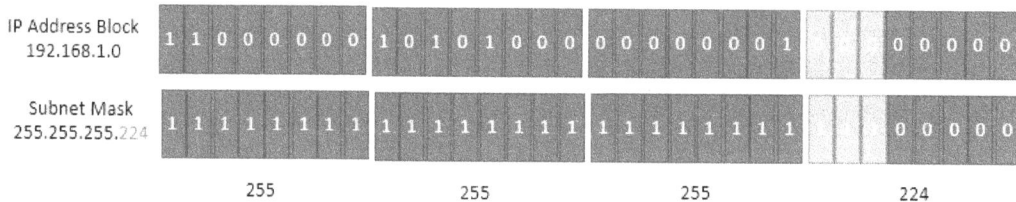

Figure 6.13 – Creating subnets resulting in a new subnet mask

Using a subnet mask of 255.255.255.224, we can see that we have eight individual subnetworks that can be created.

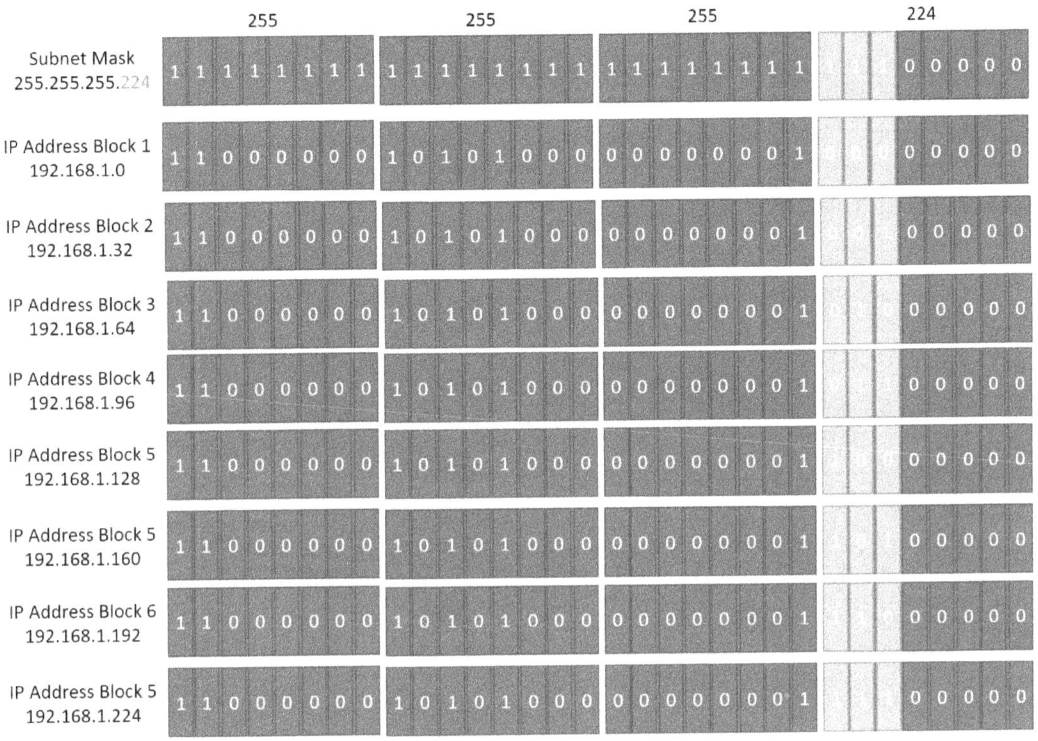

Figure 6.14 – Creation of eight subnetworks

You will note the pattern difference in the IP blocks shown in binary. Specifically, the different combination of *ones* and *zeros* in the borrowed bits define the eight different networks. The following IP address blocks are the eight subnetworks that can be created by borrowing 3 bits from the host portion of the IP address block `192.168.1.0`. Each subnetwork has 30 usable IP addresses. The eight networks are as follows:

- `192.168.1.0 (192.168.1.0 to 192.168.1.31)`

- `192.168.1.32 (192.168.1.32 to 192.168.1.63)`

- `192.168.1.64 (192.168.1.64 to 192.168.1.95)`

- `192.168.1.96 (192.168.1.96 to 192.168.1.127)`

- `192.168.1.128 (192.168.1.128 to 192.168.1.159)`

- `192.168.1.160 (192.168.1.160 to 192.168.1.191)`

- `192.168.1.192 (192.168.1.192 to 192.168.1.223)`

- `192.168.1.224 (192.168.1.224 to 192.168.1.255)`

Using subnet masking, you are then able to identify which network a particular IP address block belongs to. So, for example, the IP address 192.168.1.130 would belong to the IP block 192.168.1.128 with a subnet mask of 255.255.255.224. This is made clearer when we look at the binary representation and perform the logical AND operation, as shown in the following diagram. You can see that the results of the logical AND operation match IP address block 5:

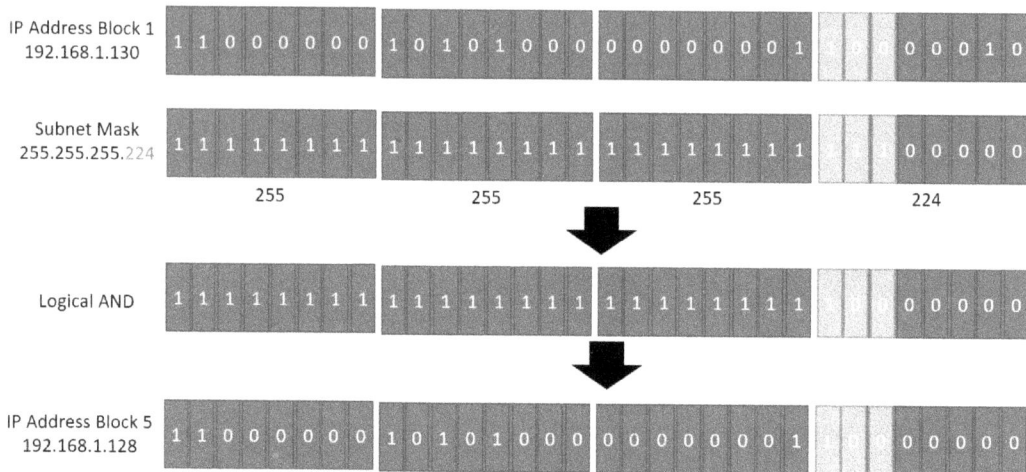

Figure 6.15 – Illustrating how IP addresses fall in a given subnetwork range

This binary depiction and logical AND operation show how IP address block 5 falls in the same network as IP address block 1, as shown in the yellow cells.

Classless Interdomain Routing (CIDR)

Instead of using subnet masks and the complexity discussed previously, we can use CIDR. With CIDR, you can create networks of different sizes like how you use subnet masks. CIDR is essentially another way to represent subnet masks but offers more flexibility. The size of your network will determine how many IP addresses you can have in that network. You can also divide your network into multiple smaller networks (subnets) by configuring specific subsets of the IP block using CIDR blocks.

When defining a network, CIDR blocks are displayed as part of the IP address block with a slash (/) followed by a decimal number between /8 and /32. For example, the IP address 192.168.1.6 could belong to a network 192.168.1.0/24. Let's look at how this works.

In a network with a CIDR block of /24, we can work out the number of IP addresses that the network can host and therefore, the number of devices that can be placed in the network. Given that the total number of bits in an IPv4 range is 32, then for a /24 CIDR, we simply subtract the 24 from 32, which gives us 8 bits. These 8 bits represent the total number of IP addresses we can have in that network. Specifically, *2^8* equals 256. So, the total number of IP addresses in network 192.168.1.0/24 is 256 IP addresses. Remember, that on any given network, the first and last IP addresses are not usable. So, the total number of usable IP addresses is 256 - 2, which equals 254.

An important point to note here is that the IP network with a CIDR block of 192.168.1.0/24 is one single network and within this network, you can host up to 254 devices that would each need an IP address. The IP address range would be from 192.168.1.1 to 192.168.1.254 (remember that the first IP 192.168.1.0 and the last IP 192.168.1.255 are not usable).

Let's look at another example. We will use the IP range of 10.0.0.0 to 10.0.255.255, which is a private range for our internal network. Let's say that you choose a network IP with a CIDR block of 10.0.0.0/16. In this network, your CIDR block is a /16. In this network, you can have a total of 65,536 IP addresses. Remember that to work out the number of IP addresses, you must subtract the value of 32 from the CIDR block notation, in this case, 16. This gives you 16 bits that can be used by devices in your network. Two to the power of *16 (2^16)* equals 65,536 IP addresses. Remember to subtract another 2 from the total number of IP addresses, which gives you a total number of 65,534 usable IP addresses. This is a very large network, and you may wish to divide this larger network into smaller subnetworks (subnets) for resource isolation and separation. Using the same network IP block, you can create subnetworks by increasing the CIDR block value. For example, in the primary network of 10.0.0.0/16, you can create several subnetworks using the CIDR block of /20. This means that you have 12 bits remaining for the hosts portion of your IP address block (32 bits – 20 bits for the network).

So, within the 10.0.0.0/16 network, you can have subnetworks with the IP block of 10.0.0.0/20. Each subnetwork would have a total of 4,096 IP addresses (*2^12* hosts bits remaining) or 4,094 usable IP addresses (subtracting the 2 IPs that cannot be used).

So far, we have looked at how IP addresses are configured and how they can be represented in both binary and decimal notation.

We also looked at how you can create a large network and subsequently, multiple smaller networks using CIDR blocks.

In the next section, we start to look at VPCs, which are virtual networks you can build on AWS similarly to how you would define networks on-premises. Many services require a VPC within which you can launch resources. VPCs also help you protect your resources through firewall technologies, which we will also discuss in this chapter.

Virtual Private Clouds (VPCs)

A VPC is a virtual network in the cloud. You choose the Region in which to create your VPC and define its network parameters such as the IP address range and any subnetworks within it, for resource isolation.

Resources deployed in your VPC can then access services on the internet or can grant inbound access from the internet, for example, if you are hosting an e-commerce web server.

AWS already provides you with a default VPC in each Region. These default VPCs are designed to get you up and running with the ability to deploy EC2 instances so that they can access the internet and, where necessary, be configured to allow direct inbound access from the internet.

You can also configure custom VPCs to suit your business requirements. When configuring a new VPC, you need to define an IP address block from one of the private IP ranges. Your VPC spans the entire Region in which you deploy it. This means that you can place workloads in different Availability Zones within the VPC to design for high availability.

All about subnets

Depending on your requirements, you can configure your non-default VPCs with multiple subnetworks (subnets).

A subnet is a subset of the VPC, and this is defined in the IP address block of the subnet. So, for example, if the IP address block of your VPC is `10.0.0.0/16`, public `Subnet-1` could be configured with a network IP address CIDR block of `10.0.1.0/24`. Each subnet you define in each VPC would also need to be configured with non-overlapping subsets of IP ranges from your primary VPC's IP address block. So, for example, if you build another subnet, say `Subnet-2`, you could use an IP address CIDR block of `10.0.2.0/24`.

The IP address CIDR block of Subnet-1 (10.0.1.0/24) and the IP address CIDR block of Subnet-2 (10.0.2.0/24) as shown in *Figure 6.8* are not overlapping. Nevertheless, both Subnet-1 and Subnet-2 belong to the same VPC's IP address CIDR block of 10.0.0.0/16. This is because the IP address CIDR blocks of both the subnets are direct subsets of the VPC's IP address CIDR block.

Subnets are also restricted to a single Availability Zone. This means you can recreate multiple subnets across multiple Availability Zones, which would host replica resources to offer high availability. If a single Availability Zone (and therefore, the subnet contained within) were to go offline, you can redirect traffic to a replica resource in the alternative subnet that resides in another Availability Zone. The following diagram shows two subnets, one in each Availability Zone, that belong to the VPC. If Subnet-1 were to go offline due to an Availability Zone outage, requests can still continue to be served by the database instance in Subnet-2.

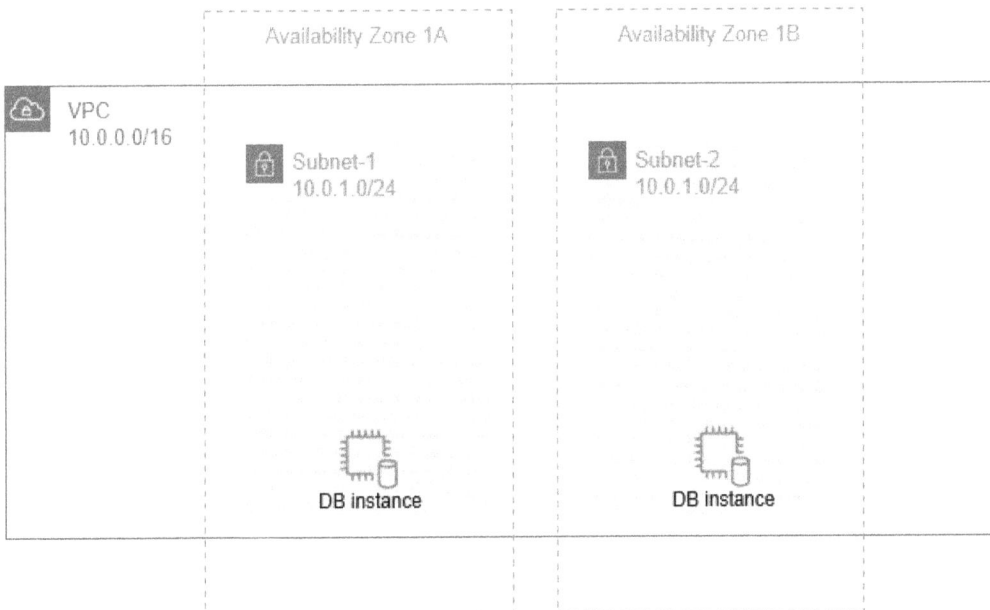

Figure 6.16 – VPCs and subnets with non-overlapping IP addresses

In the preceding diagram, we have deployed two database instances across two subnets within two Availability Zones.

Internet access

By default, your custom VPCs need to be configured with internet access if you want the EC2 instances in the VPC to send traffic to the internet. Furthermore, if you deploy servers that require direct inbound access from the internet, then you need to deploy them in subnets that have a direct route to the internet.

As depicted in *Figure 6.9*, you can define which subnetworks have direct access to the public internet, called public subnets (for example, the DMZ) and which subnets are private, called private subnets (for backend operations).

To configure a VPC with internet access, you need to deploy an internet gateway. This is a free component that can be attached to your VPC to grant the VPC internet access. In addition, you would need to configure route tables associated with the public subnet to have direct access to the internet gateway. EC2 instances, which are then deployed in the public subnet, can send traffic to the internet and be configured to receive direct inbound traffic from the internet. An additional requirement, however, is that your EC2 instances in the public subnet must also have a public IP address. This can be dynamically assigned by AWS to the EC2 instances. Note that the public IP address is dynamic, which means that if the EC2 instance is stopped and then restarted, the IP address is likely to change. In most cases, this is acceptable because you generally will place your EC2 instances behind a load balancer (a service we discuss in a later chapter). However, if you need a static public IP address – one that does not change – then you can configure your EC2 instances with an elastic IP address.

Elastic IP addresses are static and remain in your account until you release them. They are ideally suited for those EC2 instances that must always have the same public IP address over time. Elastic IP addresses can also be reassigned from one EC2 instance to another. This can be helpful if, for example, an EC2 instance fails, and you need to spin up a new server as a replacement that must have the same IP address. Elastic IP addresses, therefore, can help you design for the high availability of services by automatically being reassigned to a standby EC2 instance in case of a failure on the primary EC2 instance.

In terms of cost, it is important to remember that elastic IP addresses are free only while they are associated with an EC2 instance that is in the running state. If you have an EC2 instance in the stopped (shutdown) state, you will incur charges for that elastic IP address on an hourly basis. The following diagram illustrates how traffic from the internet can be routed to web servers deployed in public subnets of the VPC. Note that in this case, the web servers would also need public IP addresses.

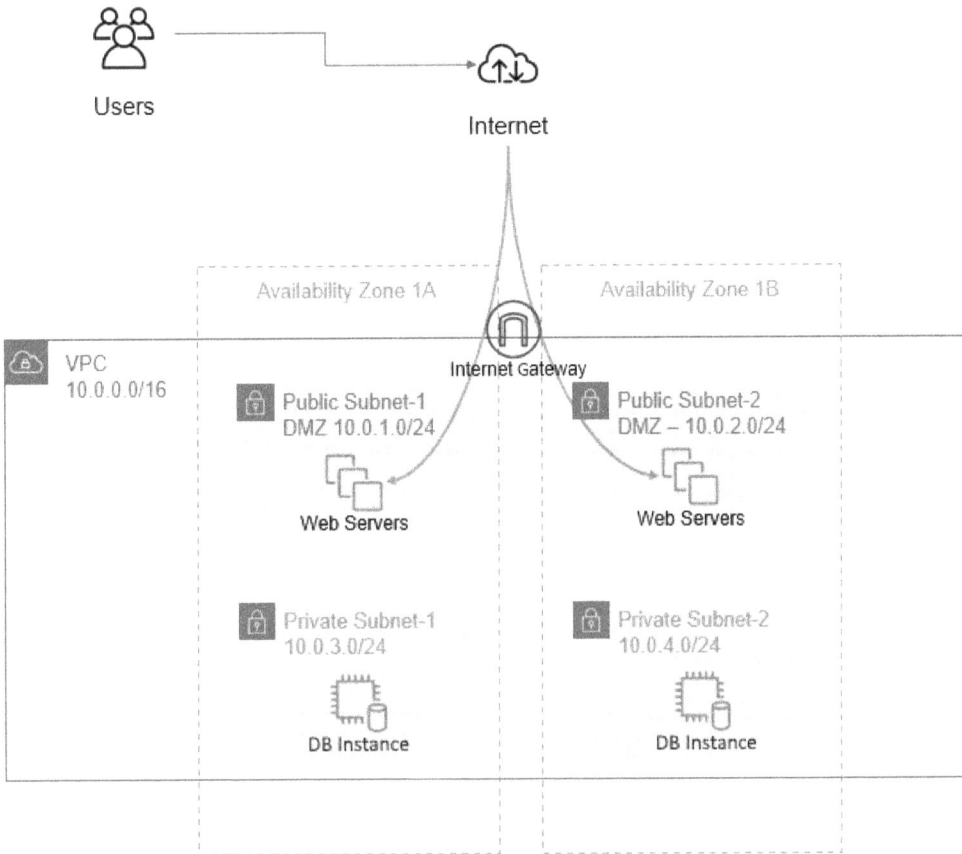

Figure 6.17 – VPCs and public and private subnets

In the preceding diagram, if Availability Zone 1A were to have a major outage, our users can be redirected to the web servers in public `Subnet-2` of Availability Zone 1B and continue operations.

VPC security

One of the primary reasons for building a VPC on AWS is to implement firewall security to ensure that access to the resources deployed within the VPC is carefully managed. For example, some resources, such as web servers, would require inbound access from members of the public on the internet. By contrast, if an application running on those web servers needed to update a backend database (for example, confirmation of customer orders), then you would want to ensure that only the web servers can make those updates to the backend database. You would not want members of the public to have direct access to the backend databases.

Amazon offers two types of security service to enable you to build a highly secure environment, ensuring that only the required level of access and traffic is permitted.

Security groups

A security group is a firewall that is designed to allow you to configure what type of traffic you permit, inbound and outbound, to your EC2 instances. When you launch an EC2 instance, you must assign at least one security group to it, which must contain the necessary rules for any inbound and outbound traffic you wish to allow. Note that you can associate up to five security groups with an instance. Security groups act at the instance level and not the subnet level.

Each VPC comes with a default security group that allows all traffic inbound, *but only where the source of that traffic is the security group itself.* This means that if you have two EC2 instances associated with the same security group, then one EC2 instance will accept traffic initiated by the other EC2 instance. In addition, all outbound traffic to any destination is permitted. This means that an EC2 instance associated with the default security group can initiate communication outbound. This is based on the idea that you can trust your EC2 instances to make any necessary outbound requests, for example, to download operating system updates.

However, security groups do not allow any traffic to inbound from other sources (other than itself) until you create necessary rules. This is to prevent any unsolicited traffic reaching your instance. So, if you are configuring a web server that needs to serve web pages on the standard HTTP protocol, then you would need to configure an appropriate inbound rule for HTTP and specify the source of the traffic in terms of an IP address range. In addition to the default security group, you can also create your own security groups.

An important feature to consider about security groups is that even though you may not have configured any inbound rules, response traffic to any outbound requests will be permitted inbound by the security group. Conversely, if you configured any inbound rules, responses to any inbound traffic that are allowed in because of those rules are permitted outbound regardless of any explicit outbound rules. This feature is what makes security groups **stateful**.

Some additional key features include the following:

- You can configure *allow* rules, but you cannot configure explicit *deny* rules.
- You can specify separate rules for inbound and outbound traffic.
- You can filter traffic based on protocols and port numbers. You can also specify sources and destinations that can be other security groups, thus offering a layered security approach.

Security groups therefore protect your EC2 instances and enable you to define what inbound traffic you will allow to the instances.

Network Access Control Lists (NACLs)

NACLs are another type of firewall service that are designed to protect entire subnets. Rather than offer protection at the instance level, they can be configured to allow or deny traffic from reaching subnets, within which you deploy your EC2 instances.

A default NACL is configured in every VPC, which is configured to allow all inbound and outbound traffic. This is acceptable because traffic must still be permitted via security groups, which, by default, block all inbound connections.

Incidentally, you can configure your subnets with custom NACLs, which, by default, block both inbound and outbound traffic, and additional configuration will be required to create any necessary inbound and outbound rules.

NACLs are also stateless, meaning that you have to configure both the inbound rules and corresponding outbound rules for traffic to flow and be responded to.

Network Address Translation (NAT)

At the start of this chapter, we discussed how a typical corporate network is usually divided into smaller subnetworks. Certain resources will be deployed in the DMZ (called public subnets on AWS) and others will be deployed in backend private subnets. By deploying resources in a private subnet, we ensured that there was no direct connection to them from the public internet. Such resources include database servers or application servers that should be accessible only via frontend web services or load balancers. This ensures that any traffic to those backend services is proxied via those frontend web services and without exposing them on the public internet.

Application servers may, however, require access to the internet, for example, to download product updates or patches. Since these servers will not have a public IP address and will be placed in private subnets, we need a mechanism for them to access the internet and to allow inbound responses to those servers. To fulfill this requirement, we can set up and configure a NAT service. As depicted in *Figure 6.10*, this NAT service needs to be placed in the public subnet of your VPC and have an elastic IP address attached to it. Your route table is then configured with a rule to allow internet-bound traffic from EC2 instances in the private subnets to access the internet via this NAT service. The NAT service relays requests from the EC2 instances in the private subnet to the internet, acting as a proxy.

Ultimately, if you need to enable EC2 instances with private IPv4 address access to the internet without directly exposing them on the internet (for example, with a public IP address), then on AWS, you can configure a NAT gateway to relay internet requests on behalf of those devices.

If you are using IPv6, as discussed earlier in this chapter, and you want to ensure that EC2 instances with IPv6 can access the internet without being directly exposed on the internet, you need to route traffic from those instances via an **egress-only internet gateway**.

VPC peering

A VPC peering connection is a private network connection between two VPCs. The service allows you to connect multiple VPCs so that instances in one VPC can access resources in another VPC. This means that traffic between VPCs over a peering connection does not traverse the public internet and offers greater levels of bandwidth, as well as security. Furthermore, the connections between the peers are highly available and there is no single point of failure.

You can create VPC peering connections between VPCs in one AWS account or between AWS accounts. Peering connections can also be configured between VPCs in the same Region or across different Regions. This means that if you have VPCs across different Regions, for example, to facilitate the deployment of resources closer to your global branch offices, those resources can access each other across Regions.

VPC transit gateway

Earlier, we discussed how VPC peering can help you connect multiple VPCs together. The problem with VPC peering, however, is that every VPC must establish a one-to-one connection with its peer. Once the connections have been established, you also need to configure your route tables in each of the peered VPCs to direct appropriate traffic across the peering connection. If you have multiple VPCs that need to connect to each other, it can become very difficult to manage the individual connections and numerous route table entries.

In the following diagram, you will note the complexity of the various peering connections to connect each VPC:

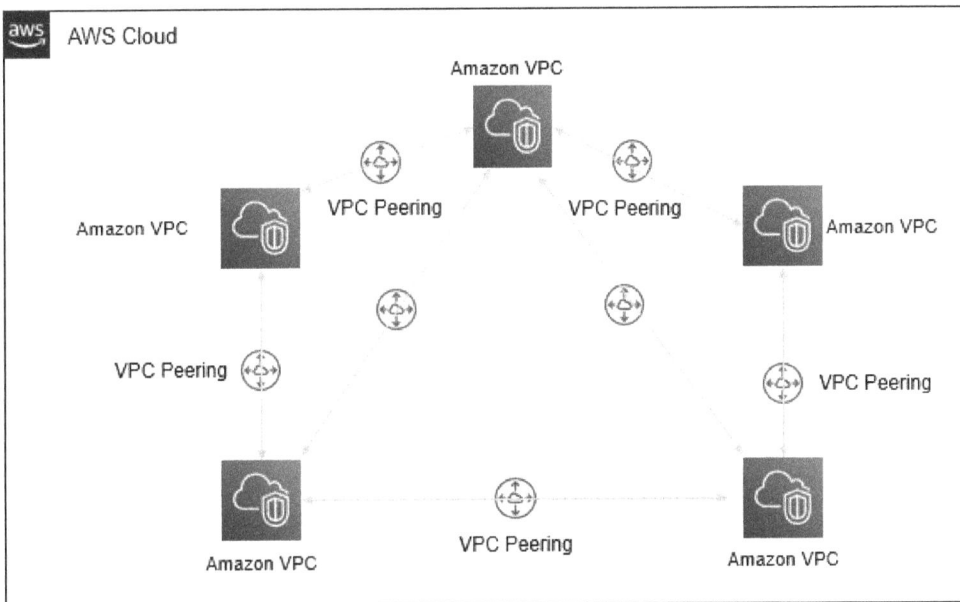

Figure 6.18 – Multiple VPC peering connections

With AWS Transit Gateway, you can connect your individual VPCs together via the gateway in a hub-and-spoke model. This greatly simplifies your network architecture, as each new VPC that is peered to the gateway needs just a single connection to be able to route traffic to the other VPCs, as long as necessary route table configurations permit it to do so.

This simplified model is depicted in the following diagram:

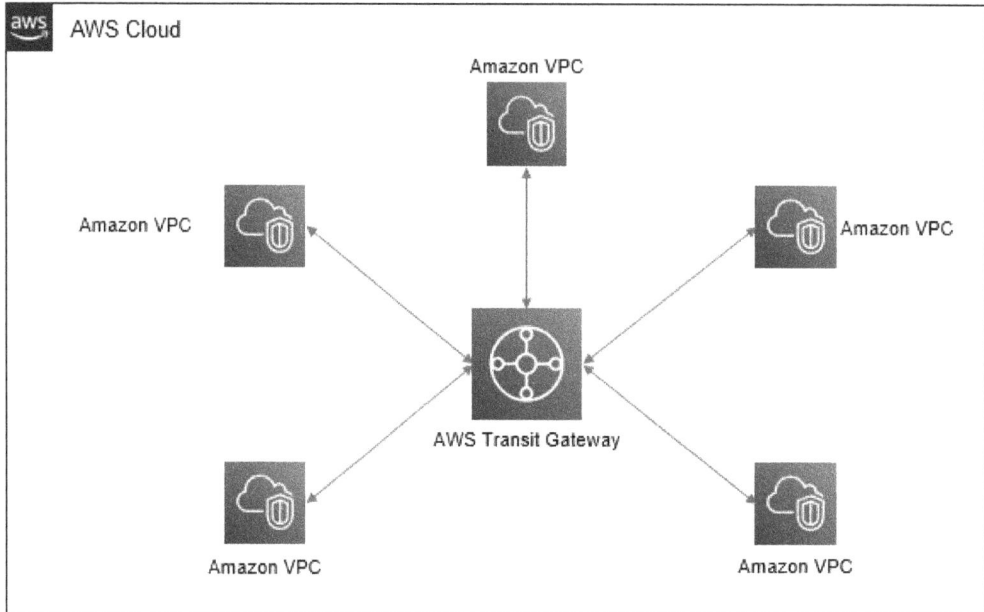

Figure 6.19 – AWS Transit Gateway

Transit gateways help to reduce the architectural design overhead when trying to connect many VPCs together. Transit gateways achieve this design by using a hub-and-spoke architectural model.

Virtual Private Networks (VPNs)

Your VPC is hosted on the AWS platform. You can deploy various resources within your VPC, which you can access via the public internet, if you configure an internet gateway and the necessary security groups and NACLs. However, you can also connect your VPC to your corporate network (on-premises or a co-location data center). This type of connection is known as VPN. A VPN is a secure encrypted site-to-site tunnel established between two endpoints over the public internet. It offers AES 128 or 256-bit **Internet Protocol security (IPsec)** encryption, which means that you can transfer data between the two endpoints securely.

To set up a VPN connection between your on-premises network and the VPC, you need to create a **Virtual Private Gateway** (**VPG**) and attach it to your VPC. You will also need to configure a customer gateway, which is a physical or virtual device located in the on-premises network that connects to the VPG over the internet. The setup is illustrated in *Figure 6.12*.

AWS supports a wide range of customer gateway devices, including Cisco, Juniper, and Check Point devices. Once the VPN connection has been established, you still need to configure your route tables on both the internal network and the VPC to direct appropriate traffic over the tunnel, as required.

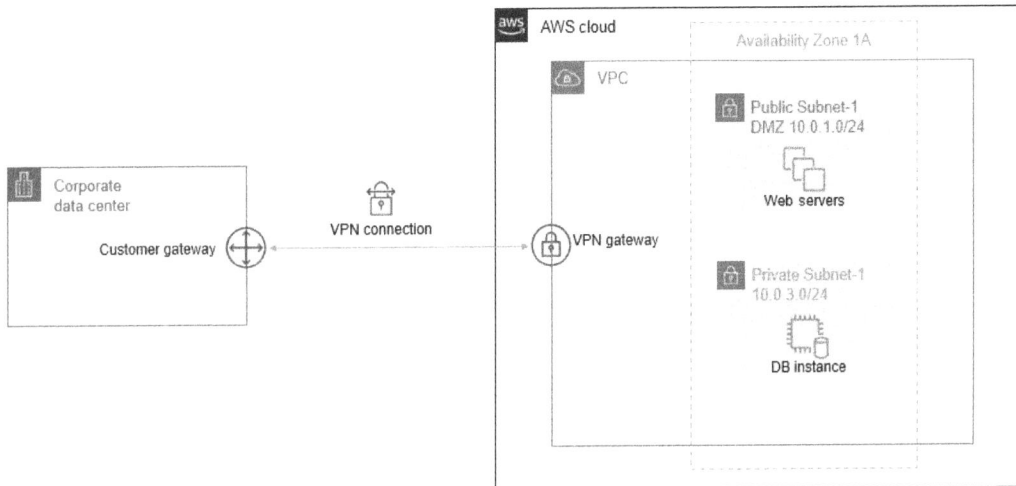

Figure 6.20 – Amazon VPNs

While VPN connections enable you to encrypt traffic between the local network and the VPC, as depicted by the diagram, the bandwidth is limited to what your internet service provider offers. Furthermore, on AWS, there is a hard limit of 1.26 Gbps. With a VPN connection you are also dependent on the routing mechanism of the public internet. If you require a more dedicated connection that bypasses the public internet altogether, then Amazon offers a service called Direct Connect, which we will look at next.

Direct Connect

Direct Connect is a service that enables you to connect your corporate data center to your VPC and the public services offered by AWS via a dedicated private connection that bypasses the internet altogether.

This is a more expensive solution than standard VPN setups due to the requirement to lay down necessary fiber optic cabling between your corporate office (or data center) and a local Amazon Direct Connect partner. Nevertheless, the service can offer dedicated links that operate at 1 Gbps, 10 Gbps, and up to 100 Gbps. Direct Connect thus gives you ultra-high-speed connectivity as well as a secure private connection.

In this section, we examined the core features and components of Amazon VPCs. VPCs enable you to build secure, isolated network areas within the AWS cloud that can be configured to allow only authorized connections into them. VPCs offer two types of firewall solutions – security groups and NACLs. The former can be configured to allow only specific types of traffic and acts as a firewall at the instance level, whereas the latter protects the entire subnet. Security groups are stateful, whereas NACLs are stateless; however, NACLs also provide the ability to deny specific types of traffic from specific sources. For example, if you host a web server and want to allow public access to the server from the internet, but also want to block access from a certain IP range (perhaps because you have discovered a potential attack from that range), NACLs can help you configure necessary deny rules as well.

You can use both NACLs and security groups to build a layered, secure architecture. NACLs will protect your entire subnets by letting you explicitly define the type of traffic that can enter the subnet. Next, you define security groups for different groups of EC2 instances that may reside in a subnet, ensuring that only the necessary traffic is permitted to connect to the application running on those EC2 instances.

We then examined features such as VPC peering, AWS Transit Gateway, VPN connections, and the Direct Connect service. In the next section, we will learn about the AWS global DNS, Amazon Route53. AWS Route53 enables you to direct traffic to resources both in your VPC and on the public internet using DNS name resolution, but also helps you shape traffic for high availability and fault tolerance.

Learning about DNS and global routing with Amazon Route53

AWS Route53 is Amazon's global DNS, which is a service to help translate human-readable names into an IP address because, ultimately, computers connect to each other over IP addresses. DNS servers across the internet host billions of such name-to-IP address records, among other types of records. When you use your favorite browser to visit a particular website, such as `example.com`, your browser sends a request to your local DNS service provider, which, if necessary, refers the query to that domain's (`example.com`) authoritative DNS server. This authoritative DNS server responds with the IP address of the website you are trying to access, and your browser is then able to establish a connection with the website (in this case, `example.com`). This process of translating domain names to IP addresses is called name resolution.

Amazon Route53 offers three primary functions:

- Domain registration
- DNS routing
- Health checks

Let's look at each of the functions next.

Domain registration

Before you can use a domain name for your website or web application that members of the public can access, you need to register a name of your choice with a domain registrar. Amazon Route53 offers complete domain name registration services. When you choose a name to register, you do so under a **Top-Level Domain** (**TLD**) such as .com, .co.uk, .org, or .net. If the name of your choice under a particular TLD is not available, you can try a different TLD. For example, if example.com is not available because someone else has already registered it, then you can try example.co.uk or example.net. There are hundreds of TLD names to choose from.

As part of the registration, you purchase a lease duration for your domain name, which can be from anywhere between 1 year and 10 years. You need to renew this lease before expiry, or you risk someone else registering the name, which will affect your ability to offer your web services using the same domain name.

Once you have registered your domain name, you then need to host it with a DNS hosting service provider. Often, the registration and hosting service is offered by the same company, in this case, Amazon Route53. The hosting service offers you the ability to create resource records for your domain, for example, configuring the domain name to point to an IP address of a web server. When users on the internet type the domain name in their browser, they are directed to your web server, which hosts your website.

There are several types of resource records you can create. Some types of resource records are displayed in the following table:

Name	Type	Value	Description
Iaasacademy.com	A - IPv4 Address	92.204.222.124	Points to server or load balancer hosting the website
Iaasacademy.com	MX-Mail exchange	iaasacademy-com.mail.protection.outlook.com	Points to email server/ service provider
Portal.Iaasacademy.com	CNAME	iaasacademy.com	Canonical name to map one domain name to another

Table 6.1 – Example resource records for iaasacademy.com

When configuring your domain names, you also need to define a zone file, which will host resource records for your domain name. It is in a zone file that you would create resource records, such as the A record, which are standard name-to-IP address resolution records. Next, we look at how zone files are created on AWS Route53.

Hosted zones

To create the necessary resource records for your domain name, you need to set up a hosted zone on Route53. A hosted zone is a container that is used to store and manage your resource records and allows you to define how traffic is routed for your domain (`example.com`) and any sub-domains (such as `portal.example.com`).

There are two types of hosted zones you can set up on Route53:

- **Public hosted zone** – This is a container that allows you to define how you want to route traffic for your domain name across the public internet. For example, when you host an S3 static website (as discussed in *Chapter 5*, *Amazon Simple Storage Service (S3)*), you configure a website endpoint name such as `http://bucket-name.s3-website-Region.amazonaws.com`. You could provide this S3 website endpoint URL to your users to access your web pages that are hosted on the S3 bucket. However, a much better option would be to create an alias record that points your company domain name (`example.com`) to the S3 website endpoint. Your users would find it much easier to remember your domain name than the original S3 endpoint name.

- **Private hosted zone** – This is a container that allows you to define how you want to route traffic for your domain name across private networks, such as one or more of your VPCs. Route53 will route traffic for the domain name to resources configured in your VPCs based on the resource records you create. For example, you could use a friendly domain name such as `devserver.example.com` that points to an EC2 instance within your VPCs, which other resources in the VPC may need to connect to. This is a much better option than having to remember IP addresses of your EC2 instances and means you do not need to hardcode those IP addresses in your application code, which can become difficult to manage.

DNS hostnames

AWS provides a DNS server (Amazon Route53 Resolver) for your VPC. This enables AWS to configure DNS hostnames for the instances you deploy in your VPC. DNS hostnames comprise a hostname and a domain name, such as `myserver.mycompany.com`. DNS hostnames enable you to create unique names for your EC2 instances. In addition, AWS provides two types of DNS hostnames:

- **Private DNS hostnames** – These resolve to the private IPv4 address of the instance. The DNS hostname can take one of the following two forms, depending on the Region:

  `ip-private-ipv4-address.ec2.internal for the us-east-1 Region`

  `ip-private-ipv4-address.region.compute.internal for other Region`

  > **Note**
  > `private-ipv4-address` is the reverse lookup IP address.

- **Public DNS hostnames** – These resolve a public DNS hostname to the public IPv4 address of the instance outside the network of the instance, and to the private IPv4 address of the instance from within the network of the instance. The DNS hostname can take one of the following two forms, depending on the Region:

  `ec2-public-ipv4-address.compute-1.amazonaws.com for the us-east-1`

  `ec2-public-ipv4-address.region.compute.amazonaws.com for other Regions.`

While Route53 offers a fully functional DNS, it also enables you to design how traffic is routed across resources placed within a Region and across Regions. Next, we look at the different routing policies offered by Amazon Route53.

Routing policies

In addition to standard name resolution services, such as pointing your domain name to a specific IP address, Amazon Route53 also offers several complex routing policies and configurations. These routing policies enable you to define various rules that offer the ability to build highly available solutions or redirect customers to resources that are closer to their location and thus reduce latency.

Let's examine the various routing policies available with Amazon Route53:

- **Simple routing policy**: The most basic and default routing policy. This resource record enables you to map a domain name to a single resource, such as an IP address of a web server or a DNS endpoint of an elastic load balancer. The policy engine will not check whether the resource is functioning and available.

- **Failover routing policy**: To offer high availability, you can host two copies of your resources ideally across different Regions. One set of resources will be designated as your primary resource and the other as a secondary resource. Route53 performs health checks (discussed later) to determine whether the primary resource is available and in the event of a failure to connect to the primary resource, Route53 will redirect all traffic to the secondary resource. This failover routing policy enables you to design your solution with an active-passive failover.

- **Geolocation routing policy**: This routing policy enables you to route traffic based on the geographical location of your users – such as continent, country, or state (in the United States). Your users' location is determined from the source of the DNS queries for your web service. This routing policy is particularly useful when you need to ensure that your content is accessible only in locations where you have distribution rights. For example, if your end users are based in Europe, you may wish to direct them to a copy of your resources deployed in the Frankfurt Region.

- **Latency routing policy**: This routing policy is particularly useful when you have resources deployed in multiple Regions and you wish to route your users' traffic to the Region that offers the lowest latency.

- **Weighted routing policy**: This routing policy enables you to route different ratios of your total traffic to different resources associated with a single domain. For example, you can choose to route three-quarters of your traffic to one copy of your resource and the remaining one-quarter to another. This is also particularly useful when you want to perform a gradual migration of your total traffic from one resource to another, such as a new version of your website. You can gradually alter the weights to migrate your total traffic over a short period of time from the old version of your website to the new version.

When creating certain routing policies, such as the failover routing policy, you also need to define health checks so that Route53 can determine whether a failover to the secondary site is required. We look at health checks in detail next.

Health checks

You can use Route53 to perform health checks against your resources, such as web servers, elastic load balancers, and S3 static websites. Depending on the results of your checks, you can then take appropriate actions and redirect traffic as necessary to ensure the high availability of services. There are three types of health checks you can perform:

- **Health checks that monitor an endpoint**: This monitors an endpoint you specify, such as an IP address or domain name.

- **Health checks that monitor other health checks**: You can choose to monitor multiple resources and determine the overall health of your collection of resources based on some minimum number of resources that are healthy. If the number of available resources drops below a specified threshold, Route53 health checks can then take appropriate action.

- **Health checks that monitor CloudWatch alarms**: You can create CloudWatch alarms that monitor your resources' metrics, for example, the number of healthy EC2 instances behind an elastic load balancer. You can then configure Route53 to monitor the same data stream that CloudWatch monitors for the alarms and will mark the resource as unhealthy if the threshold values are crossed. Route53 does not wait for the CloudWatch alarm to go into the ALARM state either, as it independently determines the health based on the metric data.

In addition to routing policies and health checks, you can also build complex routing rules, and for this, we look at traffic flow and traffic policies next.

Traffic flow and traffic policies

You may need to create complex routing of traffic using a combination of resource records and Route53 routing policies. For example, you may create configuration in which you set up latency routing policies across several Regions, which then reference weighted records in each Region for your resources. Each configuration is known as a **traffic policy**, and you use a visual editor to help you build your routing architecture, which Route53 uses to design your traffic flow. Using the visual editor, you can also create multiple versions of your traffic policy so that you can quickly adapt to changes. Note that you can only use a traffic flow to create records in public hosted zones:

- **Geoproximity routing policy**: In addition to the standard Route53 routing policies, you can also configure geoproximity routing policies (available only when you use Route53 Traffic Flow). This routing policy enables you to route traffic based on the location of your resources. You can also shift traffic from resources in one location to resources in another location using a *bias* value.

- **Route53 Resolver**: Route53 also enables you to perform DNS resolution between Route53 Resolver and DNS resolvers on your network by configuring forwarding rules. In this case, *your network* can comprise your VPC, other peered VPCs, or even your on-premises corporate data centers connected to AWS either via a VPN or AWS Direct Connect.

In this section, we learned about the Amazon DNS service offered by Route53. Route53 enables you to register domain names, create public and private hosted zones to manage your resource records, define routing policies for different use cases, and perform health checks against your resources. Route53 also enables you to build complex routing rules using multiple resource records, along with traffic flow and traffic policies.

In the next section, we look at AWS CloudFront, which is Amazon's **Content Delivery Network** (**CDN**) solution. CDNs are often used to distribute content globally from a source location and, with the help of caching services, offer reliable, low-latency access to your content.

Implementing a robust CDN with Amazon CloudFront

Amazon CloudFront is a CDN that helps you to distribute your static and dynamic digital content globally with low-latency connections. AWS CloudFront uses AWS edge locations and regional edge caches to cache content closer to your end users' locations. This means that you can host your content in one specific Region and a user who attempts to access it from another Region will retrieve the content via the edge location over the AWS backbone network. Furthermore, as content is retrieved, it is cached at a local edge location closer to the user for a period (known as a **time-to-live** or **TTL**), further improving network latency in subsequent requests for the same content.

Figure 6.21 – A typical CloudFront distribution

To configure Amazon CloudFront, you create a distribution endpoint that defines the types of content you want to serve and the source of that content. The source can be an S3 bucket, an S3 bucket configured as a static website, an Amazon EC2 instance, or an elastic load balancer, among others. As part of your configuration, you get a CloudFront URL, which you provide to your users to access your content via CloudFront. You can also use custom URLs, which allows you to define your distribution with a company-branded domain name.

When you configure your CloudFront distributions, you can choose to serve your content over HTTP and HTTPS. Some examples of types of content that can be served include the following:

- Static and dynamic content, for example, HTML, CSS, JavaScript, and images
- Video on demand in different formats, for example, Apple **HTTP Live Streaming (HLS)** and Microsoft Smooth Streaming
- Live events and conferences

CloudFront ultimately enables you to architect a CDN for your web application. Next, we look at how CloudFront is priced.

Choosing a price class for your CloudFront distribution

Amazon charges you for distributing your content via its edge locations. Rather than charge you on a per-edge location basis, however, Amazon have clubbed several edge locations across Regions into three specific price classes. The most expensive price class is where your content is accessible via all edge locations globally. This also happens to be the default price class when you create your distribution, but ultimately offers the best performance. If you are looking to reduce cost and if you know that the consumers of your content are from specific Regions, you can choose two alternative price classes:

- United States; Canada; Europe; Hong Kong, Philippines, South Korea, Taiwan, Singapore; Japan; India; South Africa; and Middle East regions – This class excludes the most expensive regions.
- United States, Canada, and Europe regions – This class is the least expensive.

In this section, we examined the services offered by the AWS CloudFront service and how it can be used to distribute your content globally over AWS backbone networks, offering lower latency and better performance.

In the next section, we provide a brief introduction to Amazon API Gateway, which allows you to architect serverless applications by creating, publishing, and managing REST, HTTP, and WebSocket APIs.

Introduction to Amazon API Gateway

Amazon API Gateway helps you design application solutions that favor the microservices architecture in place of monolith designs. Your backend developers can build a series of microservices that work with each other. For example, in an e-commerce application, you can have several microservices, such as cart-service, catalog-service, user-profile and user-session services, inventory-management-service, and more.

Without an API gateway, your frontend developer (who builds the frontend user interface) would need to be made aware of all the backend APIs and build the application to call several microservices, to provide complete functionality. Imagine, then, your backend developer later needs to refactor one of the microservices, such as splitting one microservice into two separate microservices, each with its own API. This would result in having to recode some components of the frontend user interface too.

With an API gateway, you essentially create an abstraction layer. This API gateway can be used to expose all the APIs that need to be made available to external clients to call backend services. Requests from those clients can then be routed to the various backend microservices. As per the following diagram, Amazon API Gateway acts as a "front door" for your applications to access backend data, Lambda functions, databases, and more. It handles all the incoming traffic and is capable of processing thousands of concurrent API calls.

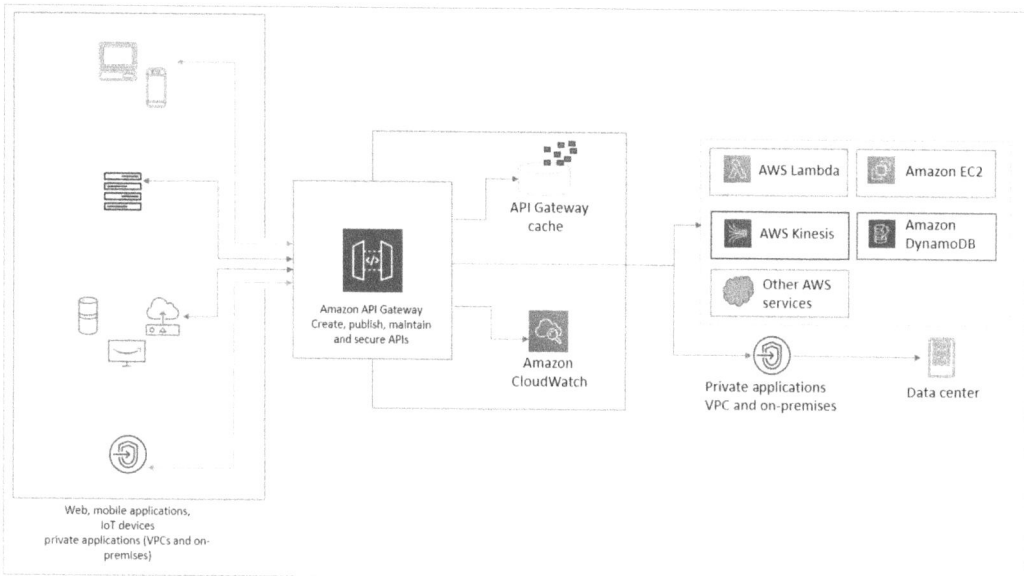

Figure 6.22 – Amazon API Gateway

Amazon API Gateway offers features that also help you protect your backend services, such as authorization, access control, and monitoring, and help protect backend resources from **Distributed Denial of Service (DDoS)** attacks.

In terms of designing your software architecture, adding this abstraction layer creates an additional hop for your clients to access backend resources. To improve application responsiveness, Amazon API Gateway offers features to optimize your APIs such as response caching and payload compression.

In this section, we provided a brief introduction to Amazon API Gateway, which is a fully managed service that enables you to create, publish, maintain, monitor, and secure APIs.

In the next section, you have the opportunity to build out your own Amazon VPC as part of the exercises in this chapter.

Exercise 6.1 – setting up a public subnet VPC

In this exercise, you will create your own custom VPC that will contain a public subnet. In later chapters, you will expand this VPC to add private subnets for different use case:

1. Log in to your AWS account as the IAM user `Alice` you created in *Chapter 4, Identity and Access Management*.

2. On the main AWS Management Console, search for `VPC` in the search box.

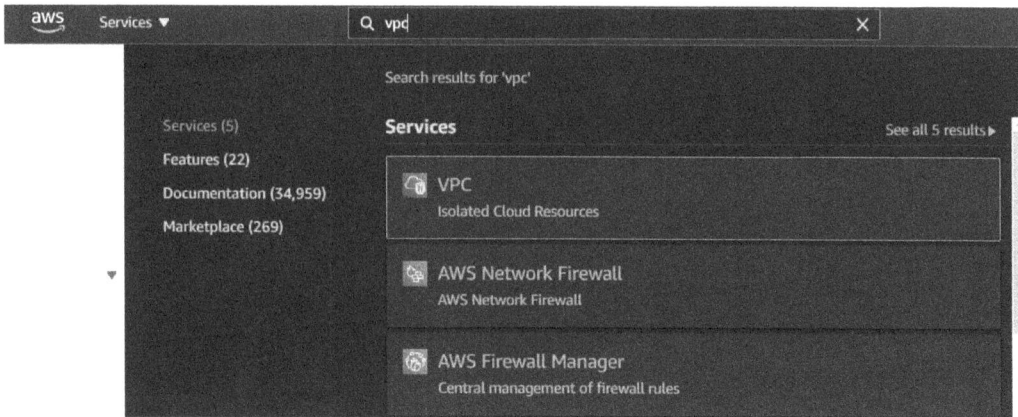

Figure 6.23 – Configuring a new VPC

3. Select **VPC** from the filtered list.

4. VPCs are Region-specific, so make sure you select the US-East-1 Region from the top right-hand corner of the screen.

5. On the main VPC console screen, click on the **Launch VPC Wizard** button. This will launch the VPC wizard.

6. Next, select the first option, **VPC with a Single Public Subnet.**

7. In *Step 2*, provide the following details for your VPC.

    For **IPv4 CIDR block**, enter the IP block of 10.0.0.0/16. This IP block represents your VPC network.

    For **VPC name**, enter the name ProductionVPC.

    The wizard gets you to create a single public subnet. Later, you will expand this VPC for future use, but for now, set the **Public subnet's IPv4 CIDR** field to 10.0.1.0/24.

    For **Availability Zone**, select **us-east-1a** from the drop-down list. Remember that while VPCs span the entire Region, each subnet you create spans a single Availability Zone. In this case, we are creating a single public subnet in **us-east-1a.**

    Next, rename the subnet name to **Public Subnet One**.

    Next, click the **Create VPC** button in the bottom right-hand corner of the screen. Review the following screenshot for the preceding steps:

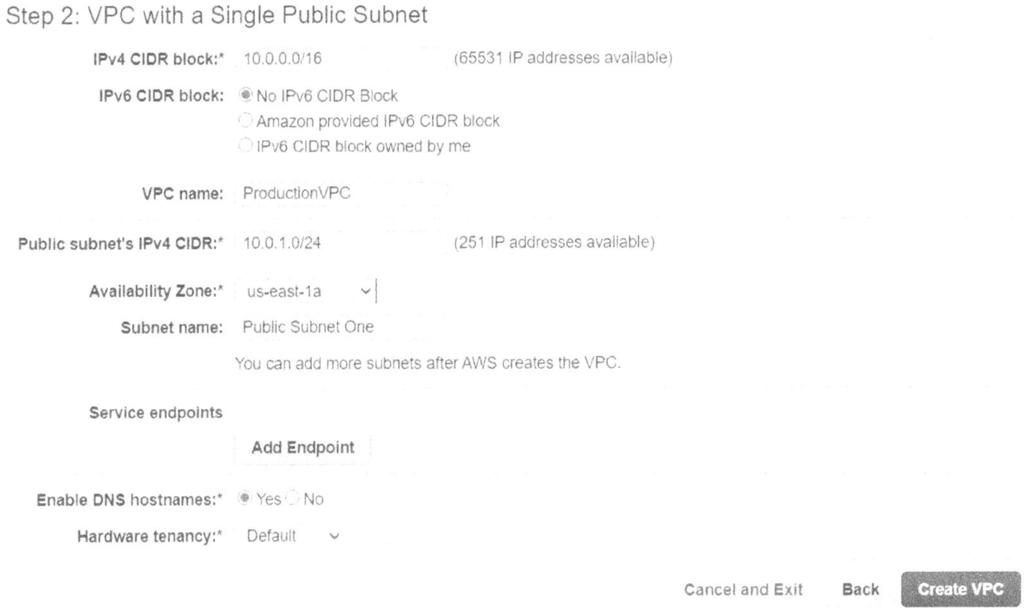

Step 2: VPC with a Single Public Subnet

IPv4 CIDR block:*	10.0.0.0/16	(65531 IP addresses available)
IPv6 CIDR block:	● No IPv6 CIDR Block	
	○ Amazon provided IPv6 CIDR block	
	○ IPv6 CIDR block owned by me	
VPC name:	ProductionVPC	
Public subnet's IPv4 CIDR:*	10.0.1.0/24	(251 IP addresses available)
Availability Zone:*	us-east-1a ∨	
Subnet name:	Public Subnet One	
	You can add more subnets after AWS creates the VPC.	
Service endpoints		
	Add Endpoint	
Enable DNS hostnames:*	● Yes ○ No	
Hardware tenancy:*	Default ∨	

Cancel and Exit   Back   **Create VPC**

Figure 6.24 – VPC with public subnet configuration page

8. The wizard runs through the parameters you specified and creates your first VPC.

9. Click **OK** on the **VPC Successfully Created** status page.

10. In the list of VPCs, you will note that your newly created **ProductionVPC** has been successfully created:

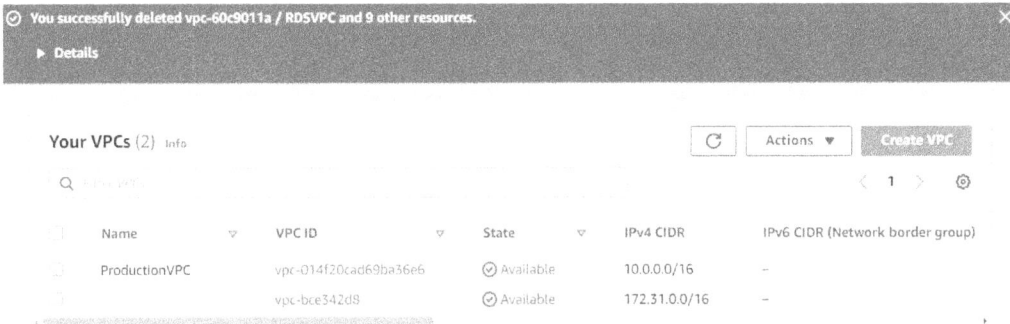

> ✓ You successfully deleted vpc-60c9011a / RDSVPC and 9 other resources.   ✕
> ▶ Details

	Name	▽	VPC ID	▽	State	▽	IPv4 CIDR	IPv6 CIDR (Network border group)
	ProductionVPC		vpc-014f20cad69ba36e6		✓ Available		10.0.0.0/16	–
			vpc-bce342d8		✓ Available		172.31.0.0/16	–

Figure 6.25 – Newly created ProductionVPC

In this section, we demonstrated how to create your first public subnet VPC using the AWS VPC wizard. Later in the training guide, you will expand the VPC to include private subnets and deploy servers within our VPCs.

Summary

This chapter examined Amazon's core network services, comprising VPCs, Amazon's DNS service offering with Route53, and Amazon's CDN solution with Amazon CloudFront.

In this chapter, you learned how computers communicate with each other via IP addresses. You also learned that using IP addressing and CIDR block ranges, you can build isolated networks and subnetworks. We then discussed how you can build the same network architecture in the cloud using the Amazon VPC service. As part of setting up your VPC, you examined the use cases to build private and public subnets and explored tools for defining security rules and limiting the types of traffic that can enter and leave your VPC. We also looked at additional VPC services that enable you to interconnect multiple VPCs using VPC peering and how to build complex connections more easily across VPCs using AWS Transit Gateway.

Next, we learned about the AWS Route53 service, a DNS offering that provides domain name registration, a route policy configuration service, and health checks. You learned that by using a routing policy, you can customize your solutions for a wide range of use cases. These include offering an active-passive failover service with the Route53 latency routing policy, and the ability to slowly migrate your users from one resource to another using the weighted routing policy.

Finally, you learned about Amazon CloudFront and how to use AWS's edge locations to distribute your digital assets and content globally from a single Region over a high-speed low-latency connection.

In the next chapter, we will learn about the **Elastic Cloud Compute (EC2)** service and how you can deploy compute services such as virtual servers running Linux or Windows operating systems in the cloud.

Questions

1. Which VPC component enables to you grant internet access to servers in the public subnet deployed in the VPC?

    A. NAT gateway

    B. Internet gateway

    C. VPC peering

    D. Security group

2. Which of the following statements are true?

    A. NACLs protect entire subnets, whereas security groups protect the individual instance.

    B. NACLs protect the individual instance, whereas security groups protect the entire subnet.

    C. NACLs enable instances in the private subnet to access the internet and act as a NAT device, whereas security groups are used to assign IAM policies to servers that need access to S3 buckets.

    D. NACLs enable instances in the private subnet to access the internet and act as a NAT device, whereas security groups are used to assign IAM policies to servers that need access to S3 buckets.

3. Which AWS service enables you to purchase and register new domain names that can be used to publish your website on the internet?

    A. Route53

    B. VPC

    C. RDS

    D. Elastic Beanstalk

4. Which AWS service enables you to distribute your digital assets such that it is cached locally to users who attempt to access this content for a time to live, and thus helps to reduce network latency?

    A. AWS CloudFront

    B. AWS CloudTrail

    C. AWS CloudWatch

    D. AWS CloudScape

5. Your organization hosts multiple AWS accounts with multiple VPCs. You would like to connect these VPCs together and centrally manage connectivity policies. Which AWS service enables you to connect multiple VPCs configured as a hub that controls how traffic is routed among all the connected networks, which act like spokes?

    A. AWS Transit Gateway

    B. AWS Global Accelerator

    C. AWS VPC Peering

    D. AWS Virtual Private Gateway

6. Which AWS service enables you to grant internet access to EC2 instances configured with IPv4, and located in the private subnet of your VPC?

    A. Egress-only internet gateway

    B. NAT gateway

    C. VPC endpoint

    D. VPN tunnel

7. You company has a primary production website in the US and a DR site in Sydney. You need to configure DNS such that if your primary site becomes unavailable, you can fail DNS over to the secondary site. Which DNS routing policy can you configure to achieve this?

    A. Weighted Routing

    B. Geolocation Routing

    C. Latency Routing

    D. Failover Routing

8. You plan to set up DNS failover using Amazon Route53. Which feature of Route53 can you use to test your web application's availability and reachability?

   A. Private DNS

   B. CloudWatch

   C. Health checks

   D. DNS ping

9. Which VPC firewall solution enables you to deny inbound traffic from a specific IP address, which can be used to prevent malicious attacks?

   A. AWS Firewall

   B. AWS Security Groups

   C. AWS **Network Access Control Lists (NACLs)**

   D. AWS CloudFront

10. Which AWS service enables you to connect your private data center to your Amazon VPC with up to 100 Gbps network connectivity?

   A. Snowball

   B. Direct Connect

   C. **Virtual Private Network (VPN)**

   D. **Virtual Satellite Network (VSN)**

7
AWS Compute Services

In the old days, traditional on-premises environments consisted of a wide range of *servers* that used to host a varied gamut of applications, websites, and other services. These *servers* were physical hardware such as your home desktop computer but were designed with more robust components (such as CPU, memory and storage) to host applications for end user consumption.

A **server** is a term given to an application running on computer hardware that grants access to a set of services, either to other applications or end users. A physical computer can host more than one "server" offering – for example, a computer may offer email services and web services on the same physical hardware.

However, traditionally, you were limited in the number of servers you could configure your physical computer to host. This was primarily because of limited memory, storage, and, more importantly, the fact that applications would conflict with one another when accessing the underlying resources. These resources could be shared libraries and runtime environments, as well as access to physical hardware such as memory, storage, and so on.

With the advent of virtualization, it then became possible to configure **Virtual Machines** (**VMs**) on a single physical host – each VM would offer one or more services and usually, these services were related to each other to avoid any conflicts. So, for example, you could have one virtual server offering email services and another offering file-sharing services.

Amazon Web Services (**AWS**) offers VMs in the cloud, which we call **Elastic Compute Cloud** or **EC2**. In this chapter, we will look at the core offering of the EC2 services, which allows us to host applications, websites, and other compute-intensive processing services on AWS. This chapter will also cover other forms of compute offerings, including a lightweight alternative to EC2, known as Lightsail, containers, and serverless offerings such as Lambda.

The following topics will be covered in this chapter:

- Introduction to **Elastic Compute Service** (**Amazon EC2**)
- Learning about Amazon **Elastic Block Storage** (**EBS**) and instance backed store
- Learning about EC2 pricing options
- Implementing Shared File Storage with **Amazon Elastic File System** (**Amazon EFS**)
- Introduction to **Virtual Private Servers** (**VPS**) with Amazon Lightsail
- Introduction to Amazon **Elastic Container Services** (ECS) and Kubernetes
- Overview of additional storage options in AWS

Introduction to Amazon EC2

Amazon EC2 is one of AWS's flagship offerings and allows you to launch and set up virtual servers in the cloud. These are very similar to setting up and configuring **VMs** in your on-premises environment, which you would configure on a hypervisor such as VMware or Hyper-V.

A hypervisor, as discussed in *Chapter 1, What Is Cloud Computing?*, is a piece of software that allows you to create virtual resources such as virtual servers. Depending on the capacity of the underlying hardware, you can then host multiple virtual servers on the same physical hardware. These virtual servers are granted access to actual physical hardware via the hypervisor, which carves out virtualized representations of the physical hardware components (CPU, memory, storage, and so on) into smaller virtual components, that are then presented to your virtual servers. You can access hypervisor software such as VMware and Hyper-V to create your virtual servers, selecting the virtual components and configuring them as required. Each virtual server, otherwise known as a VM, can then have an operating system installed and configured, along with any required applications.

Amazon EC2, however, abstracts the underlying hypervisor layer from the customer, so you do not have direct access to the hypervisor itself. Using a self-service portal or API access, you can launch virtual servers, which we term as **EC2 instances**, in your AWS account and specifically, within your **Virtual Private Cloud** (**VPC**) (discussed in the previous chapter).

Traditionally, on a VMware or Hyper-V management console, you would spin up a virtual server by specifying the exact amount of CPU and memory to configure your virtual server with.

With AWS, you choose from a wide selection of available EC2 instance types and families. An EC2 instance family refers to the generic use case of the EC2 instance types contained within the family. For example, the *general-purpose* family is designed for handling workloads that require a balance of compute, memory, and networking resources. Within this *general-purpose* family, you have several EC2 instance types that refer to the underlying hardware of the host computer used for your instance. These EC2 instance types offer different combinations of compute, memory, and storage capabilities. So, for example, within the *general-purpose* family, you have instance types such as **M5** and **T2**. The **T2** instance type runs on Intel Xeon processors, offers a burstable CPU, and is designed to offer a balance of compute, memory, and network resources. The **M5** instance type runs on Intel Xeon® Platinum 8175M processors and offers up to 384 GiB of memory and up to 25 Gbps network bandwidth using Enhanced Networking.

Within each EC2 instance type, you also have instance sizes. When selecting an EC2 instance for your application, you need to identify the family, the type, and select the size of the instance. For example, within the **T2** instance type, you have several sizes, such as `t2.micro` and `t2.xlarge`. The `t.2micro` instance size, only comes with 1 GiB of memory and 1 vCPU, whereas `t2.xlarge` comes with 16 GiB of memory and 4 vCPUs.

We will discuss the EC2 instance families, types, and sizes in more detail shortly. Suffice to say that you have access to a very large selection of EC2 instance types and sizes to meet the requirements of your applications.

In addition to your EC2 instance type, which defines CPU, memory, and other hardware-related components, you also need to attach a block storage device such as an EBS volume or an instance-store volume, as discussed in the *Learning about Amazon Elastic Block Storage and instance backed store* section in this chapter. At the very least, you will need one EBS volume to host the operating system and make the server bootable into a fully functioning server. We'll review different types of block storage for EC2 instances later in this chapter.

In addition to specifying the instance type and family, you also need to choose a particular **Amazon Machine Image** (**AMI**) to configure your EC2 instance with. AMIs are snapshots that contain base operating systems, utilities, and any applications you want to configure your EC2 instance with. We'll look at AMIs in more detail shortly.

An important point to note here is that EC2 instances are Region-specific. More appropriately, EC2 instances are tied into the Availability Zone you launch them in. When you choose to launch a new EC2 instance, you must specify the subnet of a VPC in which to launch that EC2 instance. The subnet will be tied to a given Availability Zone, as discussed in the previous chapter. So, ultimately, you dictate which Availability Zone the EC2 instances gets launched into.

The following diagram shows a high-level infrastructure architecture and the key features that support the Amazon EC2 service:

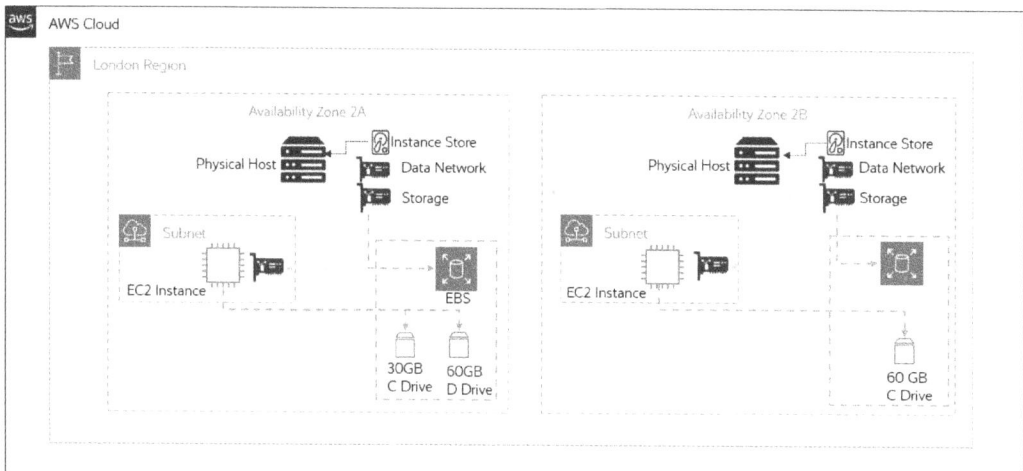

Figure 7.1 – EC2 instance components

Some key features in the preceding diagram of EC2 instances include the following:

- EC2 instances are VMs built out of the physical host located in the same Availability Zone.

- With network and storage virtualization technologies, each EC2 instance is allocated at least one elastic network interface (virtual network card or virtual **network interface card** (**NIC**)) for data. You can have additional virtual NICs for configuring multi-homed devices.

- An EC2 instance can be configured to have a portion of the instance store volume (attached locally to the host that it runs on) or EBS volumes, which are attached to a storage array in the same Availability Zone.

As shown in the preceding diagram, the EC2 instance that was deployed in **Availability Zone 2A** has a single elastic network interface and two EBS volumes attached, one as the root volume (C Drive on Windows Servers) that's 30 GB in size and another as the data volume that's 60 GB in size. We will discuss block storage volumes later in this chapter.

In the next section, we will look at AMIs. AMIs are VM images that are *like* your on-premises **virtual hard disk** (**VHD**) images. These VHDs are disk image file formats that contain the contents of a hard drive, including the operating system files as well as any applications that you wish to configure a server with. AMIs are much more than simple VHDs however, as they are also comprised of snapshots, permissions, and mappings that specify the volumes to attach to the instance when it's launched. On AWS, when you launch an instance, you need to select an AMI to configure that instance with. For example, an AMI can contain a Linux operating system image and any additional application to configure an instance as a web server.

Amazon Machine Images (AMIs)

Traditionally, when you are launching a new VM on your on-premises VMware or Hyper-V host, you would specify a particular operating system image (such as an .iso file). This would contain the operating system files necessary to install your VM. You may have also modified the machine image with any additional applications and/or utilities.

AMIs are the AWS equivalent of VM images, containing the baseline operating system and any additional applications.

AMIs are Amazon EBS snapshots or a template of the root volume (for instance, store-backed AMIs). These snapshots or templates contain the operating system and any necessary applications. In addition, AMIs contain information on block device mappings that specify the volumes to attach to the instance when it is launched.

Amazon offers several pre-configured AMIs as part of its **Quick Start AMIs** and these include standard operating systems, such as the official versions of various Linux distributions, including **Red Hat Enterprise Linux** (**RHEL**), Ubuntu, SUSE, and even Amazon's flavor called Amazon Linux. In addition, you also have access to several editions of Microsoft's Windows Server operating system and even macOS, which runs on a physical Apple Mac Mini in the backend:

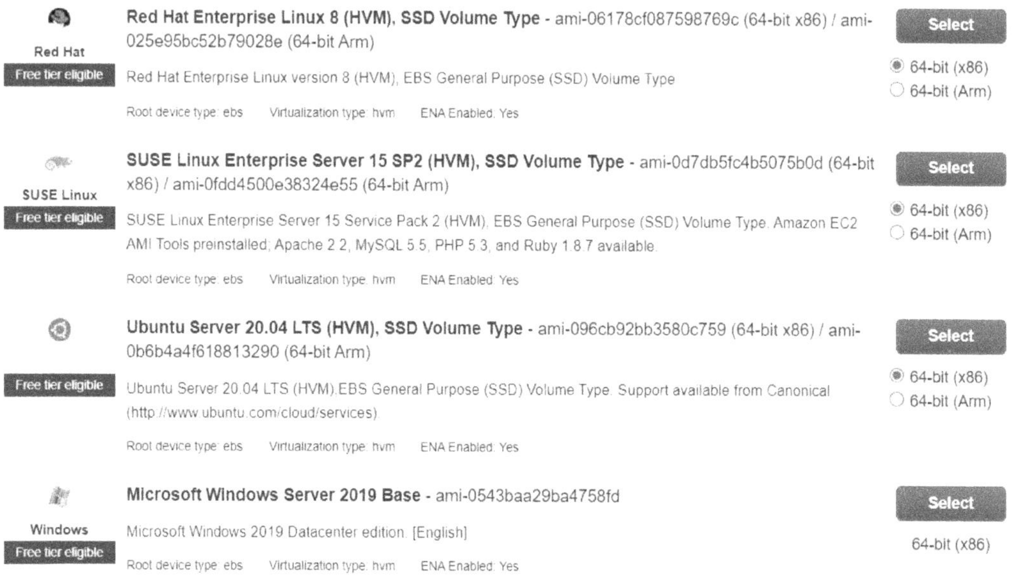

Figure 7.2 – AMIs

As shown in the preceding screenshot, for each AMI, you also have details of the release number, whether the root device type is **EBS** or **Instance**, the type of virtualization architecture, the volume type (for example, SSD), and if the AMI supports Enhanced Networking.

AMIs offered as part of the Quick Start AMIs come with any necessary licensing built into the cost of the EC2 instance that you deploy that AMI on. While you do not need a separate licensing contract to be drawn up to spin up these servers, the hourly charge you pay to Amazon will include the licensing cost.

Note, however, that the licensing that comes with any AMI is only for what is already bundled as part of the image. Any additional configurations or the installation of additional applications may carry further licensing requirements that need to be procured separately.

When you select a particular AMI as part of configuring a new EC2 instance, the image is extracted to the instance's newly attached block storage volume and made bootable. This, in turn, configures the EC2 instance as a fully functioning server.

Some additional points to be aware of about AMIs include the following:

- AMIs are Region-specific. This means that if you want to launch a particular instance configuration in a Region, the AMI must be available in that Region.

- AMIs can be copied across Regions if you need to launch a specific configuration in a Region and the required AMI is not available in that Region. For more details on how to create and copy AMIs, refer to the Amazon documentation at `https://aws.amazon.com/premiumsupport/knowledge-center/copy-ami-region/`.

In addition to the Quick Start AMIs, you have other sources to obtain these machine images. These include the following:

- **AWS Marketplace**: This is a software store managed by AWS where third-party vendors can sell their applications, often bundled as ready-to-use AMIs. Companies such as F5, Citrix, Oracle, and McAfee, among many others, sell their solutions, which can be launched with Amazon EC2 instances. Pricing is according to the owner of the AMI, plus the cost of the specific EC2 instance type you are running the AMI on. You can review the AWS Marketplace offering at `https://aws.amazon.com/marketplace`.

- **Community AMIs**: It is possible to make your AMIs public and thus make them accessible to the wider community via the Community AMI link. There are thousands of different AMIs designed to address specialized software and operating system bundles that are not generally available via the Quick Start AMIs or Marketplace. You can often find a more specific Linux distribution here such as CentOS, which is a very popular AMI that businesses use to host applications. Community AMIs are free to use but you pay for the EC2 instance charge. In addition, many AMIs are provided "as-is," with no additional support, so caution would be required when selecting a Community AMI.

Creating your own AMIs

In addition to obtaining AMIs from AWS or the Marketplace, you can create your own AMIs. This involves creating an image of a running EC2 instance you have that creates a snapshot of the EBS volume attached to the instance. You can then use the image (AMI) to launch new EC2 instances.

You generally create an AMI because you want to store a copy of a specific configuration that needs to be deployed across multiple EC2 instances in your account or even globally. Rather than having to manually configure every new instance to meet your specific requirements, you can configure one EC2 instance just the way you need it (with the right operating system patches, applications, anti-virus software, and so on) and then create an AMI of it. Then, when it comes to launching more EC2 instances with the same configuration, you can simply launch new instances with that AMI. This is also known as **prebaking an AMI** and is often used as a strategy to quickly provision new instances and minimize manual errors or misconfigurations.

In this section, we looked at AMIs and how they are used to provision new EC2 instances with an operating system and any required applications. We also discussed how you can obtain AMIs and, if necessary, create your own. In the next section, we will discuss the different types of EC2 instances you can deploy.

Exploring EC2 instance types

AWS offers a wide selection of EC2 instances that come with different virtual hardware configurations, called **instance types**. An EC2 instance type defines a particular specification of the virtual hardware components of the EC2 instance, such as the amount of processing power in terms of **Virtual CPUs (vCPUs)** and memory, the type of storage, and networking configuration. Depending on your application requirements, you can select an appropriate EC2 instance type from the wide selection available.

As we mentioned earlier, EC2 instances are broken down by *families* and within each *family*, there are *instance types*. These instance types are then broken down further by **instance size**, which comes with specific configurations of virtualized hardware. The following are the core EC2 instance families:

- **General purpose**: Designed for a balance of compute, memory, and networking resources and ideal for a wide range of workload types.

- **Compute optimized**: Designed for high-performance processing. Ideal for batch processing workloads, media transcoding, high-performance web servers, and **high-performance computing (HPC)**.

- **Memory-optimized**: Designed to deliver fast performance for workloads that process large datasets in memory.

- **Accelerated computing**: Designed with hardware accelerators, or co-processors, to perform complex functions. Ideal for floating-point number calculations, graphics processing, or data pattern matching.

- **Storage-optimized**: Designed for computing that requires high sequential read and write access to very large datasets on local storage.

Within each family, you have several EC2 instance types. So, for example, under the *general-purpose* family, you have instance types such as **T2**, **M5**, and **A1**. You can review the various instance types in the Amazon documentation at `https://aws.amazon.com/ec2/instance-types/`.

The **M5** instance type, for example, is powered by Intel Xeon® Platinum 8175M processors. It is ideal for applications that require a balance of compute, memory, and network resources. Additional features of this type include the fact that it offers up to 25 Gbps network bandwidth using Enhanced Networking and is powered by the AWS Nitro System, which is Amazon's architecture design that offers a combination of dedicated hardware and a lightweight hypervisor.

Within the **M5** instance type, there are different instance sizes, from which you get to choose an appropriate size for your specific application needs. The following is a screenshot of some of the instance sizes that fall under the M5 instance type:

Instance Size	vCPU	Memory (GiB)	Instance Storage (GiB)	Network Bandwidth (Gbps)	EBS Bandwidth (Mbps)
m5.large	2	8	EBS-Only	Up to 10	Up to 4,750
m5.xlarge	4	16	EBS-Only	Up to 10	Up to 4,750
m5.2xlarge	8	32	EBS-Only	Up to 10	Up to 4,750
m5.4xlarge	16	64	EBS-Only	Up to 10	4,750

Figure 7.3 – Sample selection of M5 instance sizes

When you review the list of instance sizes, it stands to reason that the higher the vCPU and memory, the more performance that's offered by the instance size. Depending on your application requirements, you then select the most appropriate size that matches your requirements.

Dedicated categories

In addition to the standard instances that you can deploy on AWS, you can also opt for **Dedicated Instances** and **Dedicated Hosts**. These options allow you to fulfill any compliance or regulatory requirements that state that you cannot use EC2 instances deployed on shared hosting, which is the default deployment option. Note that even though your EC2 instance may run on shared hardware with other customers, the underlying infrastructure offers isolation, and you cannot access other customers' EC2 instances unless you've been granted the necessary level of access by those other customers.

A **Dedicated Instance** is an EC2 instance that is deployed in your VPC on physical hardware that is dedicated to you and not shared with other customers.

A **Dedicated Host** is a physical host dedicated for your use alone and gives you additional control and management capabilities over how instances are placed on a physical server. In addition, dedicated hosts can help address certain third-party licensing terms based on a per-CPU core/socket basis.

In the next section, we will look at the block storage options that are available for your EC2 instances. Block storage allows you to attach block volumes to your EC2 instances, much like you would attach a hard disk to your physical servers. The root volume hosts the operating system to boot your virtual server from, and additional volumes can host data and applications for your servers.

Learning about Amazon EBS and instance backed store

Like virtual servers in your on-premises environment, EC2 instances require accessible block storage volume to host the instance operating system, data, and applications that need to run on the server. AWS offers two types of block storage options: **Elastic Block Store** and **instance store volumes**.

Amazon Elastic Block Store

Amazon Elastic Block Store offers a high-performance block storage service for your EC2 instances. These act as virtual hard drives for your virtual servers deployed in the cloud. Amazon EBS is a network storage service similar in nature to the way you would attach a storage volume from a **Storage Area Network** (**SAN**) to your VMs deployed on VMware, in an on-premises environment.

You would provision the required EBS volume, such as 8 GB, 30 GB, 1,000 GB, or whatever size you require, and attach it to your EC2 instance. EC2 instances require a root volume (or a system drive on Windows machines, such as the "C" drive) to host the operating system and certain applications. In addition, you can attach multiple EBS volumes for other purposes, such as storing data (for example, data volume).

When configuring your EC2 instance with an EBS volume, you need to select the type of volume, the amount of storage you wish to provision, and if you want to configure encryption on your EBS volume. There are different types of EBS volumes to choose from and they fall into the following categories: **Solid State Drives** (**SSDs**) and **Hard Disk Drives** (**HDDs**).

SSDs are optimized for transactional workloads involving frequent read/write operations with small I/O sizes, and they can be used to boot up your EC2 instances. These types of drives include the following:

- **gp2**: This is the default EBS volume that you can attach to an EC2 instance. The volume runs on SSDs and is suitable for general workloads such as transaction operations, low-latency interactive operations, and is suitable as boot volumes. The volume size can range from 1 GB to 16 TB, is designed to deliver up to 99.9% of durability, and has a maximum IOPS of 16,000. gp2 volumes also offer a baseline performance of 3 IOPS/GB (minimum 100 IOPS), which can burst up to 3,000 IOPS for volumes smaller than 1 TB.

- **gp3**: This is the latest version of the general-purpose SSD-based EBS volumes. Compared to gp2, gp3 volumes offer a baseline performance of 3,000 IOPS and 125 Mbps at any volume size. You can provision performance for an additional fee of up to 16000 IOPS. The size of the volume can range from 1 GB to 16 TB. gp3 volumes are ideal for running single-instance databases such as Microsoft SQL Server, Cassandra, MySQL, and Oracle DB and can be used as boot volumes.

- **io1**: This provisioned IOPS SSD offering high-performance EBS storage is ideal for critical, I/O intensive database and application workloads. io1 offers a baseline performance of 50 IOPS/GB to a maximum of 64,000 IOPS and provides up to 1,000 MB/s of throughput per volume. Note, however, that to achieve the maximum of 64,000 IOPS and 1,000 MB/s of throughput, the volume must be attached to an EC2 instance built on the AWS Nitro system. io1 volumes can also be used as boot volumes and the volume size can range from 4 GB to 16 TB.

- **io2**: The latest generation of the Provisioned IOPS SSD, this offers high performance, as well as a 100X durability of 99.999% with 10X higher IOPS to storage ratio of 500 IOPS for every provisioned GB. io2 is ideal for business-critical, I/O intensive database applications, including SAP HANA, Oracle, Microsoft SQL Server, and IBM DB2. io2 volumes can also be used as boot volumes and the volume size can range from 4 GB to 16 TB. Like io1, to achieve the maximum of 64,000 IOPS and 1,000 MB/s of throughput, the volume must be attached to an EC2 instance built on the AWS Nitro system.

- **io2 Block Express**: Currently in preview, this high-performance block storage offers 4x higher throughput, IOPS, and a capacity of io2 volumes. Designed for the most demanding applications, it offers 4,000 MB/s of throughput per volume, up to 256,000 IOPS and 1,000 IOPS/GB, as well as 99.999% durability. Volume sizes can range from 4 GB to 16 TB.

HDDs are optimized for large streaming workloads where the dominant performance attribute is throughput. These drives include the following:

- **st1**: This is a low-cost HDD volume that's ideal for frequently accessed, throughput-intensive workloads and capable of working with large datasets and large I/O sizes. Typical workloads include MapReduce, Kafka, log processing, data warehouses, and ETL jobs. st1 can burst up to 250 MB/s per TB, has a baseline throughput of 40 MB/s per TB, and the maximum throughput is 500 MB/s per volume. The volume size can range from 125 GB to 16 TB; however, it cannot be used as a boot volume.

- **sc1**: This is also known as Cold HDD and offers the lowest cost per GB of all EBS volume types. It is ideal for less frequently accessed workloads with large, cold datasets. sc1 can burst up to 80 MB/s per TB, has a baseline throughput of 12 MB/s per TB, and offers a maximum throughput of 250 MB/s per volume. The volume size can range from 125 GB to 16 TB; however, it cannot be used as a boot volume.

Some additional features of EBS volumes include the following:

- EBS volumes can be detached from one EC2 instance and attached to another. So, for example, if you have an EC2 instance with a single root volume and the instance has some sort of failure, you can detach the root volume and attach it to another EC2 instance as a data volume to extract any information as required. This makes EBS volumes very versatile and flexible.

- Data stored on EBS volumes is persistent and can exist beyond the life of an EC2 instance.

- You can also take *snapshots of your EBS volume*, which are point-in-time backups of your data stored on the volume. *The first snapshot is always a full backup and additional snapshots are incremental backups of changes*. Snapshots can then be used to create new volumes as required with the data intact. In addition, you can copy snapshots to other Regions. This means that if you need to share some data with a colleague in another Region, you simply copy the snapshot to the required Region, and it will be made available within your AWS account in that Region. You can also copy snapshots to other AWS accounts.

In this section, we looked at the different types of EBS volumes backed by both HDD and SDD technologies. We also looked at their use cases. EBS volumes are attached to your EC2 instances over a storage area network connection, usually using the ISCSI protocol. In the next section, we will look at EC2 instance store volumes. These volumes are directly attached to the physical host that your EC2 instance has been deployed on.

AWS EC2 instance store volumes

Amazon EBS volumes are storage volumes that are mounted on your EC2 instances over a network from a storage array, whereas instance store volumes are carved out of the local storage drives that are physically attached to the host that the EC2 instance runs on.

The primary benefit of attaching instance store volumes to your EC2 instance is the ultra-high throughput you get, as well as reduced latency in accessing your data. However, there is a downside to instance store volumes, namely around data persistence. The data in an instance store persists only during the lifetime of its associated instance while in the running state. If the instance is stopped, terminated, and placed into a hibernate state, then the data on the attached instance store is lost. Equally, if the underlying disk experiences some failure, then the data will be lost in such situations. The reason behind this is that when your instance experiences any of these states, every block of storage in the instance store is reset, so any data is lost.

Due to this, data on an instance store volume is considered temporary and to be used as such. It is ideal for information that changes frequently and can be used as buffers, caches, or for holding temporary content. Instance store volumes are also cheaper than EBS volumes.

> **Important Note**
> If an EC2 instance with an instance store volume is restarted, the data is not lost. This is because a reboot is not the same as a shutdown and then a cold start.

In this section, we looked at the different storage options for EC2 instances. We discussed the key features and options for Amazon EBS and identified typical use cases of where we would use which type of EBS volume. We also summarized some additional features, such as being able to encrypt your block storage or take regular snapshots as backups and for replication.

We then discussed instance store volumes and compared them with EBS, identifying use cases for instance store volumes and how data on an instance store volume is considered temporary. Instance store volumes are also known as ephemeral storage.

In the next section, we will discuss the different pricing options available to launch your EC2 instances. Depending on your use case, one set of pricing options might work out more cost-effectively than another, and this next section discusses some important concepts to be aware of.

Learning about EC2 pricing options

The Amazon EC2 service is a cloud offering that enables you to deploy virtual servers (EC2 instances) in the cloud. Traditionally, in an on-premises environment, if you needed to deploy a new physical server, you would have to make a capital investment for a few thousand dollars to procure the necessary hardware and software and then configure your server with any necessary applications.

On AWS, EC2 instances can be procured on an hourly basis, which means that you only pay for the hours that the server is running. If you turn off the server but keep it in your account (as opposed to terminating it and releasing its capacity back to AWS), then you do not pay any charges while the EC2 instance is in this stopped state. This pricing approach is what we call the **On-Demand Pricing Option** and is the default option when purchasing EC2 instances on AWS.

Let's look at these various pricing options in detail.

On-Demand Instance Pricing Option

As we've already discussed, this is the **default option** for procuring/launching an EC2 instance. You pay an hourly charge based on the instance type, the operating system (with any applications pre-installed), and the Region that you launch it in. You are charged based on the number of hours that the EC2 instance is in the **running state**. This means that you are not billed for the hours that the EC2 instance is in the **stopped state**. The **On-Demand Pricing Option** does not require any long-term commitment or upfront payments. You can increase or decrease your compute capacity depending on your application demand and you only pay per hour consumed.

The **On-Demand Pricing Option** is ideal for users who need the flexibility to consume compute resources when required and without any long-term commitment. They are ideal for test/dev environments or for applications that have short, spiky, or unpredictable workloads. An important benefit of the On-Demand pricing model is that Amazon will not interrupt your consumption in any way. So, your EC2 instances will continue to exist in your account (even in the stopped state) unless you terminate (release) them. In terms of pricing, the On-Demand Pricing Option is low cost but compared to the other pricing options, it can turn out to be the most expensive, especially if you plan to keep the instances running 24/7, 365 days a year. In such cases, alternative pricing options may be more suitable.

Reserved Instance Pricing Option

This pricing option rewards customers with a significant discount (up to 72% off the On-Demand Pricing Option) who commit to a 1-year or 3-year agreement to run specific instance types within a given Region. You can achieve very discounted hourly rates compared to the On-Demand price for the same instance. Reserved Pricing Options are not actual EC2 instances that you procure; rather, they are pricing agreements that give you the right to run a specific EC2 configuration, in a specified Region, for a specified duration (1 year to 3 years) at a specific discounted rate. Once you launch a matching EC2 instance in the given Region, the Reserved pricing model is applied instead of the On-Demand rate. There are two types of Reserved Pricing Options:

- **Standard Reserved Pricing Option**: This provides the most discount with up to 72% off the On-Demand instance price. They can be purchased on a 1-year or 3-year term and you can change the certain configurations of the EC2 instance during its life, such as the Availability Zone it is placed in, the instance size, and its network type.

- **Convertible Reserved Pricing Option**: This offers even more flexibility, including the ability to change instance families, operating systems, or even tenancies over the Reserved term. The downside of the Convertible Reserved Pricing Option is that the maximum discount is only up to 54% compared to the On-Demand Instance rate.

Here is an example of a standard 1-year reserved instance pricing for a t2.micro, Linux EC2 instance in the N. Virginia Region at the time of writing this training guide:

t2.micro

Payment Option	Upfront	Monthly*	Effective Hourly**	Savings over On-Demand	On-Demand Hourly
		STANDARD 1-YEAR TERM			
No Upfront	$0	$5.26	$0.007	38%	
Partial Upfront	$30	$2.48	$0.007	41%	$0.0116
All Upfront	$59	$0.00	$0.007	42%	

Figure 7.4 – t.2micro Reserved Instance Pricing Option in N. Virginia Region, standard 1-year term

Next, we will look at the different payment options available for EC2 instances that have been purchased under the Reserved pricing model.

Payment options for Reserved Instances

You can choose to pay all or some of the total cost of your Reserved Instance upfront and depending on how you pay for your Reserved Instances, you get varying levels of the discounted rate. They are as follows:

- **All Upfront**: This is where you pay for the entire term of the Reserved Instance upfront, right at the beginning of the contract. You do not get a monthly hourly bill and you benefit from the maximum available discount.

- **Partial Upfront**: This is where you make some upfront payment and then are charged a discounted hourly rate for the term and billed monthly.

- **No Upfront**: There is where you do not make any upfront payment, but you still get some discount compared to the On-Demand rate. You get a better discount by going with either the **Partial Upfront** or **All-Upfront** option.

An important point to note here is that while you benefit from a massive discount for your Reserved Instance, you are charged for the whole term, regardless of whether you have the EC2 instance in a running state or not. This means that for some use cases such as if you need an EC2 instance for unpredictable usage patterns and short durations, but without any interruptions, then the On-Demand pricing option might work out more cost-effectively. You would need to carefully compare the total cost of purchasing a Reserved Instance versus the On-Demand option to figure out which is more cost-effective.

Reserved Instance Marketplace

Sometimes, your business needs change, or perhaps, the project that you originally purchased your reserved instances for ends earlier than anticipated. Whatever your reasons, you can resell your Standard Reserved Instances on the **Reserved Instance Marketplace** to other AWS customers if you no longer need them. This enables you to recoup some of the costs that you may have incurred from your original purchase.

Because you are reselling your Reserved Instances, the remaining term will be less than what it was originally bought for. This also means that if you need a Reserved Instance for less than the standard contract length, you may wish to check out the Reserved Instance Marketplace to see if you can find one that fits your requirements.

Note that only Standard Reserved Instances can be listed in the Marketplace. You cannot list Convertible Reserved Instances in the Marketplace. Furthermore, to list your Reserved Instance, you must ensure that you have paid for the Reserved Instance and that you owned the Reserved Instance for longer than 30 days.

Spot Instance Pricing Option

Amazon almost always has spare compute capacity. One way to encourage the uptake of this spare capacity is by offering customers the option to launch Spot Instances at discounts up to 90% off the On-Demand price. However, Spot Instances are not suitable for all applications. They are ideal for workloads that are stateless or can be interrupted. For example, certain use cases such as data analysis, batch jobs, and background processing, can be run whenever there is available capacity, and depending on the application design architecture, those EC2 instances can be interrupted and then resumed later when capacity is available. In other words, if the application is stateless and workflows are not sensitive to interruptions, then if there are any interruptions, the application workflow simply waits for available compute capacity to continue with its operations.

You might be wondering why there would be any interruptions. Well, simply put, Amazon will try to meet supply and demand for compute capacity based on the best possible price they can get. When you purchase a Spot Instance, you are made aware of the current spot price. You would then place a maximum offer price, which needs to be higher than the prevailing spot price. Now, if demand were to increase, Amazon can increase the spot price rate and if your offer price falls below this, then another customer can be offered this capacity at the higher spot price instead, and your workloads will be interrupted. Spot Instances can also be interrupted if, for instance, supply of capacity reduces due to excessive uptake of Spot Instances by customers.

The following are some of the reasons for Spot Instance interruptions:

- **Price**: Where the spot price goes above your maximum offer price.
- **Capacity**: If there is not enough unused EC2 instances to meet the demand for On-Demand instances.
- **Constraints**: This includes the launch group or Availability Zone group, resulting in Spot Instance termination when such constraints cannot be met.

When Amazon attempts to interrupt your Spot Instance, you are provided with a warning interruption notice, which is issued 2 minutes before Amazon EC2 stops or terminates your Spot Instance. These interruption notices are made available via CloudWatch Events and as items in the instance metadata on the Spot Instance.

With regards to Amazon CloudWatch Events (discussed in *Chapter 13*, *Management and Governance on AWS*), you can create targets such as Lambda functions or Amazon **Simple Notification** (**SNS**) topics, which can process the Spot interruption notices when they happen. We will discuss SNS in detail in *Chapter 10*, *Application Integration Services*, but essentially, you can create an email subscription to an SNS topic so that when a 2-minute interruption notice is generated, this message can be forwarded to you via the SNS topic via email.

The instance metadata contains data about your EC2 instance, divided into categories; for example, hostname, IP address information, events, and security groups. It is accessible from within the EC2 instance itself, available at `http://169.254.169.254/latest/meta-data/`. The `169.254.169.254` IP address is a link-local address and is valid only from the instance.

Previously, the only behavior following an interruption from Amazon was to terminate your Spot Instance. However, you may lose data when that happens. More recently, alternative behaviors have become available, such as stopping your Spot Instance or placing the instance into a hibernate state. When an instance is placed into a hibernate state, the EBS volumes are preserved, and instance memory (RAM) is preserved on the root volume. Furthermore, the private IP addresses of the instance are also preserved. This option means that when the Spot service resumes your instance if, for example, the capacity becomes available or the spot rises fall before your maximum offer price, then your instance can continue from the state it was stopped at.

Spot Fleets

A **Spot Fleet** is a collection of Spot Instances, and optionally On-Demand instances. AWS will launch Spot Instances as part of the fleet if there is available capacity and if your maximum offer price is higher than the spot price. The Spot Fleet will attempt to maintain its target capacity if Spot Instances are terminated. What you can also do is include a few On-Demand instances as part of your Spot Fleet. This means that you always have some capacity for minimum application requirements. As demand for your application increases, Spot Instances are added to the fleet at the discounted rate, assuming there is capacity, or your offer price is higher than the spot price.

In this section, we looked at different EC2 instance pricing options. We examined the default On-Demand pricing model and compared it with the Reserved pricing model and Spot pricing options.

In the next section, we will move on to discuss another storage option for your EC2 instances, known as **Elastic File System** (**EFS**).

Implementing Shared File Storage with Amazon EFS

In our earlier discussion, we looked at Amazon EBS. These block storage volumes are directly attached to a specific EC2 instance and act as virtual hard drives for your EC2 instance. In general, an EBS volume can only be attached to one specific EC2 instance at a given time. This means that if you deploy 20 EC2 instances, each one of the instances will have one or more EBS volumes attached. This is perfectly fine if the data between those volumes does not need to be shared across those EC2 instances.

There are multiple use cases for sharing data across EC2 instances. These include file shares or data that needs to be shared across multiple applications and web servers. In those cases, using EBS volumes would create a messy architecture of having to somehow replicate data between those individual EBS volumes.

Amazon offers the **EFS** solution, which allows you to create and mount file shares across multiple EC2 instances. These instances can then update the data on the file share, which is visible to all other EC2 instances that have mounted the same share if the necessary permissions have been set.

Some key features of EFS include the following:

- Can be used by Linux-based EC2 instances as a centralized file storage solution.
- Can also be accessed from on-premises servers over a VPN or the Direct Connect service.
- EFS Volumes can grow and shrink on demand. This is unlike EBS volumes, where you need to provision storage before you can use it.
- EFS volumes are regionally-based and can be made available across Availability Zones for high availability and durability.
- Note that EFS Volumes cannot be used as boot/root volumes of your EC2 instance and cannot be used with Windows EC2 instances.

In this section, we examined an alternative storage solution for your Linux-based EC2 instance, where the requirement is to share files and data across a fleet of instances. Amazon EFS is a file-level storage solution and can be used to mount file shares on multiple EC2 instances across multiple Availability Zones in each Region.

In the next section, we will look at an alternative compute solution known as Amazon Lightsail, which allows us to easily deploy individual VPSes in the cloud.

Learning about VPSes with Amazon Lightsail

In addition to Amazon EC2, AWS also offers a more lightweight solution to deploy virtual servers preconfigured with the most common application stacks for a wide range of use cases. Rather than picking individual components of your EC2 instances, such as the instance type and storage, and then installing all the necessary application layers, Amazon Lightsail offers blueprints that will automatically configure your servers with various common use case applications and utilities required to get up and running.

A common application stack includes your chosen operating system, along with apps such as WordPress, Drupal, Plesk, LAMP, and more. Deploying Amazon Lightsail is also relatively straightforward, with guided steps to get your deployment up and running.

The best part about Lightsail is you have a fixed monthly fee based on the instance type and the associated operating system and applications that have been deployed. However, for more complex configurations and enterprise-grade production environments, you will probably still need to design and deploy EC2 instances.

Prices for Amazon Lightsail start from as little as $3.50 per month. This gives you a Linux virtual server with 512 MB memory, 1 vCPU, a 20 GB SSD disk, and up to 1 TB of data transfer. Additional configurations include opting for a static IP address, DNS management, and SSH/RDS access to your virtual server.

As your needs change over time, you can easily upgrade to EC2 instances, which involves taking a snapshot of your instance and following a step-by-step process to upgrade to EC2 using a wizard.

In this section, we reviewed Amazon's Lightsail offering, which can be used for smaller-scale application deployments and predictable pricing structures. In the next section, we will look at other compute options available on AWS.

Introduction to Amazon ECS and Kubernetes

So far, we have been looking at hardware virtualization and using hypervisors to build VMs such as EC2 instances that we can run various applications on. Different applications often have specific requirements, and many applications will not be able to run together in the same VM due to incompatibility with the underlying libraries or runtime environments.

Traditional virtualization technologies involve using bare-metal hardware, upon which you configure a hypervisor. This hypervisor, as we discussed previously, allows you to essentially *carve* out physical hardware components (CPU, memory, storage, and so on) into smaller virtual components that allow you to then deploy VMs, or in the case of AWS, EC2 instances. Each EC2 instance, however, will need to host a guest operating system (Linux or Windows, for example), shared libraries and system files, and your application.

As shown in the following diagram, VMs take up a lot of resources because they need to run the guest operating system, along with all the system files, utilities, and libraries:

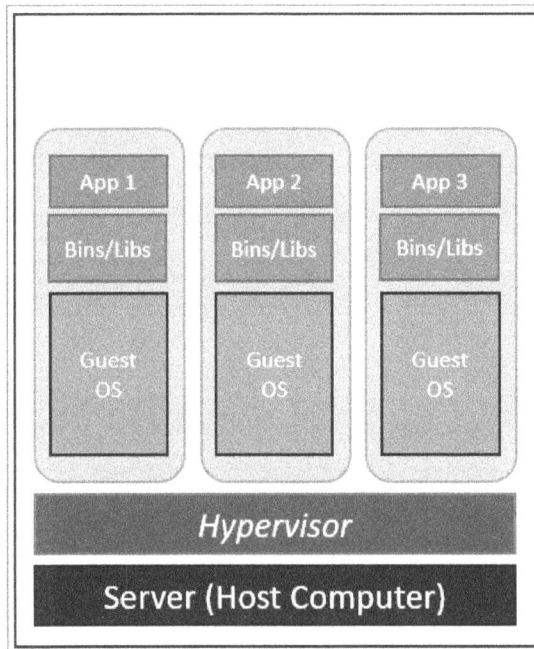

Figure 7.5 – Hypervisor architecture hosting VMs

You will note that each VM requires its own guest OS, as well as the necessary libraries, system files, and runtime environments. An alternative solution is to consider a technology known as **Docker**.

Docker is built on the concept of containerization, which is essentially **operating system virtualization**. Applications are run in isolated user spaces known as **containers**, which share the same underlying operating system. This is made possible because the container will host all the necessary dependencies required by the application, including binaries, libraries, configuration files, and runtime environments. The container is abstracted from the host operating system's container with limited access to the OS, which allows us to run several containers on the same bare-metal hardware.

Containerization also offers less overhead during startups since they share the same underlying OS rather than each container have an operating system to boot up from, which is how VMs work. The following diagram illustrates this concept:

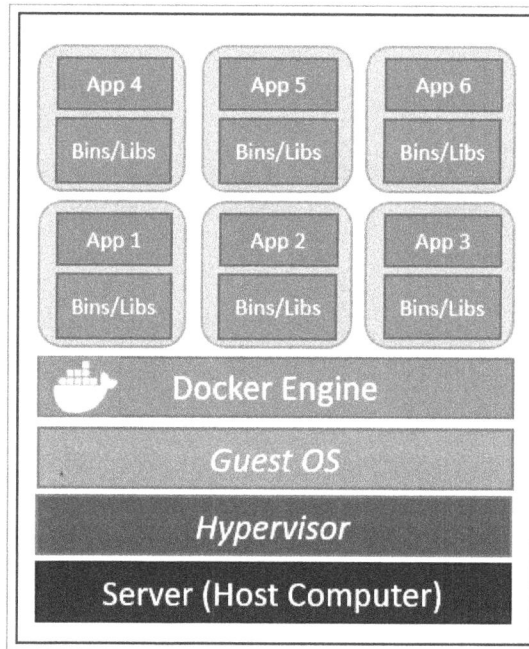

Figure 7.6 – Docker concepts

Some of the added features of using Docker include the fact that you run multiple applications on the same bare-metal hardware, all sharing the same underlying OS via the Docker engine. This keeps the code base for the application much smaller compared to a VM that also has to host the operating system.

Amazon offers **ECS**, which is a fully managed container orchestration service that allows you to deploy Docker-style applications. Amazon ECS can help you deploy and manage multiple Docker containers in the cloud rather than having to manually spin up EC2 instances and deploy the Docker solution on each.

The following diagram illustrates the core components of Amazon ECS:

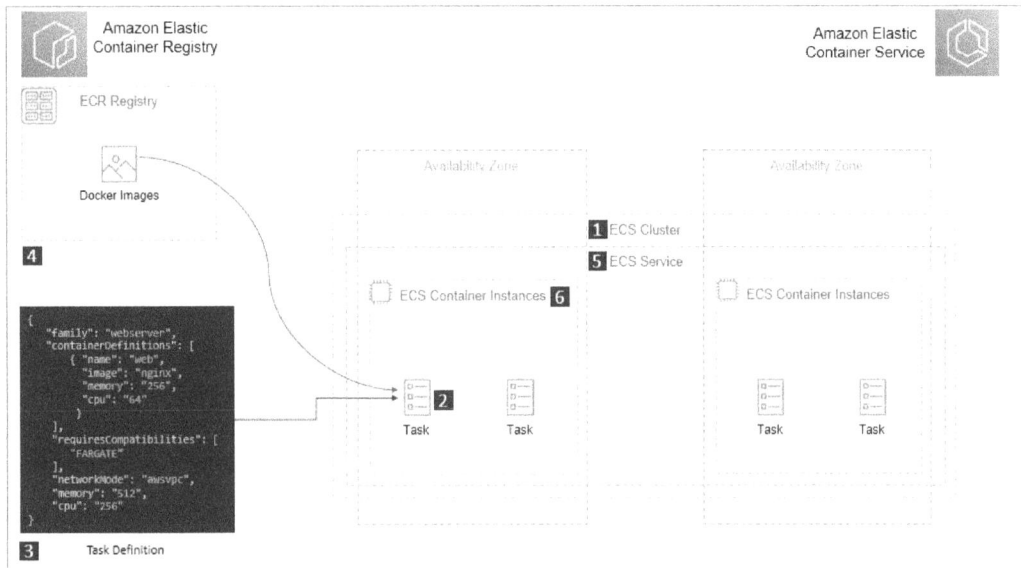

Figure 7.7 – Amazon ECS architecture

The Amazon ECS architecture comes with the following core components and features:

1. **ECS Cluster**: This is a logical grouping of tasks or services.

2. **Task** – This is a Docker container termed as a "task" in Amazon ECS. In Amazon ECS, a "task" is essentially a container.

3. **Task Definition**: Each ECS task is created from a task definition. A task definition specifies the Docker image to use, the amount of CPU and memory to use with each task or each container within a task, the launch type to use (discussed later), and the IAM role that your tasks should use, among others.

4. **Elastic Container Registry (ECR)**: This is where Docker images are stored on AWS. You can also store your images on Docker Hub or a private registry. Amazon ECR hosts your images in a highly available and high-performance architecture. When the task starts, it reviews the task definition and pulls down the Docker image required from the registry.

5. **ECS Service**: This allows you to define a specified number of instances of a task definition simultaneously in an Amazon ECS cluster. Should one of your tasks fail, the ECS service will replace it to maintain the desired number of tasks in the service.

6. **ECS Container instances**: Your tasks run on EC2 instances, within which you run your Docker container.

However, there are two types of ECS deployments, which we will discuss next.

Amazon ECS comes in two deployment options

Let's take a look at these deployment options:

- **Amazon Fargate (Fargate launch type)**: This enables you to set up your ECS environment without having to spin up an EC2 instance, provision and scale clusters, or patch and update virtual servers yourself. You simply package your application and specify your CPU, memory, and network requirements and AWS takes care of the heavy lifting for you. AWS will place the ECS tasks on the cluster, scale as required, and fully manage the entire environment for you.

- **EC2 launch type**: In some cases, you may require direct access to the underlying EC2 instances that run your container services. In this case, you must manage a cluster of EC2 instances (patch and update them) and schedule the placement of containers on the servers.

In this section, we learned about Amazon ECS. In the next section, we will look at an orchestration service known as Amazon Elastic Service for Kubernetes.

Amazon Elastic Kubernetes Service (Amazon EKS)

Kubernetes is another open source container orchestration solution. It groups containers that make up an application into logical units for easy management and discovery. Kubernetes takes care of scaling and managing your containers, taking care of failover options to ensure your application keeps running.

Amazon EKS is AWS's offering to help you deploy, manage, and scale containerized applications using Kubernetes on AWS.

In this section, we looked at the basics of containerization and how it can be used to effectively deploy and manage your application stack. We discussed the **ECS** and its two different launch types – the Fargate and EC2 launch types. We also looked at Amazon's take on Kubernetes as a container orchestration tool that uses Kubernetes.

In the next section, we will look at some additional compute solutions available from AWS.

Learning about additional compute services on AWS

In this section, we will look at some additional compute services available on AWS and their use cases.

Serverless option – AWS Lambda

So far, we have only looked at server-based computer resources except for AWS ECS Fargate. When we say serverless, we do not mean that the compute resource is running without any servers. Ultimately, servers will house the CPU that offers compute capabilities. The term serverless is just used to mean that the customer does not need to manage any actual servers as this falls within the responsibility of AWS.

AWS Lambda is a serverless offering from AWS that allows you to run code and perform some tasks. AWS Lambda is known as a **Function as a Service** (**FaaS**) solution that can be used to build an entirely serverless architecture comprised of storage, databases, and network capabilities where you do not manage any underlying servers.

To use Lambda, you must create a function using a supporting programming language, such as Python or Node.js, and upload this function to AWS Lambda. You then need to define a trigger that will execute the code. For example, you can create a Lambda function that gets triggered when you upload an image to an S3 bucket. Your function could be used to add a watermark to a copy of the image and place it in another bucket, which is then used to serve a web application that allows customers to purchase stock photos. This way, the preview images have watermarks on them, and users will need to complete a purchase before they get the original image.

The important thing to remember here is that Lambda functions are executed when they're triggered by another AWS service, so they do not run independently. Lambda functions can also be triggered on a schedule. An example of this is if you have a fleet of development servers that your dev team needs to work on. Because this is a short project, we decided to opt for On-Demand EC2 instances. However, you want to try and save on costs, so you create a Lambda function to stop the EC2 instances at 6 P.M. and another Lambda function to start up the EC2 instances again at 8 A.M. This automation sequence made possible by scheduling Lambda functions can help you save on costs.

An important point to remember is that with Lambda, you pay only for the compute time your Lambda functions consumes and there is no charge when your code is not running.

AWS Batch

AWS Batch can be used to run thousands of batch computing jobs on AWS. AWS Batch will set up and provision the necessary compute resources to fulfill your batch requests. There is no need to deploy server clusters as AWS takes care of this for you. You can schedule and execute your batch jobs across a wide range of compute services, such as EC2 and ECS. With AWS Batch, you pay for the resources that AWS creates, such as EC2 instances or ECS clusters.

AWS Outposts

It is possible to run several AWS Services such as EC2 instances, RDS databases, and host Amazon S3 storage buckets on-premises. You get to use the same AWS tools and hardware both in the cloud and on-premises, which helps build a complete hybrid solution. AWS Outposts is ideal when you want to run AWS resources with very low latency connections to your on-premises application or if you require local data processing due to any compliance and regulatory requirements.

AWS Outposts comes in two flavors:

- **VMware Cloud on AWS Outposts**: This allows you to utilize the same VMware tools to manage the infrastructure that you use with your on-premises resources.

- **AWS native variant of AWS Outposts**: This allows you to use the same AWS APIs that you use with the AWS cloud-hosted services.

You can get AWS Outposts delivered to your local on-premises location as a 42U rack and can scale from 1 rack to 96 racks to create pools of compute and storage capacity. You can also get smaller racks with 1U and 2U rack-mountable servers for locations with limited space or capacity requirements.

In this section, we looked at additional compute services on AWS, including the very popular AWS Lambda services, which offers compute capabilities delivered in a **FaaS** model.

We also discussed AWS Outposts, which allows you to host AWS infrastructure on-premises and address use cases such as very low latency connectivity or local data processing needs.

In the next section, we will look at some additional storage options available on AWS.

Understanding additional storage options in AWS

In addition to the previously discussed storage options for your compute needs, you have a couple of additional storage solutions that have been designed for specific use cases.

Amazon FSx for Lustre

Amazon FSx for Lustre is a fully managed filesystem for compute-intensive workloads and designed for applications that require high-performance and low latency connectivity, offering millions of IOPS and hundreds of gigabits per second throughput rates.

Amazon FSx for Lustre is designed to be integrated with Amazon S3. You store data in S3 and retrieve it when you need to perform compute-intensive workloads against the data. Later, you copy the data back to S3 for long-term storage.

Amazon FSx for Windows File Server

Microsoft Windows servers can be configured to offer a file-sharing solution similar to EFS. Amazon EC2 instances running Microsoft Windows may need a common filesystem to share various types of data for end users or applications. For example, if you host virtual desktops in the cloud, you may need a central file sharing solution to allow your users to share files using mapped drives.

Instead of manually building out a file share solution using EC2 instances and EBS volumes running the Windows operating system, you can use AWS FSx for Windows File Server. This is a fully managed native Microsoft Windows filesystem that offers support for the SMB protocol and Windows NTFS, **Active Directory** (**AD**) integration, and **Distributed File System** (**DFS**). By opting for FSx for Windows File Server, you benefit from a fully managed file sharing solution for all your Windows-based applications and can configure advanced file sharing solutions using DFS.

In this section, we reviewed two additional storage options for your compute needs; AWS FSx for Lustre, which is designed for high-performance computing needs offering millions of IOPS, and AWS FSx for Windows File Server, which offers a fully managed Microsoft Windows file share solution.

In the next section, we will look at how to securely access EC2 instances we deploy in our VPC with the help of **bastion hosts**. Bastion hosts are servers designed to allow secure connections from external networks such as the internet, from which you can then access the web and application servers that have been deployed in your VPC.

Securing your VPC with bastion hosts

The use of bastion hosts (or jump boxes) is used to provide secure access to EC2 instances located in the private and public subnets of your **virtual private cloud** (**VPC**). These bastion hosts (you can have one or multiple hosts deployed for redundancy) can be used as management servers and are designed to allow you to remotely connect to them via SSH or RDP protocols. Once you have established connectivity to your bastion hosts, you can then log onto backend EC2 instances acting as web or application servers.

Bastion hosts are EC2 instances that do not host any unnecessary applications, other than those required to allow connections to backend EC2 instances. These servers are hardened and secured to reduce attack surfaces and thus minimize the chances of penetration. However, you still need to manage these instances as you would any other application or web server. The onus is on you to ensure that the EC2 instances are highly secured, patched, and updated regularly. Amazon also offers an alternative solution known as **Session Manager**, which is a feature of the **AWS Systems Manager** service offering. **Session Manager** enables you to manage your EC2 instances and on-premises instances via an interactive browser shell or the AWS CLI tools, without the need to open inbound ports or maintain bastion hosts and SSH keys. It also offers a fully auditable instance management service recording instance access details. We will discuss AWS Systems Manager and Session Manager in *Chapter 13, Management and Governance on AWS*.

In the following diagram, we can see how an administrator can connect to the bastion hosts in a public subnet and from there administer backend EC2 instances running an application.

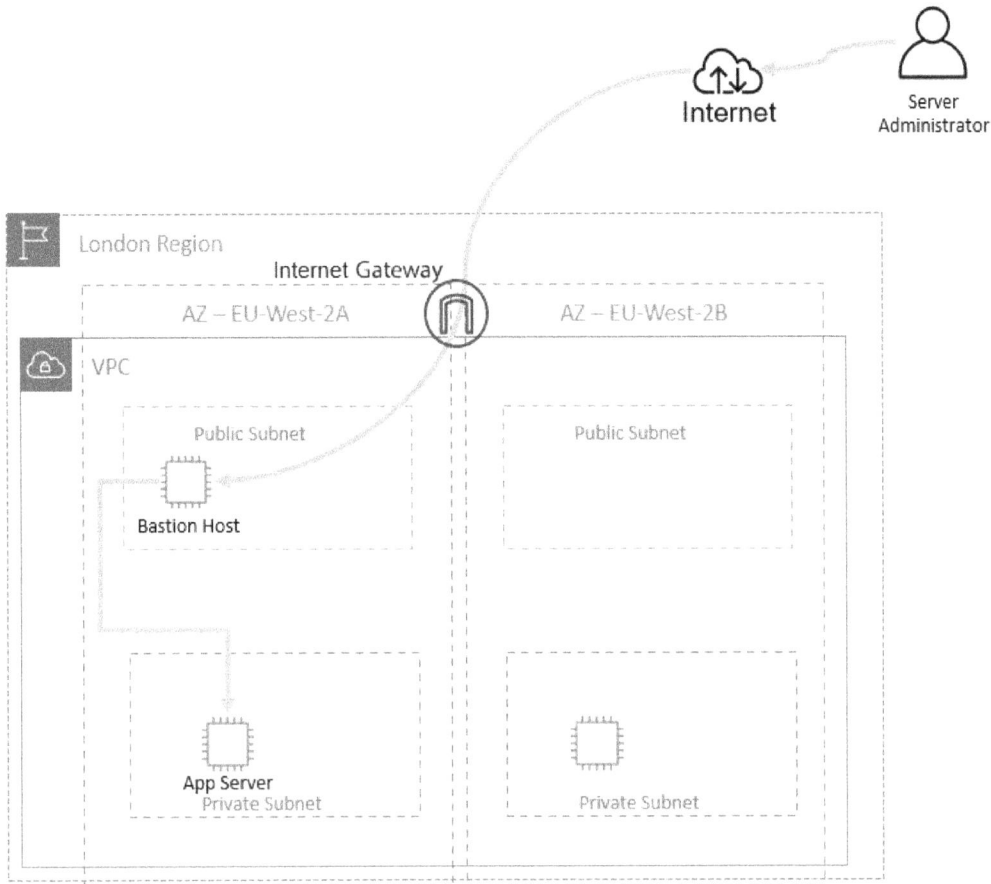

Figure 7.8 – Securing access to your VPC using bastion hosts

In the preceding diagram, the administrator connects to the bastion hosts from the internet and then from that Bastion Host, establishes a secure connection to backend EC2 instances. The security of this architecture can be improved by restricting the source of the remote traffic to the Bastion Host. For example, rather than having the remote access open to the entire internet, we can restrict the source IP address or IP range of that connection to the administrator's corporate office network IP range. This way, the administrator can only access the Bastion Host if they were working from within the corporate office's network.

In this section, we briefly looked at bastion hosts and how they can be used to securely access other EC2 instances in your VPC. Bastion hosts are standard EC2 instances themselves, but you would configure them to only serve as an entry point into your VPC for administrative access. Furthermore, you would harden the bastion hosts by installed the necessary security updates and tools, and ensuring that unnecessary applications are removed. Access to bastion hosts can also be restricted to specific source IP addresses or even only from your corporate data center via a VPN link.

Exercise 7.1 – Expanding ProductionVPC so that it includes two public subnets and two private subnets

In this exercise, we will expand the architecture of **ProductionVPC** that we built in the previous chapter so that it includes an additional public subnet in another Availability Zone and two private subnets – one in each Availability Zone. This will enable us to design an architecture that can offer high availability in case of a single Availability Zone outage.

We will be extending the VPC to fulfill our design specifications, as shown in the following diagram:

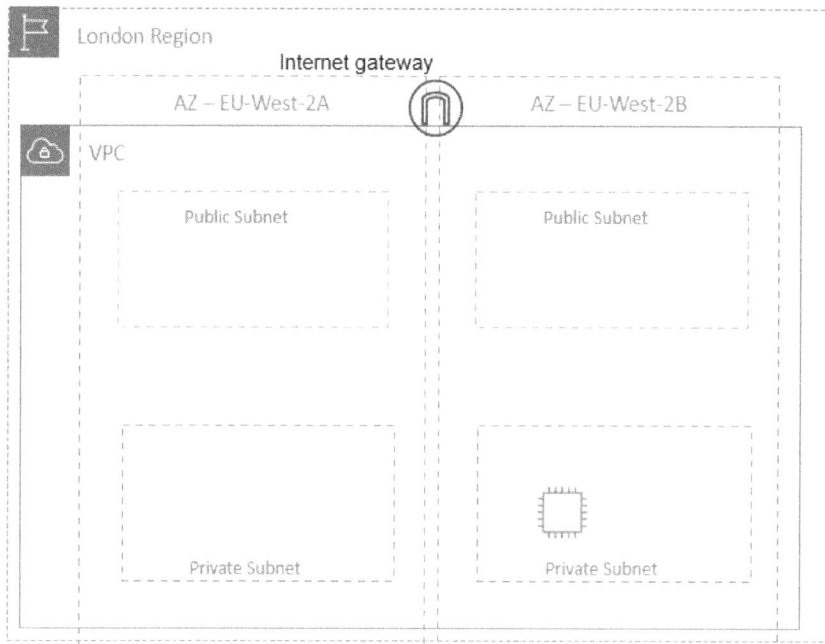

Figure 7.9 – VPC with public and private subnets across two Availability Zones

Log in to your AWS account as the root user and on the main AWS Management Console, search for VPC in the search box. Click on the VPC link to open the VPC Console. Once in the console, ensure that you are in the us-east-1 Region, where you created **ProductionVPC** in the previous chapter. If necessary, navigate to the us-east-1 Region by selecting it from the drop-down list.

Review your current VPC:

1. On the left-hand menu in **VPC console**, click on **Subnets**. You will see that there is one **Public Subnet One** that is associated with ProductionVPC, which you created in the previous chapter. You will also see other subnets, but these are associated with the default VPC in the us-east-1 Region, as per the following screenshot:

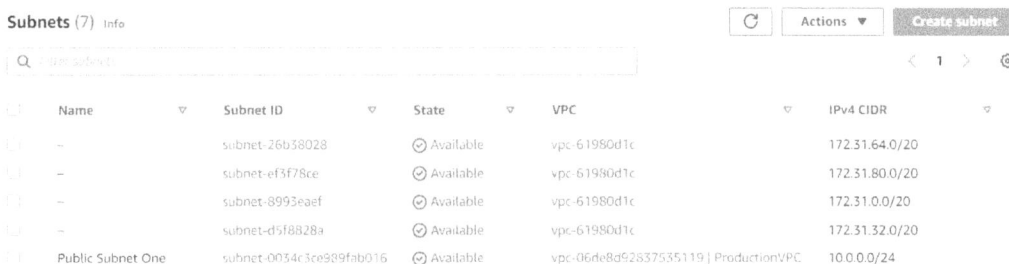

Subnets (7) Info

	Name		Subnet ID		State		VPC		IPv4 CIDR	
	–		subnet-26b38028		⊘ Available		vpc-61980d1c		172.31.64.0/20	
	–		subnet-ef3f78ce		⊘ Available		vpc-61980d1c		172.31.80.0/20	
	–		subnet-8993eaef		⊘ Available		vpc-61980d1c		172.31.0.0/20	
	–		subnet-d5f8828a		⊘ Available		vpc-61980d1c		172.31.32.0/20	
	Public Subnet One		subnet-0034c3ce989fab016		⊘ Available		vpc-06de8d92837535119 | ProductionVPC		10.0.0.0/24	

Figure 7.10 – ProductionVPC with a single public subnet

2. Next, click on the **Route Tables** link from the left-hand menu. Using the VPC Wizard, your ProductionVPC has been configured with a main route table and a public route table. From the following screenshot, you can see that the main route table associated with the ProductionVPC has a route table ID of **rtb-0d6fb017c417d8e1b**. You can tell it is the main route table because this is indicated in the **Main** column with a **Yes**. This main route table is designed to be associated with all private subnets in your VPC.

If you click on the **Main Route Table ID** link in the console, the bottom pane will provide additional information about the route table. Furthermore, if you click on the **Routes** tab of the bottom pane, you will note that currently, there is only one route: the **local route**. This local route is designed for traffic to flow within the VPC. The main route table does not offer any direct access to the internet now. This is a best practice because, for subnets that need direct access to the internet, you should ideally create a separate **public route table** and attach any *public subnets* to it:

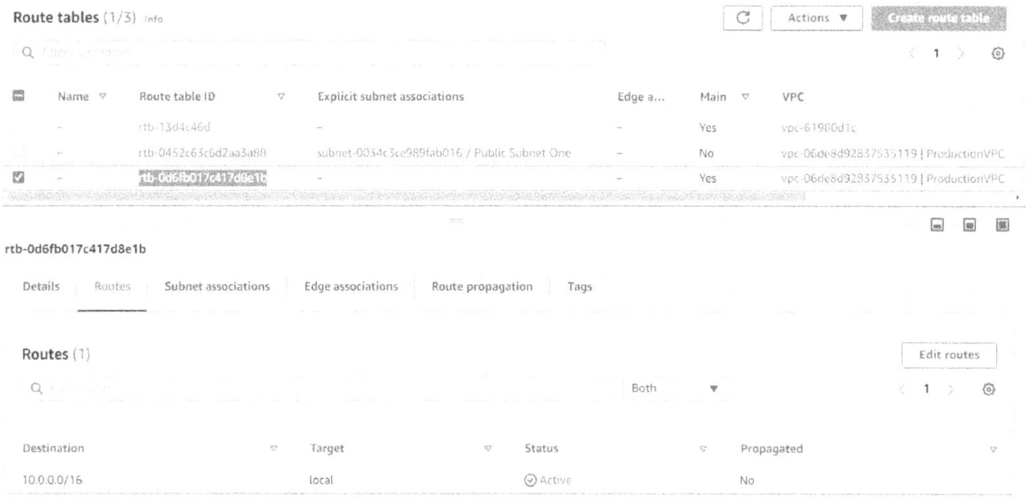

Figure 7.11 – ProductionVPC main route table configuration

3. Next, in the top pane, you will notice that the wizard has already created a public route table (with a **Route table ID** of **rtb-0452c63c6d2aa3a88**). We know this is the public route table because, in the preceding screenshot, you can see a **No** in the **Main** column. It also has one subnet associated with it: **Public Subnet One**.

4. If you click on the **Route Table ID** link of the public route table, the bottom pane will offer additional information, as per the following screenshot. Specifically, if you click on the **Routes** tab of the public route table, you will see two routes: one local route and a route to the internet. Routes to the internet are denoted with a destination of **0.0.0.0/0**. In this case, the route to the internet has a target, which is the **internet gateway**. This internet gateway gives the subnet direct access to send traffic to the internet and receive traffic from the internet if the security groups and/or **Network Access Control Lists** (**NACLs**) permit the traffic:

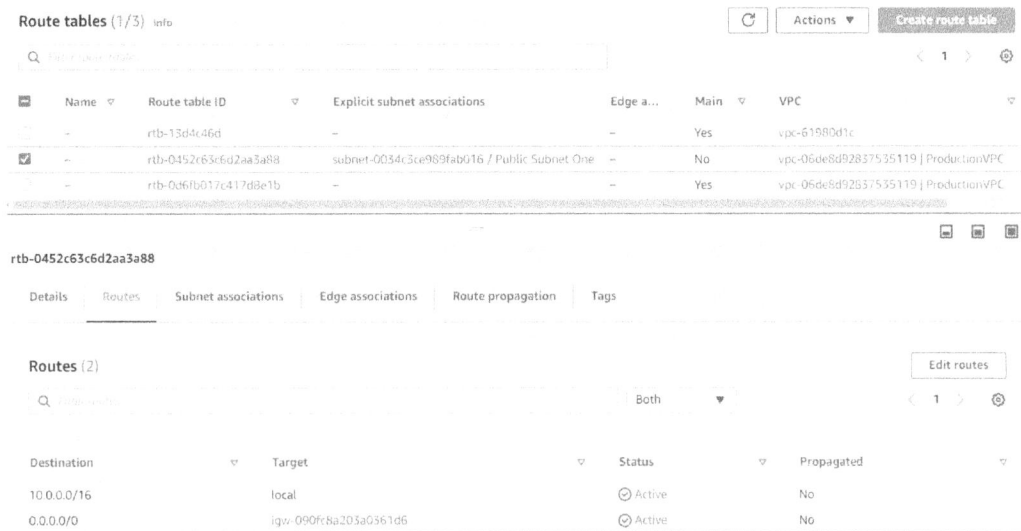

Figure 7.12 – ProductionVPC with public route table configuration

You will also notice that AWS does not create a name tag for your route tables by default. This can sometimes make it difficult to identify your resources. You can easily add name tags to make resources easy to identify. Using your mouse, simply hover over the area in the **Name** column next to the **Public Route** table until you notice an edit icon. Click on this edit icon and in the **Edit Name** text box that appears, type in Public Route Table. Perform the same action for the main route table, making sure to tag the route table with the name **Main Route Table**. This will make it easy to identify your route tables, as per the following screenshot:

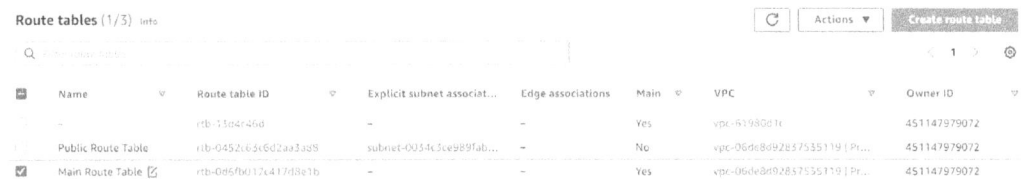

Figure 7.13 – Tagging your route tables

When you create new subnets, they are automatically associated with the main route table. This technically makes them private subnets. If you need to create a public subnet, you need to remember to disassociate the subnet from the main route table and attach it to a route table that offers public internet access via the internet gateway.

While you can configure a direct route to the internet in the main route table that sends traffic via the internet gateway, this is not advisable. You want to ensure that some subnets offer only internal private access without direct exposure to the internet. Since EC2 instances deployed as backend services require internet access to download software updates, for example, you can set up NAT gateways in the public subnets to facilitate this requirement. You would then need to configure the **Main Route** table with a route to the internet via this NAT gateway, which acts as a proxy for any internet requests from your backend EC2 instances. These backend EC2 instances do not require public IP addresses and can still communicate with the internet via the NAT gateway.

Setting up additional subnets

In this part of the exercise, we will be expanding our VPC to include multiple public and private subnets. We wish to host two public subnets and two private subnets across two Availability Zones. We wish to do this because we want to offer high availability so that if one Availability Zone fails or is offline, we can access duplicate copies of our resources in the other Availability Zone.

You already have one public subnet in your VPC. This has been configured with a name of **Public Subnet One** and has been placed in the **us-east-1a** Availability Zone as per the following screenshot:

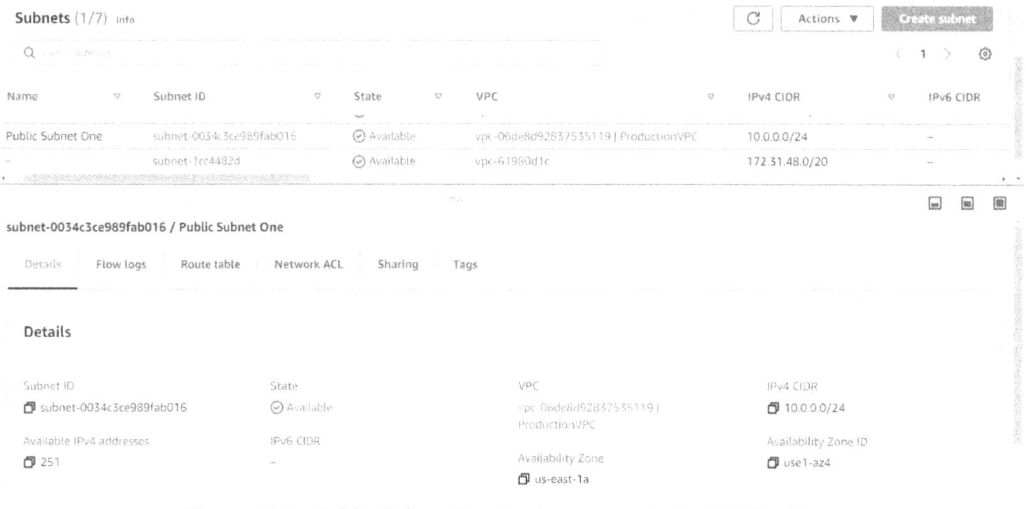

Figure 7.14 – Public Subnet One in the us-east-1a Availability Zone

As part of this exercise, we will create another public subnet in the **us-east-1b** Availability Zone:

1. From the left-hand menu in your VPC console, click on **Subnets**.

2. Click on the **Create subnet** button in the top right-hand corner.

3. In the **Create subnet** wizard that appears, choose **ProductionVPC** from the list of VPCs available.

4. Under **Subnet settings**, provide a subnet name such as `Public Subnet Two`.

5. Next, under **Availability Zone**, ensure you select **us-east-1b**. This is because we want the second public subnet to be placed in a different Availability Zone from the first public subnet.

6. For the IPv4 CIDR block, provide a block address of `10.0.2.0/24`. This CIDR block is on a separate range from the CIDR block of the first public subnet but is still a subset of the VPC's CIDR range.

7. Finally, as per the following screenshot, click on **Create subnet**:

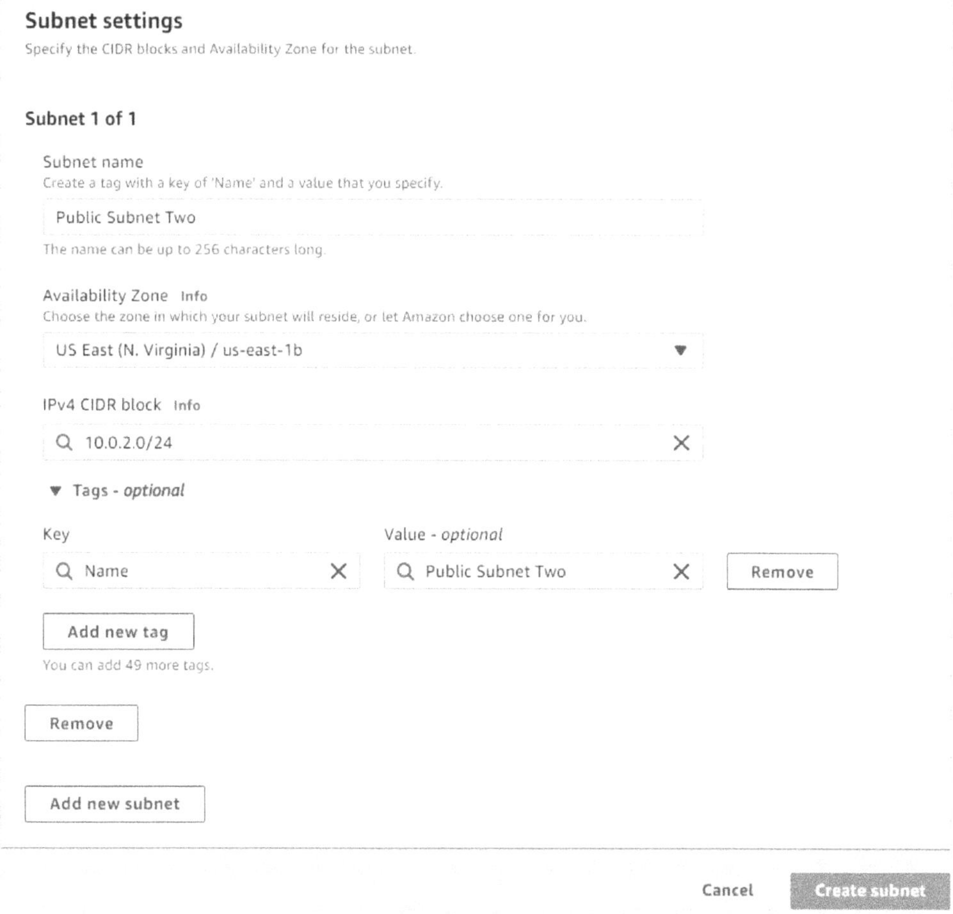

Figure 7.15 – Creating Public Subnet Two

The wizard will create the subnet and display a successful creation message.

If you click on the checkbox next to the newly created subnet, the bottom pane will provide information about the subnet. Click on the **Route table** tab. You will notice that this newly created subnet has been automatically associated with **main route table**, as shown in the following screenshot:

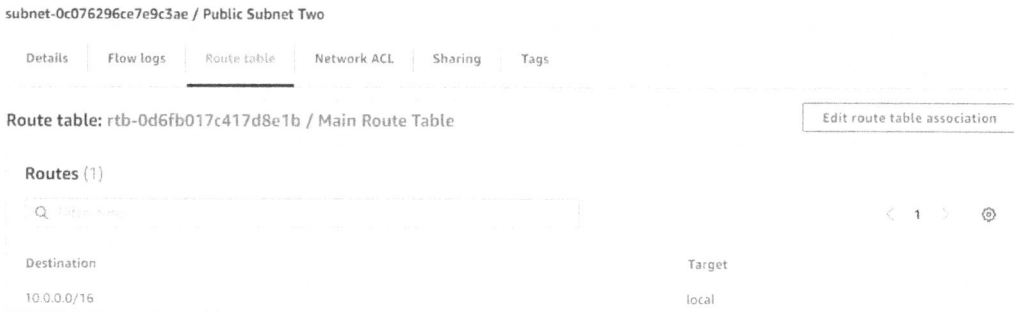

subnet-0c076296ce7e9c3ae / Public Subnet Two

| Details | Flow logs | Route table | Network ACL | Sharing | Tags |

Route table: rtb-0d6fb017c417d8e1b / Main Route Table   Edit route table association

Routes (1)

Q

Destination	Target
10.0.0.0/16	local

Figure 7.16 – Newly created subnet associated with main route table

Because we want this subnet to be a public subnet, we need to change its association from the main route table to the public route table, as follows:

1. Click on the **Edit route table association** button, as shown in the preceding screenshot.

2. In the **Edit route table association** dialog box that appears, select **Public Route Table** from the **Route table ID** drop-down list, as shown in the following screenshot:

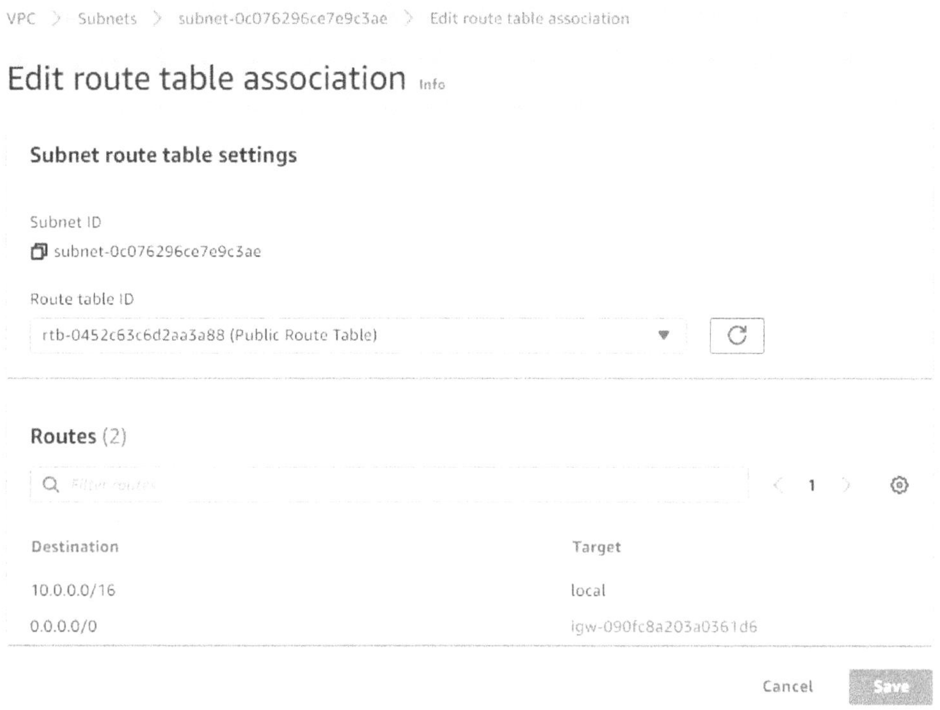

VPC > Subnets > subnet-0c076296ce7e9c3ae > Edit route table association

Edit route table association Info

Subnet route table settings

Subnet ID

⬡ subnet-0c076296ce7e9c3ae

Route table ID

rtb-0452c63c6d2aa3a88 (Public Route Table)   ▾   ↻

Routes (2)

| Q Filter routes | | < 1 > ⚙ |

Destination	Target
10.0.0.0/16	local
0.0.0.0/0	igw-090fc8a203a0361d6

Cancel   Save

Figure 7.17 – Edit route table association

3.   Click **Save**.

At this point, both public subnets have been correctly configured. Next, we'll create two private subnets, one in **us-east-1a** and the other in **us-east-1b**.

Creating private subnets

Follow these steps to create the aforementioned private subnets:

1.   While still in the VPC console, click on the **Subnets** link from the left-hand menu.

2.   In the **Create subnet** dialog box that appears, select **ProductionVPC** from the **VPC ID** drop-down list.

3.   Next, in the **Subnet settings** section, type `Private Subnet One` in the **Subnet name** text box.

4.   Under **Availability Zone**, select **us-east-1a** – this is the same zone where we placed **Public Subnet One**. This way, any frontend web resources in **Public Subnet One** can access any backend resources in **Private Subnet One**, allowing those resources to be in the same Availability Zone.

5. For the IPv4 CIDR block, provide the IP CIDR range of `10.0.3.0/24` – this CIDR block does not conflict with any of the other subnets and is still a subset range of the main VPC's IP range.

6. Finally, click the **Create subnet** button at the bottom of the page, as per the following screenshot:

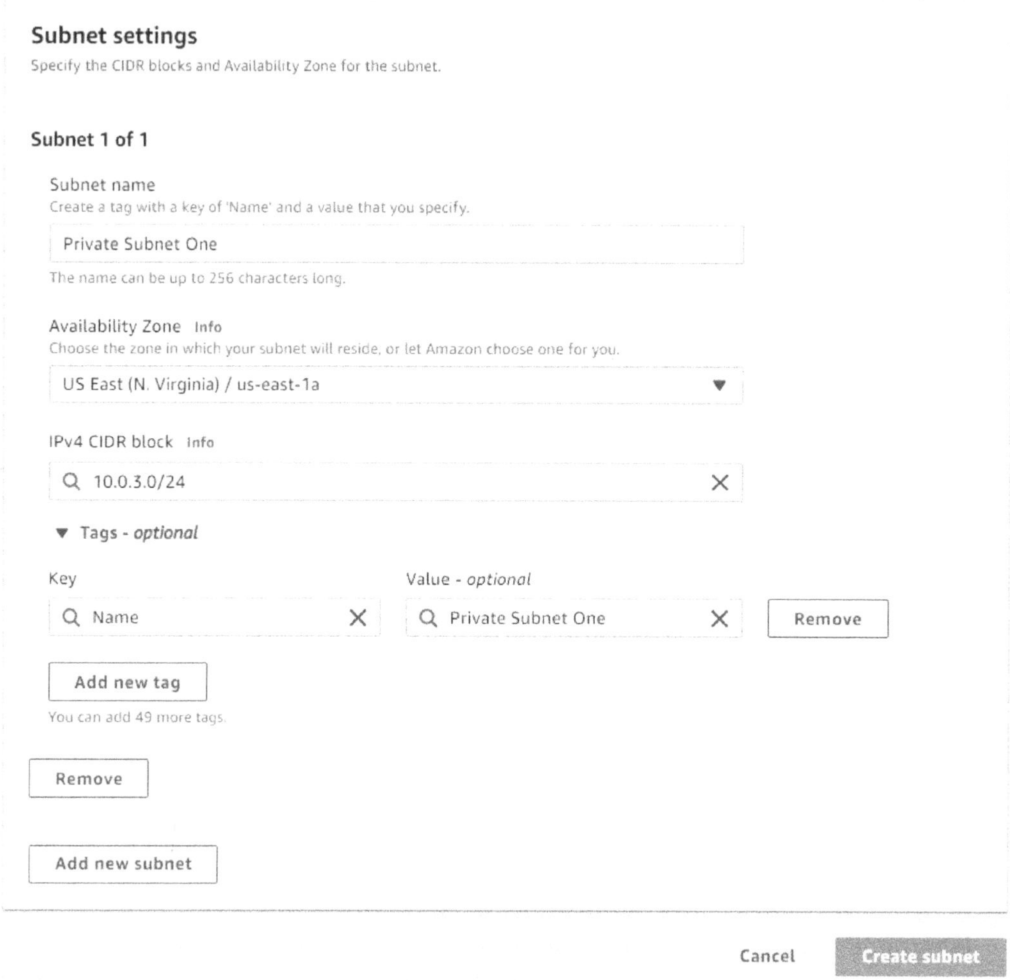

Figure 7.18 – Private Subnet One settings

You will receive a success message. Next, we will perform the same steps we did previously, but this time to create the second private subnet in **us-east-1b**:

1. Click **Create subnet**.

2. Select **ProductionVPC** from the VPC ID drop-down list.

3. For the subnet name, type in `Private Subnet Two`.

4. Under **Availability Zone**, select **us-east-1b**.

5. For the IPv4 CIDR block, type in `10.0.4.0/24` as the CIDR block range for this subnet.

6. Click the **Create subnet** button at the bottom of the page.

At this point, you have a VPC configured with two public subnets and two private subnets.

> **Note**
> You do not need to associate the two private subnets to the main route tables manually, as this is done for you by default.

To check this, perform the following steps:

1. Select **Route Tables** from the left-hand menu.

2. From the top pane, click on the checkbox next to the main route table.

3. In the bottom pane, you will notice that both private subnets are associated with the main route table (not explicitly) but by default instead, as shown in the following screenshot:

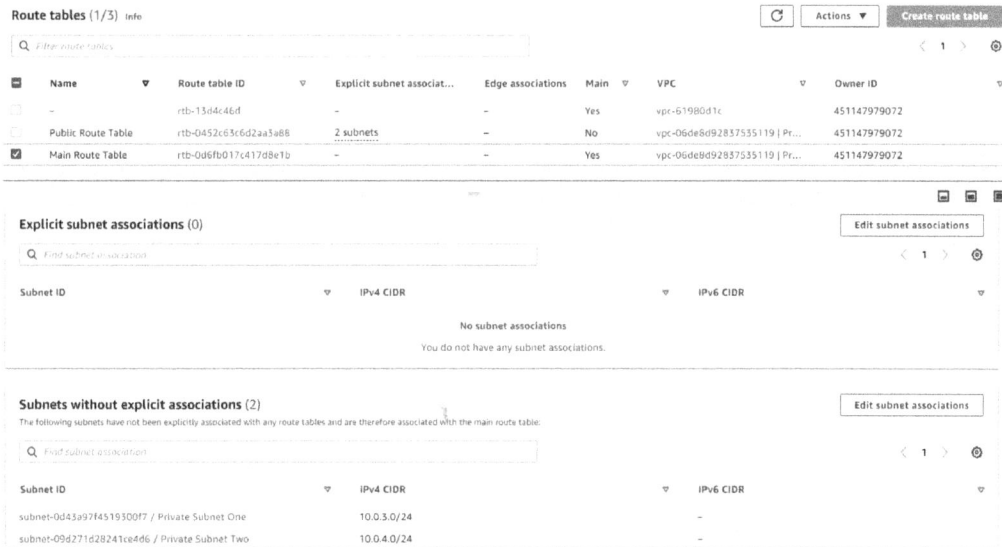

Figure 7.19 – Private subnet route table associations

In this exercise, we expanded our **ProductionVPC** to include two public subnets and two private subnets across two Availability Zones. This will enable us to deploy duplicate workloads in each Availability Zone to design for high availability.

In the next exercise, we will deploy our first custom security group in **ProductionVPC**. This security group will be used to define what types of traffic we will allow inbound to an EC2 instance that will be deployed in the third exercise.

Exercise 7.2 – Creating a Bastion Host security group

In this exercise, we will create a custom security group that will be used by an EC2 instance. This will act as a Bastion Host, as previously discussed in this chapter. Let's get started:

1. Log in to your AWS account and navigate to the VPC console. Ensure you are in the **us-east-1** Region. From the left-hand menu, confirm that **ProductionVPC** is available in this Region.

2. Select the **Security Groups** link from the left-hand menu, which is located under the **Security** category, as shown in the following screenshot:

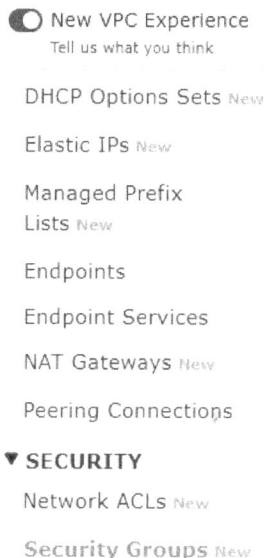

Figure 7.20 – Selecting Security Groups from the VPC console

3. Click on the **Create security group** button on the top right-hand corner of the screen.

4. Under **Basic details**, provide a name for your security group; for example, `BastionHost-SG`.

5. Next, provide an appropriate description, such as `Allow Remote Access to Bastion Host Server`.

6. Under **VPC**, make sure your select **ProductionVPC** from the drop-down list.

7. Next, under the **Inbound rules** section, click the **Add rule** button.

8. Select **RDP** for the type of rule. **RDP** is a remote access protocol that's used to connect to Microsoft Windows servers and desktops. RDP uses the TCP (6) protocol and operates on port `3389`.

9. Under the **Source** column, ensure that **Custom** is selected and in the text box, type in `0.0.0.0/0`. This IP block represents all external networks, including the internet.

10. Next, click on the **Create security group** button on the bottom right-hand corner of the page.

You will receive a successful creation message to confirm that the security group has now been created. We will use this security group in the next exercise to ensure that we can connect to our EC2 instance over the **Remote Desktop Protocol** (RDP), which operates on port `3389`.

Exercise 7.3 – Launching an EC2 instance

In this exercise, we will launch a Windows-based EC2 instance in **Public Subnet One** of our **ProductionVPC**. We will use this EC2 instance as a Bastion Host, allowing us to configure other EC2 instances in the VPC:

1. Log in to your AWS account and from the **Services** drop-down list, select **EC2** under the **Compute** category, as shown in the following screenshot:

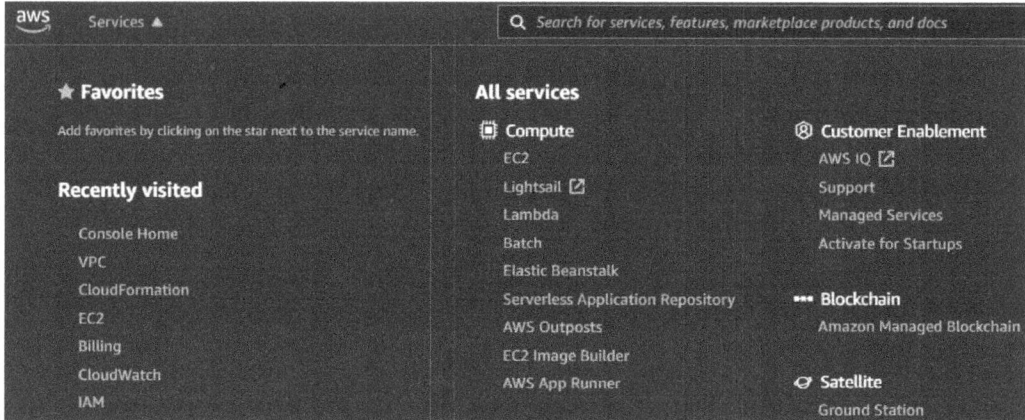

Figure 7.21 – Accessing the EC2 console

2. From the **Regions** list in the top right-hand corner, ensure that you are in the **US East N.Virgina (us-east-1)** Region.

3. On the **EC2** dashboard, you will note that there are 0 Instances in the running state.

4. Click on the **Instances (running)** link, as shown in the following screenshot. This will bring up the **Instances** console:

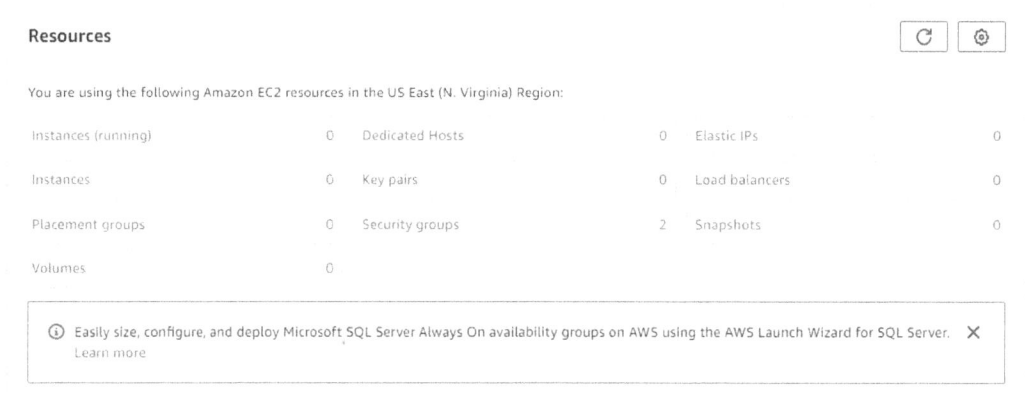

Figure 7.22 – EC2 dashboard

5. From the **Instances** console, select the **Launch instances** button from the top right-hand corner of the screen.

6. You will now be presented with **Step 1: Choose an Amazon Machine Image (AMI)**. From the list of available AMIs, select **Microsoft Windows Server 2016 Base - ami-05ce3abcaf51f14b2**.

7. In **Step 2: Choose an Instance Type**, ensure that the checkbox next to the **t2.micro** instance is selected. This instance type is available as part of your Free Tier offering.

8. Next, click the **Next: Configure Instance Details** button in the bottom right-hand corner of the screen.

9. In **Step 3: Configure Instance Details**, choose the following options:

   ▪ For **Network**, select **ProductionVPC**.

   ▪ For **Subnet**, select the **Public Subnet One** subnet.

   ▪ For **Auto-assign Public IP**, select **Enable**.

   ▪ Leave all the remaining settings as their default values and click the **Next: Add Storage** button in the bottom right-hand corner of the screen.

10. In **Step 4: Add Storage**, accept the default root volume size and click on the **Next: Add Tags** button in the bottom right-hand corner of the screen.

11. In **Step 5: Add Tags**, provide a key-value combination to tag the EC2 instance with a name, as shown in the following screenshot. For **Key**, type in Name, and for **Value**, type in Windows-BastionSrv:

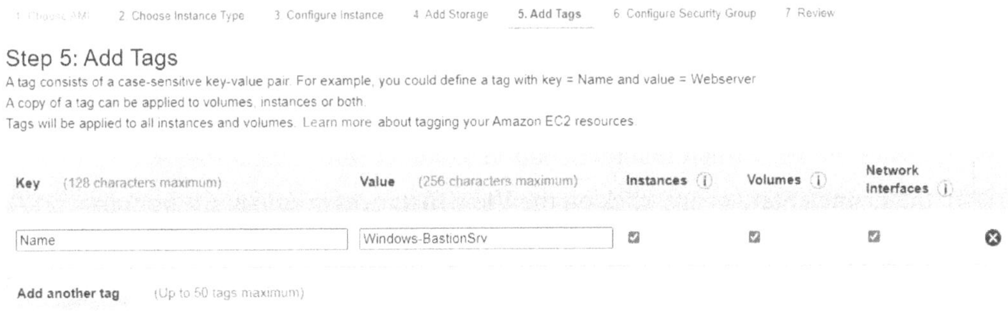

| 1. Choose AMI | 2. Choose Instance Type | 3. Configure Instance | 4. Add Storage | 5. Add Tags | 6. Configure Security Group | 7. Review |

Step 5: Add Tags

A tag consists of a case-sensitive key-value pair. For example, you could define a tag with key = Name and value = Webserver
A copy of a tag can be applied to volumes, instances or both.
Tags will be applied to all instances and volumes. Learn more about tagging your Amazon EC2 resources.

Key (128 characters maximum)	Value (256 characters maximum)	Instances ⓘ	Volumes ⓘ	Network Interfaces ⓘ	
Name	Windows-BastionSrv	☑	☑	☑	❌

Add another tag   (Up to 50 tags maximum)

Figure 7.23 – Step 5: Add Tags

12. Next, click the **Next: Configure Security Group** button in the bottom right-hand corner of the screen.

13. In **Step 6: Configure Security Group**, click on the **Select an existing security group** option under the **Assign a security group** heading.

14. Next, from the list in the **Security Group ID** column, select the security group ID that corresponds to the newly created **BastionHost-SG**, which we set up in the previous exercise.

15. Finally, click on the **Review and Launch** button in the bottom right-hand corner of the screen.

16. You will then be presented with **Step 7: Review Instance Launch**. Review the settings you have defined and then click on the **Launch** button in the bottom right-hand corner of the screen.

17. At this point, you will be presented with another dialog box, requesting you to either select an existing EC2 key pair or create a new key pair.

Key pairs are cryptographically encrypted public/private keys that are used to encrypt the credentials required to authenticate against the EC2 instance operating system so that you can remotely access them.

For Microsoft Windows-based EC2 instances, the public key of the keys pair is used to encrypt the *Administrator* password. You then use the private key to decrypt the password to remotely access the Windows machine. For Linux-based EC2 instances, the private key can be used to establish a **Secure Socket Shell** (**SSH**) connection. Establishing SSH connections to your Linux-based EC2 instance allows you to work on the server remotely using the Linux shell interface. Let's take a look:

1. In the **Select an existing key pair or create a new key pair** dialog box, select the option to **Create a new key pair**, from the drop-down list.

2. Next, provide a key pair name. For example, I am naming my key USEC2Keys.

3. Click on the **Download Key Pair** button, which will download the private key in .pem format into your load Downloads folder. Make sure that you copy and save the keys in a safe location or on the desktop for easy access.

4. Next, click on the **Launch Instances** button to launch your EC2 instance.

5. On the **Launch Status** page, click on the **View Instances** button at the bottom right-hand corner of the page.

At this point, you will be redirected to the instances console and will be able to see your EC2 instance, as per the following screenshot:

Figure 7.24 – Successfully launched the Windows BastionSRV EC2 instance

It will take a few minutes for the server to fully launch and become accessible. Look out for the **Status Check** column, as per the preceding screenshot, and ensure that AWS has completed all its checks; you will receive a **2/2 checks passed** message in the **Status check** column.

You can now remotely connect to the Windows Server using the Microsoft Remote Desktop client. If you are performing these labs from a laptop or desktop using Microsoft Windows, then the **remote desktop client** will already be installed on your machine. Simply click the **Start** button and search for the `Remote Desktop Connection` app. If you are performing these labs on a macOS-based computer, you will need to download the Microsoft Remote Desktop client from the Apple Play Store.

To connect to your new Windows Bastion Host server, follow these steps:

1. From the EC2 instances console, click on the checkbox next to the EC2 instance you have just launched.

2. In the bottom pane, make a note of the EC2 instance's public IP address.

3. Next, with the EC2 instance still selected, click on the **Actions** menu from the top right-hand corner of the screen, and then on the **Security** sub-menu. From here, select the **Get Windows password** link, as per the following screenshot:

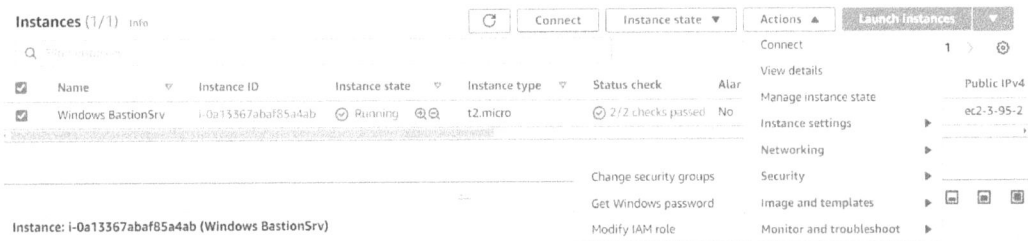

Figure 7.25 – The Get Windows password option for your EC2 instance

4. You will be presented with the **Get Windows password** page. Here, you need to browse for the EC2 key pairs you previously downloaded using the **Browse** button. Select the key pair from the location where you stored a copy of it; you should also find that it is still in the `Downloads` folder.

5. Upon selecting the key, you will note that the text box below the **Browse** button gets populated with the encrypted key, as per the following screenshot:

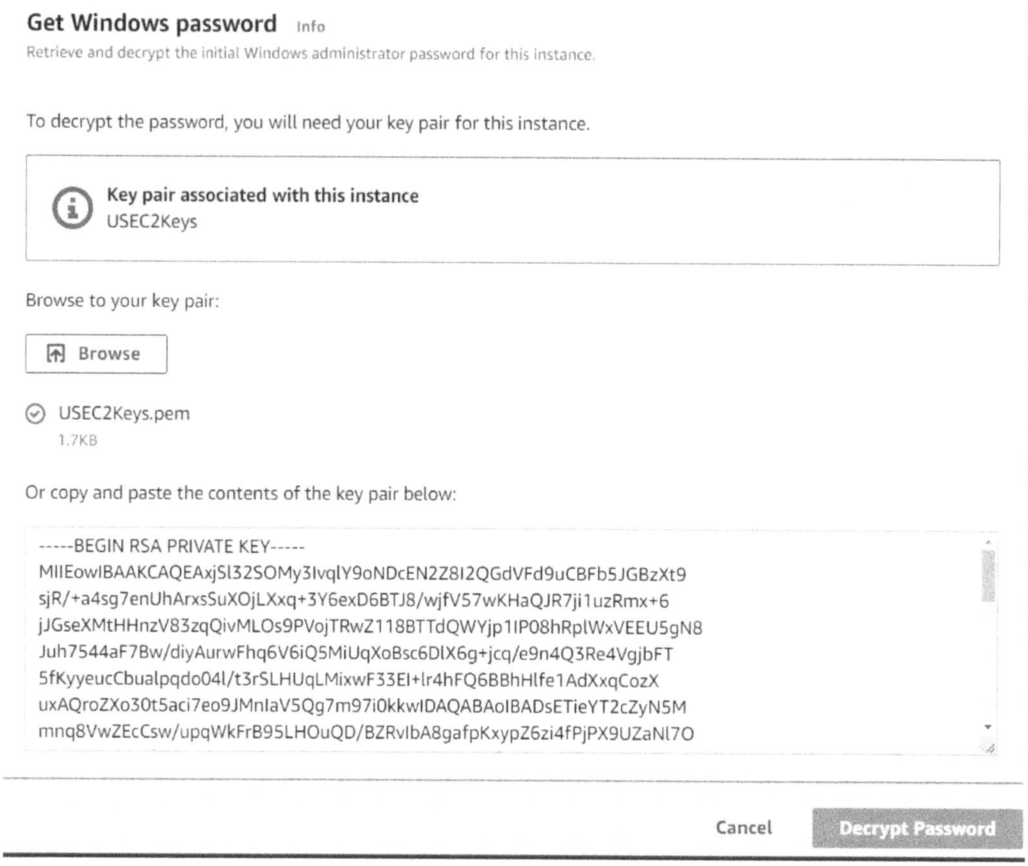

Get Windows password   Info

Retrieve and decrypt the initial Windows administrator password for this instance.

To decrypt the password, you will need your key pair for this instance.

> (i) **Key pair associated with this instance**
> USEC2Keys

Browse to your key pair:

[Browse]

⊘ USEC2Keys.pem
  1.7KB

Or copy and paste the contents of the key pair below:

```
-----BEGIN RSA PRIVATE KEY-----
MIIEowIBAAKCAQEAxjSl32SOMy3IvqlY9oNDcEN2Z8I2QGdVFd9uCBFb5JGBzXt9
sjR/+a4sg7enUhArxsSuXOjLXxq+3Y6exD6BTJ8/wjfV57wKHaQJR7ji1uzRmx+6
jJGseXMtHHnzV83zqQivMLOs9PVojTRwZ118BTTdQWYjp1IP08hRplWxVEEU5gN8
Juh7544aF7Bw/diyAurwFhq6V6iQ5MiUqXoBsc6DlX6g+jcq/e9n4Q3Re4VgjbFT
5fKyyeucCbualpqdo04l/t3rSLHUqLMixwF33EI+lr4hFQ6BBhHlfe1AdXxqCozX
uxAQroZXo30t5aci7eo9JMnIaV5Qg7m97i0kkwIDAQABAoIBADsETieYT2cZyN5M
mnq8VwZEcCsw/upqWkFrB95LHOuQD/BZRvIbA8gafpKxypZ6zi4fPjPX9UZaNl7O
```

Cancel   **Decrypt Password**

Figure 7.26 – Decrypting your key pairs

6. Click on the **Decrypt Password** button in the bottom right-hand corner of the screen; you will be provided with the Windows Administrator password. Make a note of this password.

7. Next, launch your **Remote Desktop Connection** application.

8. In the **Remote Desktop Connection** app, type in the public IP address of your Windows Bastion Server in the text box, next to the **Computer:** field, as per the following screenshot:

Figure 7.27 – Remote Desktop Connection client

9. Click the **Connect** button.

10. You will be prompted to provide your security credentials in the Windows Security dialog box, as per the following screenshot:

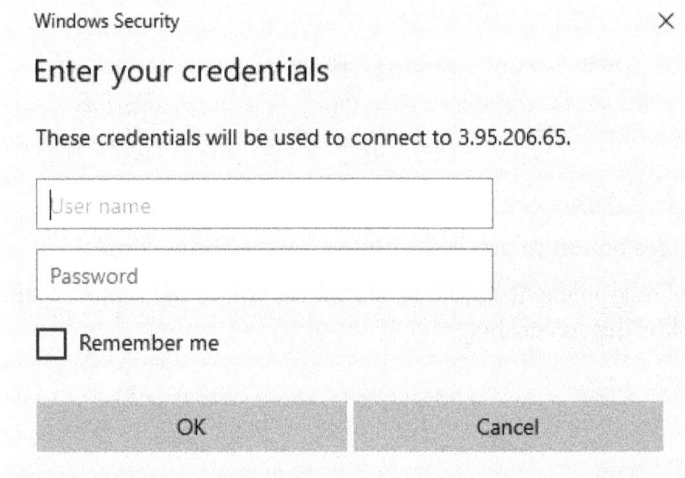

Figure 7.28 – Windows Security – Enter your credentials

11. For **Username**, type in `Administrator`.

12. For **Password**, type in the password you decrypted earlier, and then click **OK**.

13. If the password has been correctly typed in, you will be prompted with a **Remote Desktop Connection** security prompt, informing you that the connection to the remote computer cannot be authenticated due to its security certificate. This warning message can be ignored and you can proceed to log in to the server, as per the following screenshot:

Figure 7.29 – RDP certificate warning dialog box

14. Click on the **Yes** button to proceed with the remote connection.

15. The Remote Desktop client should now connect you to the remote Windows Server, as per the following screenshot:

Figure 7.30 – Remote Bastion Host Server

In this exercise, you were able to launch a new EC2 instance that will act as our Bastion Host server. You will be able to remotely connect to the Windows EC2 instance using the RPD client and can now perform any operation, as required, on this server.

> **Important Note**
>
> As part of ending this exercise, you should terminate your EC2 instances to ensure you do not go over the billing alarm threshold you configured in *Chapter 4, Identity and Access Management*. To terminate your EC2 instances, from the EC2 dashboard, click on **Instances** from the left-hand menu. Next, select the checkbox next to the **Windows BastionSrv** EC2 instance you launched. Then, from the top right-hand menu, click on the **Instance state** drop-down menu and select **Terminate instance**. You will be prompted to confirm that you wish to terminate your EC2 instances. Go ahead and click the **Terminate** button.

In the next exercise, we will demonstrate the Amazon ECS service while focusing on the Fargate launch type with a simple example.

Exercise 7.4 – Launching an application on Amazon Fargate

In this exercise, you will launch a task on ECS, which is essentially a Docker container:

1. Log in to your AWS account as **Alice** and in the top search bar in the AWS Management Console, type in ECS.

2. From the search results, select **Elastic Container Service**.

3. You will be presented with the ECS splash screen, as per the following screenshot:

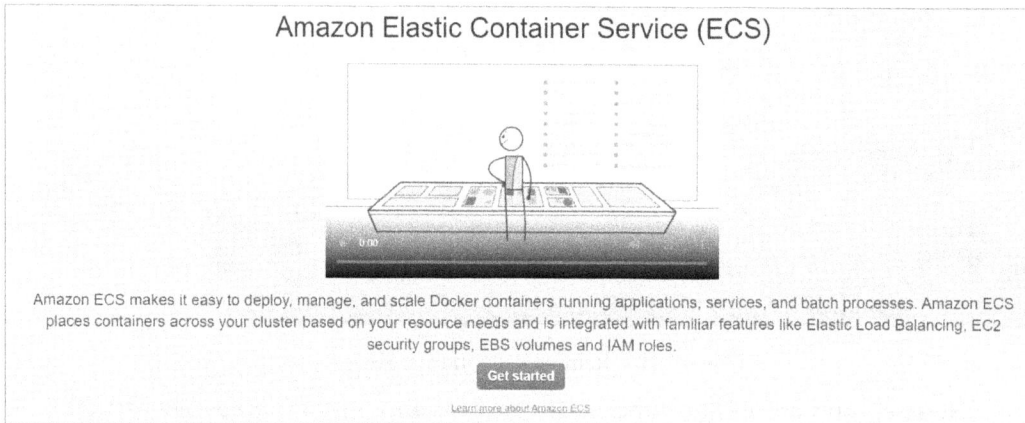

Figure 7.31 – Amazon ECS

4. From the left-hand menu, select **Clusters**.

5. In the right-hand pane, click the **Create cluster** button.

6. You will be prompted to select the cluster template. You will be deploying a Fargate cluster, so go ahead and select the **Network only** option design for use with either AWS Fargate or an external instance capacity.

7. Click the **Next step** button at the bottom of the page.

8. On the next screen, name your cluster MyCluster.

9. Next, click the **Create** button at the bottom of the page.

10. You will see a notification once the cluster has been created. Click the **View Cluster** button.

11. Next, from the left-hand menu, click on **Task Definition**.

12. In the right-hand pane, click the **Create new Task Definition** button.

13. You will then be prompted to select the launch type's compatibility, as per the following screenshot:

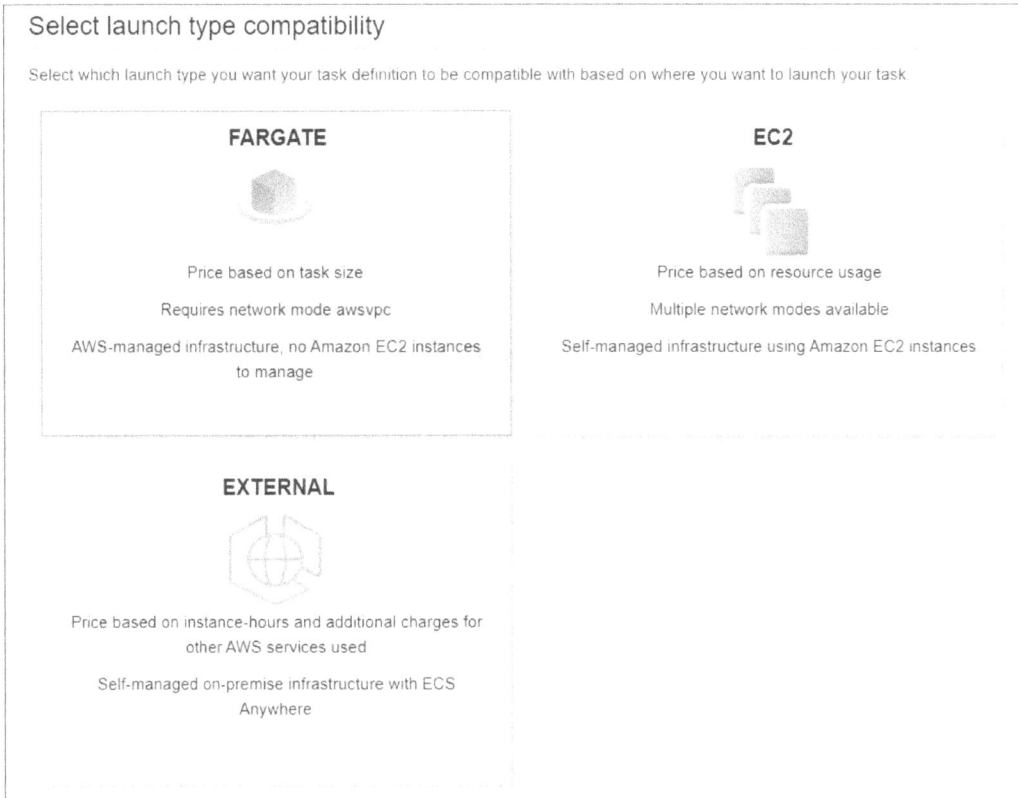

Figure 7.32 – ECS task definition

14. Select the Fargate type and click the **Next step** button at the bottom of the page.

15. Provide a name for your task definition; for example, `fargate-task`.

16. Scroll further down and under **Task size**, set **Task memory (GB)** to **1GB**.

17. Next, set **Task CPU (vCPU)** to **0.5 vCPU**.

18. Next, under **Container Definitions**, click the **Add container** button.

19. You will be selecting an existing container here, so in the **Add container** dialog box, for **Container name**, type in `nginx`.

20. Similarly, for **Image**, ensure you type in `nginx`, as per the following screenshot:

Add container

▼ Standard

             Container name*   nginx   🛈

                       Image*   nginx   🛈

Figure 7.33 – Fargate launch container

21. Next, for **Port mappings**, type in `80` for the **Container** port.

22. Scroll toward the bottom and click the **Add** button. This will take you back to the **Create new Task Definition** page. Scroll toward the bottom and click on the **Create** button.

23. After a few seconds, you should find that your task definition has been created.

24. Click on the **View task definition** button. This will take you back to the ECS dashboard.

25. Next, click on the **Clusters** link from the left-hand menu. You will see that your **MyCluster** cluster is now available. Click on the **MyCluster** link.

26. Next, click on the **Task** tab.

27. Click the **Run new Task** button.

28. On the **Run Task** page, ensure that **Launch type** is set to **Fargate**.

29. You will find that the task definition has been pre-populated with your task. If not, select it from the drop-down arrow.

30. For **Number of tasks**, ensure that only `1` is set as we will only run one task.

31. Under **VPC and security groups**, select **ProductionVPC**.

32. For **Subnets**, select **Public Subnet One**.

33. For **Security groups**, click the **Edit** button next to the provided security group name.

34. You will be prompted to create a new security group, as per the following screenshot:

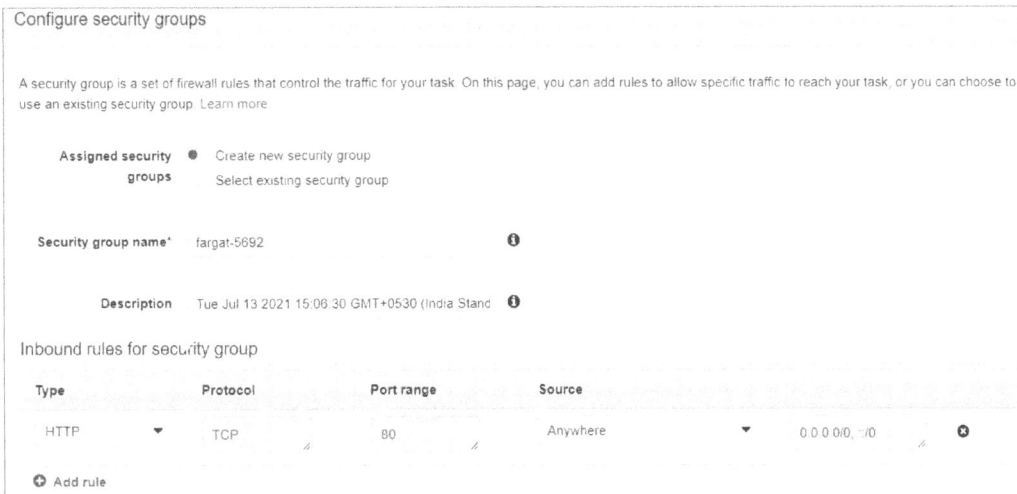

Figure 7.34 – Fargate task security group

35. Accept the default to create an inbound rule that allows port 80 from the internet. Click the **Save** button at the bottom of the page.

36. Next, ensure that **Auto-assign public IP** is set to **ENABLED**.

37. Click the **Run Task** button at the bottom of the page.

38. You will be taken back to the ECS dashboard.

39. Within a few seconds, you should find that your task is now in the **RUNNING** status, as per the following screenshot:

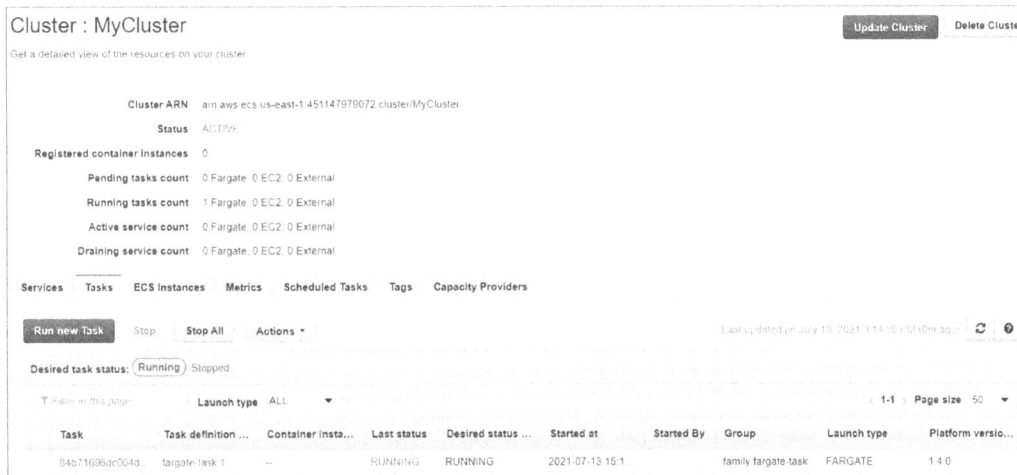

Figure 7.35 – Fargate running task

40. Next, under **Task**, click on the task ID link. This will bring up the **Details** page of the task.

41. Make a note of the **Public IP** address and copy and paste it into a new browser tab.

42. You should then find that you can connect to the nginx web page, as per the following screenshot:

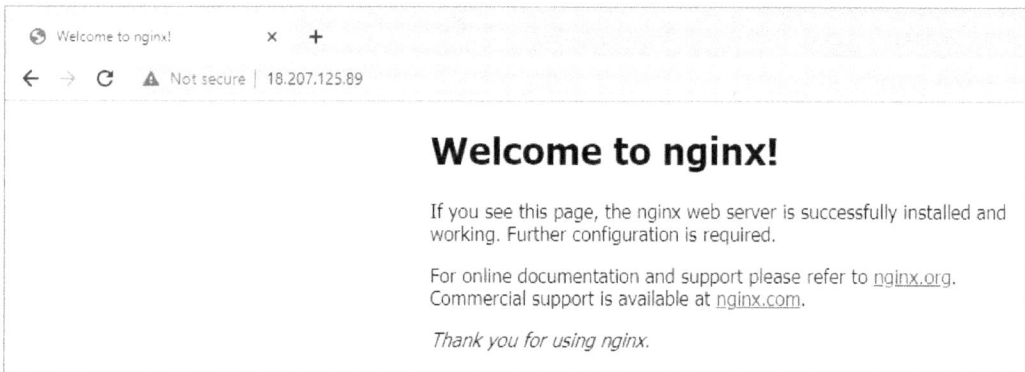

Figure 7.36 – nginx application running on Fargate

43. You have just deployed your first Fargate task!

44. To clean up, go back to your ECS dashboard. From the left-hand menu, click the **Clusters** link.

45. In the right-hand pane, click the **MyCluster** link.

46. Next, from the **Task** tab, select the checkbox next to your task.

47. Finally, click on the **Stop** button. You will be prompted to confirm your action. Click the **Stop** button in the dialog box that appears.

48. This will stop your task. You will also receive a notification, stating that the task was successfully stopped. You do not need to worry about the cluster as you only pay for the tasks on Fargate.

In this exercise, we demonstrated how you can launch a simple nginx Docker container on ECS using the Fargate launch type. Next, we will summarize this chapter.

Summary

In this chapter, we looked at a wide range of AWS compute solutions, starting with their flagship solution, Amazon **EC2**. We discussed AMIs, which are machine images that contain the base operating system, any applications and the state of patches, and updates that are used to launch your EC2 instance as a fully-fledged server. AMIs are made available via the Quick Start AMIs, Community, and Marketplace, and you can also create your own AMIs based on your corporate requirements.

We examine the different EC2 instance types and families and how an instance type is effectively a description of associated hardware and software specifications that your EC2 instance offers. We also looked at storage options for your EC2 instance, with EBS, EFS, and FSx as possible solutions.

Then, you learned how to choose a particular EC2 Pricing Option, depending on your application needs, and identify opportunities for cost-effective deployment by combining the On-Demand, Reserved, and Spot Instance options.

In the next chapter, we will look at databases on AWS. Almost every application requires some form of database to host structured data. Amazon offers both relational and non-relational databases to help design modern web and mobile applications. Amazon also offers additional database solutions for specific niches such as social networking platforms, data warehousing solutions, and databases designed for hosting highly sensitive data.

Questions

1. Which AWS EC2 pricing option can help you reduce costs by allowing you to use your existing server-bound software licenses?

    A. On-Demand

    B. Reserved

    C. Spot

    D. Dedicated Hosts

2. Which AWS EC2 pricing option enables you to take advantage of unused EC2 capacity in the AWS cloud and can offer up to a 90% discount compared to On-Demand prices?

    A. Spot Instances

    B. Reserved Instances

    C. On-Demand Instance

    D. Dedicated Hosts

3. Which of the following is true with regards to the benefits of purchasing a Convertible EC2 Reserved Instance? (Choose 2 answers)

    A. You can exchange a Convertible Reserved Instance for a Convertible Reserved Instance in a different Region.

    B. You can exchange one or more Convertible Reserved Instances at a time for both Convertible and Standard Reserved Instances.

    C. To benefit from better pricing, you can exchange a No Upfront Convertible Reserved Instance for an All Upfront or Partial Upfront Convertible Reserved Instance.

    D. You can exchange All Upfront and Partial Upfront Convertible Reserved Instances for No Upfront Convertible Reserved Instances.

    E. You can exchange one or more Convertible Reserved Instances for another Convertible Reserved Instance with a different configuration, including instance family, operating system, and tenancy.

4. Which feature of the AWS EC2 service helps prevent accidentally terminating an EC2 instance by preventing the user from issuing a termination command either from the console or CLI?

    A. Enable "termination protection"

    B. Enable "termination protect"

    C. Enable "prevent termination"

    D. Enable "protect EC2"

5. Which storage solution enables you to share a common filesystem across multiple Linux-based EC2 instances that can be used to support applications that require access to data with very low latency connectivity?

    A. EFS

    B. EBS

    C. S3

    D. NTFS

6. Which type of IP address offering from AWS gives you a static, publicly routable address that will not change, even if you stop and restart an EC2 instance that it is associated with?

    A. Public IP address

    B. Private IP address

    C. Elastic IP address

    D. Regional IP address

7. Which AWS service enables you to configure multiple Windows-based EC2 instances to share and access a common storage solution that is based on using the industry-standard SMB protocol and eliminate the administrative overhead of managing Windows file servers?

    A. Amazon FSx for Windows File Server

    B. Amazon Elastic File System

    C. Amazon Elastic Block Store

    D. Amazon DFS Volumes for Windows File Server

8. Which of the following types of EBS volumes can be used as boot volumes for your EC2 instances? (Select 2 answers)

    A. General Purpose SSD (gp2)

    B. Provisioned IOPS SSD (io1)

    C. Throughput Optimized HDD (st1)

    D. Cold HDD (sc1)

    E. FSx for Windows

9. Which of the following AWS services enables you to quickly launch a web server with a pre-configured WordPress installation pack, offers predictable monthly pricing, comes with integrated certificate management, and provides free SSL/TLS certificates?

    A. AWS Lightsail

    B. AWS EC2

    C. AWS RDS

    D. AWS Elastic Beanstalk

10. Which AWS service can be used to run a piece of code that can create thumbnails of images uploaded to one Amazon S3 bucket and copy them to another S3 bucket?

    A. AWS SNS

    B. AWS Lambda

    C. AWS RDS

    D. AWS Snowball

8
AWS Database Services

Most applications need to store, access, organize, and manipulate data in some way. Normally, the data would reside externally to the actual application in what we call a *database* for several reasons, including efficiency improvements. Databases are designed to do more than simply store data, however. Depending on the type of database, data can be organized and stored in a structured or semi-structured manner, offer high-speed access to the data, and give you the ability to perform queries and scans against the data. Data can also be combined from different *tables* within the database to help you create complex analytics and reporting. Typical examples of where you would use a database include storing customer records and their orders for your e-commerce website, storing a product listing catalog, and storing temperature information from your home IoT thermostat devices. AWS offers three primary database solutions and several others for specific application types.

In this chapter, we will cover the following topics:

- Managed databases versus unmanaged databases
- Introduction to database concepts and models
- Introduction to **Amazon Relational Database Service** (**Amazon RDS**)
- Learning about Amazon DynamoDB (a NoSQL database solution)
- Understand the use cases for Amazon Redshift and data warehousing
- Understanding the importance of in-memory caching options with Amazon Elasticache
- Learning about additional database services for specific niche requirements
- **Database Migration Service** (**DMS**)

In this chapter, you will learn about the various managed databases solutions offered by AWS and launch your very first Amazon Relational Database service running the MySQL engine. Later in this book, we will configure a database to store data that's been uploaded via a web application.

Technical requirements

To complete the exercises in this chapter, you will need to access your AWS account and be logged in as our fictitious administrator, **Alice**, using her IAM user credentials.

Managed databases versus unmanaged databases

Traditionally, in an on-premises setup, you would configure a server (physical or virtual) with a base operating system and then proceed to install the database software on it. Because the primary purpose of a database is to store data, you would also need to ensure that you had adequate storage attached to your server. Due to the importance of the data store, you would take additional security measures to protect the data and ensure you had adequate backups and copies of the data (ideally stored offsite in another location) in case of disasters.

On AWS, you can set up an **Elastic Compute Cloud** (**EC2**) instance and install your database, such as **Microsoft SQL Server** or **Oracle**, in the same manner to serve your frontend web and application servers as required. In this case, you take full ownership of managing the database, provisioning the required amount of **Elastic Block Store** (**EBS**) volumes for storage, and ensuring adequate backups are made. You also need to design for high availability and performance.

Alternatively, AWS also offers **managed database solutions**. AWS takes care of provisioning your database instances, where you specify certain parameters to ensure the required capacity for your application. AWS will also provision and manage the required storage for your database, as well as perform all backups and replications as required. Ultimately, you get a fully managed solution where AWS takes care of almost every configuration option you choose, except for ensuring that your application is optimized for the chosen database solution.

Learning about additional database services for specific niche requirements

In this section, we introduced you to the concept of unmanaged databases and how traditionally, we would have to install our database software on a physical or virtual server. However, hosting a database on a server carries additional administrative efforts. While on AWS, you can install a database on an EC2 instance, it makes more sense to consider using AWS managed database offerings such as **Amazon RDS** as this reduces the management burden on the customer. In the next section, we will introduce you to database concepts and models.

Introduction to database concepts and models

Today, there are several types of database models, but the most common are *relational* and *non-relational* models. Relational databases have existed for years and allow you to efficiently manage your data with the ability to perform complex queries and analyses. However, they have certain restrictions, such as the fact that you need to define the database schema (its structure) before you can add data, and changing this later can be difficult. Non-relational databases offer a lot more flexibility and are used for many modern-day web and mobile applications. Let's look at the key differences.

Relational databases

A **relational database** is often compared to a spreadsheet, although databases offer more capabilities than just letting you analyze data using complex calculations and formulas. Like a spreadsheet, a relational database can be composed of one or more `tables`. Within each table, you have rows and columns – columns define `attributes` for your data and rows contain individual `records` in the database. Each row in your table then contains data that relates to the attributes that were defined in the columns. So, for example, in a customer's table, you can have columns such as `First Name` and `Last Name`, then your rows will contain data related to those columns comprised of your customer's first and last names.

Another important factor to consider with **relational databases** is the need to define your database schema before you add data to the database. For example, if you have a column called `First Name` and a second column called `Date of Birth`, then you need to define the type of data you will permit in each of those columns prior to adding any data; for the `First Name` column, the type of data will be `string`, whereas for the `Date of Birth` column, you will define the type of data as `date`.

An important column (`attribute`) that must exist in a relational database is the `Primary Key` field. Each record must have a primary key that is unique across the whole table. This ensures that each record within the table is unique, allowing you to easily query specific records in the table. As shown in the following table, the customer records table has a primary key called `CustomerID`:

Primary Key	Customer-ID	First Name	Last Name	Flat/House	Street	Town
	Cust001	Penny	Smith	11	Hoxton Grove	Richmond
	Cust002	Mark	Edwards	18	Cavendish Ave	Harrow
	Cust003	Luke	Skywalker	33	Hazelwood Ave	Morden
	Cust004	Jean Luke	Picard	22	Delta Drive	Knightsbridge

Figure 8.1 – Customer contact table

A database can also host multiple tables for specific record sets. Rather than storing all the records within a single table, you can have separate tables for related data. So, for example, in one table, you can have your *customer contact* details and in another, you can have your *customer order* details. In most cases, the tables will have some relationship with other tables in the same database. In this example, the tables are related to specific customers; one for their contact details and another for their orders.

The purpose of separating different sets of data into separate tables is to allow for better management, performance, and to avoid duplicate data. For example, if you have a single table to host customers' contact details as well as their orders, then you would have multiple records relisting the customers' contact information for every single order they placed. By separating the orders from the contact details into separate tables, we can avoid this duplication of data and improve performance. The following is an example of a customers' orders table:

Customer-ID	Order-ID	Order Date	Order Amount
Cust002	Omega001	12 April 2019	$200.75
Cust002	Omega002	15 June 2019	$250.99
Cust002	Omega003	17 August 2019	$330.99

Figure 8.2 – Customer order table

In the preceding two tables, note that we avoided replicating the data by separating the customer contact details table from the orders table. If we have all the data in one table, then we would have multiple columns with the same pieces of information repeated, such as First Name and Last Name for every order placed by the same customer.

The tables in a database can relate to each other, and we need some form of connection between the two tables to effectively structure the data. In the previous example, rather than repeating all customer address details in the orders table, we simply include the Customer-ID column, where we identify which customer the order relates to. Remember that the Customer-ID column is the primary key of the **Customer Contacts Table**, so each Customer ID uniquely identifies a customer.

Ultimately, we can now query the database by combining data from both columns using the Customer ID as a reference point. We can then produce a report from a query to list all the orders that have been placed by a customer, whose Customer-ID is Cust002. In this report, we can list the customer contact details, which will be extracted from the first table, and the list of orders placed from the second table, where Customer-ID is Cust002. This report could then be sent to the customer as a statement of account.

Relational databases enable you to perform such complex queries, analyses, and reports from large datasets. Performance is directly correlated to the types of queries you need to perform and the volume of data you host. Often, this means that your infrastructure may need to be upgraded from time to time to cope with demanding applications.

Relational databases use the **Structured Query Language** (**SQL**), which is a language designed to create databases and tables, as well as manage the data within the database using queries. SQL is written out as statements that you then issue to the database to perform a specific action. For example, the SELECT statement enables you to query specific data, whereas the WHERE statement enables you to restrict your SELECT query to match a specific condition. For the *AWS Certified Cloud Practitioner* exam, you are not expected to know how to use the SQL language.

Relational databases are also known as **Online Transaction Processing** (**OLTP**) **databases**. **OLTP databases** are designed for adding, updating, and deleting small amounts of data in a database regularly. Typical examples include a *customer orders* database for an e-commerce website or a student's database for a university.

Non-relational (NoSQL) databases

With relational databases, you store data in a structured format of a defined schema in tables. Each column of a table (known as an attribute) will only hold one type of data and this needs to be predefined. You usually query multiple tables of related data and combining queries across your tables can yield required pieces of information.

However, the problem with relational databases is the lack of flexibility since data needs to be structured. Furthermore, the more tables you have across your database, the more complex the queries tend to be, and the more resources are required to run and manage the databases. Relational databases also do not lend themselves well when trying to perform thousands of reads and writes to the database per second.

In contrast, non-relational databases do not follow the traditional relational approach to storing data. Non-relational database data is stored using different models, depending on the type of data being stored. These are as follows:

- **Key-value stores**: This is a collection of key-value pairs contained within an object.

- **Document data stores**: This is typically a **JavaScript Object Notation** (**JSON**) format document (although other formats can be used) that's used to store data in a structured or semi-structured form. Data can be comprised of nested attributes of key-value pairs. All the documents in the store are not required to maintain identical data structures and this offers greater levels of flexibility.

- **Columnar data stores**: Data is organized into cells grouped by columns rather than by rows. Furthermore, reads and writes are carried out using columns rather than rows.

There is no requirement to predefine the schema of the database, and this creates a lot of flexibility because you can freely add fields (attributes) to a document without the need to define them first. Therefore, you have documents with different numbers of fields. For example, one document listing a customer's details could include their name, address details, order history, and credit card information, while another document could include a list of their favorite products.

Non-relational databases were developed as an alternative to relational databases where the flexibility of the schema was required, as well as to handle very large data stores that required thousands of reads/writes per second, something that relational databases have traditionally found difficult to do. Non-relational databases can cope with this kind of load because a query does not have to view several related tables to extract the results. Furthermore, a non-relational database can handle frequent changes to the data.

Like relational databases, though, non-relational databases do require you to have at least one primary key field (attribute) and this is the only attribute required. Beyond this, your database tables are effectively schemaless. The primary key is used to ensure that each record of the database is unique.

In this section, we reviewed the primary differences between relational and non-relational databases. We looked at the use cases for both types of database solutions and compared the key differences between the two. On AWS, both relational databases and non-relational databases are offered. In the next section, we will look at the services that are offered in detail.

Introduction to Amazon RDS

Amazon RDS offers traditional relational databases as fully managed services on the AWS platform. Ideal for transactional database requirements, also known as **OLTP**, AWS offers six different database engines, as follows:

- MySQL
- PostgreSQL
- MariaDB
- Microsoft SQL server
- Oracle
- Amazon Aurora

Another term you might have heard of is **Relational Database Management System (RDBMS)**. An RDBMS performs functions to **create, read, update, and delete** (**CRUD**) data from the database using an underlying software component, which we call the database engine.

An important point to understand here is that when you choose to set up an Amazon RDS database, you are setting up a *database instance* with a chosen engine to run on that instance. You can then create one or more databases supported by that engine on your database instance. This means you can have several databases running on an individual database instance.

Furthermore, on Amazon RDS, when you set up a database instance, you specify hardware capabilities in the form of CPU and memory allocation. The type of instance will also determine the maximum storage bandwidth and network performance that the instance can offer. AWS offers three different types of instance classes with varying virtual hardware specifications and is designed for various uses cases. These are as follows:

- **Standard classes (includes m classes)**: These classes offer a balance of compute, memory, and network resources, and they are ideal for most application requirements. Standard classes offer the following specs:

  - Between 2 and 96 vCPUs

  - Up to 384 GB of memory

- **Memory-optimized classes (includes r and x classes)**: These classes are ideal for most demanding applications that require greater levels of memory and are optimized for memory-intensive applications. Memory-optimized classes offer the following specs:

  - Between 4 and 128 vCPUs

  - Up to 3,904 GB of memory

- **Burstable classes (includes t classes)**: These classes are designed for nonproduction databases and provide a baseline performance level, with the ability to burst to full CPU usage. Burstable classes are ideal for database workloads with moderate CPU usage that experience occasional spikes. Burstable classes offer the following specs:

  - Between 1 and 8 vCPUs

  - Up to 32 GB of memory

The following screenshot shows the different **DB instance class** options you can select from:

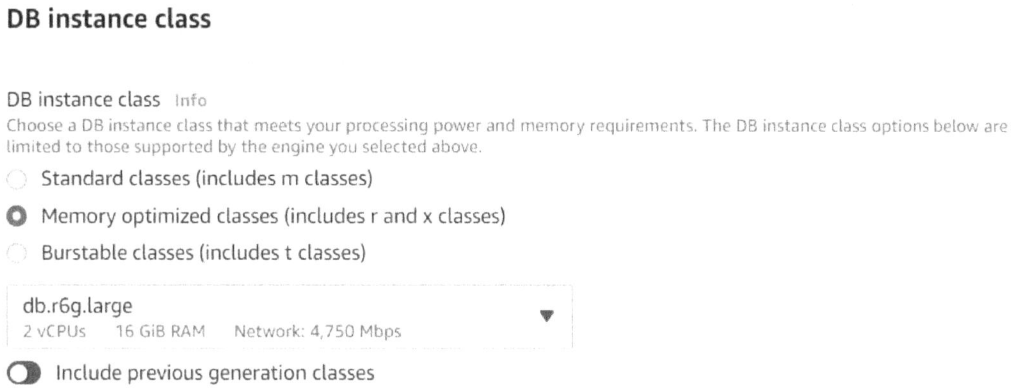

DB instance class

DB instance class Info
Choose a DB instance class that meets your processing power and memory requirements. The DB instance class options below are limited to those supported by the engine you selected above.

○ Standard classes (includes m classes)

● Memory optimized classes (includes r and x classes)

○ Burstable classes (includes t classes)

db.r6g.large
2 vCPUs   16 GiB RAM   Network: 4,750 Mbps   ▼

◐ Include previous generation classes

Figure 8.3 – Database instance class options

In addition to the compute resource, Amazon RDS also requires storage capabilities to host all the required data. The storage platform runs on Amazon **EBS**, so it is decoupled from the actual database instance class. This allows you to upgrade the storage volumes without necessarily having to upgrade the instance class
and vice versa, so long as compatibility is maintained. The volume's throughput is determined by the instance types chosen, as well as the **input/output operations per second** (**IOPS**) that the EBS volume supports. AWS offers the following different storage options for your databases:

- **General Purpose SSD**: Designed for standard workloads and ideal for most databases, General Purpose SSD volumes offer between 20 GiB to 64 TiB of storage data for MariaDB, MySQL, PostgreSQL, and Oracle databases, and between 20 GiB to 16 TiB for Microsoft SQL Server.

  The number of IOPS that's achieved is dependent on the size of the storage volume, with a baseline I/O performance of 3 IOPS per GiB (minimum 100 IOPS). The larger the volume size, the higher the performance; for example, a 60 GiB volume would give you 180 IOPS.

  General Purpose SSDs also offer bursts in performance for volumes less than 1 TiB in size for extended periods. This means that smaller volumes will get an additional performance boost when required and you do not need to allocate unnecessary storage for short-term occasional spikes. Bursting is not relevant for volume sizes above 1 TiB, however.

- **Provisioned IOPS SSD**: AWS recommends using Provisioned IOPS SSDs for production applications that require fast and consistent I/O performance. With Provisioned IOPS SSDs, you specify the IOPS rate and the size of the volume. Like General Purpose SSDs, you can allocate up to 64 TiB of storage, depending on the underlying database engine you use. Provisioned IOPS SSD does not offer any bursting, however.

- **Magnetic**: AWS also offers magnetic storage volumes for backward compatibility. They are not recommended for any production environments and are limited to 1,000 IOPS and up to 3 TiB of storage.

Ultimately, you can upgrade your storage if required, but this will usually require a short outage, typically of a few minutes (for Magnetic, this can take much longer), so this must be planned for.

Deploying in Amazon VPCs

Amazon RDS is a regional service, which means that you need to select the Region you want to deploy your database instance in first. Amazon RDS database instances can only be deployed in a VPC, and like EC2 instances, in a specified subnet. Since a subnet is always only associated with a single Availability Zone, this also means that if there is an Availability Zone failure, your database will not be accessible. With Amazon RDS, you can only deploy a **single master database instance**. This type of instance can perform both read and write operations to the database. Amazon does offer various solutions in case the master database instance fails and we look at these options later in this chapter.

Deploying your RDS database in a VPC gives granular control over how the databases is going to be accessed and allows you to configure various network security components such as the private IP addressing range you will use, security groups to protect your RDS instance, and **Network Access Control Lists** (**NACLs**) to protect traffic in the subnet that will host the database.

Deploying your RDS database in a VPC also means that you can configure various architectures from which to access that database. The following are some scenarios that you could configure:

- **A DB instance in a VPC that's accessed by an EC2 instance in the same VPC**: In this configuration, you would need to configure the security group associated with the RDS database to accept inbound traffic from the security group associated with the EC2 instance on the relevant database port; for example, port 3306 for a MySQL RDS instance.

- **A DB instance in a VPC that's accessed by an EC2 instance in a different VPC**: In *Chapter 6, AWS Networking Services, VPCs, Route53, and CloudFront,* we discussed that you could connect two VPCs using a service known as VPC peering. Instances in either VPC can then communicate with each other over that peering connection using private IP addresses as though they were within the same network. Once the peering connection has been established, you would then need to configure the necessary rules for your security groups to enable traffic to flow between the EC2 instance and your RDS database. VPC peering can allow you to peer connections between VPCs in the same Region, cross-Region, and even across AWS accounts.

- **A DB instance in a VPC that's accessed by a client application through the internet**: While nothing is stopping you from placing your RDS database instance in a public subnet of a VPC, this is not considered a best practice for production environments. Databases are considered backend services that contain critical and maybe sensitive information. They should always be placed in the private subnet of the VPC. Placing your RDS database in a public subnet should only really be done for testing purposes or specific use cases.

- **A DB instance in a VPC that's accessed by a private network**: With a VPC in place, you can set up a VPN tunnel or a Direct Connect service between your on-premises network and the VPC. This allows you to place your RDS database in a private subnet of the VPC, and still be able to access it from your corporate offices via the VPN tunnel or Direct Connect service.

For production environments, always place your RDS database instances in the private subnet(s) of your VPC. The architecture shown in the following diagram illustrates one such best practice methodology. Here, the RDS database has deployed a private subnet, dubbed a **database subnet**. To access the database via a standard web application, traffic from the internet is routed via an **Elastic Load Balancer** (**ELB**) (discussed in detail in *Chapter 9, High Availability and Elasticity on AWS*) and distributed to web servers placed in another set of private subnets within your VPC. Your web servers then connect to the RDS database to perform any data operations such as adding, updating, or deleting records as required by the application. Traffic can flow based on the rules that have been defined by your **Network Access Control Lists** (**NACLs**) and security groups:

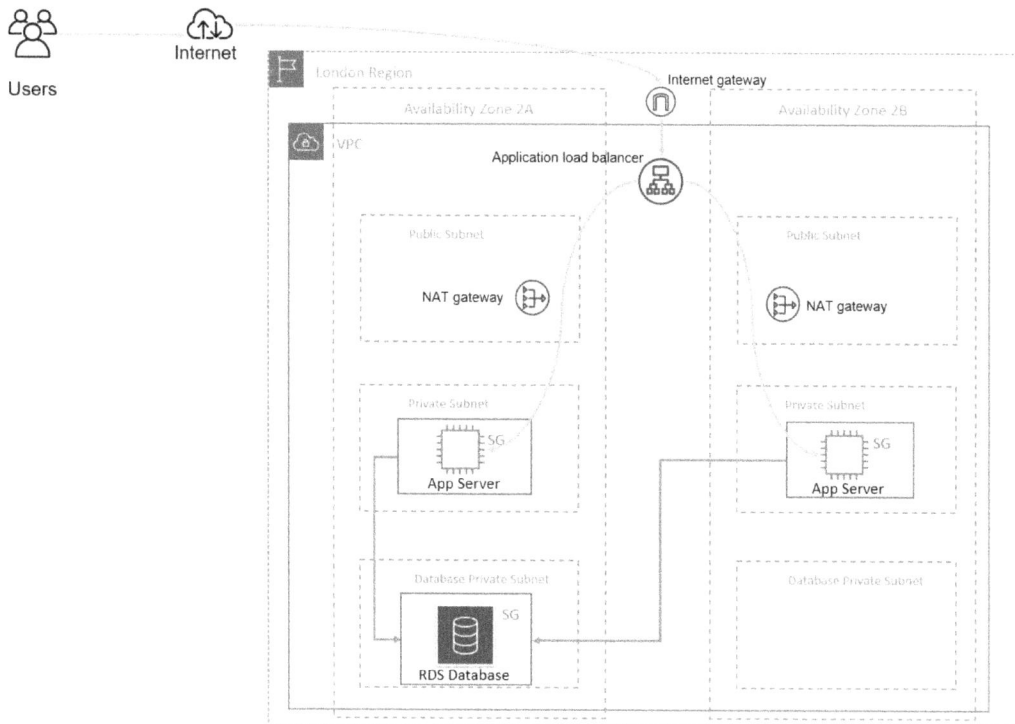

Figure 8.4 – Amazon RDS deployed in a VPC in private subnets

In the preceding diagram, users from the internet can access the database via the web/app server rather than have direct access to the database. Users will connect to the web/app servers via an **Application Load Balancer** (**ALB**) (which distributes traffic among healthy EC2 instances in the fleet). The web/app server will have a process in place to send database operations requests to the RDS database in the backend private subnet.

Note that the RDS database will only accept traffic from the EC2 instances that are attached to the appropriate security groups. This ensures that if the EC2 instances are replaced or if additional EC2 instances are attached to the fleet, and if they are attached to the same security sroup, they will be able to communicate with the RDS database.

> **Note**
> Your applications connect to the backend database via an RDS DNS endpoint name, rather than the database instance's specific IP addresses. This allows you to easily manage failovers in the event of a disaster, which we will discuss next in this chapter.

Backup and recovery

Your database is going to be very important to you and ensuring that you can recover from failures, data loss, or even data corruption is going to be an important factor in your design architecture. AWS enables you to address your disaster recovery and business continuity concerns, and Amazon RDS comes with several configuration options to choose from.

What are your RPO and RTO requirements?

When deciding on a strategy to protect your database on AWS from unexpected failures or data corruption, you need to consider what configuration options are going to meet your organization's expectations for recovery. If your business hosts critical data that needs to be recovered fast in the event of a failure, you need to design an architecture that will support this requirement. To help you identify how critical your recovery is going to be, you need to determine your **Recovery Point Objective** (**RPO**) and **Recovery Time Objective** (**RTO**). These two parameters will help you design a recovery strategy that meets your business requirements:

- **RTO**: This represents the time (in hours) it takes to recover from a disaster and return to a working state. The time taken will involve provisioning a new database instance, performing a restore job, and any other administrative or technical tasks that need to be completed.

- **RPO**: This represents how much data (again, as a measure of time, and generally in hours) you will lose in the event of a disaster. The shorter the RPO, the less data you are likely to lose in the event of a failure.

If your organization stipulates that it can only afford an RTO of 2 hours and an RPO of 4 hours, this means that you need to recover from failure to a working state within 2 hours and the maximum amount of data you can afford to lose (probably because you can create that data) is 4 hours' worth.

Based on your RPO and RTO levels, you can then choose a disaster recovery strategy that fits your requirements. For example, if your RPO is set to 4 hours and your recovery strategy was based on restoring older backups of your database, then you should be performing a backup of your database every 4 hours.

High availability with Multi-AZ

For Amazon RDS database engines running MariaDB, MySQL, PostgreSQL, Oracle, Microsoft SQL, and Amazon Aurora, AWS offers high availability and failover support using the **RDS Multi-AZ** solution. Multi-AZ is an architectural design pattern where a primary (master) copy of your database is deployed in one Availability Zone and a secondary (standby) copy is deployed in another Availability Zone. Data is then **synchronously** replicated from the master copy to the standby copy continuously. Normally, with relational databases, only one database can hold the *master* status, meaning that data can be both written to and read from it. In the case of Multi-AZ deployments, this is still true, and the standby copy of the database simply receives all the changes that have been made to the master synchronously. However, you cannot write or read from the standby directly.

If the master copy of your database fails, then AWS will perform a failover operation to the standby copy of the database. The standby copy will be promoted to become the new master, and the previous master will be terminated and replaced with another standby copy. Replication will then be initiated in the opposite direction. During failover, your application may experience a brief outage (about 2 minutes), but then will be able to reconnect to the database (the standby copy that has been promoted to the new master) and continue operating:

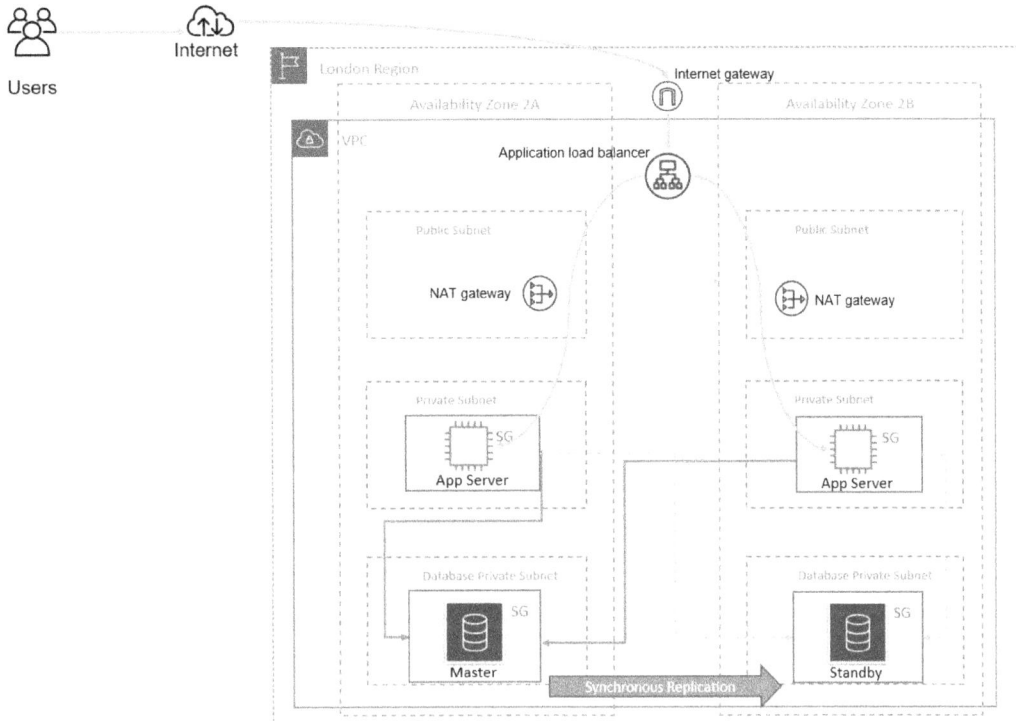

Figure 8.5 – Amazon RDS configured with Multi-AZ

Failover can be triggered for several reasons other than Availability Zone outages, including patching the master database or upgrades of the instance. You can also perform a failover test to ensure that the configuration has been set up correctly by performing a reboot of the master database and requesting a failover operation on reboot.

With Multi-AZ, you can reduce your RTO and RPO levels drastically. Because existing data has already been replicated to a standby copy in another Availability Zone, failover happens in a matter of minutes and data loss is minimized.

However, in certain circumstances, Multi-AZ alone as a DR strategy may not be enough. For example, what happens if there is data corruption on the master copy of your database? That corrupted data will be replicated across to the standby too.

Backup and recovery

AWS also offers options to perform regular backups of your database, which you can use to perform point-in-time restores. AWS offers two options here: *automatic backups* and *manual snapshots*.

Automatic backups

AWS offers a fully managed automatic backup service free of cost up to the total size of your database. While performing automatic backups, the first snapshot that's created will be a full backup; subsequent snapshots will be incremental, ensuring that only changes to the data are backed up. Some additional features of automatic backups include the following:

- **Backup window**: Automatic backups are performed during a predefined window that is configurable for the customer. The default backup time allocated is 30 minutes but you can change this as well. Furthermore, if the backup requires more time than what's been allotted to the backup window, the backup continues after the window ends, until it finishes.

- **Backup retention**: Amazon RDS offers a retention period of 1 day to 35 days. This means that you can store up to 35 days of backups that you can restore from. You can also set the backup retention to 0 days, which essentially translates to disabling backup operations. If you disabled automatic backups at the time of launching your RDS instance, you can enable this later by setting the backup retention period to a positive non-zero value.

- **Automatic backups and transactional logs**: When you configure your RDS database for automatic backups, the backup process runs at the predefined backup window. In addition, AWS will also capture transaction logs (as updates to your DB instance are made) up to the last 5-minute interval, which is uploaded to Amazon S3. AWS can then use the daily backups as well as the transaction logs to restore your DB instance to any second during the retention period, up to LatestRestorableTime, which is typically the last 5 minutes.

Manual snapshots

In addition to automatic backups, you can also create manual snapshots of your database, which can provide additional protection. You can then use manual snapshots to restore your DB instance to a known state as frequently as you like.

Manual snapshots can be very useful if you plan to make a major change to your database and would like an additional snapshot before making that change.

Automatic and manual backups are particularly useful if you need to restore due to data corruption. Remember that even if you have Multi-AZ enabled, any data corruption on the master copy will be replicated across to the standby, and having older backups can enable you to revert your database to a time before that data corruption occurred.

Cross-Region snapshots

You can copy your snapshots across Regions to improve the availability of your backups even further, in the event of a regional outage or disaster. You can also configure the replication of your automatic backups and transactional logs across to another Region. Amazon RDS initiates a cross-Region copy of all snapshots and transaction logs as soon as they are ready on the DB instance.

I/O suspense issue

An important behavior pattern to be aware of is that a brief I/O suspension (usually lasting only a few seconds) is experienced when your backup process initializes. In scenarios where you have a single Database instance deployed, this results in a brief outage when connecting to the database. This means that if your backup operations are taking place during business hours, then users may experience some interruption during the backup process.

To work around this problem, you can choose to ensure that your backup processes take place outside of business hours, or better still, follow Amazon's recommendations for deploying a Multi-AZ deployment for database, particularly for MariaDB, MySQL, Oracle, and PostgreSQL engines. This is because, in a Multi-AZ configuration, the backup is taken from the standby copy of the database and not the master. Note that for Microsoft SQL Server, I/O activity is suspended briefly during backup, even for Multi-AZ deployments.

Horizontal scaling with read replicas

Traditionally, relational databases do not scale well horizontally due to their architecture. A relational database can normally only have one master copy (the copy that you can write data to), which means that if you experience a failure on the master copy, you need to resort to restoring data from backups. AWS offers Multi-AZ as a means to overcome this single point of failure by enabling you to create a standby copy of the database that has data synchronously replicated to it.

However, you cannot use a standby copy to perform write or read queries since your standby copy is only accessible in the event of the failure of the master copy. If the master copy of your database fails, your standby copy gets promoted to become the new master copy of the database, upon which you will be reading and writing to it.

When it comes to scaling horizontally, where you can have multiple nodes of your database, AWS offers an option to scale read copies of your database using a feature called **read replicas**. AWS RDS can use its built-in replication functionality for Microsoft SQL, MySQL, Oracle, and PostgreSQL to create additional read replicas of the source DB instance. Data is then replicated from the source DB to the replicas using *asynchronous* replication. This can help reduce the load on your master copy by redirecting read queries to the read replicas instead. Application components that only need to read from the database can be routed to send requests to the read replicas, allowing your master copy to focus on those applications that need to write to the database:

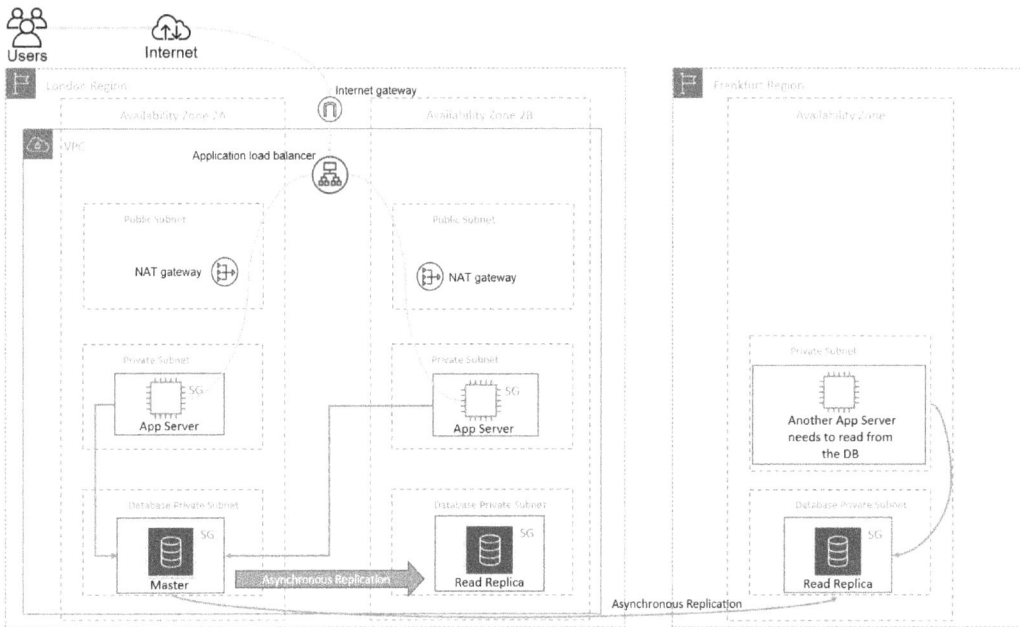

Figure 8.6 – AWS RDS with read replicas

Read replicas can also be configured for cross-Region replication, as depicted in the preceding diagram. (The exception is for the Microsoft SQL Server engine, which doesn't allow Multi-AZ read replicas or cross-Region read replicas.) This means that you can maintain read copies of your database in a different Region that can be used by other applications that may only require read access to the data. Storing read replicas across Regions can also help you address any compliance or regulatory needs that stipulate that you need to maintain copies of your data at a considerable distance.

You can also set the read replica as Multi-AZ, which will enable you to use the read replicas as a DR target. This means that if you ever need to promote the read replica to a standalone database, it will already be configured for Multi-AZ. This feature is available for MySQL, MariaDB, PostgreSQL, and Oracle engines. Finally you can add up to five read replicas to each DB instance.

Furthermore, in the event of a major disaster, a read replica can also be promoted to become the master copy, after which it becomes independent of the original master copy of the database.

In this section, we examined Amazon RDS and the key offerings by its managed RDS. We reviewed the various database engines on offer, concepts related to high availability and scalability, as well as backup and recovery.

In the next section, we will briefly look at one AWS RDS offering, specifically Amazon Aurora. While Amazon Aurora is an RDS database solution from AWS, it offers several enhanced capabilities.

A brief introduction to Amazon Aurora

Amazon Aurora is AWS's proprietary MySQL- and PostgreSQL-compatible database solution and was designed for enterprise-grade production environments. Amazon Aurora comes with a vast array of features that enable you to design your database solution with high availability, scalability, and cost-effective deployments to suit a variety of business needs.

Amazon Aurora is architected to offer high resilience, with copies of the database placed across a minimum of three Availability Zones, It is up to five times faster than standard MySQL databases and three times faster than standard PostgreSQL databases.

The service offers *fault tolerance* and *self-healing storage capabilities* that can scale up to 128 TB per database instance. Amazon Aurora also offers the ability to host up to 15 low latency read replicas. Let's review some of the key features of Amazon Aurora.

Amazon Aurora DB clusters

Amazon Aurora is deployed as **DB clusters** that consist of one or more **DB instances** and a *cluster volume*. This cluster volume spans multiple Availability Zones, within which copies of the cluster data are stored.

The Aurora DB cluster is made of up two types of DB instances, as follows:

- **Primary DB instance**: This instance supports both read and write operations, and it performs all of the data modifications to the cluster volume. You have *one* Primary DB instance.

- **Aurora Replica**: You can have up to 15 Aurora Replicas in addition to the primary DB instance. Aurora Replicas connect to the same storage volume as the primary DB instance but are only used for read operations. You use Aurora Replicas as a failover option if the primary DB instance fails. You can also offload read queries from the primary DB instance to the replicas.

With regard to the architecture, DB instances (compute capacity) and cluster volume (storage) are decoupled, as illustrated in the following diagram.

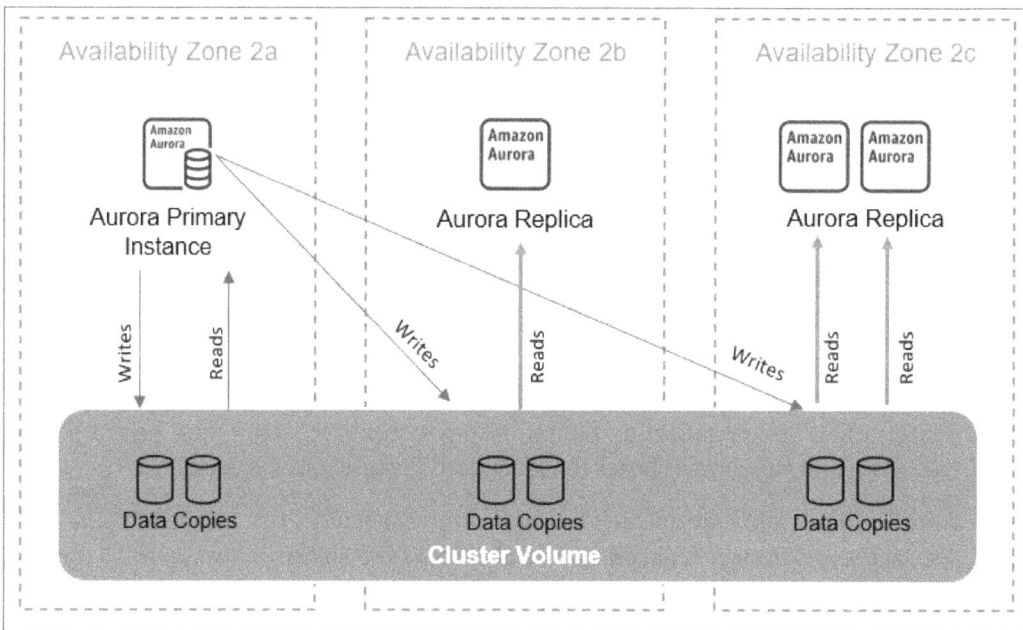

Figure 8.7 – Amazon Aurora DB cluster architecture

This decoupling of the computer capacity and storage also means that even a single DB instance is still a cluster due to the fact storage volumes are spread across multiple storage nodes, across multiple Availability Zones.

With regards to provisioning your DB instances, you have a choice of two instance classes. These are memory optimized (designed for memory-intensive workloads) and burstable performance (which provides a baseline performance level with the ability to burst to full CPU usage).

While the standard Amazon Aurora deployment seems somewhat similar to deploying an Amazon RDS database, where you choose the compute capacity and underlying storage, AWS also offers a serverless alternative, which we will look at briefly next.

Amazon Aurora Serverless

Amazon Aurora Serverless (version 1) is an on-demand autoscaling configuration for Amazon Aurora. The DB Cluster automatically scales compute capacity up and down based on your requirements. The serverless alternative automatically starts up, scales compute capacity to match your application's usage, and shuts down when it's not in use. Furthermore, the cluster volume is always encrypted.

In terms of use cases, Amazon Aurora Serverless is ideal for applications with unpredictable workloads. Another use case is where you have a lightweight application that experiences peaks for 30 minutes to several hours a few times a day or perhaps at regular intervals throughout the year. Examples include budgeting, accounting, and reporting applications.

In this section, we were briefly introduced to the Amazon Aurora Service. In the next section, we will examine Amazon DynamoDB, which is AWS's non-relational (NoSQL) database offering.

Learning about Amazon DynamoDB (NoSQL database solution)

Amazon offers a fully managed non-relational database solution called Amazon DynamoDB. Unlike AWS's relational database offerings (excluding Amazon Aurora, which also has a serverless offering, as discussed earlier), you do not need to worry about provisioning the right DB instance with the right specification for your application. DynamoDB is offered as a serverless solution because you do not need to define any database instance configuration, such as CPU or memory configuration. Amazon manages the underlying infrastructure that hosts the DynamoDB service.

DynamoDB is a regional service just like Amazon RDS, but it comes with higher levels of scalability and high availability. You do not need to provision a single DB instance in one Availability Zone as you do with a single instance of an Amazon RDS database. Instead, when you provision a DynamoDB table, Amazon provisions the database and automatically spreads the data across several servers to handle your throughput and storage requirements. All data is stored on **solid-state disks** (**SSDs**) and the underlying storage is replicated across multiple Availability Zones.

The architecture of DynamoDB means that it can be used for use cases that are similar Amazon RDS's, although they are more ideal for applications that can have millions of concurrent users and where the application needs to perform thousands of reads and writes per second.

Tables, items, and attributes

Let's look at the core components of a DynamoDB database:

- **Tables**: Like Amazon RDS databases, your data is stored in tables. So, you can have a customers table that will host information about your customers and their orders. Each table will also have a unique primary key, which is crucial for uniquely identifying every record in the table. Records are known as items in DynamoDB Tables.

- **Items**: Items are like records in Amazon RDS databases. A table can have one or more items, and each item will be a unique combination of attributes that define that item. Items can be up to 400 KB in size and can contain key-value pairs called attributes.

- **Attributes**: An attribute is like a column heading or a field in an Amazon RDS database. Attributes help define the items in your table. So, in a Customers Table, the attributes could be First-Name or Last-Name and so on.

Unlike Amazon RDS, you do not need to predefine the schema of the table. This offers greater flexibility as your table evolves. Other than the primary key, you can add new attributes and define them to expand the table at will.

Furthermore, the items don't have to have a value for all attributes – so, for example, you can have a table that contains customer address details, their orders, and their favorite dessert for a restaurant. Some items may have a value set against the favorite dessert, while some may not, and this is perfectly fine.

However, attributes need to have data types defined. The following are the options that are available here:

- **Scalar**: This only has one value and it can be a number, string, binary, Boolean, or null.

- **Set**: This represents multiple scalar values and can be a string set, number set, or binary set.

- **Document**: This is a complex structure with options for nested attributes. You can use JSON formatted documents and retrieve data you need without having to retrieve the entire document. There are two subtypes of the document data type:

  - **List**: An ordered collection of values

  - **Map**: An unordered collection of name-value pairs

In this section, we examined the different components of a DynamoDB database. In the next section, we will learn how to provision the required capacity for our database requirements.

Provisioning capacity for DynamoDB

When it comes to provisioning your database, all you need to provide is the parameters that define the **read capacity units** (**RCUs**) and **write capacity units** (**WCUs**). These values enable Amazon to determine the underlying infrastructure to provision to host your database and your throughput levels.

Depending on your RCU and WCU, DynamoDB will provision one or more partitions to store your data in and use the primary key to distribute your items across multiple partitions. Spreading data across multiple partitions enables DynamoDB to achieve ultra-low latency reads and writes, regardless of the number of items you have in the table.

The two options that are available for provisioning capacity are as follows:

- **On-Demand**: DynamoDB will provision capacity based on your read and write requests and provision capacity dynamically. This option is ideal when you have unpredictable application traffic and unknown workloads.

- **Provisioned**: You specify the number of reads and writes per second that are required by your application. This is ideal if you have predictable application access patterns. You can always also enable *auto-scaling* to automatically adjust to traffic changes.

In this section, we looked at Amazon DynamoDB and AWS's non-relational database solution. We discovered how DynamoDB is suitable for modern web applications that require thousands of reads and writes per second, as well as how DynamoDB is built for this purpose specifically.

In the next section, we will look at Amazon's data warehousing solution, known as Amazon Redshift.

Understanding the use cases for Amazon Redshift and data warehousing

A data warehousing solution is a specialized database solution designed to pull data from other relational databases and enable complex querying and analytics to be performed across different datasets. For example, you can combine data across customer orders, inventory data, and financial information to analyze product trends, demands, and return on investments.

Clients of Amazon Redshift include **business intelligence** (**BI**) applications, reporting, and analytics toolsets.

Online Analytical Processing (OLAP)

Amazon Redshift is designed for analytics and is optimized for scanning many rows of data for one or multiple columns. Instead of organizing data as rows, Redshift transparently organizes data by columns; it converts the data into columnar storage for each of the columns. Let's look at what this means.

In a traditional database, data for each record is stored as rows. The columns represent the attributes of your data, and each row will contain field values for the relevant columns.

Let's look at a table we saw earlier and see how the data is stored in blocks on disk:

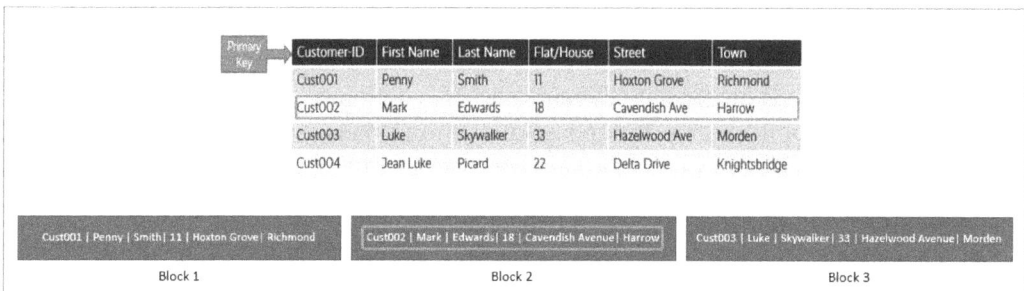

Primary Key	Customer-ID	First Name	Last Name	Flat/House	Street	Town
	Cust001	Penny	Smith	11	Hoxton Grove	Richmond
	Cust002	Mark	Edwards	18	Cavendish Ave	Harrow
	Cust003	Luke	Skywalker	33	Hazelwood Ave	Morden
	Cust004	Jean Luke	Picard	22	Delta Drive	Knightsbridge

Cust001	Penny	Smith	11	Hoxton Grove	Richmond	Cust002	Mark	Edwards	18	Cavendish Avenue	Harrow	Cust003	Luke	Skywalker	33	Hazelwood Avenue	Morden
Block 1						Block 2						Block 3					

Figure 8.8 – Data stored in blocks on disk

Note that the data is stored sequentially for each column that makes up the entire row in blocks on the disk (*block 1, 2, 3, and so on*). If the record size is greater than the block size, then the record is stored across more than one block. Similarly, if the record size is smaller than the block size, then the record may consume less than the size of one block. Ultimately, this way of storing data leads to inefficiencies in the use of storage.

Having said that, in a traditional relational database, most transactions will involve frequent read and write queries for a small set of records at a time, where the entire record set needs to be retrieved.

Now, let's look at how Redshift stores data. Using the same customer data table, each data block stores values of a single column for multiple rows, as per the following diagram:

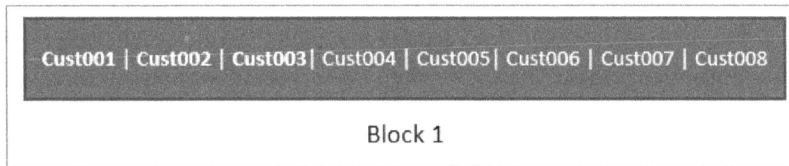

Cust001 | Cust002 | Cust003 | Cust004 | Cust005 | Cust006 | Cust007 | Cust008

Block 1

Figure 8.9 – Data stored on Amazon Redshift

Amazon Redshift converts the data as it is added into columnar storage. With this architecture, Amazon Redshift can store column field values for as many as three times the number of records compared to traditional row-based database storage. This means that you only consume a third of the I/O operations when it comes to reading column field values for a given set of records, compared to row-wise storage. Furthermore, because the data that's stored in blocks will be of the same type, you can use a compression method design for the columnar data type to achieve even better I/O and reduce the overall storage space. This architecture works well for data warehousing solutions because, by its very nature, your queries are designed to read only a few columns for a very large number of rows to extract data for analysis. In addition, queries require a fraction of the memory that would be required for processing row-wise blocks.

Ultimately, Redshift is designed to host petabytes of data and supports massively parallel data processing for high-performance queries.

Redshift architecture

The Redshift architecture is built on a cluster model that is comprised of the following:

- **Leader node**: A single node that manages all communications between client applications and *compute nodes*. The leader node carries out all operations such as the steps required to carry out various complex queries – the leader node will compile code and distribute these to compute nodes.

- **Compute nodes**: Up to `128` compute nodes can be part of a Redshift cluster. The compute nodes execute the compiled code that was provided by the leader node and sends back intermediate results for the final aggregation. Each compute node will have its own dedicated CPU, memory, and disk type, which determines the node's type:

    - **Dense compute nodes**: These can store up to 326 TB of data on magnetic disks.

    - **Dense storage nodes**: Can store up to 2 PB of data on **solid-state disks** (**SSDs**).

    - **RA3 instances**: This next generation of Nitro-powered compute instances come with *managed storage* (unlike the other previous node types). You choose some nodes based on your performance requirements and only pay for the managed storage that you consume. This architecture has the compute and storage components decoupled. Furthermore, data storage is split, whereby local SSD storage is used for fast access to cached data and Amazon S3 is used to use longer-term durable storage that scales automatically. You will need to upgrade your dense compute and dense storage nodes if you wish to make use of managed storage.

In this section, we introduced you to the Amazon Redshift service, a cloud-hosted data warehousing solution designed for **OLAP** operations. In the next section, we will look at another feature of Amazon Redshift known as the Redshift Spectrum service, which allows you to directly query data held in Amazon S3.

About Redshift Spectrum

Another solution from Amazon Redshift is the **Redshift Spectrum** service, which allows you to perform SQL queries against data stored directly on Amazon S3 buckets. This is particularly useful if, for instance, you store frequently accessed data in Redshift and some infrequent data in Amazon S3. Rather than import the infrequent data into Redshift, which will only be queried occasionally, storing them in Amazon S3 and using Redshift Spectrum will be more cost-effective.

It is also important to note that data in S3 must be structured and you must define the structure to enable Redshift to consume it.

Understanding the importance of in-memory caching options with Amazon Elasticache

Often, you will find yourself accessing a set of data regularly, which is what we term as frequently accessed data. Every time you run a query on the database, you consume resources to perform the query operation and then retrieve that data. Overall, this can add additional load to your database and may even affect performance as you constantly write new data to the database.

As part of your overall application architecture, you should consider using in-memory caching engines offered by AWS to alleviate the load on your primary databases. **Amazon Elasticache** is a web service that offers in-memory caching in the cloud. By caching frequently accessed data on Amazon Elasticache, applications can be configured to retrieve frequently accessed data from it rather than make more expensive database calls.

AWS offers two in-memory caching engines, as follows:

- **Amazon Elasticache for Redis**: This is built as a cluster, which is a collection of one or more cache nodes, all of which run an instance of the Redis cache engine software. Redis is designed for **complex data types**, offers Multi-AZ capabilities, encryption of data, and compliance with FedRAMP, HIPAA, and PCI-DSS, as well as high availability and automatic failover options.

- **Amazon Elasticache for Memcached**: This is designed for **simple data types**. Here, you can run large nodes with multiple cores or threads and scale out. It should be used where you require object caching.

In this section, we learned about the various AWS Elasticache services, which offer in-memory caching capabilities for our applications. In-memory caching can be used to alleviate the load on your primary databases by caching frequently access data rather than having to run expensive queries repeatedly. In the next section, we will look at some additional databases offered by AWS that address specific niche market requirements.

Learning about additional database services for specific niche requirements

In addition to Amazon RDS and DynamoDB, AWS also offers additional databases that meet the requirements of specific niche applications. In this section, we will take a look at two of those databases: **Amazon Neptune** and **Amazon Quantum Ledger Database** (**QLDB**).

Introduction to Amazon Neptune

Amazon Neptune is a fully managed graph database service and a type of NoSQL database. Graph databases are designed to store data as nodes (person, place, location, and so on) and directions. Each node would have some property and nodes have relationships between them. So, for example, **Alice** lives in **London**, and in **London**, there is a resident called **Alice**. This is a simple example, but you can start to imagine how complex your nodes and their relationships can become. These kinds of complex relationships between the nodes are just as important as the data itself and are ideal for a graph database solution. Traditional relational databases would require you to define complex joins between tables and even then, this would result in inefficiencies when trying to extract data.

Amazon Neptune support well-known graph models such as Property Graph and W3C's RDF and their respective query languages such as Apache TinkerPop, Gremlin, and SPARQL. Amazon Neptune is a highly available database solution that offers point-in-time recovery and continuous backups to Amazon S3 with Availability Zone replication. Typical use cases for Amazon Neptune include applications such as fraud detection, knowledge graphs, drug discovery, and network security.

Amazon QLDB

Some types of data are highly sensitive and maintaining data integrity is of paramount importance. Examples of this include bank transaction records, where you need to track the history of credits and debits, or insurance claim applications that require you to maintain a verifiable history of the claim process. Another example is that of having to trace the movement of parts in a supply chain network and being able to prove the journey those items took to reach the customer.

Although you can use relational databases to host these ledger types of data, you would need to build in an audit trail, which can be very cumbersome and prone to errors. Furthermore, because data stored in relational databases is not inherently immutable, it can become difficult to verify if data has been altered or deleted.

An alternative solution is to build a **blockchain** network. Blockchain frameworks such as Hyperledger Fabric and Ethereum enable you to build decentralized databases where the data stored is immutably and is cryptographically verifiable. However, blockchain networks are very complex and are designed on a decentralized model where you have multiple nodes that need to verify each record before it is committed to the database.

Amazon **QLDB** is a fully managed ledger database that enables you to store immutable records with cryptographically verifiable transaction logs in a centralized database model. Amazon QLDB can maintain a history of all data changes. The following are the key benefits of Amazon QLDB:

- **Immutable and transparent**: It enables you to track and maintain a sequence transaction log (journal) of every single change you make to your data. With QLDB, this transaction log is immutable, which means it cannot be altered or deleted. QLDB tracks each application data change and maintains a complete and verifiable history of all changes over time.

- **Cryptographically verifiable**: You can use a SHA 256 cryptographic hash function to generate secure output files of your data's change history. This is also known as a *digest*, which acts as proof of any changes that have been made to your data. This can validate the integrity of all your data changes.

- **Easy to use**: It uses a flexible document data model. You can use a SQL-like query language to query your data known as PartiQL. Amazon QLDB transactions are ACID-compliant.

- **Serverless**: Amazon QLDB is a fully managed database service with no need to provision database instances or worry about capacity restraints. You start by creating a ledger and defining your tables. At this point, QLDB will automatically scale as required by your application.

In this section, we examined a couple of additional database solutions designed for niche applications. In the next section, we will look at the database migration services offered by AWS, enabling you to move your on-premises databases to the cloud.

Database Migration Service

Amazon offers a **Database Migration Service** (**DMS**) that can be used to migrate data from one database to another. Often, this is used as part of an on-premises to cloud migration strategy, where you need to migrate database services located in your data center to your AWS account in the cloud. AWS DMS offers support for both homogeneous migrations, such as from MySQL to MySQL or Oracle to Oracle, as well as heterogeneous migrations between engines, such as Oracle to Microsoft SQL Server or Amazon Aurora.

An important point to be aware of is that, while migrating, you can continue to use your source database, which minimizes downtime for your business operations. In addition, you can also use DMS to perform continuous data replication from your on-premises environment to the cloud to offer high availability or disaster recovery capabilities.

Exercise 8.1 – Extending your VPC to host database subnets

In *Chapter 7*, *AWS Compute Services*, you expanded your VPC to include both private subnets and public subnets. Generally, you would only host services in a public subnet that would need direct exposure on the internet. Examples include the bastion host server we deployed earlier in *Chapter 7*, *AWS Compute Services* (which we will discuss in the next chapter).

Most applications are deployed across tiers – so, for example, you can have a web tier, an application tier, and a database tier. These different tiers are designed to separate different components of your application stack, allowing you to create a degree of isolation, as well as benefit from a layered security model. In *Chapter 7*, *AWS Compute Services* , as part of *Exercise 7.1 – Expanding ProductionVPC so that it includes two public subnets and two private subnets*, you also configured two private subnets across two Availability Zones to host your application servers. In this example, the application tier and web tier are the same. However, in many real-world scenarios, they would be separate.

In this exercise, you will be extending your VPC to add an additional tier, known as the database tier, within which you will be able to launch an Amazon RDS database. Like EC2 instances, Amazon RDS needs to be deployed in a VPC.

In the following diagram, you can see that your VPC now has three tiers – a public (DMZ) tier to host bastion host servers, NAT gateways, and Elastic Load Balancers, an application tier comprised of the **Private Subnet One – App** and **Private Subnet Two – App** subnets, and finally, a database tier comprised of the **Private Subnet Three – Data** and **Private Subnet Four – Data** subnets. Note that the subnets are spread across two Availability Zones to enable you to offer high availability of services in the event of an Availability Zone failure. We will discuss high availability in more detail in the next chapter:

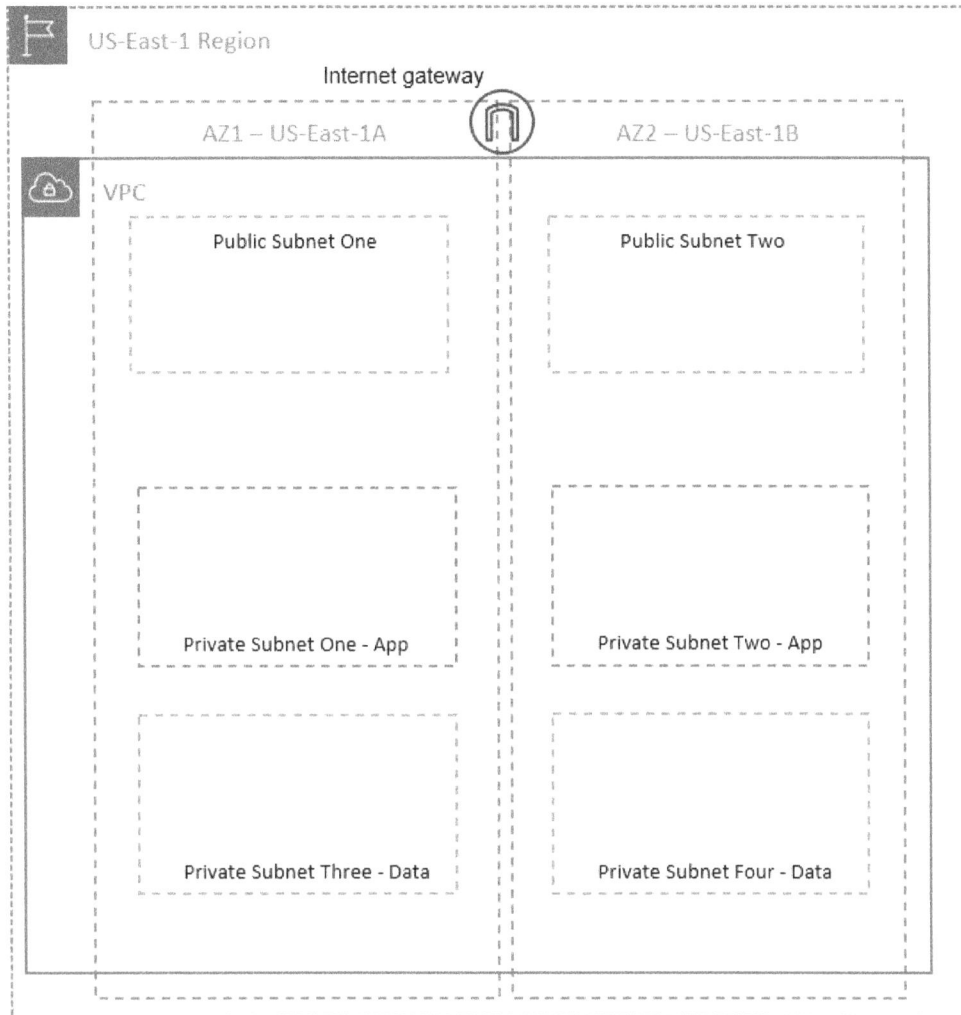

Figure 8.10 – Extending the VPC to include a database tier

Let's start by extending your VPC so that it includes our database tier:

1. Log back into your AWS account as our administrator, **Alice**.

2. Navigate to the **VPC** dashboard and ensure you are in the **US-East-1** Region.

3. In the left-hand menu, click on **Subnets**.

4. Next, click the **Create subnet** button in the top right-hand corner of the screen.

5. You will be presented with the **Create subnet** wizard page.

6. Under **VPC ID**, select **ProductionVPC** from the drop-down menu.

7. In the **Subnet settings** section, under **Subnet 1 of 1**, provide a name for your first database subnet. For this exercise, name your subnet **Private Subnet Three – Data**.

8. Under **Availability Zone**, select the **us-east-1a** Availability Zone.

9. Next, for **IPv4 CIDR block**, type in 10.0.5.0/24.

10. Next, rather than create this subnet and repeat the wizard to create the second database subnet, simply click on the **Add new subnet** button, as per the following screenshot:

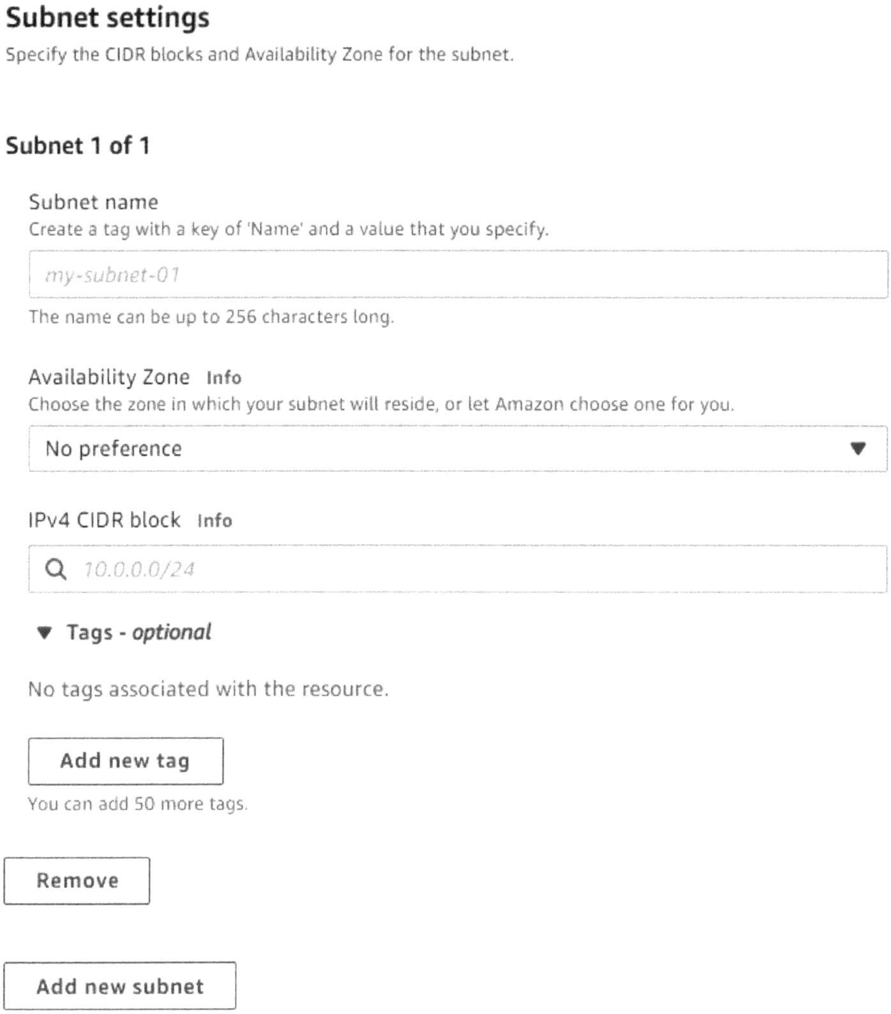

Subnet settings

Specify the CIDR blocks and Availability Zone for the subnet.

Subnet 1 of 1

Subnet name

Create a tag with a key of 'Name' and a value that you specify.

```
my-subnet-01
```

The name can be up to 256 characters long.

Availability Zone   Info

Choose the zone in which your subnet will reside, or let Amazon choose one for you.

```
No preference                                                              ▼
```

IPv4 CIDR block   Info

```
Q   10.0.0.0/24
```

▼ Tags - *optional*

No tags associated with the resource.

```
Add new tag
```

You can add 50 more tags.

```
Remove
```

```
Add new subnet
```

Figure 8.11 – Creating multiple subnets

11. A new subsection, **Subnet 2 of 2**, will appear, allowing you to create an additional subnet in the same wizard. Under **Subnet name**, type in **Private Subnet Four – Data**.

12. For **Availability Zone**, select the **us-east-1b** Availability Zone from the drop-down list.

13. For **IPv4 CIDR block**, type in `10.0.6.0/24`.

14. Click the **Create subnet** button at the bottom of this page.

AWS will successfully create two new subnets, which you will use to host your Amazon RDS database. In the right-hand menu, click on **Subnets** to view all the subnets now associated with your **ProductionVPC**, as per the following screenshot:

Name	Subnet ID	State	VPC	IPv4 CIDR	IPv6 CIDR	Availa
Public Subnet Two	subnet-0c076296ce7e9c3ae	⊘ Available	vpc-06de8d92837535119	Pr...	10.0.2.0/24	–	251
Public Subnet One	subnet-0034c3ce989fab016	⊘ Available	vpc-06de8d92837535119	Pr...	10.0.0.0/24	–	251
Private Subnet Two - App	subnet-09d271d28241ce4d6	⊘ Available	vpc-06de8d92837535119	Pr...	10.0.4.0/24	–	251
Private Subnet Three - Data	subnet-0b8982e1596b9f0e3	⊘ Available	vpc-06de8d92837535119	Pr...	10.0.5.0/24	–	251
Private Subnet One - App	subnet-0d43a97f4519300f7	⊘ Available	vpc-06de8d92837535119	Pr...	10.0.3.0/24	–	251
Private Subnet Four - Data	subnet-0bcb73d4c74af47fa	⊘ Available	vpc-06de8d92837535119	Pr...	10.0.6.0/24	–	251

Figure 8.12 – Subnets in ProductionVPC

Now that you have created an additional two subnets for your **ProductionVPC**, you can proceed with the next part of this exercise. Like EC2 instances, Amazon RDS databases require you to configure the necessary security groups that will permit traffic to the database instances.

In our layered security model, we wish to ensure that only our application servers will be able to communicate with the databases in the backend. In this part of the exercise, you will create a new security group that will be configured to allow database-relevant traffic from any application servers you deploy later. To do this, you must configure an inbound rule on the new database security group to accept traffic on port `3306` for MySQL traffic from the security group of the application servers, specifically from the **AppServers-SG** security group. Let's get started:

1. Ensure you are currently in the **VPC** dashboard. Then, from the left-hand menu, click on **Security Groups**.

2. Click the **Create security group** button in the top right-hand corner of the screen.

3. For **Security group name**, type in `Database-SG`. Under description, type in `Allow MYSQL traffic from AppServer-SG`.

4. Under **VPC**, ensure you select **ProductionVPC** from the drop-down list.

5. Next, in the **Inbound rules** section, click on the **Add rule** button.

6. Under **Type**, select **MySQL/Aurora**.

7. Ensure that **Source** is set to **Custom**, and in the search box, start typing sg-.
    You should see that a list of all security groups shows up in a list. Select the
    AppServers-SG security group.

8. Provide an optional description if required.

9. Click on the **Create security group** button in the bottom right-hand corner
    of the screen.

AWS will now confirm that the security group has been created successfully.

In this exercise, you extended your VPC to host two additional private subnets that
we will use to host our Amazon RDS database. You also created a new security group,
Database-SG, which will be associated with our Amazon RDS database instance. It will
also allow MySQL/Aurora traffic on port 3306 from any EC2 instance that is associated
with the **AppServer-SG** security group.

In the next exercise, we will configure an Amazon RDS database subnet group that will be
used to inform Amazon RDS of which subnets it can deploy our databases to.

Exercise 8.2 – Creating a database subnet group

Before you can launch an RDS database in your VPC, you need to define a DB subnet
group. A **DB subnet group** is a collection of two or more subnets within the VPC where
you want to deploy your database instance. When creating your DB subnet group, at least
two subnets must be selected in the VPC that are associated with two separate Availability
Zones in a Region. Amazon RDS uses the subnet group's IP address CIDR block to assign
your RDS database instance(s) with an IP address.

Amazon RDS can then deploy the database instance on one of your chosen subnets that is
part of the group. In the case of a Multi-AZ deployment, the master copy will be deployed
in one subnet in a particular Availability Zone, while the standby copy will be deployed in
another subnet that is hosted within another Availability Zone.

Note that the subnets in a DB subnet group are either public or private, but they cannot be a mix of both public and private subnets. Ideally, you want to configure private subnets as part of your subnet group because you want to deploy any backend databases in the private subnets of your VPC. Your databases should only be accessible from web/application servers and not directly from the internet.

To set up the DB subnet group, follow these steps:

1. Ensure that you are logged into your **AWS Management Console** as the IAM user **Alice**.

2. From the top left-hand menu, click on the **Services** drop-down arrow and select **RDS** located under the **Database** category. This will take you to the Amazon RDS dashboard.

3. Ensure that you are in the **us-east-1** Region and from the left-hand menu, click on **Subnet groups**.

4. Next, in the main pane of the screen, click the **Create DB Subnet Group** button.

5. On the page that appears, you will need to define your DB subnet group details:

    - Provide a name for your DB subnet group; for example, `ProductionVPC-DBSubnet`.

    - For the description, type in `DB Subnet Group to host RDS Database in Production VPC`.

    - Under **VPC**, select **ProductionVPC** from the drop-down menu.

    - Next, under **Availability Zone**, choose the Availability Zones that include the subnets you want to add. For this exercise, select the checkboxes next to **us-east-1a** and **us-east-1b**.

6. Next, under **Subnets**, select the subnets you created earlier for your RDS database. When you click on the drop-down list next to **Subnets**, you will notice that the subnet names are not visible; instead, the subnets are identified by their subnet IDs. You can determine the correct subnets by cross-referencing the subnet ID against the subnets you created in the VPC dashboard, which you can access in another browser window. Alternatively, you can ensure you select the correct subnets by comparing their relative IPv4 CIDR blocks. So, in our example, the private DB subnets are in the `10.0.5.0/24` and `10.0.6.0/24` Ipv4 CIDR blocks, as per the following screenshot:

VPC
Choose a VPC identifier that corresponds to the subnets you want to use for your DB subnet group. You won't be able to choose a different VPC identifier after your subnet group has been created.

ProductionVPC (vpc-06de8d92837535119)   ▼

us-east-1a

☐ subnet-0034c3ce989fab016 (10.0.0.0/24)
☑ subnet-0b8982e1596b9f0e3 (10.0.5.0/24)
☐ subnet-0d43a97f4519300f7 (10.0.3.0/24)

us-east-1b

☑ subnet-0bcb73d4c74af47fa (10.0.6.0/24)
☐ subnet-09d271d28241ce4d6 (10.0.4.0/24)
☐ subnet-0c076296ce7e9c3ae (10.0.2.0/24)   Zones.

Select subnets   ▲

subnet-0b8982e1596b9f0e3 (10.0.5.0/24) ✕

subnet-0bcb73d4c74af47fa (10.0.6.0/24) ✕

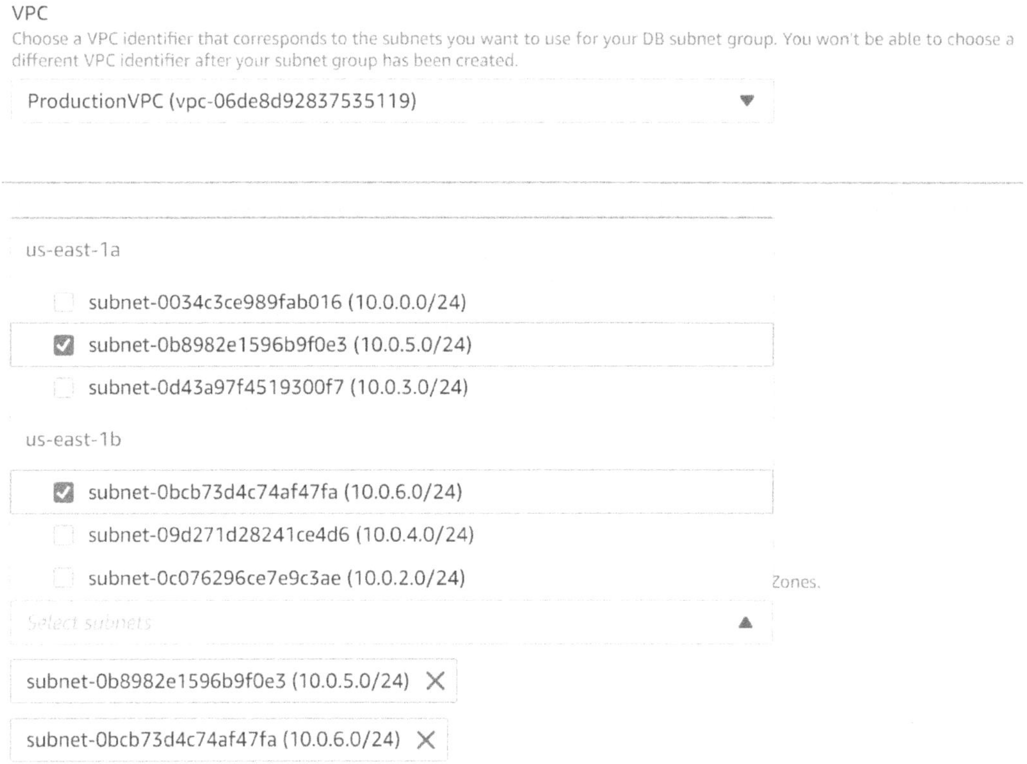

Figure 8.13 – Creating database subnet groups

7. Next, click the **Create** button at the bottom right-hand corner of the screen.

AWS will create your DB subnet group using the details you provided, as per the following screenshot:

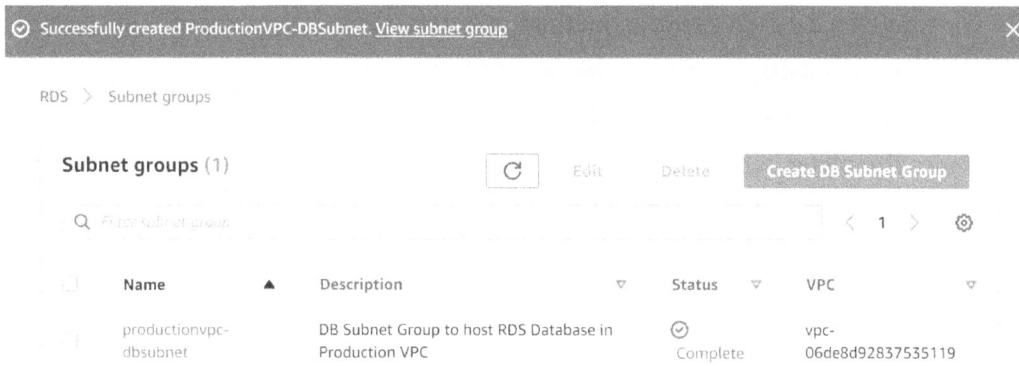

Figure 8.14 – Successfully creating a database subnet group

In this exercise, you learned about RDS DB subnet groups, which allow you to define a minimum of two subnets across two Availability Zones, where Amazon RDS can deploy your RDS DB instance when you choose to launch your database.

In the next exercise, we will launch our RDS database in **ProductionVPC**. We will also use this database to host the backend data of our web application, which we will then deploy in the fourth exercise of this next chapter.

Exercise 8.3 – Launching your Amazon RDS database in ProductionVPC

In this exercise, you will launch an Amazon RDS MySQL database in the DB subnet group of **ProductionVPC**. Let's get started:

1. Ensure that you are logged into your AWS account as the IAM user **Alice**.

2. Navigate to the Amazon RDS dashboard.

3. From the left-hand menu, select **Databases**.

4. On the right-hand side of the pane, click the **Create database** button.

5. Next, you will be presented with the **Create database** wizard, where you will need to define various parameters of your VPC. Amazon offers the t2.micro database instance running the MySQL engine as part of the Free Tier offering, which comes with the following features for up to 12 months:

   - 750 hours of Amazon RDS in a Single-AZ db.t2.micro instance.

   - 20 GB of General Purpose storage (SSD).

   - 20 GB for automated backup storage and any user-initiated DB snapshots.

6. For **Choose a database creation method**, select the option next to **Standard create**.

7. Next, for the database engine option, select **MySQL**.

8. Leave the **Edition** and **Version** settings as-is.

9. Under **Templates**, select the **Free Tier** option.

10. Next, you need to provide some settings:

   - For the DB instance identifier, type in `productiondb`.

   - Under **Credential Settings**, leave the **Master** username set to **admin** and provide a password of your choice. Make sure that you note this password down; otherwise, you will not be able to connect to the database.

11. Under **Database instance class**, leave the settings as-is.

12. Under **Storage**, leave the settings as-is except for **Storage autoscaling**, where you should *disable* the option for **Enable storage autoscaling**.

13. Under **Availability & durability**, you will note that the option to enable **Multi-AZ** is grayed out. This is because Multi-AZ is not available in the Free Tier.

14. Next, under **Connectivity**, do the following:

   - Ensure that you select **ProductionVPC** from the drop-down list under **Virtual private cloud (VPC)**.

   - Note that the subnet group has been pre-populated with the DB subnet group we created earlier, which in this case is `productionvpc-dbsubnet`.

   - Under **Public Access**, select **No**.

   - Next, under **VPC Security Group**, ensure that **Choose Existing** is selected. Then, from the drop-down list under **Existing VPC security groups**, select the **Database-SG** security group you created earlier.

   - Next, under **Availability Zone**, you can select an AZ of your choice or leave it as the **No preference** option.

   - Next, expand the **Additional configuration** option and ensure that the database port is set to `3306`.

15. Under **Database authentication options**, ensure that **Password authentication** is enabled.

16. Under **Additional configuration**, do the following:

- Type in the same database name as you did earlier for **DB instances identifier**. In our example, the name will be `productiondb`.

- Leave **DB parameters group** and **Options group** as-is.

- Under **Backups**, ensure that **Enable automatic backups** is enabled and then set **Backup retention period** to **1 Day**.

- Under **Backup window**, select the **No preference** option. For real-world applications, you may wish to set the backup window to a period outside of normal business hours.

- Under the **Maintenance** subheading, leave the settings as-is.

- Finally, click the **Create database** button at the bottom right-hand corner of the screen.

Your RDS database will take a few minutes to launch. As part of the launch process, an initial backup will also be performed. Once the database has been successfully launched and ready to use, you will see that its **Status** will be set to **Available**, as per the following screenshot:

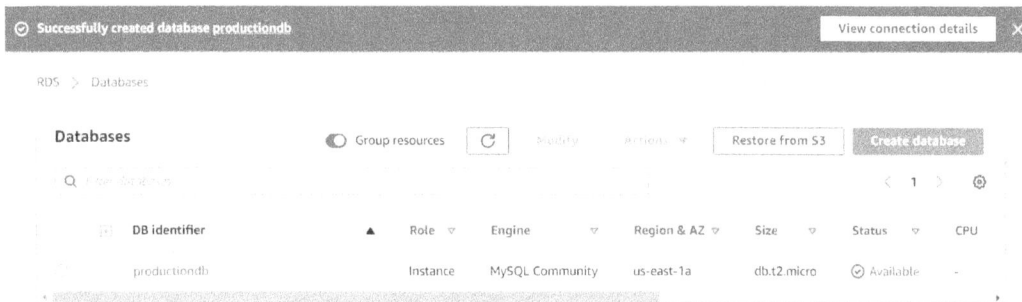

Figure 8.15 – RDS database created successfully notification

In the next exercise, you will learn how to deploy a DynamoDB table.

Exercise 8.4 – Deploying an Amazon DynamoDB table

In this exercise, you will deploy a very simple DynamoDB table. Let's get started:

1. Ensure that you are logged into your AWS account as the IAM user known as **Alice**.

2. Next, navigate to the DynamoDB dashboard. You can search for DynamoDB from the top search box of **AWS Management Console**.

3. If this is the first time you have visited the **DynamoDB console** page, you will be presented with a splash screen.

4. Click the **Create table** button.

5. Provide a name for your table in the text box next to **Table name**; for example, Recipes.

6. In the **Primary key** field, enter RecipeName and ensure that the type is set to **String**.

7. Under **Table settings**, uncheck the box next to **Use default settings**.

8. In the **Read/write capacity mode** section, select the **On-demand** option.

9. Click the **Create** button at the bottom of the page. DynamoDB will create a new table for you in a few seconds, as per the following screenshot:

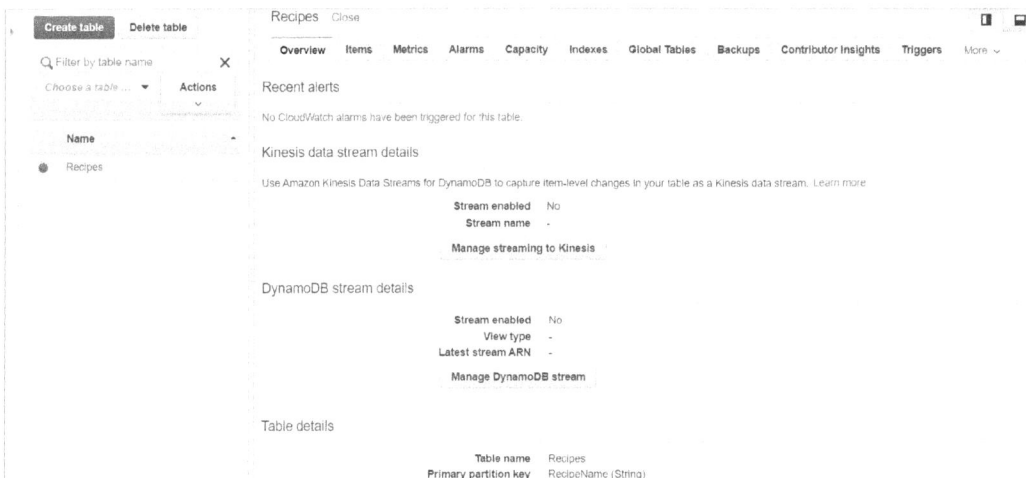

Figure 8.16 – DynamoDB table – Recipes

10. Click on the **Items** tab.

11. You can start adding items in the **Items** tab. Click the **Create item** button.

12. You will see a dialog box in which you can add a new item (record) to your database.

13. In the text box next to **RecipeName String**, enter Vegan Sausage Rolls.

14. Click on the **Save** button.

15. Note that the new item has been added and that the value of the primary key for this item is the name of the recipe, Vegan Sausage Rolls.

16. Click the **Create item** button again.

17. In the text box next to **RecipeName String**, enter Vegan Peri Peri Burger.

18. Click on the *plus* sign and select the **Append** drop-down list. From the list of options provided, select the **StringSet** type. A new field will be created for you. In the **field** box, enter an attribute (field name); for example, Ingredients. You will also notice that an additional entry appears below this **StringSet**, which is where you would input the values for the field you just created. Click on the *plus* sign next to the *empty array* line and select **Append**.

19. In the **Value** box, type in the following words, followed by pressing *Enter* on your keyboard after each word – Lettuce Tomato Cucumber. Click on a different part of the screen to update the values, as per the following screenshot:

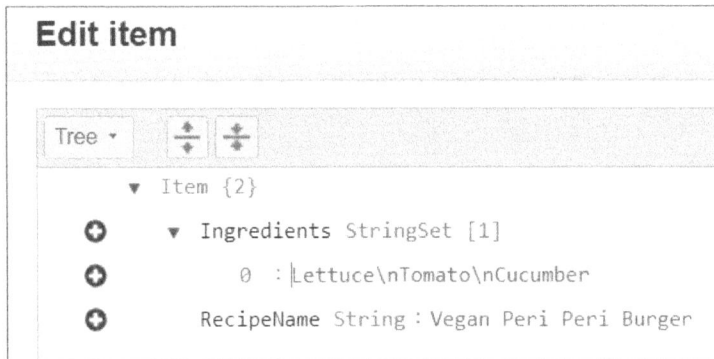

Figure 8.17 – DymanoDB item entry

20. Click on the **Save** button.

21. At this point, your table has been updated with a new record. You will see the two items in your table. The **Vegan Sausage Roll** item only has one field with a value in it, namely the primary key. The **Vegan Peri Peri Burger** item has two fields associated with it, which are the primary key and an attribute called **Ingredients**. Review the following screenshot for reference:

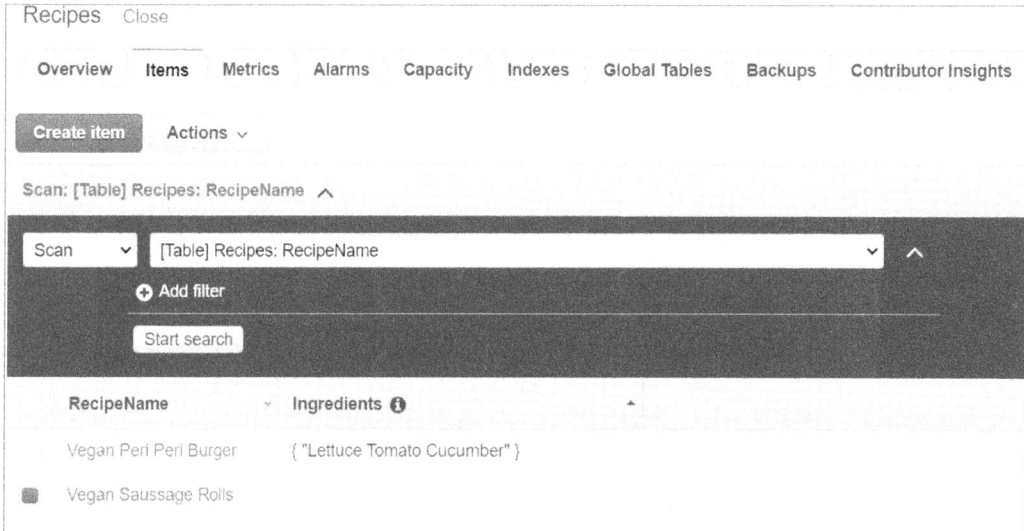

Figure 8.18 – DynamoDB Recipes table

As you can see, DynamoDB offers lots of flexibility in not requiring a rigid schema definition before inputting data.

Next, we will conclude by summarizing this chapter.

Summary

In this chapter, we learned about the various database services offered by Amazon, comprising both relational and non-relational databases services. You learned how AWS enables you to quickly deploy new RDS databases and offers full management of your database as a service, rather than you having to provision EC2 instances that you will install database software on.

Amazon RDS comes with six engines – MySQL, PostgreSQL, Microsoft SQL, Oracle, MariaDB, and Amazon Aurora. Amazon RDS is a regional service and must be deployed in your VPC. You have options to configure for high availability using services such as Multi-AZ and backup and restore strategies. You can also scale out read copies of your RDS database to offload read queries away from the primary master copy of your database.

Amazon Aurora comes with a lot more features and addresses some of the limitations of traditional RDS engines out of the box, including features such as self-healing and high availability.

We then looked at Amazon DynamoDB, which is a non-relational database designed for the modern web, mobiles, and IoT applications that can read and write thousands of requests per second. Amazon DynamoDB is offered as a completely serverless solution – you do not need to provision database instances or storage yourself. You simply specify **WCUs** and **RCUs** and AWS will provision the underlying infrastructure for you.

In addition, we looked at two in-memory caching engines offered by the Amazon Elasticache service – Redis and Memcached – and compared which engine to use in what scenario.

Finally, we examined AWS DMS3, which offers both homogenous migrations such as Oracle to Oracle migrations and heterogenous migration such as Oracle to Microsoft SQL type migrations. AWS DMS can be used to migrate on-premises databases to the cloud very easily.

In the next chapter, we will discuss concepts related to high availability and scalability. In addition, we will carry out various lab exercises that will enable you to learn how to combine the various core services we have learned about so far. You will do this by deploying a multi-tier application architecture.

Questions

1. A company plans to migrate its on-premises MySQL database to Amazon RDS. Which AWS service should they use for this task?

    A. Amazon Snowball

    B. AWS Database Migration Service (AWS DMS)

    C. AWS VM Import/Export

    D. AWS Server Migration Service

2. Which of the following is the primary benefit of using an Amazon RDS database instead of installing a MySQL-compatible database on your EC2 instance?

    A. Managing the database, including patching and backups, is taken care of by Amazon.

    B. Managing the database, including patching and backups, is taken care of by the customer.

    C. You have full access to the operating system layer that the RDS database runs on.

    D. You can choose which drive and partition to install the RDS database on.

3. AWS RDS supports six database engines. From the following list, choose *three* engines supported by Amazon RDS.

    A. Microsoft SQL

    B. Oracle

    C. MySQL

    D. FoxPro

    E. Db2

4. You are building an application for a wealth asset management company that will be used to store portfolio data and transactions of stocks, mutual funds, and forex purchased. To that end, you need a backend database solution that will ensure a ledger-like functionality because they want to maintain an accurate history of their applications' data, for example, tracking the history of credits and debits for its customers. Which AWS database solution would you recommend for this business requirement?

    A. Amazon RDS

    B. Amazon DynamoDB

    C. Amazon QLDB

    D. Amazon Redshift

5. Which AWS database solution enables you to build a complete data warehousing
    solution, capable of handling complex analytic queries against petabytes
    of structured data using standard SQL and industry-recognized business
    intelligence tools?

    A. AWS DynamoDB

    B. AWS Redshift

    C. AWS Neptune

    D. AWS Pluto

6. You are looking to host a production-grade enterprise relational database solution
    that offers high-end features such as self-healing storage systems that are capable of
    scaling up to 128 TB per database instance. Which of the following AWS database
    solutions fulfills the requirement?

    A. Amazon DynamoDB

    B. Amazon Aurora

    C. Amazon Redshift

    D. Amazon Neptune

7. Which AWS feature of Amazon Redshift enables you to run SQL queries against
    data stored directly on Amazon S3 buckets?

    A. Redshift DaX

    B. Athena

    C. Redshift Spectrum

    D. Redshift Cache

8. Which AWS service enables you to migrate an on-premises MySQL database to an Amazon RDS database running the Oracle Engine?

    A. AWS Cross-Region Replication

    B. AWS SMS

    C. AWS DMS

    D. AWS EFS

9. You are running a single RDS DB instance. Which configuration would you recommend so that you can avoid I/O suspension issues when performing backups?

    A. Configure RDS read replicas.

    B. Configure RDS Multi-AZ.

    C. Configure RDS Cross Region Backup.

    D. Configure DynamoDB DaX.

9
High Availability and Elasticity on AWS

Most applications follow a design pattern that comprises several layers—such as the network layer, the compute layer, and the storage and database layers. We call this a multi-tier application. So, for example, you can have a three-tier application stack comprising a web services layer that offers frontend web interface access, an application layer where perhaps all data processing happens, and a backend database layer to store and manage data.

In this chapter, we start to bring together the various core **Amazon Web Services** (AWS) services we have learned about so far to design and architect a complete **end-to-end** (E2E) solution. Furthermore, in previous chapters, we have only deployed single resource instances of various AWS services—for example, a single **Elastic Compute Cloud** (EC2) instance to offer compute capability, or as in the previous chapter, where we deployed a single Amazon **Relational Database Service** (RDS) database instance in the public subnet of our **virtual private cloud** (VPC).

In real-world scenarios, you generally need to incorporate components that will help you achieve **high availability** (**HA**) and **scalability**. We can increase our application's availability by having more than one EC2 instance serving the same application or website. This way, if one of the EC2 instances fails, users can continue to access the services offered by being directed to other healthy EC2 instances in the fleet, and away from those that are unhealthy or in a failed state. We can also ensure that we place our EC2 instances across multiple **Availability Zones** (**AZs**), ensuring that if one AZ fails or just simply goes offline, users can be redirected to healthy EC2 instances in another AZ.

Similarly, we need to offer solutions that are scalable. AWS offers services that are capable of automatically scaling out when required; for example, when we notice an increase in traffic, we can add more EC2 instances to cope with the load. On the flip side, the same AWS service can automatically scale in when demand drops, allowing us to save on the unnecessary expense of running underutilized servers in the cloud.

Finally, we also need to consider the global availability of our services. Many companies have global customers, and while many AWS services are designed for regional availability and scalability options, other AWS services can help us deliver global availability and even offer resilience against regional outages.

In this chapter, we discuss the following key concepts:

- Introduction to vertical and horizontal scaling concepts
- Overview of the **Open Systems Interconnection** (**OSI**) model
- Distributing web traffic with Amazon **Elastic Load Balancing** (**ELB**)
- Implementing elasticity with AWS Auto Scaling
- Designing multi-Region HA solutions

Technical requirements

To complete this chapter and the exercises within, you need to have access to your AWS account and be logged in as **Alice**, our **Identity and Access Management** (**IAM**) user (administrator) that we created in *Chapter 4, Identity and Access Management*.

Introduction to vertical and horizontal scaling concepts

When you deploy a given EC2 instance in your VPC, you need to choose an instance type and one or more associated **Elastic Block Store** (**EBS**) (or instance store) volumes of specific sizes. Your EC2 instance will always need one root volume and one or more data volumes based on your application requirements.

However, from time to time, you may need to upgrade your original configuration—perhaps you need more memory or more **central processing units** (**CPUs**) to cope with the load on your server. You may be running out of storage space and therefore need to increase the amount of storage on your EBS volumes. When upgrading to an instance of a higher specification, we call this **vertical scaling**. To perform most upgrades this way, you generally need to stop processing application requests, and most of the time, you may first need to shut the EC2 instance down.

The actual upgrade can take anything from a few minutes to a few hours, depending on what you are upgrading. For example, upgrading the instance type usually involves shutting down the server, modifying the instance type, and restarting it again, as shown in the following screenshot:

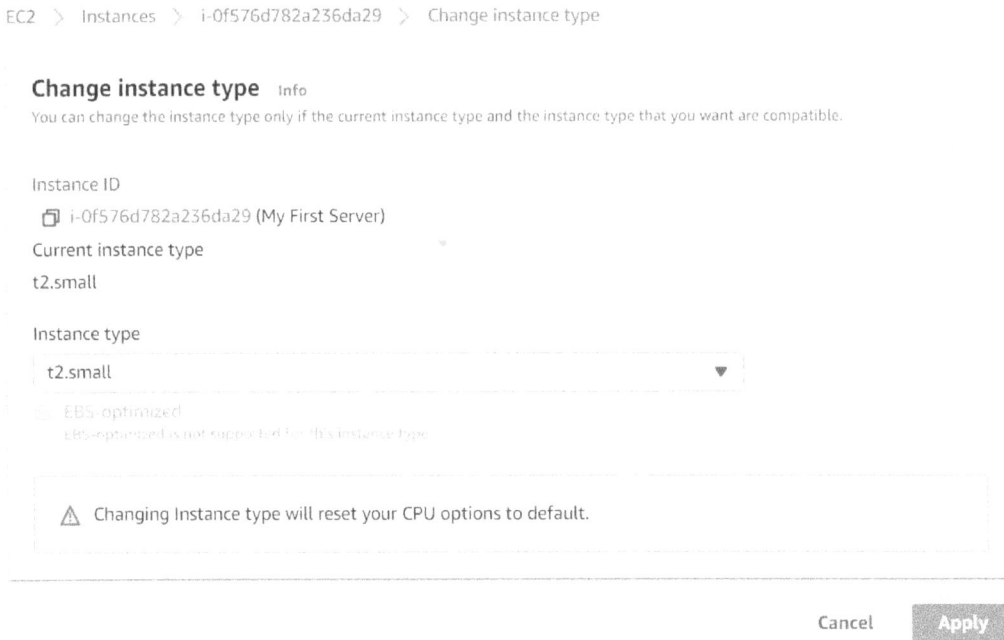

EC2 > Instances > i-0f576d782a236da29 > Change instance type

Change instance type Info
You can change the instance type only if the current instance type and the instance type that you want are compatible.

Instance ID
⬚ i-0f576d782a236da29 (My First Server)
Current instance type
t2.small

Instance type
t2.small ▾

EBS-optimized
EBS-optimized is not supported for this instance type

⚠ Changing instance type will reset your CPU options to default.

Cancel   Apply

Figure 9.1 – Changing EC2 instance type: vertical scaling

In the preceding screenshot, you will select the drop-down arrow next to **Instance type** to select a higher-specification EC2 instance type. Once selected, simply click on **Apply** and start up the EC2 instance again. The EC2 instance is started with the upgraded specifications.

Similarly, you can also upgrade the storage volume attached to the server or attach additional volumes. You can modify existing volumes to increase the storage size or change the type of storage from **General Purpose SSD (gp2)** to **Provisioned IOPS (io1)**. When upgrading your storage, AWS needs to perform some optimization tasks, and this will take some time, depending on the size of the volume.

Vertical scaling does have its limitations, however. It cannot offer HA; so, if there is a problem with the EC2 instance and it fails, you will need to provision a new EC2 instance as a replacement.

Rather than having a single EC2 instance host your application, you could consider hosting multiple EC2 instances with the same application offering. This way, if one EC2 instance fails, customers can be redirected to another EC2 instance that is in a healthy state.

Often, you need more than one EC2 instance participating as a fleet to cope with demand and offer HA in case of failures of any instance. AWS offers a service called Auto Scaling (which we look at in detail later in this chapter), which can automatically launch (or terminate) an EC2 instance to cope with load based on performance parameters such as average CPU utilization across a fleet of servers. This ability to then launch additional EC2 instances serving the same application is known as **horizontal scaling**.

With horizontal scaling, you can add more EC2 instances to your fleet when demand increases and terminate unnecessary instances when demand drops.

This requires careful architectural design, as the application needs to be aware that it is being run from multiple EC2 instances. For example, if you have two copies of your WordPress blog running from two EC2 instances, content data is usually stored on the local storage attached to a single EC2 instance—in this case, the EBS volumes. If you remember from *Chapter 7, AWS Compute Services – EC2, Lightsail,* EBS volumes can only be attached to one EC2 instance at a time.

Important Note

Amazon EBS Multi-Attach is a new feature that enables you to attach a single Provisioned IOPS SSD (io1 or io2) volume to multiple instances located in the same AZ. However, it has several limitations and does not necessarily replace the use case for **Elastic File System (EFS)** volumes. For additional information, refer to `https://docs.aws.amazon.com/AWSEC2/latest/UserGuide/ebs-volumes-multi.html`.

This means that any blog articles you write will be stored on one EC2 instance, and the other EC2 instance running WordPress will not be aware of this content, resulting in inconsistencies between the two servers. One way to handle this is to offer some ability to share data between the EC2 instances, such as having the application data hosted on an EFS volume that can act as a file share for multiple EC2 instances. The WordPress application will also have to be configured to store all blog content and related media on this central EFS volume instead, as shown in the following diagram:

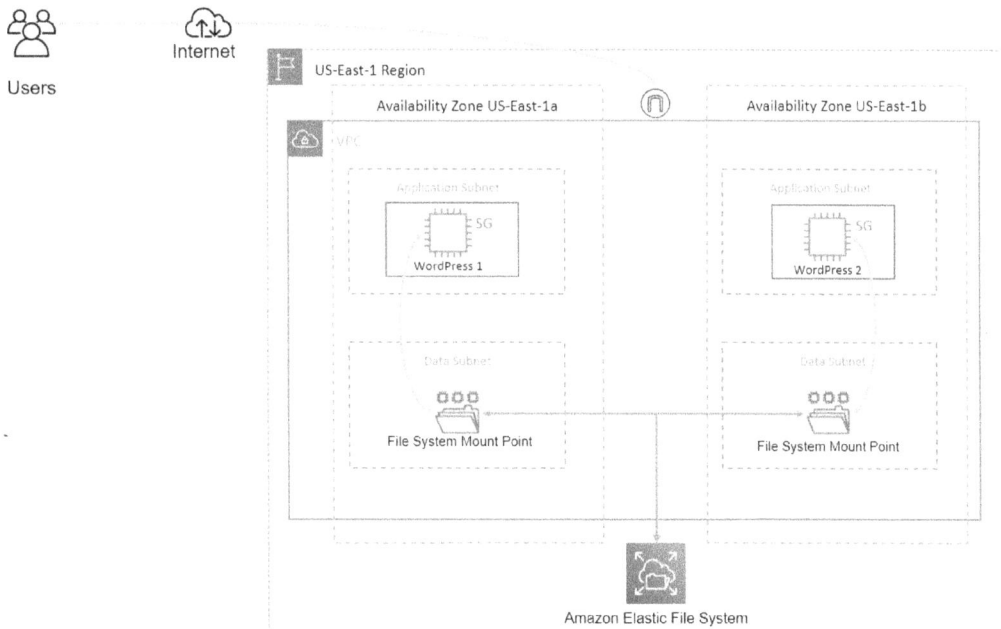

Figure 9.2 – Enabling horizontal scaling at the application layer

In this section, we compared vertical scaling and horizontal scaling scalability options on AWS. Vertical scaling refers to an in-place upgrade to add more CPUs, memory, or storage. Vertical scaling does not offer any HA because if the EC2 instance fails, you cannot fail over to another instance. Horizontal scaling is designed to add more nodes (compute or otherwise) to your fleet and can help reduce the overall load on an individual instance. With horizontal scaling, you can also offer HA so that if one node fails, traffic can be redirected to other healthy nodes.

Next, we introduce the **OSI model**, which is a reference model for how applications communicate over a network and how network traffic flows across from the physical layer of network cabling and Wi-Fi through to the application itself. Having a broad understanding of this reference model will help you appreciate and assist with troubleshooting communication issues between your applications across networks.

Overview of the OSI model

The flow and distribution of network traffic across various devices and applications are defined by a concept known as the **OSI model**. Published in 1984, this model provides a visual description of how network traffic flows over a particular network system

There are **seven** layers to the OSI model that flow from top to bottom, with layer 7 being at the top and layer 1 at the bottom. The OSI model is used as a reference point for various vendors to express which layer of network communication the product they are offering works on. This is because different hardware and software products operate at different layers.

The OSI model also assists in identifying network problems. When analyzing the source of any network issue, identifying whether only a single user is affected or whether a network segment or the entire network is down can help identify potential equipment or mediums that are experiencing the fault.

The following diagram illustrates the seven layers of the OSI model:

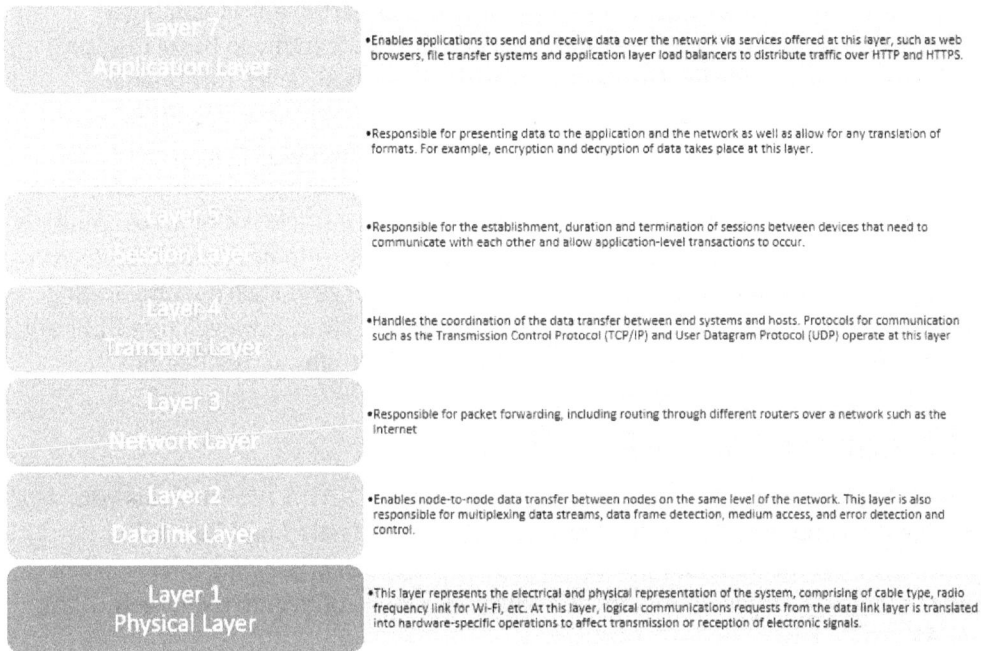

Figure 9.3 – OSI model

In the next section, we look at ELBs, which form a crucial part of the overall design architecture for HA and horizontal scalability.

Distributing web traffic with Amazon ELB

When you have more than one EC2 instance that works as part of a fleet hosting a given application, you need a mechanism in place to distribute traffic to those instances in a manner that spreads the load across the fleet. At a very basic level, this is what Amazon ELBs are designed to do. Amazon ELBs distribute traffic across multiple targets, which can be EC2 instances, containers, **Internet Protocol** (**IP**) addresses, and even Lambda functions. They can handle varying traffic for your application, evenly distributing the load across those registered targets either in a single AZ or across multiple AZs within a given Region. This also means that ELBs can assist in designing architecture that offers HA and fault tolerance, as well as working with services such as Auto Scaling to deliver automatic scalability features to your applications. Note, however, that ELBs are regional-based only, so you cannot use an ELB to distribute traffic across Regions.

Load balancers and VPCs

Amazon load balancers are designed to work with your VPC. AWS recommends you enable more than one AZ for your load balancer. Amazon **Elastic Load Balancing** (**ELB**) then creates **load balancer nodes** in the AZs you specify. The load balancer then distributes traffic to its nodes across the AZs. Its nodes then connect to the targets in the relevant AZs.

An important point to note here is that your internet-based clients need to only connect to the load balancer, which then distributes traffic to targets in your VPC. This means you no longer need to place any targets such as EC2 instances in the public subnet unless you have a specific reason to do so. You can place your web servers, for instance, in a private subnet, and because they are registered to the load balancer, traffic will be routed to them. Furthermore, this also means that your web servers can function with just private IP addresses, reducing the overall attack surface by not having any public IP addresses, as depicted in the following diagram:

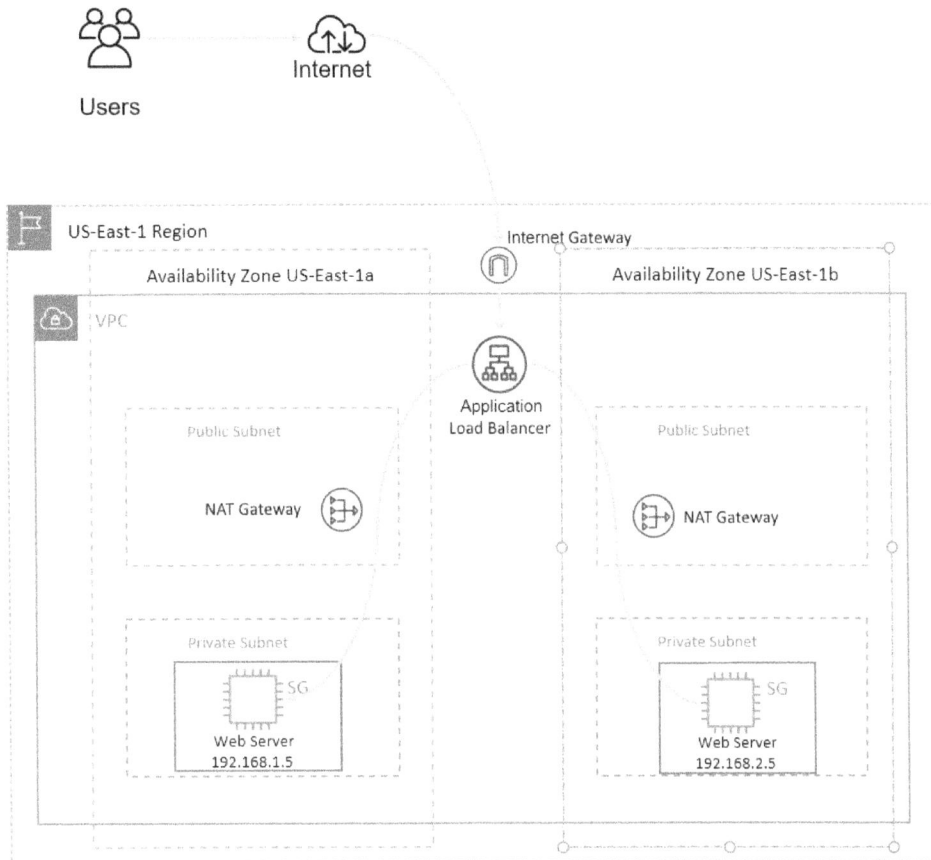

Figure 9.4 – Amazon ELB VPC configuration

As seen in the preceding diagram, the web servers will receive traffic from the **Application Load Balancer** (**ALB**) (we look at the types of load balancers next). The ALB will distribute this traffic using the default round-robin method based on the number of AZs enabled for the load balancer.

In addition, when you create a load balancer, you need to specify whether you are creating an *internet-facing* load balancer or an *internal* load balancer. These are outlined in more detail here:

- **Internet-facing load balancer**—This has a publicly resolvable **Domain Name System** (**DNS**) name, enabling it to route requests from internet-based clients. The DNS name resolves client requests to public IP addresses of the load balancer nodes for your load balancer. In the preceding diagram, we can see an example of an internet-facing load balancer that accepts traffic from clients on the internet.

- **Internal load balancer**—The nodes of an internal load balancer only have private IP addresses, and its DNS name is resolvable to the private IPs of the nodes. This means that internal load balancers can only route requests from clients that already have access to the VPC.

Internal load balancers are particularly useful when you are designing a multi-tier application stack—for example, when you have multiple web servers that receive traffic from your internet-based clients and then need to send on that traffic for processing to application or database servers distributed across multiple AZs in private subnets, via another load balancer. In this case, the web servers would be registered against the internet-facing load balancers and the application/database servers would be registered against the internal load balancers, as depicted in the following diagram:

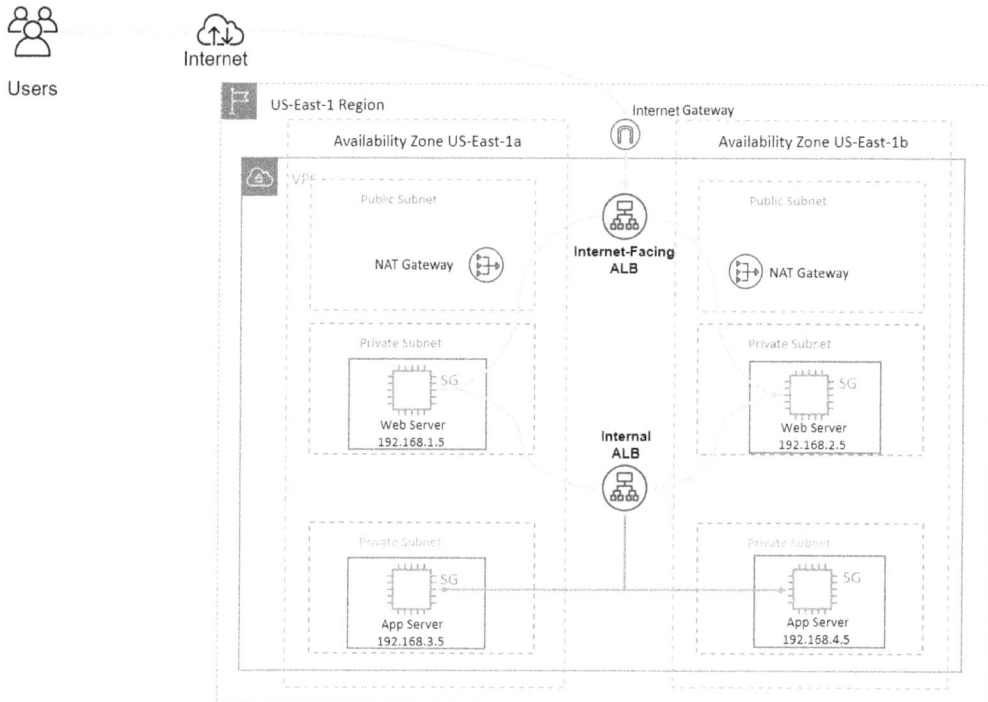

Figure 9.5 – Internal versus internet-facing ELBs

Another important point to note is that for traffic to be accepted by an ELB, you need to configure security groups, specifying inbound rules that the port, protocol, and source of the traffic for that load balancer can accept. In addition, the security groups associated with your targets must also be configured to allow inbound traffic from the load balancers. You can usually do this by specifying the source of the traffic as being the security group associated with the load balancer itself.

Amazon offers four types of ELBs, as listed here:

- ALB
- **Network Load Balancer (NLB)**
- **Gateway Load Balancer (GWLB)**
- **Classic Load Balancer (CLB)**

Let's look at these individually in some detail next.

ALB

We will start with the Amazon **ALB**, which is the most common type of load balancer to use for most applications. ALBs are designed to act as a single entry point for clients to connect to your applications running on targets such as a fleet of EC2 instances. It is recommended that your EC2 instances are based across multiple AZs, increasing the overall availability of your application in the case of a single AZ failure.

ALBs are designed to distribute traffic at the application layer (using **HyperText Transfer Protocol (HTTP)** and **HTTP Secure (HTTPS)**). The application layer is also known as the *seventh layer of the OSI model*. ALBs are therefore ideal for ensuring an even distribution of traffic to your web applications across the internet.

Some key benefits of ALBs include the following:

- Support for path-based routing, allowing you to forward requests based on the **Uniform Resource Locator (URL)** in the request
- Support for host-based routing, allowing you to forward requests based on the URL in the request
- Support for routing requests to multiple applications on a single EC2 instance
- Support for registering Lambda functions as targets, as well as containerized applications, and much more

An ALB has a configuration component called a **listener**. This listener service allows you to define rules on how the load balancer will route requests from clients to registered targets. These rules consist of a priority, actions, and any conditions that once met will enable the action to be performed. Your listener should have at least one default rule, and you can have additional rules. These rules also define which protocol and port to use to connect to your targets.

When configuring an ALB, you also need to configure one or more **target groups** that will contain one or more **targets**, such as EC2 instances. You can have multiple target groups with an ALB, and this feature enables you to define complex routing rules based on the different application components. So, for example, if you have a corporate website that also hosts an e-commerce section to it, you can split the traffic so that one target group contains targets that host just the corporate pages of your website (for example, www.mycompany.com) and another target group contains targets that host the actual e-commerce portion of your website (for example, shop.mycompany.com). Another feature of ALBs is that you can also register a target with multiple target groups.

This approach allows you to better manage traffic to your website and use different targets to ensure better performance and management. As an example, if your e-commerce portion is having a major upgrade that will take that portion of the site down for a few hours, users can still visit the corporate portion of your website to access the latest information or services offered, as illustrated in the following diagram:

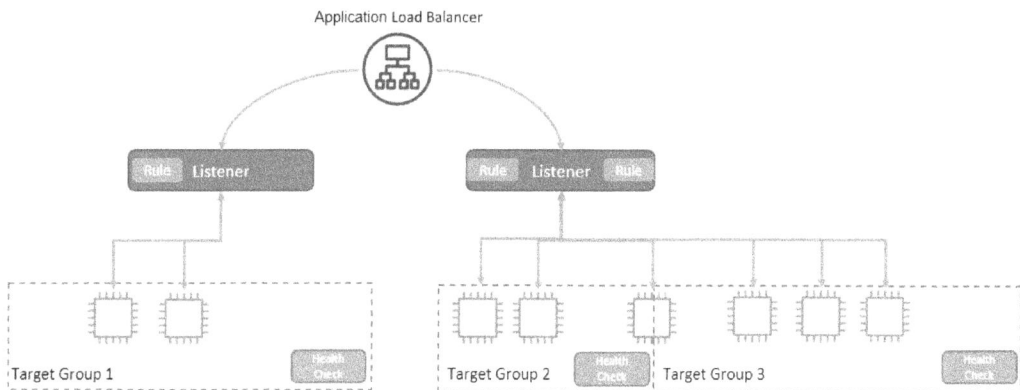

Figure 9.6 – AWS ALB with multiple listener rules

In the preceding diagram, you will note that we have three target groups, and one target is part of two groups. **Target Group 1** could be our primary corporate web pages, whereas **Target Group 3** could be our e-commerce site. Each listener will also contain a default rule, and the listener on the right contains an additional rule that routes requests to another group.

Health checks

In addition to defining listener rules to route traffic to appropriate targets within target groups, ALBs also perform health checks against your targets to determine whether they are in a healthy state. If a target such as an EC2 instance is not responding within a predefined set of requests based on the health check settings, it is marked as *unhealthy* and the ALB stops sending traffic to it, redirecting traffic to only those targets that are in a *healthy* state.

Using this approach, end users are always directed to EC2 instances (or any other targets) that are functioning and responding to the ALB, reducing their chances of experiencing outages.

Traffic routing

When it comes to routing traffic to individual targets in a target group, ALBs use **round robin** as the default method, but you can also configure routing based on the **least outstanding requests** (**LOR**) routing algorithm.

Previously, we mentioned that you use the ALB to split traffic across two target groups in our example of an e-commerce store. The first target group will host EC2 instances that offer access to the corporate website (www.mycompany.com), and the second target group could host the e-commerce portion (shop.mycompany.com). This is also known as **host-based routing**, which allows you to define target groups and listener rules based on the hostname of your domain (www versus shop). In addition to host-based routing, ALBs can be used to route traffic for the following use cases:

- **Path-based conditions**—This allows you to direct traffic based on different sections of your web application to different target groups. Traffic is then routed based on that path of the URL. For example, you can send one portion of traffic to mycompany.com/store for users looking to purchase products, and another portion to mycompany.com/blog for users looking to read articles about the latest market trends.

- **Host-header conditions**—This allows you to route traffic based on fields in the request URL—for example, query patterns or by source IP address.

- **Multiple applications on a single EC2 instance**—This allows you to register more than one application on an EC2 instance using different port numbers.

- **Support for registering targets by IP address**—This allows you to also redirect traffic from your ALB to your on-premises servers using private IP addressing over **virtual private network** (**VPN**) tunnels or Direct Connect connections.

- **Support for registering Lambda functions as targets**—This allows you to configure Lambda functions as targets. Any traffic forwarded will invoke the Lambda function, passing any content in **JavaScript Object Notation** (**JSON**) format.

- **Support for containerized applications**—This includes **Elastic Container Service** (**ECS**), where you can schedule and register tasks with a target group.

In this section, we discussed how traffic can be routed with ALBs and saw different use cases. Next, we take a look at some security features of ALBs.

ALB and WAF

Amazon also offers several security tools that we will examine in detail in a later chapter. One such tool is called the **Web Application Firewall** (**WAF**), which helps protect against common web exploits such as SQL injections and **cross-site scripting** (**XSS**). Amazon ALBs offer WAF integration to help you protect your applications from such common web attacks.

NLB

NLBs are designed to operate at the fourth layer of the OSI model and be able to handle millions of requests per second. NLBs are designed for load balancing of both **Transmission Control Protocol** (**TCP**) and **User Datagram Protocol** (**UDP**) traffic and maintain ultra-low latencies. With NLBs, you can preserve the client's source IP, allowing your backend services to see the IP address of the client, which may be a requirement for the application to function.

NLBs also offer support for static IP addresses and elastic IP addresses, the latter allowing you to configure one fixed IP per AZ. Again, this may be a requirement for your application, and hence you would need NLBs.

NLBs do not inspect the application layer and so are unaware of content types, cookie data, or any custom header information.

Some key benefits of NLBs include the following:

- Ability to handle volatile workloads and handle millions of requests per second.

- Support for static IP addresses for your load balancer and one elastic IP address per subnet.

- You can register targets by IP address—this allows you to register targets outside the VPC such as in your on-premises environment.

- Support for routing requests to multiple applications on a single instance using multiple ports.

- Support for containerized applications such as those running on Amazon ECS.

In this section, we discussed NLBs and learned about their various use cases, particularly when you need to support millions of requests per second and operate at the fourth layer of the OSI model over TCP and UDP protocols. In the next section, we look at GWLBs, designed to enable you to distribute traffic across various software appliances offered on the AWS Marketplace.

GWLB

Before allowing traffic to enter your VPC, you may wish to perform security inspections and analysis of that traffic to block any kind of suspicious activity. Often, you could deploy your own security tools on EC2 instances to inspect that traffic to deploy third-party tools procured from the AWS Marketplace such as firewalls, **intrusion detection systems/ intrusion prevention systems** (**IDSes/IPSes**), and so on. Managing traffic being routed via these third-party tools is made easier with the help of Amazon **GWLBs**.

Amazon GWLBs can manage the availability of these third-party virtual appliances and act as a single entry and exit point for all traffic destined for these services. This enables you to scale the availability and load-balance traffic across a fleet of your virtual appliances. GWLB operates at the third layer of the OSI model (the network layer) and exchanges application traffic with your virtual appliances using the **Generic Network Virtualization Encapsulation** (**GENEVE**) protocol on port 6081. Traffic is sent in both directions to the appliance, allowing it to perform stateful traffic processing.

CLB

Amazon CLB is a previous-generation ELB designed to operate at both layer 4 and 7 of the OSI model but without the extended features offered by the ALB or the level of throughput you can expect from an NLB. CLBs enable you to distribute traffic across EC2 instances that are in a single AZ or across multiple AZs. They are ideal for testing and for non-production environments, or if your existing application is running in the **EC2-Classic network mode**.

In this section, we looked at Amazon ELBs, which enable you to centrally distribute incoming traffic across multiple targets such as a fleet of EC2 instances offering access to a web application. Amazon ELBs can also perform health checks against your targets and redirect traffic away from unhealthy targets to ones that are responding and in a healthy state. This reduces the chances of your end users experiencing any kind of outage by inadvertently being connected to an instance that is not healthy.

Amazon ELBs ultimately help you design your architecture for HA and support scalability features in conjunction with Amazon Auto Scaling, which we will discuss in the next section.

Implementing elasticity with Amazon Auto Scaling

One of the most amazing services on AWS is the ability to automatically scale your workloads when demand increases and then scale back in when demand drops. This service is offered as part of various core technologies—for example, computing services such as EC2 and database services such as DynamoDB.

Automatic scaling in response to a particular condition such as an increase in demand (for example, when average CPU utilization across your fleet of EC2 instances goes above a threshold such as 70%) can help provision additional capacity when it is required most. However, you are not stuck with the new size of your fleet. You can configure Auto Scaling so that if demand drops below a specific threshold value, it will terminate EC2 instances and therefore *save on costs*. Let's look at Auto Scaling for EC2 instances in detail next.

Auto Scaling is a regional service, and you can scale across AZs within a given Region, allowing you to launch EC2 instances across AZs for **HA** and **resilience**.

In the following diagram, we can see that two additional EC2 instances were added to a fleet across AZs **2A** and **2B**. This was due to the average CPU utilization rising above 80%. Once the new instances are part of the fleet, the average CPU utilization should start to fall as the load on the application is spread across six instances now instead of the original four:

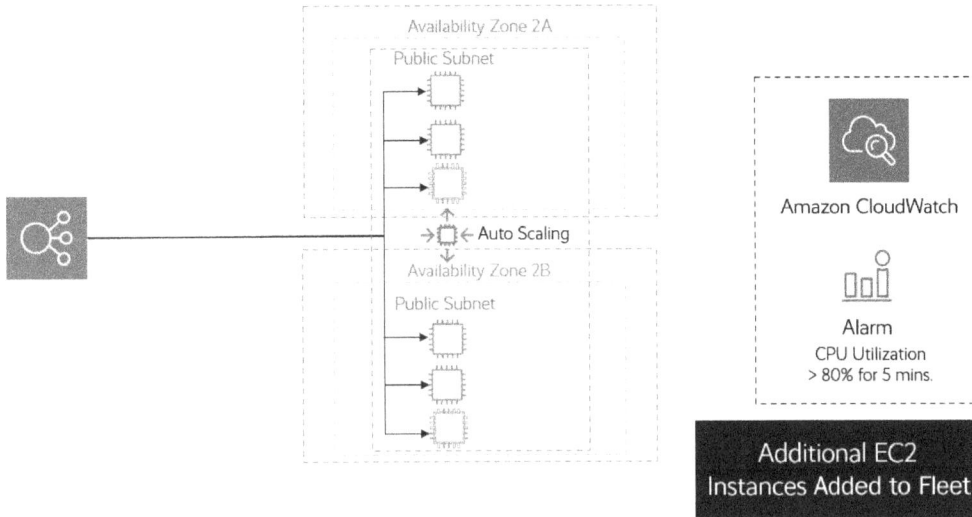

Figure 9.7 – Auto Scaling service example

Amazon Auto Scaling helps you provision necessary EC2 instances on-demand and terminate them when the demand for your resources drops to or below a certain threshold. With Amazon Auto Scaling, you do not need to carry out complex capacity planning exercises. Next, we look at some core components of the Amazon Auto Scaling service.

Auto Scaling groups

When you configure Auto Scaling, you define a collection called an Auto Scaling group. This Auto Scaling group will monitor and manage your fleet of EC2 instances. As part of the configuration, you need to define the following:

- **Minimum number of EC2 instances**—This is the minimum size of the group, and Auto Scaling will ensure that the number of EC2 instances in your fleet never drops below this level. If an instance fails, taking the total count below this value, then the Auto Scaling service will launch additional EC2 instances.

- **Desired number of EC2 instances**—If you specify the desired capacity (usually because you know that at this value, your users have optimal experience), then the Auto Scaling service will always try to ensure that you have the number of EC2 instances equal to the desired capacity. Note that your desired number can be the same as the minimum number, which would mean that the Auto Scaling service ensures that you always have this minimum number of EC2 instances.

- **Maximum number of EC2 instances**—This is the maximum size of the fleet. You need to specify the maximum size that you would want to scale out to. This also has the effect of ensuring that Auto Scaling does not deploy more than the maximum number of instances if, say, a bug in the application running on those instances causes unnecessary scale-outs.

In the following diagram, we can see how Amazon Auto Scaling will provision your desired capacity of EC2 instances and can then scale out to the maximum number of EC2 instances as per your Auto Scaling group configuration:

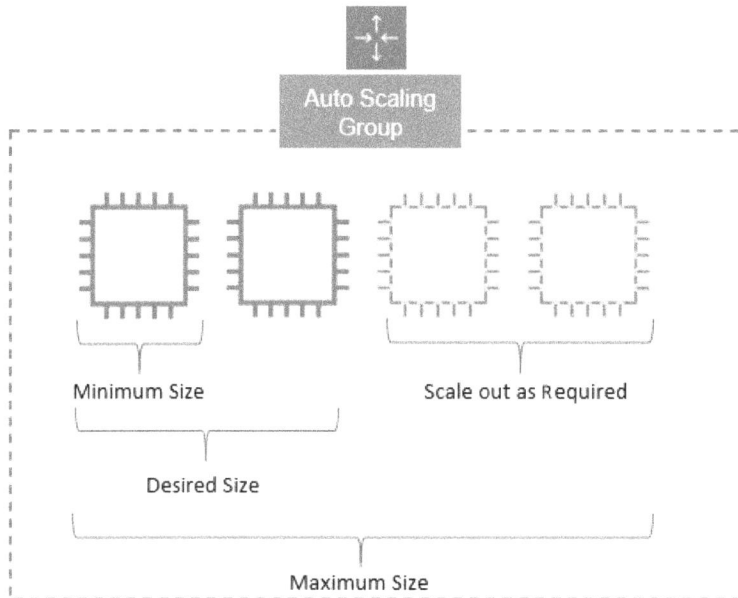

Figure 9.8 – Auto Scaling groups

The Auto Scaling service will launch and/or terminate EC2 instances as part of the group based on the parameters you define and then scale out or scale back in based on the scaling policies you set. You also define how health checks are made against the EC2 instances in the group—this can be either with the Auto Scaling service performing health checks itself or using the health check services of the ELB that the fleet of EC2 instances is registered to. Depending on the health check results, your scaling policies will be triggered accordingly.

Configuration templates

To set up AWS Auto Scaling, you need to configure either a **Launch Template** or a **Launch Configuration**. Configuration templates enable you to define specifications of the EC2 instances to launch within the group. So, for example, the template will define the **Amazon Machine Image (AMI) identifier (ID)**, instance type, key pairs, security groups, and block device mappings.

Within the configuration template, you can also define scripts to be run at launch time, known as **bootstrapping**, which will allow you to automatically configure the EC2 instance to participate in the fleet of existing instances, where possible. For example, at the launch of an EC2 instance, you can configure it with Apache Web Services so that it can function as a web server. These scripts can be defined in the **user data** section of the template and can be written in **Bash** for Linux **operating systems (OSes)** or **PowerShell** for Windows OSes. They are described in more detail here:

- **Launch Configuration**—A Launch Configuration is a basic template where you specify information for instances such as AMI IDs, instance types, key pairs, and security groups. You can associate your Launch Configuration with multiple Auto Scaling groups, but you can only launch one specific Launch Configuration for an Auto Scaling group at a time. Furthermore, once you have defined the parameters of a Launch Configuration, you cannot change it and you will have to re-create it if you need to modify it in any way.

  Launch Configurations are the original way of defining configuration templates for your Auto Scaling groups and while still available, are now no longer recommended by Amazon. Instead, you are advised to use Launch Templates, which offer more features and flexibility and we'll look at next.

- **Launch Templates**—Similar to Launch Configurations in that you define the specifications for your EC2 instances, Launch Templates also offer additional features, including the following:

  - Configuring multiple versions of a template

  - Hosting a default template and templates with variations for different use cases

  - Launching both Spot and On-Demand Instances

  - Specifying multiple instance types and multiple Launch Templates

  Launch Templates also enable you to use newer features of EC2, such as newer **EBS volumes** (gp3 and io2), **EBS volume tagging**, **elastic inference**, and **Dedicated Hosts**. You cannot use Launch Configurations to set up Dedicated Hosts.

Launch Templates are the preferred option when configuring your templates for EC2 instances to be launched as they provide a lot more flexibility. Next, we look at different scaling options to suit different use cases.

Scaling options

The final configuration component of your Auto Scaling service is to determine your scaling policy. Scaling refers to the automatic addition or termination of your compute capacity to meet the demands of your end users and the load on your application. Scaling actions are triggered by an event—for example, the average CPU utilization across your fleet of servers has gone above 80% for the last 20 minutes and users will start to experience poor performance. Based on this event, Auto Scaling can be configured to launch one or more EC2 instances to bring the average CPU utilization down to below 60%.

Depending on your business use case, you have several scaling options to choose from, as outlined here:

- **Always maintain current instance levels**—Your scaling options can be configured to maintain the number of EC2 instances in the fleet, which is where you do not scale in or out; instead, AWS Auto Scaling simply replaces any failed or unhealthy EC2 instances to maintain the fleet size.

- **Scale manually**—You can change the minimum, maximum, and desired number of instances and Auto Scaling will make the necessary modifications to reflect your change.

- **Scale based on schedule**—You can configure Auto Scaling to automatically launch new EC2 instances or terminate existing ones at predefined schedules, whereby you specify a date and time for the scaling action to take place. For example, a large payroll company could scale out the number of EC2 instances running a payroll application in the third week of the month when clients need to submit all their payroll data before a specific deadline. During other weeks, the payroll company can operate on a small number of instances and still offer the optimal client experience.

- **Dynamic scaling (scale on-demand)**—This is ideal when you are not able to predict demand. Dynamic scaling will be triggered when an event occurs, such as CPU utilization rising above a predefined threshold value for a period of time. Likewise, if demand drops, you can then automatically scale back in. There are three different forms of dynamic scaling on offer, outlined as follows:

- **Target tracking scaling policy**—Auto Scaling will launch or terminate EC2 instances in the fleet based on a target value of a specific metric. So, for example, if you know that average CPU utilization of 45% is ideal for the end user experience and anything above this threshold affects performance, then you can set your target tracking scaling policy for CPU utilization to 45%. If demand on your application increases, causing this metric to rise, then additional EC2 instances are launched. Likewise, if demand drops, causing the metric to fall much below 45%, Auto Scaling can terminate EC2 instances. You can think of a target-tracking scaling policy as a home thermostat where you try to maintain an ideal room temperature at home.

- **Step scaling**—Here, an increase or decrease in capacity is based on a series of *step adjustments*, where the size of the breach of threshold specified determines the amount of scaling action.

- **Simple scaling**—This is where capacity is increased or decreased based on a single scaling metric.

- **Predictive scaling**—A more advanced form of scaling that uses **load forecasting**, **scheduled scaling actions**, and **maximum capacity behavior**. Maximum capacity behavior enables you to override the maximum number of instances in the fleet if the forecast capacity is higher than this maximum capacity value.

In this section, we looked at the AWS Auto Scaling service, which enables you to automatically scale your compute resources (and other resources such as databases) across multiple AZs in a given Region. You can scale out as well as scale back in to cope with demand and ensure that you always have the right number of resources to offer the best end user experience. By automatically scaling back in when demand is low, you can also ensure effective cost management.

In the next section, we move to examine how we can offer global HA and fault tolerance of the AWS resources that power your application.

Designing multi-Region HA solutions

In *Chapter 6, AWS Networking Services – VPCs, Route53, and CloudFront*, we looked at Amazon Route 53, which offers DNS and traffic routing policies to help design highly available and resilient architectures incorporating configurations that increase performance and security best practices. We also looked at how Amazon CloudFront can help cache content locally closer to end users, which reduces latency and improves overall performance.

While Amazon Auto Scaling and ELB services help you offer HA and scalable services within a given Region on their own, there is no provision for global availability of services. If you were to host your application in a single Region alone and if that Region were to fail, your end users would not be able to access your applications until the Region came back online and resources made available.

Services such as Route 53 and CloudFront, however, enable you to extend your application's availability to be even more resilient on a global scale. In this section, we look at one such option to offer global availability if your primary Region experiences a major outage and where you perhaps have a global customer base.

Specifically, Amazon Route 53 offers several routing policies, one of which is known as a failover routing policy, which helps you design an active/passive configuration for your application availability.

In the following diagram, we deploy two copies of the application across two AZs. We then configure Amazon Route 53 with a **failover** routing policy, where the primary version of your site is based in the London Region and the secondary site is in the Sydney Region. The following diagram and associated key points highlight the proposed architecture:

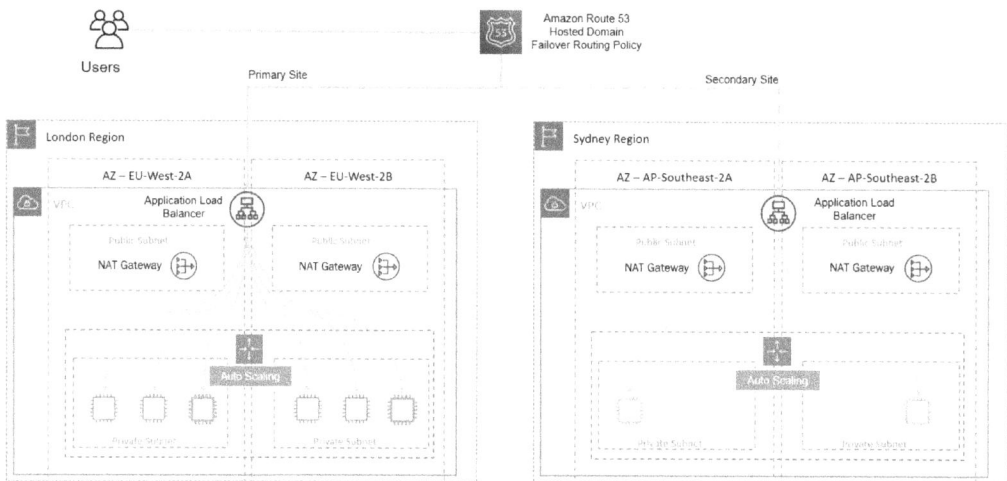

Figure 9.9 – Route 53 configured with failover routing policy, enabling an active/passive solution for application architecture

In the preceding diagram, we have a **primary site based in the London Region** with the following deployment:

- EC2 instances are deployed as part of an Auto Scaling group. In the London Region (our primary site), we have a desired/minimum capacity of four EC2 instances that can expand to a maximum of six EC2 instances to support demand as required.

- Traffic is distributed via the ALB to the EC2 instances across two AZs in the London Region. The source of that traffic is routed via Route 53 from end users on the internet.

- Route 53 performs health checks against the primary site. Health checks are run against each EC2 instance via the ALB, and if the primary site is reachable, Route 53 continues to direct traffic to the primary site only.

- If there is a regional outage or if the Auto Scaling group fails to maintain healthy EC2 instances behind the load balancer, Route 53 marks the site as unhealthy and performs a failover.

- During failover, Route 53 redirects all traffic to the secondary site in Sydney, shown in *Figure 9.9* as the dotted yellow traffic lines on the right.

- While traffic is being redirected to the secondary site, which may start off with a minimum number of instances, Auto Scaling can scale out the number of nodes in the fleet to cope with demand up to the maximum specified number of instances.

The preceding example and diagram illustrate how we can combine regional services such as Auto Scaling and ALBs along with global services such as Route 53 to design a highly available and resilient application architecture.

In this section, we learned about designing application solutions that offer multi-regional HA options using both ELBs and Route 53 services specifically.

In the next section, we examine a series of hands-on exercises that will help you configure the various services you have learned about so far, incorporating IAM, **Simple Storage Service (S3)**, VPCs, EC2, RDS, ELBs, and Auto Scaling, to build a two-tier application solution. This two-tier application solution will comprise a web/application tier and a backend database tier.

To complete the upcoming exercises, it is vital that you have completed all previous exercises in all the previous chapters. Furthermore, to complete the exercise, you will need to access the source code files of the application, which are available at the *Packt Publishing* GitHub repository: `https://github.com/PacktPublishing/AWS-Certified-Cloud-Practitioner-Exam-Guide`

Finally, as you carry out each exercise, you will be provided with some background details to help you appreciate the architecture and reasons behind the deployment. All exercises need to be done in the us-east-1 Region, which is where you have already built your **production VPC** and host your **RDS database**.

For all exercises, ensure that you are logged in to your AWS account as our IAM user **Alice**.

Extended exercises – setting the scene

The upcoming exercises are based on the following scenario. You work for a fictitious company called **The Vegan Studio**. The company is in the hospitality industry. Specifically, the company runs a chain of cafes and restaurants across the **United States (US)**, serving only vegan dishes for those looking to indulge in meat-free cuisine. The company employs over 4,000 employees across its business, and keeping everyone engaged and feeling part of a large family is something the business takes great pride in.

Every year, they run several contests for their employees to participate in. This year, they are running a *My Good Deed for the Month* contest. A web application has been designed by one of the developers, which you need to now deploy in a highly available and scalable manner on AWS in the us-east-1 Region. The contest will run for a month and all employees are encouraged to submit a statement of any good deeds they carried out. Five winners will be chosen from a list of entries and awarded a special hamper prize. Participants must back up their good deeds with evidence if requested (just to be sure no one is fooling around)!

Next, we will walk you through a series of exercises to deploy the application designed by your developer with HA and scalability features.

> **Important Note**
>
> Some AWS services, such as ELBs and **Network Address Translation (NAT)** gateways, are chargeable. We suggest you complete all the exercises in reasonably quick succession and then perform the cleanup exercise at the end. Overall, the cost should not be more than $5. To ensure costs are kept to a minimum, we will not be configuring the RDS database you deployed in the previous chapter with Multi-AZ.

The following exercises will make use of a multi-tier application design that will be deployed in the **production VPC** that you already built in the previous chapters. Recall that the VPC comprises both public and private subnets, spanning across two AZs. From the previous chapters, you have already built a foundation architecture as per the following diagram:

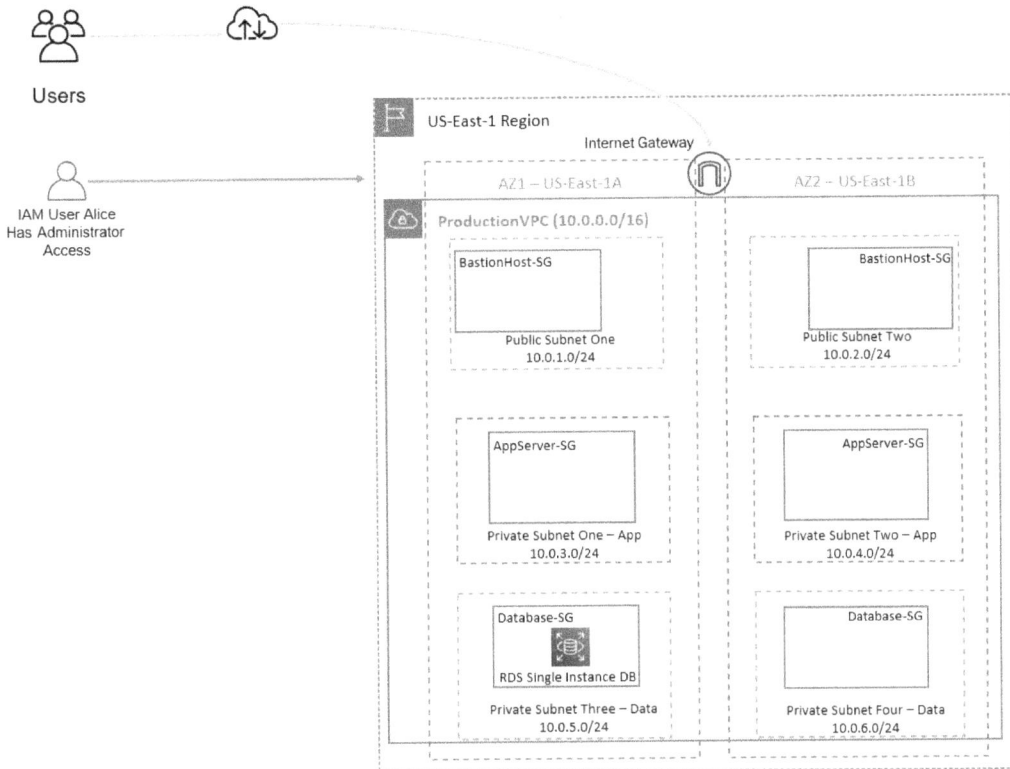

Figure 9.10 – Production VPC architecture prior to deploying the "Good Deed of the Month" contest application

As per the previous diagram, your current architecture is comprised of the following key AWS services and resources:

- A VPC created in the `us-east-1` Region with public and private subnets across two AZs. The private subnets have been designed to support a two-tier application solution comprising a web/application tier and a database tier.

- The public subnets will normally be used to deploy bastion hosts for remote administration and NAT gateways. For the upcoming series of exercises, we will not be deploying any bastion hosts as this is not required for the labs in these exercises. However, we will amend the bastion host security group to allow inbound **Secure Shell (SSH)** connections if you wish to later deploy bastion hosts. Furthermore, because we will be deploying Linux servers, remote administration requires SSH access on port 22.

- The application tier private subnets do not currently have any EC2 instances deployed.

- The database tier private subnets currently host a single instance MySQL RDS database in the us-east-1a AZ.

Through the upcoming exercises in this chapter, we will build on the architecture to design a fully functional application solution with HA and scalability features.

Exercise 9.1 – setting up an Amazon S3 bucket to host source files

In this exercise, you will first create an Amazon S3 bucket that will be used to host your source files for your application. You need to first download the source file, which is available in a ZIP folder format, and extract its contents into a new folder or onto the desktop of your computer for easy access.

The first step is to prepare your source code files. Your source code files contain a database connection file that will need to be amended to the specific RDS database you configured in the previous chapter. Follow these next steps:

1. Once you unzip the downloaded folder, you can see the contents of the main `vegan-php-files` folder, as per the following screenshot:

Desktop › AWS Certified Cloud Practitioner - Packt › Chapter 9 › vegan-php-files › v5

Name	Date modified	Type	Size
css	25-06-2021 03:50	File folder	
fonts	25-06-2021 03:50	File folder	
images	25-06-2021 04:43	File folder	
js	25-06-2021 03:50	File folder	
action	16-10-2020 17:27	PHP File	1 KB
db	16-10-2020 17:39	PHP File	3 KB
health	24-06-2021 14:05	Chrome HTML Do...	1 KB
index	25-06-2021 04:40	Chrome HTML Do...	6 KB
script	16-10-2020 17:28	JavaScript File	2 KB

Figure 9.11 – vegan-php-files source code

2. You will note that in the v5 directory, there is a file called db that is a **PHP: Hypertext Preprocessor (PHP)** file. This file contains default database connection string details that you will first need to amend before you upload the source code to your S3 bucket. Specifically, you will need to provide the RDS database connection details, which include the RDS endpoint DNS name, master username, password, and database name. *Recall that you made a note of these values in the last chapter.*

3. In a notepad or text editor tool, open the db.php file from the v5 folder.

4. Within the PHP file, you will need to edit the values of the placeholders with the appropriate database connection values. In the following screenshot, you will see the placeholders:

```php
<?php
define('DB_SERVER', 'Enter the RDS Database Endpoint DNS name here');

define('DB_USERNAME', 'Enter your RDS Database Username, usually its Admin');

define('DB_PASSWORD', 'Enter your RDS Database Password');

define('DB_DATABASE', 'Enter your RDS Database Name');
```

Figure 9.12 – db.php file

5. You will need to replace the placeholders with the connection details to your database, making sure to place all the values within single quotation marks. Do not make any other changes to the code. Here is a screenshot of where you can obtain these values after you create your database. The database endpoint is visible on the main **Connectivity & security** tab, and you will find the username and database name in the **Configuration** tab. Note that the password is not visible, as you should have made a note of it when launching the database instance:

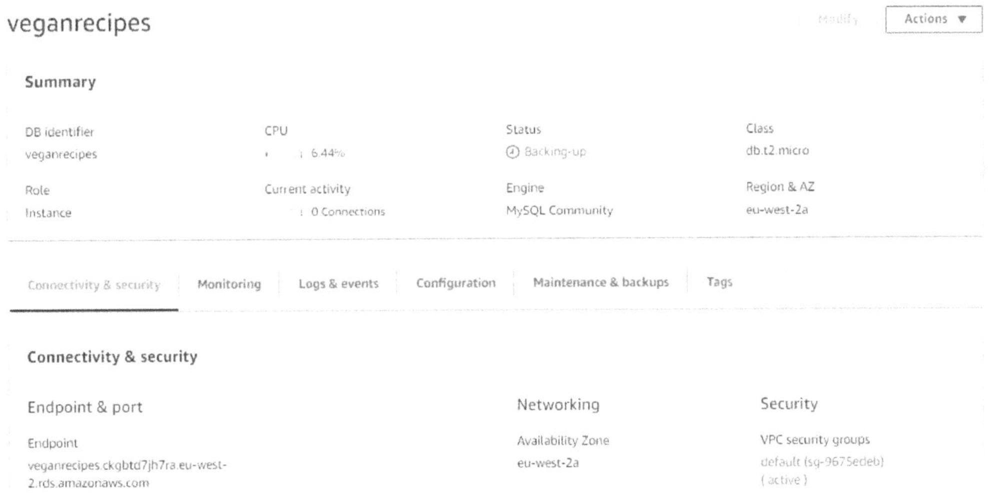

Figure 9.13 – Amazon RDS database settings

6. Save the file in its original location.

7. Log in to your AWS Management Console and navigate to the Amazon S3 dashboard.

8. In the left-hand menu, click **Buckets**.

9. Click **Create bucket** in the right-hand pane of the dashboard.

10. For the **Bucket name** field, provide any name of choice. You will need to choose a unique name as bucket names are obtained on a first-come, first-served basis. For example, my bucket name is `vegan-good-deed`. Ensure that the Region selected is the `us-east-1` Region.

11. Scroll to the bottom of the page, leaving all settings at their default values, and click the **Create bucket** button. Your Amazon S3 bucket will be created, and you will be redirected back to the list of available buckets.

12. Click on the bucket you just created, and you will be redirected to the **Objects** listing page, where you will note that there are currently no objects, as per the following screenshot:

vegan-good-deed

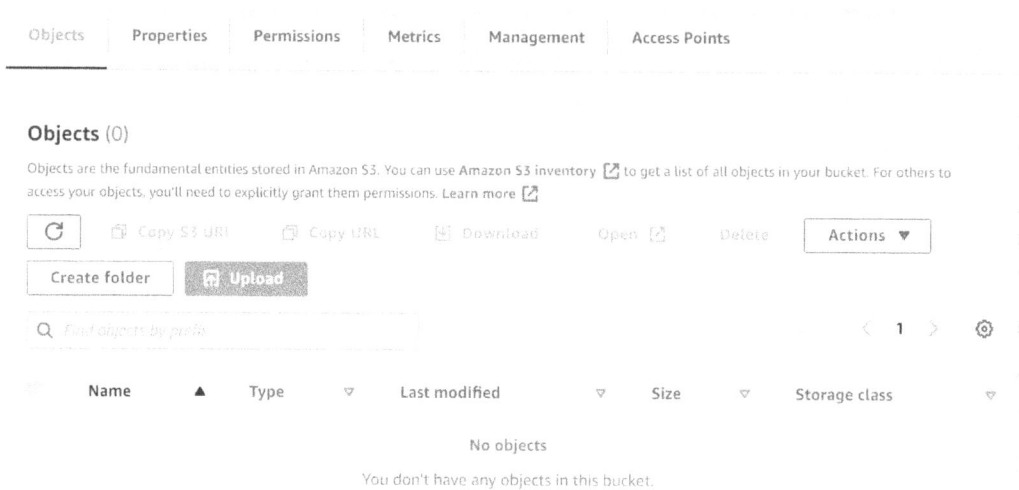

Objects	Properties	Permissions	Metrics	Management	Access Points

Objects (0)

Objects are the fundamental entities stored in Amazon S3. You can use Amazon S3 inventory to get a list of all objects in your bucket. For others to access your objects, you'll need to explicitly grant them permissions. Learn more

C	Copy S3 URI	Copy URL	Download	Open	Delete	Actions ▼

Create folder	Upload

Q Find objects by prefix   < 1 >   ⚙

Name ▲	Type ▽	Last modified ▽	Size ▽	Storage class ▽

No objects

You don't have any objects in this bucket.

Figure 9.14 – New bucket creation

13. Click the **Upload** button.

14. Next, you want to try to resize your browser page with the S3 bucket **Upload** page visible, and next to it resize the `vegan-php-files` folder so that you can easily drag and drop all the folders and files into the S3 bucket's **Object** area, as per the following screenshot. You need to ensure that the folder hierarchy is maintained for the application to work:

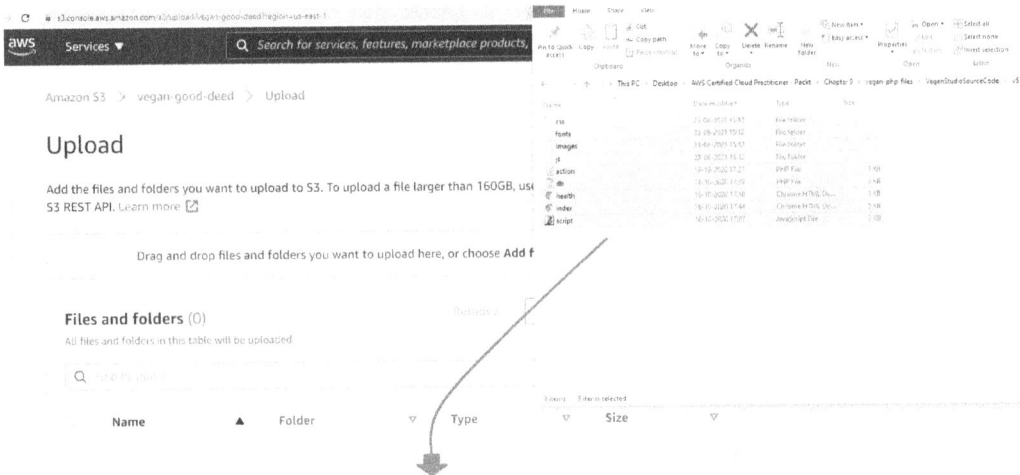

Figure 9.15 – Copying files and folders to the S3 bucket

15. Your Amazon S3 **Upload** page will provide a summary of files and folders to be uploaded. You will need to then click on the **Upload** button at the bottom of the page.

16. Once all the files and folders have been uploaded, you receive an **Upload succeeded** message.

Now that your source code and files for your application have been uploaded, we will move on to the next exercise. As part of this series of exercises, you will need to configure your EC2 instances to download the source code files for the application. Using Bash scripts at the time of launching your EC2 instances, you will download the source code from the Amazon S3 bucket and place it in the appropriate folders within the EC2 instances to serve the application.

Because your EC2 instance would need to have permissions to access the previous S3 bucket we created and download the source code, we need to configure an IAM role that your EC2 instance will use to authenticate to Amazon S3.

Exercise 9.2 – creating an IAM role

In this exercise, you will create an IAM role that your EC2 instances will use to authenticate and access the source code files in your Amazon S3 bucket. Proceed as follows:

1. Ensure that you are logged in to your AWS account and navigate to the IAM dashboard.

2. Click on **Roles** from the left-hand menu.

3. Click **Create role**, as per the following screenshot:

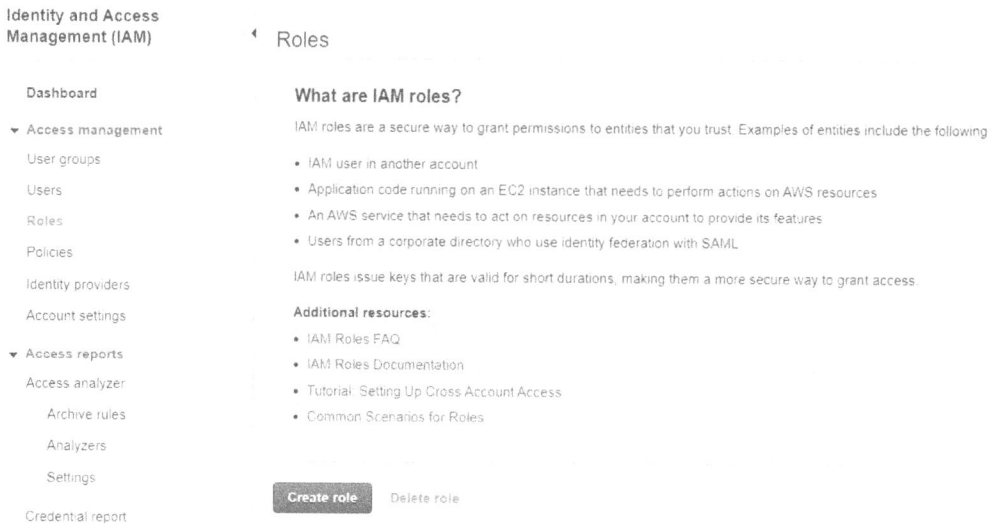

Figure 9.16 – Creating an IAM role

4. In the **Select type of trusted entity** field, click the **AWS services** option, and under **Choose a use case**, select **EC2** under **Common use cases**.

5. Click the **Next: Permissions** button at the bottom of the page.

6. On the **Attach permissions policies** page, filter the list by searching for S3. Next, select the AmazonS3ReadOnlyAccess policy and click the **Next: Tags** button, as per the following screenshot:

Create role   1   ②   3   4

▾ Attach permissions policies

Choose one or more policies to attach to your new role

Create policy   ⟳

Filter policies ⌄   Q s3   Showing 8 results

	Policy name ▾	Used as
▸	AmazonDMSRedshiftS3Role	None
▸	AmazonS3FullAccess	Permissions policy (1)
▸	AmazonS3OutpostsFullAccess	None
▸	AmazonS3OutpostsReadOnlyAccess	None
✓ ▸	AmazonS3ReadOnlyAccess	Permissions policy (1)
▸	IVSRecordToS3	None
▸	QuickSightAccessForS3StorageManagementAnalyticsReadOnly	None
▸	S3StorageLensServiceRolePolicy	None

▸ Set permissions boundary

* Required   Cancel   Previous   **Next: Tags**

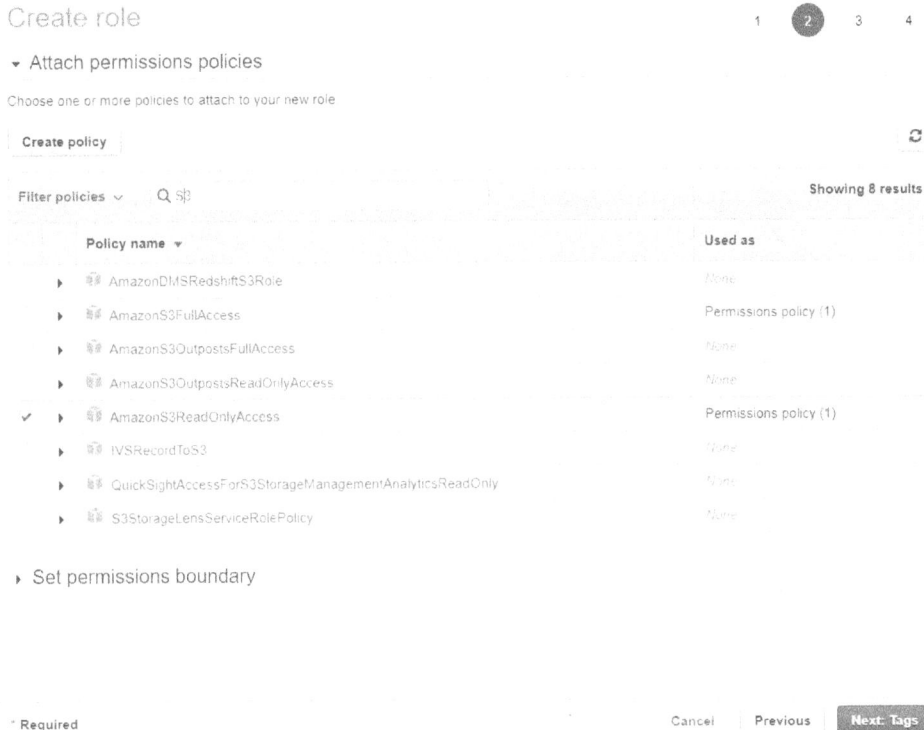

Figure 9.17 – Creating an IAM role (continued)

7. Set a key-value pair to tag your role with a key of **Name** and a value of EC2-to-S3-Read-Access. This allows us to easily identify the role. Click the **Next: Review** button on the bottom right-hand corner of the page.

8. In the **Review** section, provide a name for your role, such as EC2-to-S3-Read-Access, and a description.

9. Finally, click the **Create role** button.

AWS will now create your IAM role, and you will need to reference this role when we launch our EC2 instances in a later chapter.

At this stage, you have the following AWS services configured:

- An Amazon VPC with public and private subnets across two AZs. You have a public subnet to host bastion hosts and NAT gateways, and four private subnets—two for your web/application servers located in the **web/application tier** and another two for your **database tier**.

- An RDS database that will store all data such as the *good deeds of the month* that the employees of The Vegan Studio will submit.

- An Amazon S3 bucket with the source code files configured to point to the RDS database.

- An IAM role to allow your EC2 instances to download the source code files from the S3 bucket.

When you deploy your application, you will install Apache Web Services and host your application files on EC2 instances. Specifically, you will be deploying two EC2 instances that will be placed across two AZs. To distribute traffic across those EC2 instances, you will need to configure an ALB that will be configured to accept inbound HTTP (port 80) traffic from the internet and distribute them to your EC2 instances. In the next exercise, you will need to configure your ALB.

Exercise 9.3 – configuring an ALB

In this exercise, you will be configuring an ALB that will be used to accept inbound traffic from your users on the internet and distribute them across the EC2 instances you deploy later in this chapter.

ALBs, as discussed earlier in this chapter, can be used to distribute web and application traffic using HTTP and HTTPS protocols. You will configure an internet-facing load balancer so that you can accept inbound requests from the internet.

CLBs and ALBs require you to also configure a security group within which you define which traffic would be permitted inbound to those load balancers. Therefore, the first step is to revisit the VPC dashboard and create a new security group for your ALB, as follows:

1. Navigate to the VPC dashboard and ensure you are still in the us-east-1 Region.

2. From the left-hand menu, click on **Security Groups**.

3. Click the **Create security group** button in the top right-hand corner of the screen.

4. Provide the security group with a name such as ALB-SG and a description such as Allow inbound HTTP traffic from Internet.

5. Ensure that you select **ProductionVPC** from the VPC drop-down list.

6. Click the **Add rule** button under **Inbound rules**.

7. For **Type**, select **HTTP** from the drop-down list, and set the source to **Custom**, with a **classless inter-domain routing (CIDR)** block of 0.0.0.0/0. This source denotes the public internet.

8. Provide an optional description and then click on the **Create security group** button on the bottom right-hand corner of the screen.

AWS will now create your security group, which we will use to configure our ALB.

> **Important Note**
> ELBs do not fall under the Free Tier offering from AWS, and you must ensure
> you delete them once you have completed all the labs.

Now that we have configured a security group, we can move on to configuring our load
balancing service. However, your ALB requires a **target group** to send traffic to. The target
group will be used to register the EC2 instances that will accept traffic from the ALB. So,
the first step is to create your target group, as follows:

1. Navigate to the EC2 dashboard and ensure that you are in the us-east-1 Region.

2. From the left-hand menu, click on **Target Groups**, under the **Load Balancing** menu.

3. From the right-hand pane, click on **Create target group**.

4. Next, you are presented with a two-step wizard. In *Step 1*, select **Instances** and then
    scroll further down to provide a **Target group name** value. I have named my target
    group Production-TG.

5. Under **Protocol**, ensure that **HTTP** is selected and the port is set to 80.

6. Next, under **VPC**, ensure you select Production-VPC.

7. Scroll further down till you reach the **Health checks** section.

8. Next, set the **Health check protocol** to **HTTP**.

9. For the **Health check path** field, type in /health.html as per the following
    screenshot:

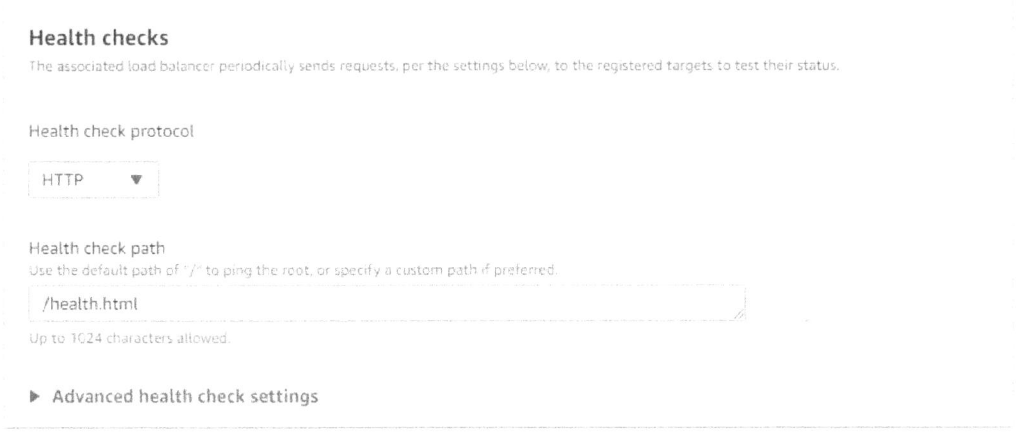

Health checks

The associated load balancer periodically sends requests, per the settings below, to the registered targets to test their status.

Health check protocol

| HTTP ▼ |

Health check path

Use the default path of "/" to ping the root, or specify a custom path if preferred.

| /health.html |

Up to 1024 characters allowed.

▶ Advanced health check settings

Figure 9.18 – Load balancer target group health checks

10. Next, expand the **Advanced health check settings** field.

11. Set the **Port** value to **Traffic Port**.

12. Set the **Healthy threshold** value to 3.

13. Set the **Unhealthy threshold** value to 2.

14. Next, set the **Timeout** value to 2.

15. Finally, set the **Interval** value to 10 seconds.

16. Click the **Next** button at the bottom of the page.

17. This will take you to *Step 2*, where you would normally register any EC2 instances. However, as we have not launched any EC2 instances yet, you can ignore this step and simply click on the **Create target group** button at the bottom of the page.

    Next, now that we have the **target group** configured, we can launch our ALB. ELBs are configured in the EC2 management console or via the **command-line interface** (CLI).

18. From the left-hand menu, click on **Load Balancers** under the **Load Balancing** category.

19. Next, click on the **Create Load Balancer** button at the top of the screen in the right-hand pane.

20. Click on the **Create** button in the **Application Load Balancer** section of the page, as per the following screenshot:

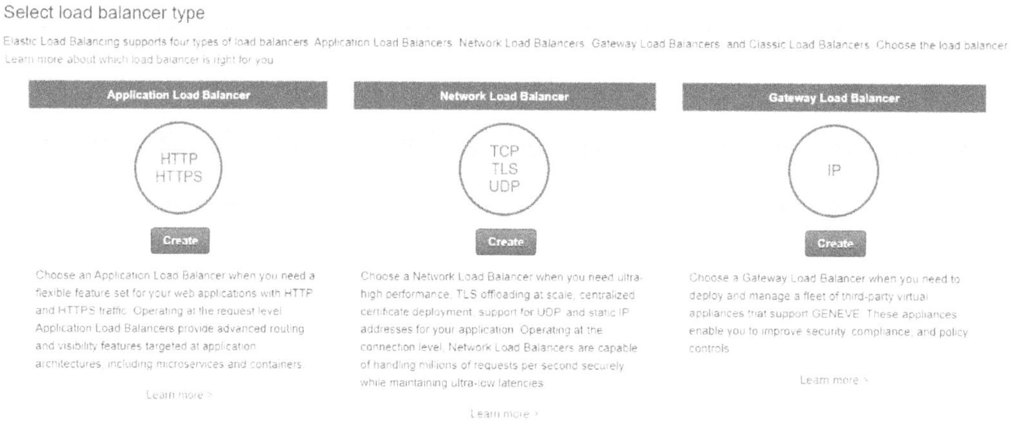

Figure 9.19 – Selecting Application Load Balancer as load balancer type

21. In **Step 1: Configure Load Balancer**, proceed as follows:

    A. Set the name of the load balancer to `Production-ALB`.

    B. Ensure that the **Scheme** field is set to **Internet facing** and that **IP address type** is set to **ipv4**.

    C. Next, under **Network mapping**, select `Production-VPC` under the **VPC** heading.

    D. Under **Mappings**, you need to select which AZs will be enabled for the ALB.

    E. Select the checkboxes next to both the `us-east-1a` and `us-east-1b` AZs.

    F. In the **Subnet** drop-down list for the `us-east-1a` AZ, select the **Public Subnet One** subnet.

    G. Next, in the **Subnet** drop-down list for the `us-east-1b` AZ, select the **Public Subnet Two** subnet.

    AWS will then deploy the ALB *nodes* in these public subnets, routing incoming traffic from the internet to the EC2 instances in the private subnets that we registered as targets for the load balancer. Internet-facing load balancers should be created in subnets that have been configured with an internet gateway such as in this case: the public subnets of your VPC.

22. Next, under **Security Groups**, select the `ALB-SG` security group from the drop-down list. You can also delete the **default** security group that was pre-selected by clicking on the **X** sign next to the group. This is because we only want to associate the load balancer with the `ALB-SG` security group.

23. In the **Listener** section, ensure that the **Protocol** field is set to **HTTP** and the **Port** field is set to **80**. Next, under **Default Action**, select the `Production-TG` target group you created earlier from the drop-down list.

24. Finally, scroll further down and click on the **Create load balancer** button.

25. You will receive a confirmation message stating that the load balancer has been created successfully. Click on **View load balancers**, which will take you back to the list of load balancers deployed, and you should find your `Production-ALB` load balancer in the list. After a few moments, the status of the load balancer should change from **Provisioning** to **Active**.

At this point, we have now configured our ALB. Let's go ahead and look at our architectural diagram to see how our configuration is coming along:

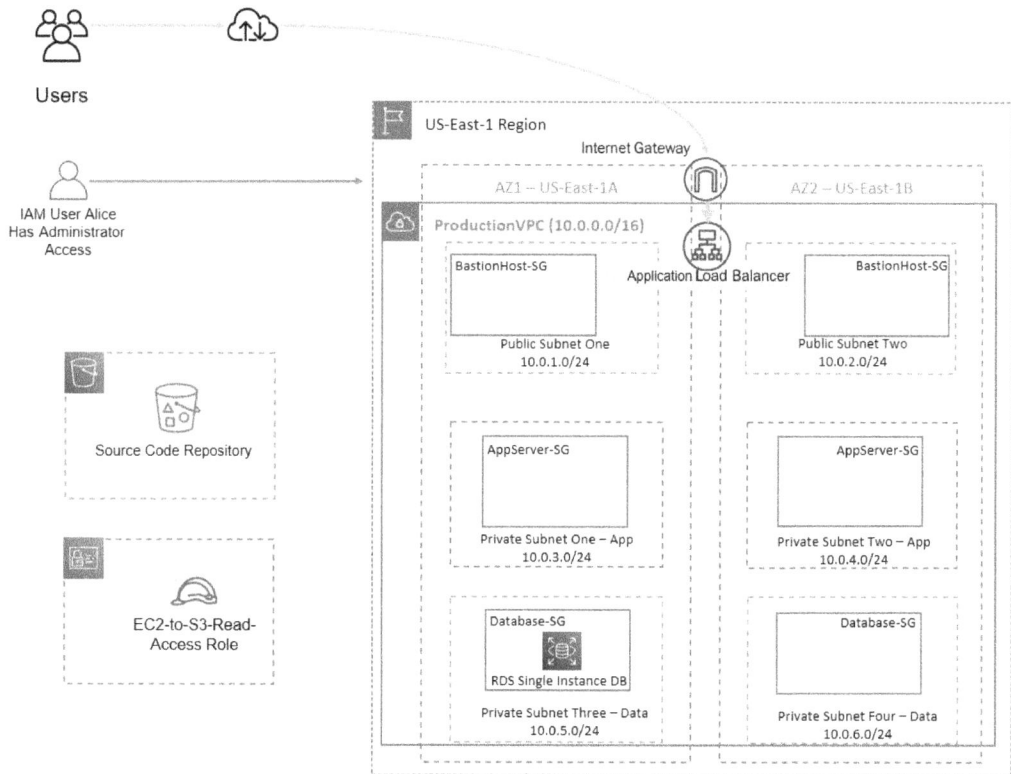

Figure 9.20 – Production VPC architecture after configuring S3 bucket, IAM role, and ALB

For traffic to be allowed inbound to the application servers, we need to ensure that the security groups associated with those servers have been correctly configured. Specifically, the AppServers-SG security group must allow traffic on the HTTP protocol (port 80) from the ALB we deployed in the previous exercise.

Furthermore, in *Chapter 7*, *AWS Compute Services*, you configured the AppServers-SG security group to accept traffic from the BastionHost-SG security group. This was to enable inbound traffic on **Remote Desktop Protocol** (**RDP**) (port 3389), which enables the Windows Remote Desktop client to perform remote access operations. Although we will not be deploying any bastion hosts in the remaining exercises in this chapter, we will amend the inbound rule on the AppServers-SG security group such that the protocol and port used to accept traffic from the BastionHost-SG security group will be set to the SSH protocol on port 22. This is because we will be deploying Linux EC2 instances to host our application, and any remote management of Linux servers requires you to configure SSH access.

Exercise 9.4 – amending the Production-VPC security group

In this exercise, we will amend the RDP inbound rule in the `AppServers-SG` security group such that it is configured to accept traffic on the SSH protocol (port 22) from the `BastionHost-SG` security group. Next, we will add a new rule to accept traffic on the HTTP protocol (port 80) from the ALB's security group, `ALB-SG`. Finally, we will amend the `BastionHost-SG` security group such that it is configured to accept traffic on the SSH protocol (port 22) from the internet. This is useful if you later wish to perform any remote administration of your Linux servers.

Amend the `BastionHost-SG` security group, as follows:

1. Navigate to the VPC dashboard and ensure that you are in the `us-east-1` Region.

2. From the left-hand menu, click on **Security Groups**.

3. In the middle pane, select the checkbox next to the **Security group ID** value associated with the `BastionHost-SG` security group.

4. In the pane below, click on **Inbound rules** and then click on the **Edit inbound rules** button.

5. Next, delete the existing **RDP** rule by clicking on the **Delete** button on the far right of the page.

6. Click the **Add rule** button.

7. For the type, select **SSH** from the drop-down list. Next, ensure that the **Custom** option is selected in the **Source** column, and in the search box next to it, type in `0.0.0.0/0`.

8. Finally, click on the **Save rules** button in the bottom right-hand corner of the page.

   Amend the `AppServers-SG` security group.

9. Click on the **Security Groups** link from the left-hand menu again to see all your security groups in the VPC.

10. In the middle pane, select the checkbox next to the **Security group ID** value associated with the `AppServers-SG` security group.

11. In the pane below, click on **Inbound rules** and then click on the **Edit inbound rules** button.

12. Next, delete the existing **RDP** rule by clicking on the **Delete** button on the far right of the page.

13. Click the **Add rule** button.

14. For the type, select **SSH** from the drop-down list. Next, ensure that the **Custom** option is selected in the **Source** column, and in the search box next to it, start by typing in `sg-`. You will notice that a list of your security groups will become visible. Select the `BastionHost-SG` security group from this list.

15. Next, click the **Add rule** button again.

16. For the type, select **HTTP** from the drop-down list. Next, ensure that the **Custom** option is selected in the **Source** column, and in the search box next to it, start by typing in `sg-`. You will notice that a list of your security groups will become visible. This time, select the `ALB-SG` security group from the list.

17. Finally, click on the **Save rules** button in the bottom right-hand corner of the page.

We will not need to amend the `Database-SG` security group because this has already been configured to only accept traffic from the `AppServers-SG` security group using the MySQL port `3306`.

Recall from the architectural diagram in *Figure 9.20* that the web/application EC2 instances are going to be placed in a private subnet. Our The Vegan Studio employees will be able to access the *Good Deed of the Month* contest application on those EC2 instances via the ALB. However, the EC2 instances will need access to the internet to download updates as well as the source code files stored on the Amazon S3 bucket.

Remember that, unlike the public subnet, the private subnet does not grant direct access to the internet. Any EC2 instance in the private subnet would need to direct internet-bound traffic via an AWS NAT gateway, as discussed in *Chapter 6, AWS Networking Services – VPCs, Route53, and CloudFront*.

In the next exercise, we will deploy a NAT gateway.

Exercise 9.5 – deploying a NAT gateway

In this exercise, we will deploy a NAT gateway in the **Public Subnet One** subnet of our production VPC. Ideally, you want to deploy multiple NAT gateways in each public subnet across the AZs you have resources in to avoid a **single point of failure** (**SPOF**). However, for the purposes of this lab, we will use a single NAT gateway.

In addition, you will need to configure your **main route table** with a new route that will allow outbound traffic to the internet via this NAT gateway.

We will start this exercise by first allocating an elastic IP address for our AWS account, which is a requirement to configure a NAT gateway. To do this, follow these steps:

1. Navigate to your VPC dashboard and ensure that you are in the `us-east-1` Region.

2. NAT gateways require an elastic IP address, and so you will need to allocate one first to your AWS account. From the left-hand menu, click on **Elastic IPs**. In the right-hand pane, click the **Allocate Elastic IP address** button.

3. You will be presented with the **Allocate Elastic IP address** page. Ensure that **Amazon's pool of IPv4 addresses** is selected and then click the **Allocate** button.

AWS will allocate an elastic IP address from its pool of available addresses for your AWS account. Next, you will need to set up your NAT gateway, as follows:

1. From the left-hand menu, click on **NAT Gateways**.

2. In the right-hand pane, click the **Create NAT gateway** button.

3. On the **Create NAT gateway** page, proceed as follows:

    A. Provide a name for your NAT gateway—for example, `Production-NAT`.

    B. Next, from the **Subnet** drop-down list, select the **Public Subnet One** subnet.

    C. Next, from the drop-down list under **Elastic IP allocation ID**, select the elastic IP you allocated to your account moments ago.

    D. Finally, click the **Create NAT gateway** button at the bottom of the page.

The NAT gateway will take a couple of minutes to be provisioned. Once ready, the NAT gateway state will be set to **Available**, as per the following screenshot:

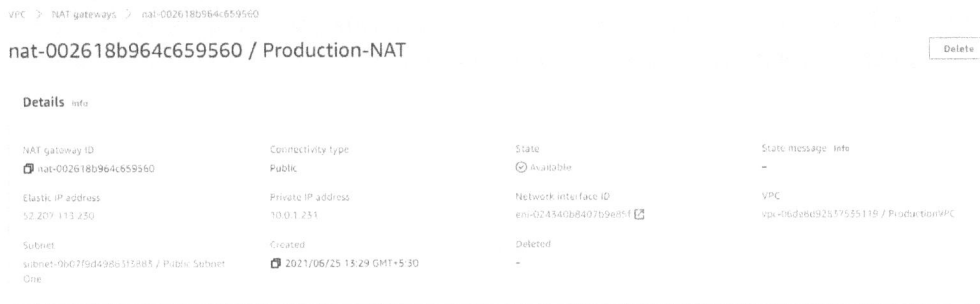

Figure 9.21 – NAT gateway

Now that you have deployed your NAT gateway, you will need to configure your main route table with a route to the internet that uses the NAT gateway, as follows:

1. From the left-hand menu in the VPC dashboard, select **Route Tables**.

2. Click on the checkbox next to **Main Route Table**.

3. In the bottom pane, click on the **Routes** tab.

4. Next, click the **Edit routes** button on the far right-hand side of the page.

5. You will be presented with the **Edit routes** page. Click the **Add route** button.

6. Under the **Destination** column, provide the destination as 0.0.0.0/0.

7. Next, click on the **Target** search box to open up a list of potential targets. Select the **NAT Gateway** target, and AWS will display available NAT gateways associated with this VPC. You should find the Production-NAT NAT gateway in the list. Go ahead and select this.

8. Finally, click on the **Save changes** button.

Your main route table has now been configured with a route to the internet that will use the NAT gateway.

Now that you have configured your NAT gateway and the main route table correctly, we can proceed with deploying our EC2 instances that will host the *Good Deed of the Month* application in the next exercise.

Exercise 9.6 – deploying your application servers with Amazon Auto Scaling

In this exercise, we will configure the Amazon Auto Scaling service to define a **Launch Configuration** for our deployment, which will include a script to configure our EC2 instances with the Apache web service and download the application source files from the Amazon S3 bucket. As part of the exercise, you will also create an EC2 instance profile that will be used to contain the IAM role you created earlier and allow the EC2 instance to assume that role.

The EC2 instances will also be provisioned as targets in the Production-TG target group we created earlier in the ALB exercise. The Production-ALB ALB will then be able to distribute inbound traffic from our The Vegan Studio employees on the internet to those EC2 instances, enabling them to submit any good deeds they carried out for review by our panel.

In addition, we will configure Auto Scaling policies to always ensure that we always have two running EC2 instances, one in each private subnet, across the two AZs, us-east-1a and us-east-1b. In terms of health checks, these will be performed both at the EC2 level and via the ALBs using the health check parameters you defined in *Exercise 9.3* earlier.

Creating an Auto Scaling Launch Configuration

As part of this exercise, you will need access to a Bash script that we have included in the GitHub repository `https://github.com/PacktPublishing/AWS-Certified-Cloud-Practitioner-Exam-Guide`. In the `vegan-php-files.zip` file you downloaded earlier, which you unzipped, you will find a file called `userdata-script` in the top-level folder. You will need to amend this script to match your configuration. Open the script file in a notepad or text editor application and change the last line of the script, replacing `[Source Bucket]` with the actual name of your bucket. So, for example, if your bucket name is `vegan-good-deed`, then the last line should be changed from `aws s3 cp s3://[Source Bucket] /var/www/html -recursive` to `aws s3 cp s3://vegan-good-deed /var/www/html -recursive`. Make sure to save the file.

Next, we look at the steps required to set up our AWS Auto Scaling Launch Configuration, as follows:

1. Navigate to the EC2 dashboard and ensure that you are in the `us-east-1` Region.

2. From the left-hand menu, select **Launch Configurations** from the **Auto Scaling** category.

3. In the right-hand pane, click on the **Create Launch Configuration** button.

4. You will be presented with the **Create Launch Configuration** page.

5. Provide a name for your Launch Configuration—for example, `Production-LC`.

6. Next, you need to search for the Amazon Linux 2 AMI. It might be difficult to find the AMI in the new **user interface** (**UI**). To identify the AMI ID, open another browser window to access your AWS account and navigate to the EC2 dashboard. Click on **Instances** from the left-hand menu and then click on the **Launch instances** button in the far right-hand corner of the screen. You will find a list of quickstart AMIs. From this page, *make a note of the AMI ID for the Amazon Linux 2 instance*. Ensure that the AMI ID is for the `64-bit x86` architecture, which I have highlighted as per the following screenshot:

Amazon Linux 2 AMI (HVM), SSD Volume Type - ami-0ab4d1e9cf9a1215a (64-bit x86) / ami-0d296d66f22f256c2 (64-bit Arm)

Amazon Linux
Free tier eligible

Amazon Linux 2 comes with five years support. It provides Linux kernel 4.14 tuned for optimal performance on Amazon EC2, systemd 219, GCC 7.3, Glibc 2.26, Binutils 2.29.1 through extras. This AMI is the successor of the Amazon Linux AMI that is approaching end of life on December 31, 2020 and has been removed from this wizard.

Root device type: ebs   Virtualization type: hvm   ENA Enabled: Yes

Figure 9.22 – AMI ID for Amazon Linux 2 instance

7. Back in the previous browser window where you are configuring your Auto Scaling Launch Configuration, click on the drop-down arrow under **AMI** and paste in the AMI ID you copied previously in the search box. You should then be able to find the relevant AMI to use. Make sure you select this AMI.

8. Next, under **Instance type**, click on the **Choose instance type** button, and in the search box, type in t2.micro. You can then select the t2.micro instance type from the filtered list. Go ahead and click the **Choose** button.

9. Next, under the **Additional configuration** option, click the drop-down arrow under **IAM instance profile** and select the EC2-to-S3-Read-Access instance profile that contains the IAM role you created earlier.

10. Next, expand the **Advanced details** section.

11. Under **User Data**, ensure that **As Text** is the selected option and, in the textbox provided, go ahead and paste in a copy of the Bash script file you amended a few moments ago.

12. Next, under **IP address type**, ensure that you select **Do not assign a public IP address to any instances**. This is because the EC2 instances are going to be launched in the private subnets and will not require a public IP address.

13. Leave the settings in the **Storage (volumes)** field at their default values.

14. Next, under **Security groups**, click on the **Select an existing security group** option, and from the list of available security groups, select the security group ID associated with the AppServers-SG security group.

15. In the **Key pair (login)** section, select **Choose an existing key pair** from the **Key pair options** drop-down list.

16. In the drop-down list under **Existing key pair**, ensure that you select the key pair you created earlier. In my example, this is the USEC2Keys key pair.

17. Next, tick the box to acknowledge that you have access to the private key file that you downloaded earlier in *Chapter 7, AWS Compute Services*.

18. Finally, click on the **Create Launch Configuration** button at the bottom of the screen.

At this point, you have successfully created your first Auto Scaling Launch Configuration, as per the following screenshot:

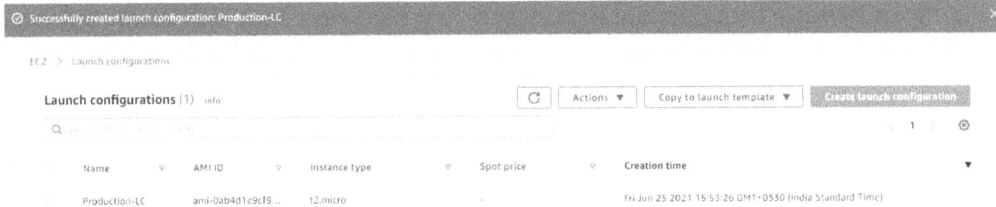

Figure 9.23 – Amazon Auto Scaling Launch Configuration

Now that you have created your Launch Configuration, you can proceed to configure your Auto Scaling groups.

Configuring Auto Scaling groups

As part of creating your Auto Scaling groups, you can define Auto Scaling policies. Because we will not be performing any real load testing on our application servers, we will simply configure our Auto Scaling policy to ensure that we always have a minimum of two EC2 instances across the two AZs. Proceed as follows:

1. From the left-hand menu of the EC2 dashboard, click on the **Auto Scaling Groups** link under **Auto Scaling**.

2. Click the **Create an Auto Scaling group** button in the right-hand pane of the screen.

3. In **Step 1, Choose Launch Template or configuration**, provide a name to identify your Auto Scaling group—for example, `Production-ASG`.

4. In the next section of the screen, you will have an option to select a Launch Template from a drop-down list. However, instead of a Launch Template, we have configured a Launch Configuration. To access your Launch Configuration, click on the **Switch to Launch Configuration** link on the far right-hand side of the screen.

5. Next, under **Launch Configuration**, select the `Production-LC` Launch Configuration you created earlier.

6. Click the **Next** button to move on to *Step 2*.

7. In **Step 2, Configure settings**, select `Production-VPC` from the **VPC** drop-down list.

8. In the drop-down list under **Subnets**, ensure that you select both the **Private Subnet One - App** and **Private Subnet Two - App** subnets, as per the following screenshot:

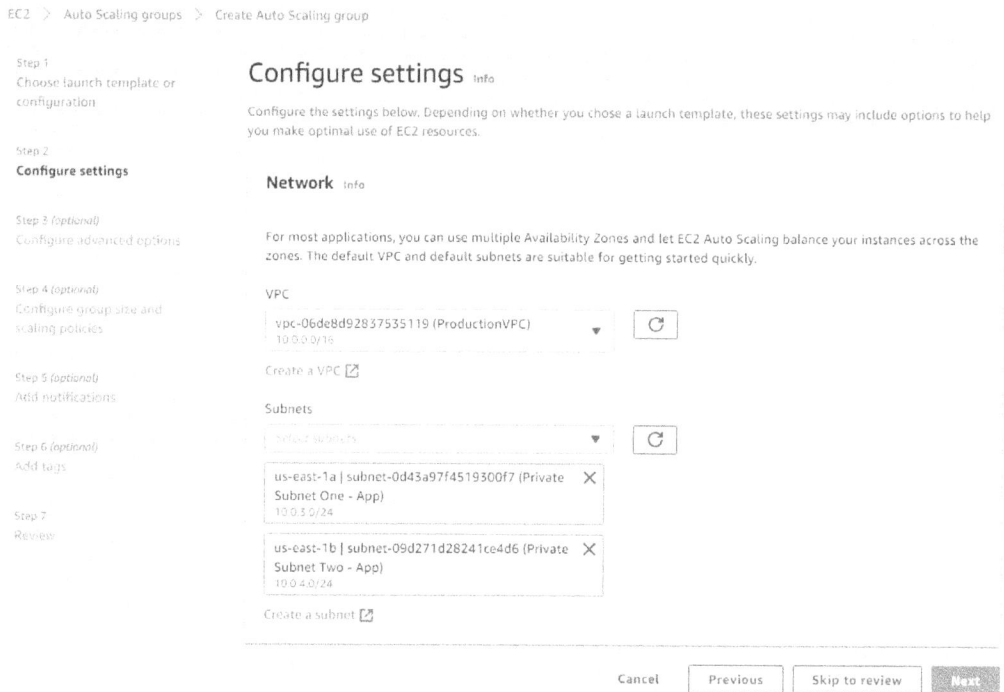

Figure 9.24 – Auto Scaling group subnet selection

9. Click the **Next** button.

10. In **Step 3, Load balancing – optional**, we will be using the ALB you created earlier. Select the **Attach to an existing load balancer** option.

11. Next, ensure that **Choose from your load balancer target groups** is selected under the **Attach to an existing load balancer** section.

12. From the drop-down list under **Existing load balancer target groups**, select the `Production-TG` target group that is associated with the `Production-ALB` ALB.

13. Next, under the **Health checks – optional** section, select the **ELB** checkbox. This is to enable ELB health checks in addition to the EC2 health checks.

14. Click the **Next** button at the bottom of the screen.

15. In **Step 4, Configure group size and scaling policies**, under **Group size**, set the **Desired**, **Minimum**, and **Maximum** capacity values to 2 each. We want to always maintain two EC2 instances in our fleet.

16. Under **Scaling policies – optional**, ensure that **None** is selected, and then click the **Next** button.

17. In **Step 5, Add notifications**, do not add any notifications and click on the **Next** button.

18. In **Step 6, Add tags**, click the **Add tag** button. Specify a key-value pair to set the name of the servers you launch such that the **Key** field is set to Name and the **Value** field is set to Production-Servers.

19. Click the **Next** button to continue.

20. You are then presented with a **Review** page. Review the configuration settings that you have defined to make sure you followed the preceding series of steps correctly. When you are satisfied, go ahead and click the **Create Auto Scaling group** button at the bottom of the page.

AWS will then start configuring your Auto Scaling group and proceed to launch two EC2 instances based on the parameters of the groups. The EC2 instances will be configured as per the configuration you defined in the Launch Configuration earlier.

Once Auto Scaling has completed the deployment of your EC2 instances, you will be able to see the details of your deployment, as per the following screenshot:

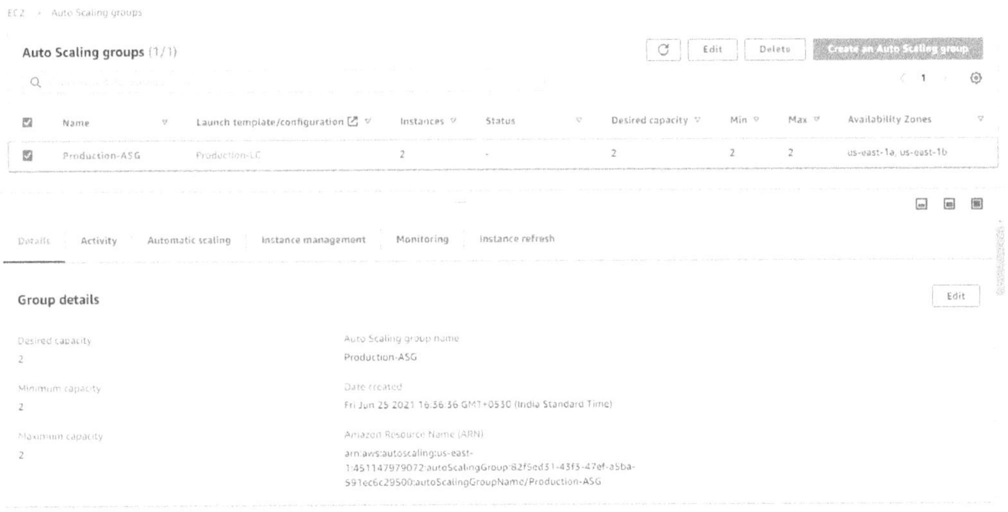

Figure 9.25 – Auto Scaling group deployment completed

At this point, your application has now been deployed across two EC2 instances. If you click on the **Activity** tab, you will see the AWS Auto Scaling service launched two EC2 instances in response to the fact that the minimum and desired capacity were not met before the launch of those EC2 instances. The Auto Scaling service will always try to ensure you have the desired number of EC2 instances in your fleet. Next, we will review the deployment and access the application.

Reviewing your deployment and accessing your application

You can check whether the Auto Scaling service has correctly deployed your application. Specifically, you can check whether two EC2 instances have been deployed and registered with your ALB. Furthermore, you can also check whether the ALB has marked those EC2 instances as healthy, indicating that the health checks have passed as well.

Here are the steps to perform these checks and then access the application:

1. In the EC2 dashboard, click on the **Target Groups** link from the left-hand menu under **Load Balancing**.

2. In the right-hand pane, click on the `Production-TG` target group that you created earlier.

3. On the details page of the `Production-TG` target group, you will note that two EC2 instances have been launched and both are in a **healthy** state, as per the following screenshot:

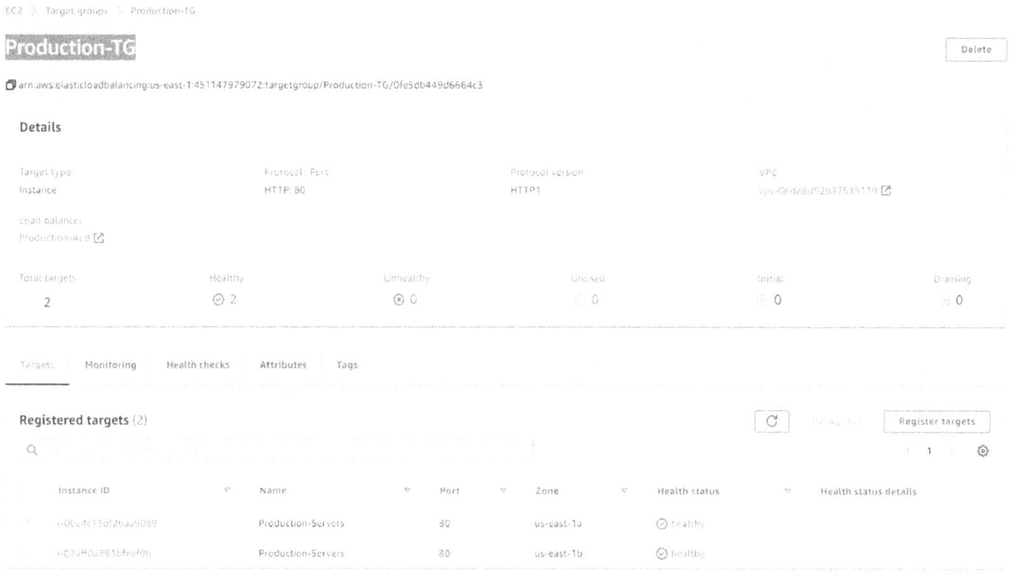

Figure 9.26 – Healthy EC2 instances registered to load balancer target group

4. Next, click on the **Instances** link from the left-hand menu.

5. You will notice that two instances with the name `Production-Servers` have been launched, with one EC2 instance in the `us-east-1a` AZ and the other in the `us-east-1b` AZ, as per the following screenshot:

	Name	Instance ID	Instance state	Instance type	Status check	Alarm status	Availability Zone
	Production-Servers	i-02aff0a893bfecf0b	⊘ Running	t2.micro	⊘ 2/2 checks passed	No alarms +	us-east-1b
	Production-Servers	i-00e4c73bf26aa9089	⊘ Running	t2.micro	⊘ 2/2 checks passed	No alarms +	us-east-1a

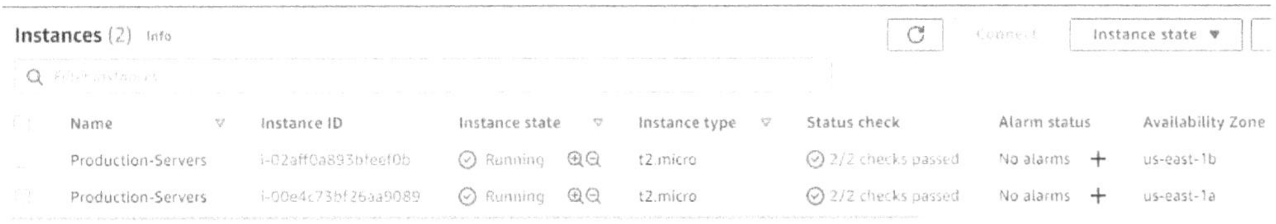

Figure 9.27 – Auto Scaling group successfully launched two EC2 instances

6. Next, we can access our application. From the left-hand menu, click on the **Load Balancers** link under **Load Balancing**.

7. In the right-hand pane, you will find your ALB details, as per the following screenshot:

	Name	DNS name	State	VPC ID	Availability Zones
	Production-ALB	Production-ALB-1998282540.us-east-1.elb.amazona...	Active	vpc-06de8d92837535119	us-east-1a, us-east-1b

Load balancer: | Production-ALB

Description   Listeners   Monitoring   Integrated services   Tags

Basic Configuration

Name	Production-ALB
ARN	arn:aws:elasticloadbalancing:us-east-1:451147979072:loadbalancer/app/Production-ALB/4e6bae1279c9580c
DNS name	Production-ALB-1998282540.us-east-1.elb.amazonaws.com (A Record)
State	Active
Type	application
Scheme	internet-facing
IP address type	ipv4
	Edit IP address type

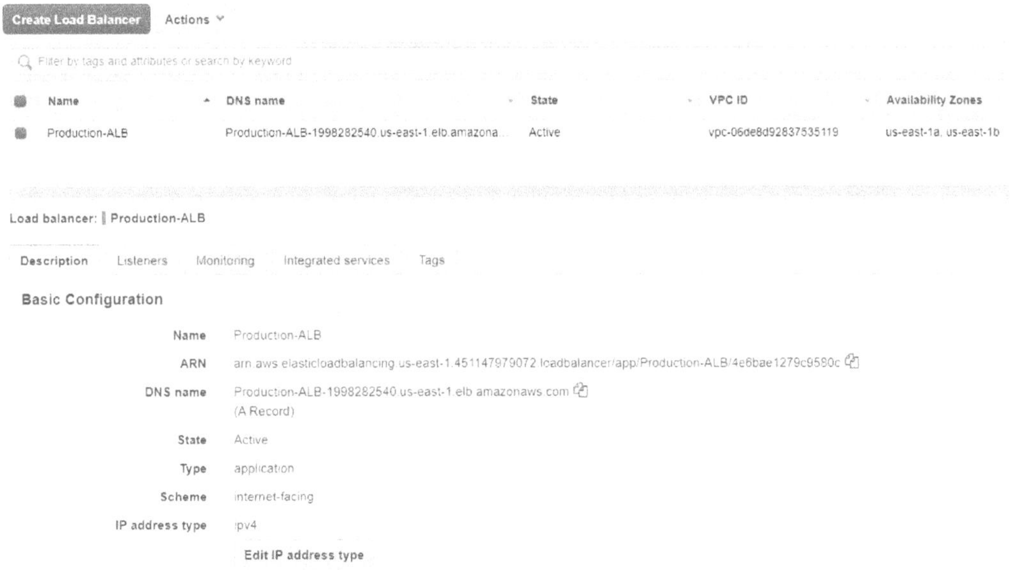

Figure 9.28 – ALB details for Production-ALB

8. In the bottom pane, you will find a **DNS name** link for your ALB. Copy this URL and paste it into a new browser window. If you have successfully completed all of the previous exercises, you will be able to access the **Good Deed of the Month Contest** web application, as per the following screenshot:

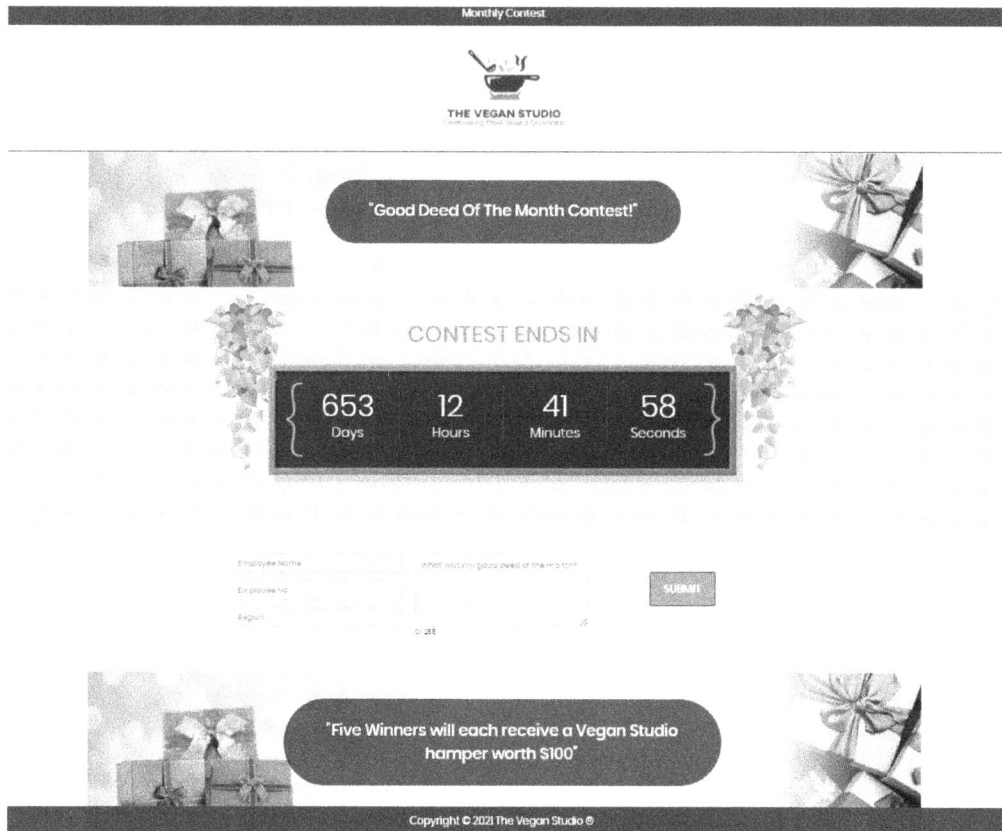

Figure 9.29 – Good Deed of the Month Contest web application

9. You can test your application by entering details of some potential good deeds you have done yourself. Once you have filled in the form in the middle of the web page, click the **SUBMIT** button.

10. You will note that when the web page reloads after you click on the **Submit** button, your *good deed of the month* is read back from the MySQL RDS database and presented on the page. If you submit more entries, these are also reported back. This demonstrates how the application can write to and read from the backend RDS database.

Let's take another look at the application architectural diagram to see how you have built this multi-tier solution:

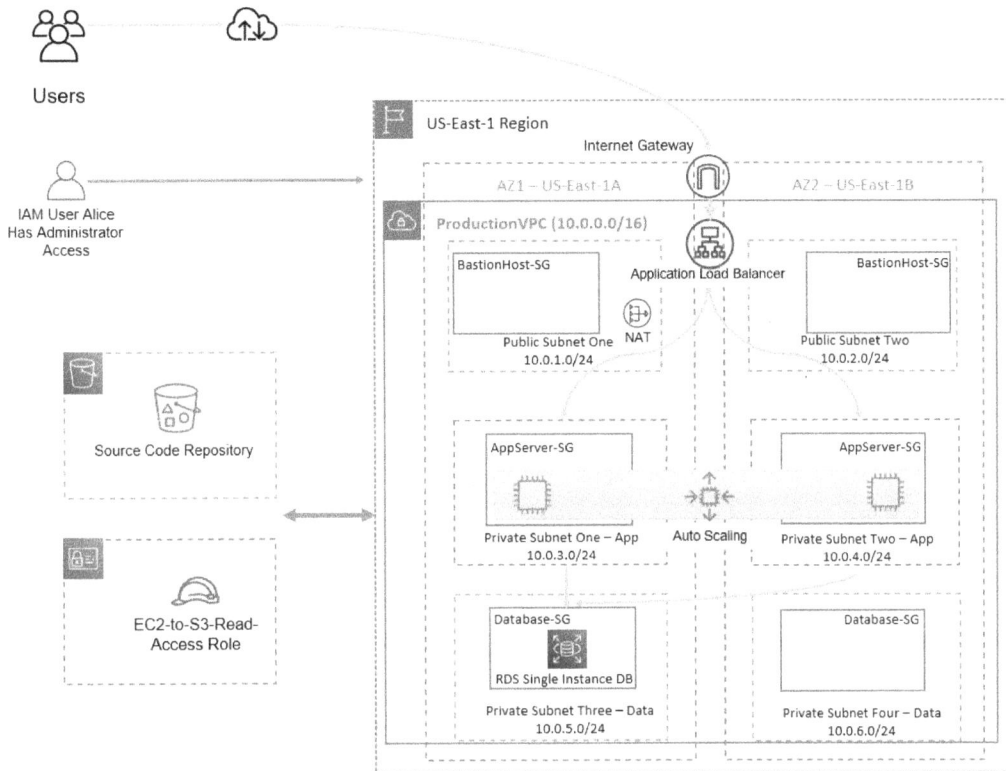

Figure 9.30 – Multi-tier application architecture

The key components of your architecture include the following:

- A VPC in the us-east-1 Region that consists of public and private subnets across the us-east-1a and us-east-1b AZs. The VPC consists of six subnets in total: two public subnets to host your ALB nodes and NAT gateway, two private subnets to host your application tier, and another two private subnets to host the database tier.

- You also have an Amazon S3 bucket to host all the application source code and files.

- An IAM role that will allow your EC2 instances to gain authorization to read and download the application source code from the S3 bucket.

- An RDS database deployed in the us-east-1a AZ as a single database instance. Ideally, you would want to configure your database with Multi-AZ for HA.

- Two EC2 instances deployed using an Auto Scaling group and Launch Configuration. The Launch Configuration contains the necessary **bootstrapping** script to set up and configure the EC2 instances as web servers and automatically serve the application to your users. Furthermore, the Auto Scaling group automatically registers any EC2 instances deployed to the ALB target group. The target group runs health checks against the EC2 instances, marking them as healthy or unhealthy based on the health checks defined.

- Finally, the application code has the necessary database connection to access the backend RDS database and store any application data. Note that storing database connection details within the application is not considered best practice, and AWS offers several options such as the **AWS Systems Manager Parameter Store** or **AWS Secrets Manager** to manage such pieces of sensitive data. To keep this lab simple, we stored the database connection details within the application code.

Next, we look at how to test the AWS Auto Scaling service by simulating a failure of an EC2 instance.

Testing the Auto Scaling service

In this part of the exercise, you will stop an EC2 instance to simulate failure. When the EC2 instance is in a stopped state, it will not respond to the load balancer health checks. The load balancer will then mark the EC2 instances as unavailable. This will send a notification to the Auto Scaling group, confirming that there are fewer than two EC2 instances in the group, which is less than the desired capacity. Auto Scaling should then replace the instance and new server. Let's proceed with simulating this failure of an EC2 instance, as follows:

1. In the EC2 dashboard, click on the **Instances** link from the left-hand menu.

2. Next, in the right-hand pane of the screen, you will note that you have two EC2 instances running. Select the instance that is in the us-east-1b AZ, as per the following screenshot:

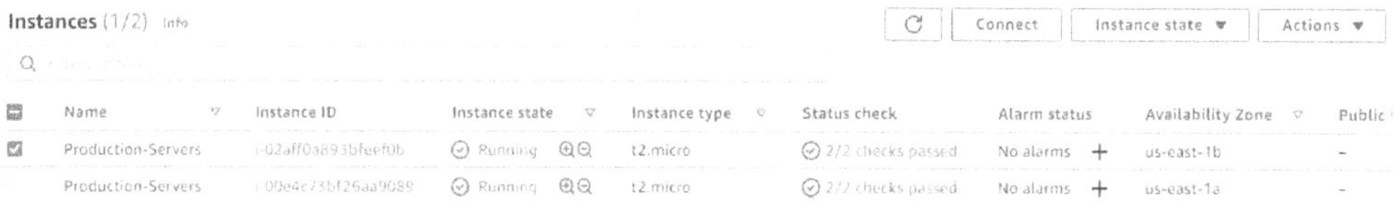

Figure 9.31 – EC2 instances in Running state

3. From the **Instance state** drop-down list at the top right-hand corner of the screen, select **Stop instance** while ensuring the EC2 instance in the us-east-1b AZ is selected.

4. You will be prompted with a dialog box to confirm whether you want to stop the selected instance. Go ahead and click the **Stop** button.

5. AWS will then perform a shutdown of your EC2 instance, which will take a couple of minutes. Wait until the EC2 instance is in a **Stopped** state, and then proceed to click on the **Auto Scaling Group** link under the **Auto Scaling** category from the left-hand menu.

6. Next, in the right-hand pane, click the `Production-ASG` Auto Scaling group.

7. Next, click on the **Activity** tab.

8. You will find additional activities that clearly show that the Auto Scaling service terminated the stopped EC2 instance. This is because, in a stopped state, it cannot respond to health checks. This is then followed by the launch of a new EC2 instance to replace the one that got terminated (see the following screenshot), in order to maintain our desired capacity at two instances as per the Auto Scaling group configuration. You will note that the Auto Scaling service will not try to restart the stopped instance. The Auto Scaling group will use the same Launch Configuration to configure the server with the application and register it to the ALB's target group, as illustrated here:

Status		Description		Cause		Start time		End time	
Successful		Launching a new EC2 instance: i-0cb11660eda954fa0		At 2021-06-27T05:39:20Z an instance was started in response to a difference between desired and actual capacity, increasing the capacity from 1 to 2.		2021 June 27, 11:09:23 AM +05:30		2021 June 27, 11:09:55 AM +05:30	
WaitingForELB ConnectionDraining		Terminating EC2 instance: i-02aff0a893bfeef0b - Waiting For ELB Connection Draining		At 2021-06-27T05:39:00Z an instance was taken out of service in response to an EC2 health check indicating it has been terminated or stopped.		2021 June 27, 11:09:00 AM +05:30			

Figure 9.32 – Auto Scaling activity history

9. Now that the Auto Scaling service has replaced your EC2 instance, you can visit your application via the ALB URL to confirm that your application is still functioning as expected. Note that when one of the EC2 instances was stopped, the application was still accessible via the ALB URL because traffic would have been forwarded onto the other EC2 instance that was still running in the `us-east-1a` AZ.

Congratulations! Well done on completing the series of exercises to get to this stage. You have now learned how to design and architect a multi-tier application solution using a combination of AWS services to help you build an HA and scalable application.

In the next exercise, you will perform a cleanup operation to terminate unwanted resources so that you do not incur any further charges.

Exercise 9.7 – cleanup

In this exercise, you will terminate the various resources you deployed in the previous exercises. The first step is to delete the Auto Scaling group, which will terminate your EC2 instances. If you try terminating the EC2 instances manually, then the Auto Scaling group will simply launch new ones. Proceed as follows:

1. From the EC2 dashboard, click on **Auto Scaling Group** from the left-hand menu under **Auto Scaling**.

2. From the right-hand pane, select the `Production-ASG` Auto Scaling group. Click the **Delete** button and confirm the delete request by typing in `delete` in the textbox and clicking the **Delete** button.

3. Next, click on **Launch Configuration** from the left-hand menu under the **Auto Scaling** service.

4. Next, select the `Production-LC` Launch Configuration, and from the **Actions** menu, click **Delete Launch Configuration**. Confirm the delete request.

5. Next, click **Load Balancers** under the **Load Balancing** menu.

6. From the right-hand pane, select the `Production-ALB` load balancer, and from the **Actions** drop-down list, click **Delete**.

7. Next, click on **Target Groups** under **Load Balancing** in the left-hand menu. In the right-hand pane, select the `Production-TG` target group, and from the **Actions** drop-down list, click **Delete** and delete the target group.

Your load balancer and the Auto Scaling group have been removed from your account. Next, navigate to the Amazon RDS console, as follows:

1. From the left-hand menu, click on **Databases**.

2. In the right-hand pane, select the database that you created earlier in *Chapter 8, AWS Databases Services*.

3. From the **Actions** drop-down list, click **Delete**.

4. Uncheck the **Create final snapshot?** box and click the acknowledgment box that states that upon deletion, automated backups, including system snapshots and **point-in-time recovery** (**PITR**), will no longer be available. Next, type `delete me` in the confirmation textbox and click the **Delete** button. Your Amazon RDS database will now be deleted.

5. Next, we should also remove the database subnet group created previously. From the left-hand menu, click on **Subnet groups**.

Extended exercises – setting the scene   357

6. In the right-hand pane, select the database subnet group you created previously and click the **Delete** button.

   Now that your database has also been deleted, we can delete the VPC. Navigate to the VPC console.

7. Before we can delete the VPC, you need to delete the NAT gateway. From the left-hand menu, click on **NAT Gateways**. In the right-hand pane, select the `Production-NAT` NAT gateway, and from the **Actions** drop-down list, click **Delete NAT gateway**. You will then be presented with a dialog box to confirm the deletion. Type `delete` in the confirmation box and click the **Delete** button.

8. Next, you need to check whether there are any network interfaces still attached to your VPC. Usually, you will find that the Amazon RDS network interface (`RDSNetworkInterface`) may still be attached to the `Database-SG` security group. If that is the case, you will first need to delete this interface before you can delete the VPC, as follows:

   A. Navigate to the EC2 console and select **Network Interfaces** from the left-hand menu.

   B. Check whether there are any interfaces still attached to your VPC by cross-referencing the VPC ID with your `Production-VPC` security group ID. Select the network interface and then, from the **Actions** menu, click the **Delete** button. The following screenshot shows the attached network interface:

Figure 9.33 – Network interfaces attached to your VPC

9. Navigate back to the VPC console. Next, from the left-hand menu, click on **Your VPCs**.

10. From the right-hand pane, select the `Production-VPC` security group, and then, from the **Actions** drop-down list, select **Delete VPC**.

11. You will be presented with a list of all components of your VPC that will be deleted. Confirm your delete request by typing `delete` into the confirmation textbox and then clicking the **Delete** button.

Your VPC should now get deleted. Within the VPC console, there is still one more component you need to delete, and that is the elastic IP address you allocated to your AWS account. This is because elastic IP addresses are only free if they are associated with running instances (or in our case, the NAT gateway). Proceed as follows:

1. From the left-hand menu, click on **Elastic IPs**.

2. In the right-hand pane, select the IP address you allocated to your AWS account, and from the **Actions** drop-down list, click the **Release Elastic IP addresses** link. Next, in the **Release Elastic IP addresses** dialog box, click the **Release** button.

At this point, your elastic IP address has been released back to AWS. You will not incur any charges on unused elastic IP addresses in your account.

This completes your cleanup exercise for this chapter, and you can now rest assured that you will not incur any further costs associated with this lab.

> **Important Note**
> You still have an Amazon S3 bucket that hosts all the source code for the application you deployed in this chapter. While you could delete that resource, we advise you to keep the bucket as we will be using it for the exercises in the next chapter.

Next, we provide a summary of this chapter and the key concepts to remember for the exam.

Summary

In this chapter, you learned about the differences between vertical scaling and horizontal scaling. We discussed options to increase an EC2 instance's specification and capacity. We then examined the AWS ELB service and how it can be used to evenly distribute incoming application traffic across a fleet of EC2 instances. You learned about the different types of ELBs and their use cases—ALBs, NLBs, GWLBs, and CLBs. We discussed how, using ELB, you can distribute the placement of EC2 instances that power your application across multiple AZs, thereby offering HA of services in case of AZ failures or outages.

Next, we examined how we can automatically scale out (add more EC2 instances to our fleet of servers that support an application) using the Amazon Auto Scaling service. Auto Scaling can help us scale out when demand increases and equally scale back in when demand drops, ensuring that you always have the right number of EC2 instances to provide the best **user experience** (**UX**) for your application.

Both ELB and Auto Scaling, however, are Regional services only. This means using these two services alone cannot offer global resilience. To offer HA across Regions, we discussed how we can use Route 53 and other global services such as CloudFront. Route 53 offers several routing policies to make it possible to distribute traffic to application servers spread across the globe and offers options to build complete **disaster recovery** (**DR**) and business continuity solutions for your business.

In the next chapter, we look at a number of AWS services designed to help you build applications on AWS that move away from traditional monolith architectures in favor of modern decoupled architecture and microservices.

Questions

1. You are planning on developing a website in multiple languages such that you have one fleet of EC2 instances that serves the English version of your site and another fleet that serves the Spanish version of your site. For each language version, you will be configuring URLs with different paths such that the English version of your site will contain /en/ in the path and the Spanish version will contain /es/.

    Which type of load balancer would you use to route traffic to ensure users connect to the site in their desired language?

    A. CLB

    B. NLB

    C. ALB

    D. Path-based load balancer

2. You are building a multi-tier architecture with web servers placed in the public subnet and application servers placed in the private subnet of your VPC. You need to deploy ELBs to distribute traffic to both the web server farm and the application server farm. Which type of load balancer would you choose to distribute traffic to your application servers?

    A. Internet-facing

    B. Internal load balancer

    C. Dynamic load balancer

    D. Static load balancer

3. Which ELB is ideal for handling volatile workloads and can scale to millions of requests per second?

   A. ALB

   B. NLB

   C. CLB

   D. Premium load balancer

4. Which configuration feature of the AWS Auto Scaling service enables you to define a maximum number of EC2 instances that can be launched in your fleet?

   A. Auto Scaling group

   B. Auto Scaling Launch Configuration

   C. Auto Scaling max fleet size

   D. Auto Scaling policy

5. When an ELB detects an unhealthy EC2 instance, which action does it perform regarding distributing incoming traffic?

   A. It continues to send traffic to the failed instance.

   B. It terminates the failed instance so that it is not part of the ELB target group.

   C. It only sends traffic to the remaining healthy instances.

   D. It restarts the unhealthy EC2 instance.

6. Which service does an AWS ALB integrate with to protect your applications from common web attacks?

   A. WAF

   B. Shield

   C. Inspector

   D. **Key Management Service (KMS)**

10
Application Integration Services

AWS offers a suite of services that enable you to build architectures that enable communication between the different components of your application in a bid to move away from monolith designs. These integration services facilitate design patterns for distributed systems, serverless applications, and decoupled applications.

Ultimately, decoupling your application from traditional all-in-one monolith architectures ensures a reduced impact when making changes. It also facilitates easier upgrades and new features being released faster.

In this chapter, we will look at several services that offer integration capabilities. These include messaging solutions between application components using a queuing service, notification services, which can be used for **application-to-application** (**A2A**) notifications or **application-to-person** (**A2P**) type notifications, event-driven workflow designs, and coordinating multiple services into serverless workloads.

In this chapter, we will cover the following topics:

- Understanding notification services such as Amazon **Simple Notification Service** (**SNS**)

- Decoupling your application architecture with Amazon **Simple Queue Service** (**SQS**) and Amazon MQ

- Designing event-driven workflows to connect your application data with various AWS services using EventBridge
- Coordinating multiple AWS services into serverless workloads with Amazon Step Functions and Amazon **Simple Workflow Service** (**SWF**)

Technical requirements

To complete the exercises in this chapter, you will need access to your AWS Free Tier account, as well as permissions to access the various AWS services. You will also need access to the VPC you built in *Chapter 6, AWS Networking Services – VPCs, Route53, and CloudFront.*

Understanding notification services such as Amazon SNS

Amazon SNS is a push-based messaging and notification system that can be used to allow one application component to send messages to other application components or directly to end users.

Amazon SNS uses a publisher/subscriber model where one application component will act as a publisher of messages and the other application components will consume those messages as subscribers. Amazon SNS allows you to design high throughput, many-to-many messaging between distributed systems, microservices, and event-driven applications.

Let's look at an example. Suppose you want to be notified if any of your IAM users upload an object to a particular Amazon S3 bucket that they have access to. To achieve this, you can configure **S3 event notifications** to send out an alert whenever the s3:ObjectCreated:* action occurs. This notification can be sent to an **SNS topic** (discussed later), which you subscribe to using your email address. This way, every time your users upload a new object to your S3 bucket, Amazon SNS will send out a notification to you via email. This is an example of A2P messaging using SNS.

Let's look at another example. Suppose you host an S3 bucket that allows end users to upload images in a default format, and you have a requirement to convert those images into multiple formats. To achieve this requirement, you can use Amazon Lambda, which, as we discussed in *Chapter 7, AWS Compute Services*, is a serverless compute solution that allows you to run code in response to an event or trigger.

For this specific example, you can configure your S3 notification service to send a message to an Amazon SNS topic and have the Lambda function subscribe to that topic. The message can include information about the new image that has been uploaded and can trigger your Lambda function to access the image in the bucket, create different formats of the image, and save them in another S3 bucket. This automated process is an example of A2A messaging using SNS.

Next, we will take a look at a key component of the Amazon SNS services, specifically SNS endpoints.

Amazon SNS endpoints

As we mentioned previously, Amazon SNS is a push-based messaging solution, enabling one or more publishers to push messages to one or more subscribers. With Amazon SNS, your subscribers need to use a supported endpoint type. These endpoints are depicted in the following diagram:

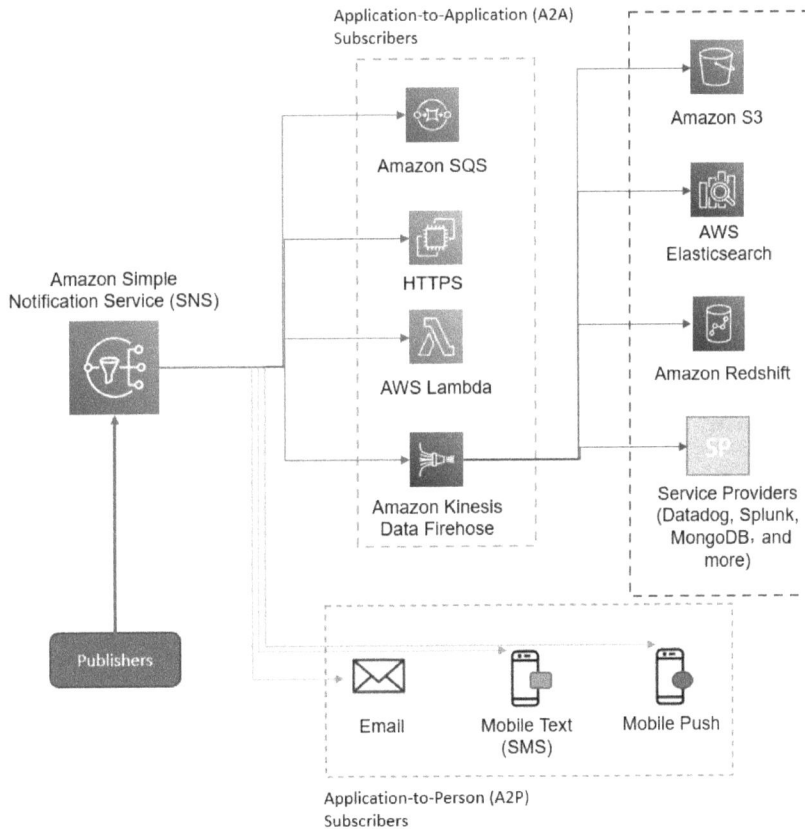

Figure 10.1 – Amazon SNS subscriber endpoints

Amazon SNS A2A endpoints include Amazon SQS, HTTP/S endpoints, AWS Lambda, and Amazon Kinesis Firehose. Data from Amazon Kinesis Firehose can then be offloaded and stored in Amazon S3 buckets, AWS Elasticsearch, and Amazon Redshift, as well as other third-party service providers.

Amazon SNS A2P endpoints include email, mobile text messages, and mobile push endpoints.

Amazon SNS also ensures high levels of message durability. Messages are stored and replicated on multiple devices across geographically separated servers and data centers.

Amazon SNS topics

At the heart of the Amazon SNS service is the SNS topic feature, which is a logical access point that acts as a communication channel between your publishers and subscribers. Before you can send out messages to your subscribers, you need to create a topic. Your publisher needs to be made aware of which SNS topic to send messages to and your end clients must subscribe to the topic to be able to receive those messages.

In the following diagram, we can see that an application component allows us to upload objects to an Amazon S3 bucket (**1**). Amazon S3 can be set up with an event notification service that pushes out a notification, stating that an upload took place to an Amazon SNS topic. In this diagram, an admin has subscribed to the SNS topic. Any notifications resulting from the objects being uploaded to the S3 bucket are then sent to the admin:

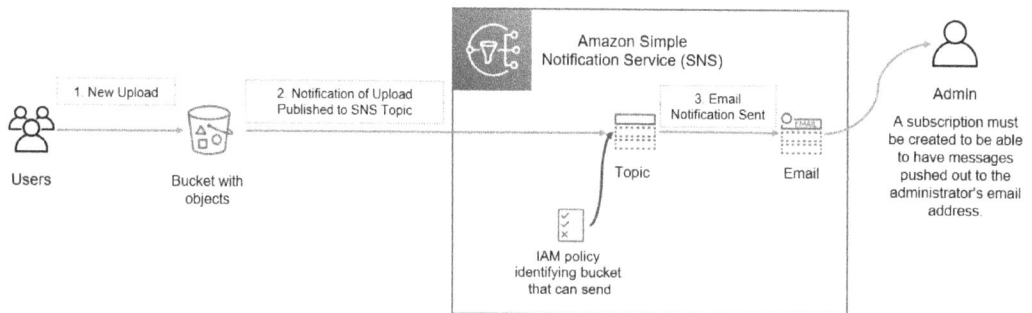

Figure 10.2 – Example – configuring an SNS notification for an S3 event notification

Your publishers also need permission to be able to publish messages to the topic. In the previous example, where we wanted to send a notification to an administrator every time a new object was uploaded to an S3 bucket, you would also need to create permissions that grant the S3 bucket the ability to send messages to the SNS topic. You can do this by configuring an IAM policy that specifies which bucket can send messages to the topic and attach it directly to the SNS topic as an inline policy.

Subscribers to your topic will then have messages pushed out to them whenever a new message is published by the publisher. In the case of the preceding example, the publisher is the S3 notification service, and the subscriber is your administrator's email address.

Note that when you create a topic, you need to assign it a name. This can be up to 256 characters in length and can contain hyphens (-) and underscores (_). Amazon SNS will assign an **Amazon Resource Name (ARN)** to the topic you create, which will include the service name – in this case, `sns` – the Region, the AWS account ID, and the topic name. So, for example, an SNS topic called `new-recipe-upload-alert`, created in the London Region, with an AWS account ID of `123456789789` will have an ARN of `arn:aws:sns:eu-west-2:1234567890123789: new-recipe-upload-alert`.

Next, we will look at the topics you can create on Amazon SNS, which will depend on the application's use case.

Standard and FIFO topics

When configuring Amazon SNS, you create a **standard topic** by default. Standard topics are used when the message's delivery order is not going to affect your application in any way and where duplicating messages will not create any issues in your workflow. All supported delivery protocols support standard topics.

In addition, you can create **FIFO topics**. These are designed to ensure strict message ordering and prevent message duplication. Note that only the Amazon SQS endpoint (specifically, the Amazon SQS FIFO queue) can subscribe to a FIFO topic. We will discuss the Amazon SQS service later in this chapter.

Amazon SNS Fanout scenario

A key feature offered by Amazon SNS is the ability to replicate messages pushed out to an SNS topic across multiple endpoints. This is known as the **Fanout scenario** and it allows parallel asynchronous processing.

Let's look at an example. Let's say that you are a theatre company and that you sell tickets for your various performances. You are required to process online payments from customers and issue them their tickets. At the same time, you are also required to store information on all sales in AWS's data warehousing solution, which is offered by Amazon Redshift. One way to design this architecture is depicted in the following diagram:

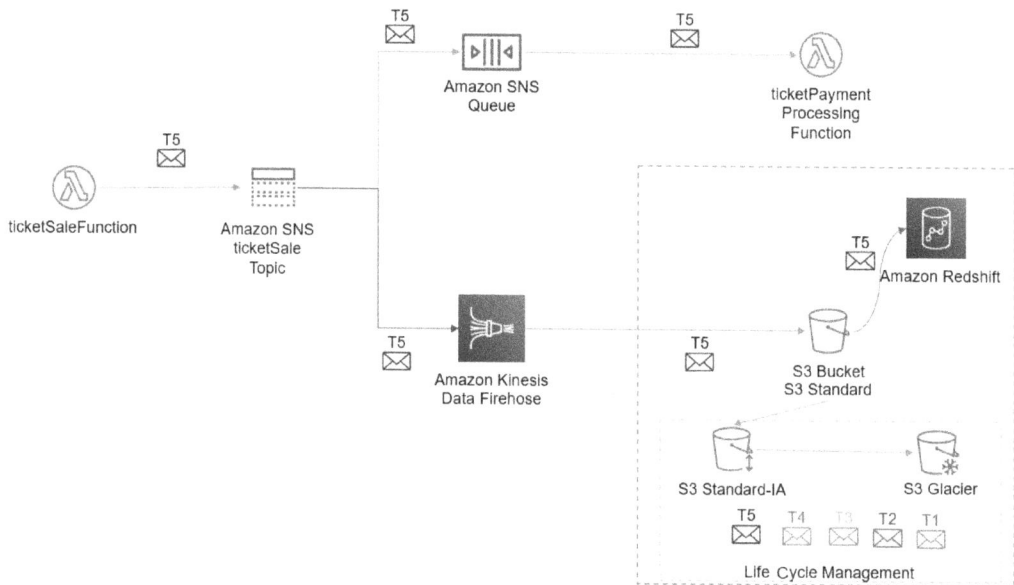

Figure 10.3 – Example of an Amazon SNS Fanout scenario

In the preceding diagram, incoming ticket sales are sent to an SNS topic, which then gets replicated to an SQS queue and an Amazon Kinesis Data Firehose stream. Any messages that are sent to the SQS queue are processed by the payment function to complete the sale transaction. Additional queues may be added to the architecture for order fulfillment and customer notification.

Next, the same message is processed by Amazon Kinesis Data Firehose, which feeds the data into an Amazon Redshift cluster. Note that to stream data from Amazon Kinesis Firehose to Redshift, you need to deliver the data into an S3 bucket and then issue an Amazon Redshift `COPY` command to load the data into your Amazon Redshift cluster. We will discuss Amazon Kinesis in the next chapter.

At the same time, messages will continue to reside in the S3 bucket, which can be archived using the life cycle management process, as we discussed in *Chapter 5, Amazon Simple Storage Service (Amazon S3)*. This can help address any compliance requirements to store historic information on all ticket sales.

Amazon SNS pricing

Amazon SNS is a managed service with no upfront cost. You pay based on usage and this is based on the type of topic that's used; that is, standard topics or FIFO topics. Standard topics are charged based on the number of API requests made per month and the number of deliveries to the various endpoints. For example, mobile push notifications are charged at $0.50 per million notifications after you have exhausted your free tier threshold of 1 million notifications.

The maximum payload size for your messages is 256 KB. Except for SMS messages, you are billed for every 64 KB chunk as one request. So, a payload size of 256 KB is equal to four requests. Furthermore, if you need to send messages that are larger than 256 KB, you can use the Amazon Extended SNS Client Library, which allows you to send payloads via the Amazon S3 service. When you do this, additional Amazon S3 storage costs are incurred.

With regards to FIFO topics, you are charged based on the number of published messages, subscribed messages, and their respective amount of payload data.

In this section, we looked at the Amazon push-based messaging solution offered by Amazon SNS. Amazon SNS works based on a publisher/subscriber model and enables you to design and architect A2A messaging and A2P messaging. Amazon SNS can help you build integration between different application components, allowing you to design distributed systems, microservices, and serverless architectures.

You also learned about some of the core features of Amazon SNS, including standard and FIFO topics, as well as the Fanout scenario concept.

In the next section, we will look at another messaging integration service known as Amazon SQS. Amazon SQS is a pull-based messaging solution and lends itself well to designing decoupled architectures, enabling you to migrate away from monolith application architectures.

Decoupling your application architecture with Amazon SQS and Amazon MQ

Amazon SQS is another fully managed messaging integration solution that enables you to decouple your application components into distributed systems and facilitate the design and architecture of microservices. One of the primary advantages of using a queuing system such as Amazon SQS is the ability to move away from monolithic application designs. In a monolithic design, where all the components of your applications are dependent on each other and always need to be available to each other, you often suffer from frequent failures and outages. A queueing system such as Amazon SQS can help the different components of your application work independently and queues can hold messages in the form of requests/tasks until capacity becomes available. With asynchronous processing and the ability for different components to scale independently, you benefit from higher levels of availability, where each component can scale as needed without impacting the overall workflow.

In the following diagram, we can see how Amazon SQS can be used to queue messages between various components of your application and achieve a decoupled architecture (also known as loose coupling):

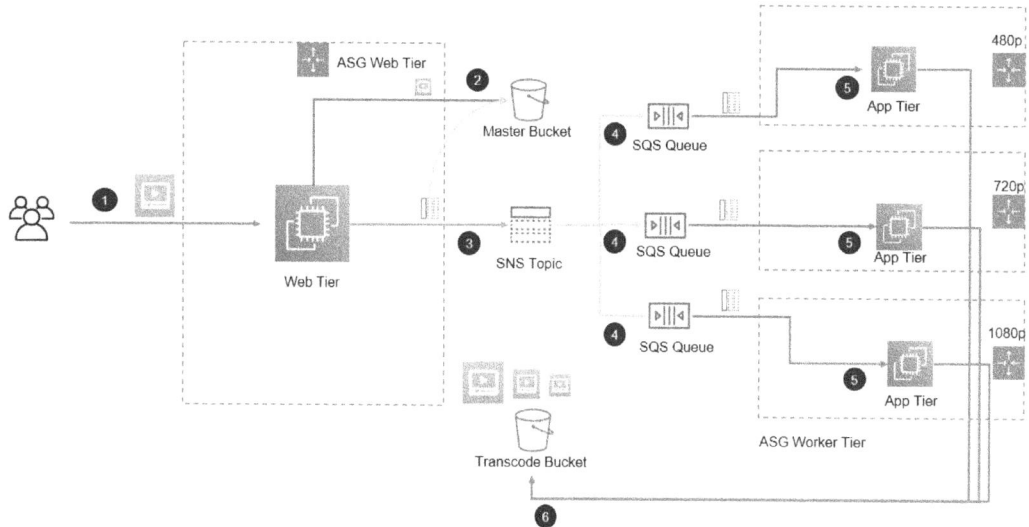

Figure 10.4 – Amazon SQS use case example

In the preceding diagram, a media transcoding example makes use of both the Amazon SNS and Amazon SQS services to convert raw videos uploaded by users into various formats and resolution sizes. In this architecture, we have different auto-scaling groups that are provisioning a fleet of servers, with each fleet responsible for converting the videos into a specific format. Separate queues are created to handle messages destined for the different fleets of servers to process. Here is a quick breakdown of the workflow:

1. Users upload videos via a frontend web server farm that is part of an auto-scaling group designed to scale out and scale in based on demand.

2. The videos are uploaded to a master bucket.

3. At the same time, an SNS notification is sent out to multiple Amazon SQS queues in a fanout configuration (refer to the SNS Fanout scenario discussed earlier in this chapter).

4. Each SQS queue holds messages for the appropriate app server to pull when capacity is available.

5. The relevant app servers retrieve the messages from the appropriate SQS queue, which identifies the videos that need to be processed in the master bucket. The app servers then retrieve the raw videos from the master bucket.

6. The app servers convert the format and resolution of the videos and upload the completed videos in the correct format into the transcode bucket.

Whereas Amazon SNS offers a push-based message notification solution, Amazon SQS is a fully managed pull-based message queue system that will also retain the messages for a short duration (the default is set to 4 days but this can be configured to a maximum duration of 14 days).

This means that if you have backend services that need to process lots of messages in the queue from frontend web requests, you can retain those messages until your backend services can process new messages in the queue. Amazon SQS increases the overall fault tolerance of your application solutions, allowing your decoupled application components to run independently.

Amazon SQS queue types

Amazon SQS offers two types of queues designed to help address different use cases. These are discussed next.

Amazon SQS standard queues

Standard queues support a nearly unlimited number of API calls per second, (SendMessage, ReceiveMessage, or DeleteMessage) and are designed for messages to be delivered at least once. However, this does mean that on an odd occasion, duplicate copies of the message could be delivered. In addition, messages may not be delivered in the order in which they were introduced into the queue. So, your application must be able to cope with messages that are not delivered in the order in which they entered the queue, as well as the occasional duplicate message:

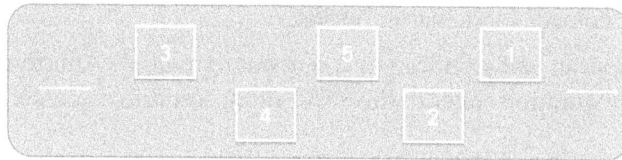

Figure 10.5 – Amazon SQS standard queue

Standard queues are particularly useful when you need to process vast amounts of transactions per second. However, note that there is a 120,000 quota for the number of inflight messages for a standard queue. A typical example of where you might use standard queues is, for example, when you need to process a high number of credit card validation requests for an e-commerce application.

Amazon SQS FIFO queues

FIFO stands for **first-in first-out**. FIFO queues are designed to preserve the order of your messages, as well as ensuring only one-time delivery with no duplicates.

FIFO queues only offer throughput at a rate of 300 transactions per second. This means that they cannot offer unlimited throughput; however, high throughput of messages can be offered by using a process known as batching, which offers support for 3,000 transactions per second, per API method (SendMessageBatch, ReceiveMessage, or DeleteMessageBatch). These 3,000 transactions represent 300 API calls, each with a batch of 10 messages.

In addition, as depicted in the following diagram, messages are delivered in the order in which they were introduced into the queue and Amazon SQS will preserve this order:

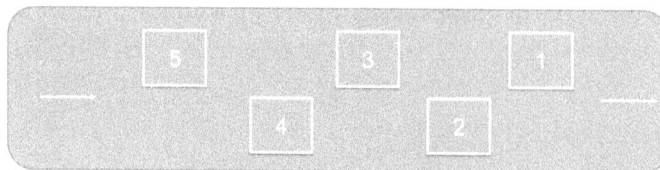

Figure 10.6 – Amazon SQS FIFO queue

So, FIFO queues are ideal for those applications where the order of events is important, such as when you're making sure that user-entered commands are run in the right order. FIFO queues also ensure no duplicates are created more than once, such as processing payment transactions. However, they are not ideal where exceptional levels of scaling might be required or where you need to process greater than 3,000 transactions per second (with batching).

Amazon SQS pricing and security

To ensure sensitive data is protected, you can encrypt messages using Amazon **Key Management Service** (**KMS**). In terms of pricing, there are no upfront costs. You pay based on the number and content of requests and the interactions with Amazon S3 and AWS KMS. Note that as part of the free tier, you also get the first 1 million requests for free every month.

In this section, we looked at another managed messaging solution offered by Amazon SQS. Amazon SQS is a pull-based message queuing solution that allows you to decouple your application components, enabling them to work independently of each other. Messages are stored in either standard or FIFO queues and application components retrieve these messages as required when there is available capacity.

In the next section, we will look at Amazon MQ, which is a message broker service designed for Apache ActiveMQ and other message brokers.

Amazon MQ

A message broker is a piece of software designed to help you facilitate communications between application components to exchange information. Message brokers allow different services to communicate with each other directly, even if those services are written in different languages or run on different platforms.

Many organizations have existing message brokering services within their on-premises environments that support their on-premises applications. One such service is **Apache ActiveMQ**, which is probably one of the most popular Java-based message brokers.

When clients are looking to migrate their on-premises applications, you must consider where such third-party message brokering services are being consumed. AWS offers a service known as Amazon MQ, which is a fully managed message broker service that provides compatibility with popular message brokers.

Amazon recommends using Amazon MQ to migrate applications from existing message brokers where compatibility with APIs such as **JMS** or protocols such as **AMQP 0-9-1**, **AMQP 1.0**, **MQTT**, **OpenWire**, and **STOMP** is required.

If, as part of the migration, you are looking to completely rearchitect your application layer, then you may wish to consider Amazon SNS and Amazon SQS instead, as you do not require third-party message brokers. Amazon recommends these services for new applications that can benefit from nearly unlimited scalability and simple APIs.

In this section, we looked at Amazon MQ, which enables customers to easily migrate to a message broker in the cloud and offers compatibility with existing messaging brokers such as Apache ActiveMQ and RabbitMQ. With Amazon MQ, you reduce your overall operational overhead when provisioning, configuring, and maintaining message brokers that depend on connectivity with APIs such as JMS or protocols such as AMQP 0-9-1, AMQP 1.0, MQTT, OpenWire, and STOMP.

In the next section, we will look at event-driven workflow services and the services offered by Amazon to help architect event-driven solutions for your applications.

Designing event-driven application workflows using AWS EventBridge

Amazon EventBridge is a serverless *event bus service* that allows you to stream real-time events from your applications, SaaS-based services, and AWS services to a variety of targets. These targets can include AWS Lambda, Kinesis, an HTTP/S endpoint, or another event bus service in another account. Amazon EventBridge helps you create application architectures where you need to react and perform some action against those events that are generated.

Events can be generated when there is a change in the state of a given resource, such as when an EC2 instance changes its state from a running state to a stopped state. Another example of an event is when your auto-scaling group launches or terminates an EC2 instance. Additional functionality, as required by your application architecture, can be created by reacting to such state changes.

With EventBridge, you set up rules that define matching incoming patterns or events. When an event occurs, as defined by the rule, it can be sent to a target for further processing. For example, an event that resulted in a critical server being stopped (by accident) can be sent to a Lambda function to have it restarted automatically.

EventBridge can also be configured to trigger events at a defined schedule. For example, if you have a large fleet of EC2 instances that are used to test new applications, and you generally run your tests every weekday from Monday to Friday during normal business hours, then there is no need to have the servers running outside of those business hours. You can set up a scheduled event to trigger a Lambda function that stops the servers at 6 P.M. Monday to Friday and restarts them every weekday at 8 A.M. This ensures that outside of normal business hours, your servers are in the stopped state. If these are on-demand EC2 instances, you do not get charged while those servers remain stopped.

Amazon EventBridge is an updated solution from a previous version known as Amazon CloudWatch Events. With CloudWatch Events, you were limited to a default event bus that enabled you to route all AWS events, as well as custom events. However, with the new Amazon EventBridge, you can introduce custom event buses in addition to the default event bus. Custom event buses can be created exclusively for your workloads and enable you to control access to events that are limited to a set of AWS accounts or custom applications. In addition, you can use content-based filtering and advanced rules for routing events. EventBridge can handle more processing, reduce the load on downstream events, and use partner event sources such as Zendesk, PagerDuty, and Datadog.

In the following diagram, we can see how EventBridge works at a high level. We can see the sources of events, the types of buses that can be used, and the potential targets for those events:

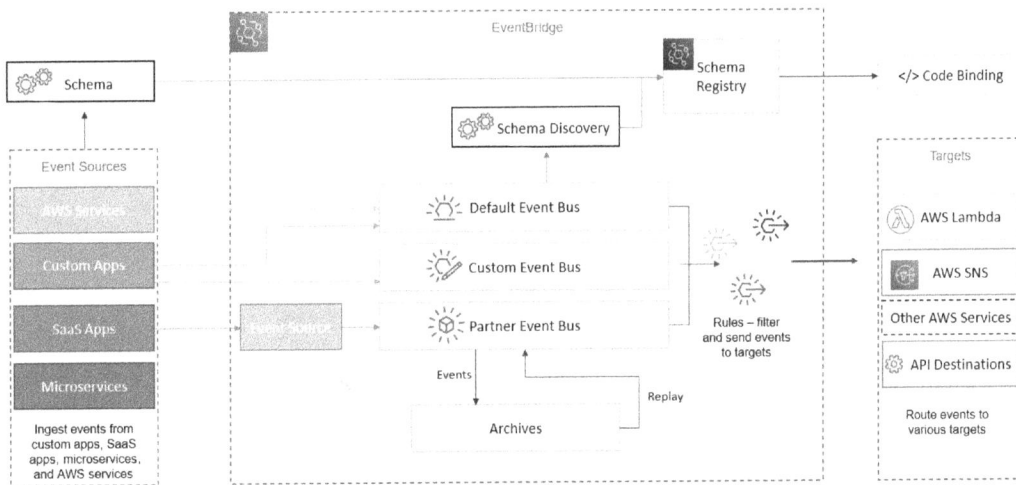

Figure 10.7 – How EventBridge works

The following are the key concepts and components of Amazon EventBridge:

- **Events**: As mentioned earlier, events represent a change in the state of a given environment or resource. This can be a state change in an application, an AWS resource, or even a SaaS partner service or application.

- **Rules**: A rule enables you to match an event to a target for processing. You can have a single rule route to a single target or a parallel processing route to multiple targets. There is no ordering of how the rules are processed but you can customize the JSON that's sent to the target to ensure only data of interest is passed on to the target.

- **Targets**: A target can process these events and perform some action. Targets include Lambda functions, SNS topics, ECS tasks, and SQS queues. Events are passed on to targets in JSON format.

- **Event buses**: Event buses can receive events. You have a default bus in your AWS account that is used to receive events from AWS services, but you can also create custom events for your custom applications. Partner event buses can be used to receive events from partner SaaS applications and services, which are then directed to your AWS account.

In this section, we looked at Amazon EventBridge, which allows you to stream real-time events from your applications, SaaS-based services, and AWS services to a variety of targets for processing. Targets can include AWS Lambda functions, Kinesis streams, ECS tasks, and SQS queues, among others. In addition, you can configure Amazon EventBridge to handle scheduled events that are triggered on a defined schedule. Amazon EventBridge offers more flexibility and advanced features compared to the previous CloudWatch Events services.

In the next section, we will look at task-oriented integration services such as Amazon Step Functions and Amazon SWF.

Coordinating multiple AWS services into serverless workloads with Amazon Step Functions and Amazon SWF

In this section, we'll look at two different AWS services that enable you to design task-based workflows between your application components. The first service we will look at is AWS Step Functions, while the second will be AWS SWF.

AWS Step Functions

Applications tend to have several components that make up individual workflows and processes. Each workflow represents an element of the application that then leads on to the next to provide a complete end-to-end solution. Amazon Step Functions enables you to define these workflows as a series of state machines that contain "states" that make up the workflow. These states make decisions based on input, perform some action, and produce an output to other states.

States can be any of the following types:

- **Success or fail state**: Where the execution stops with a success or failure
- **Wait state**: Where the state waits for a timeout period or a scheduled time
- **Parallel state**: Where the state performs parallel branches of execution
- **Map state**: Where the state accesses a list of items such as a list of orders
- **Choice state**: Where the state chooses between branches of execution
- **Task state**: A state that focuses on carrying out a specific task and may call other AWS services, such as Lambda functions, to perform the task

The state machine coordinates work through the different states and uses the task state to perform the actual work. Steps Functions helps you visualize your workflow as a series of event-driven steps, as well as the state of each step in your workflow to make sure that your application runs in a defined order.

Amazon Step Functions makes use of the **Amazon States Language** (**ASL**). This is a JSON-formatted structured language that helps you define your state machines, including states such as **Task** states, which perform certain actions. ASL is used to define how states transition from one state to the next, as in the case of the **Choice** state, or when you need to stop execution with an error, as in the case of a **Fail** state, and so on.

With Step Functions, you can also introduce human interaction, particularly where manual intervention is required. Let's look at an example of a credit card application process. You are likely to have several steps that form part of the application process. In the following diagram, we can see those steps in detail:

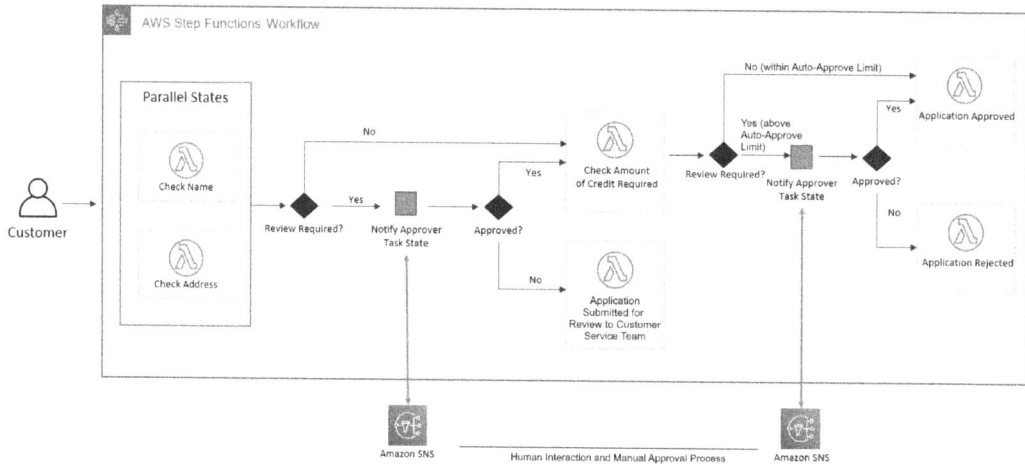

Figure 10.8 – Example of a credit card application workflow

In the preceding diagram, we have a Step Functions workflow process that illustrates how a potential customer could apply for a new credit card. The initial workflow would involve signing up for a credit card, which would require the customer's details to be verified. The next step could involve enabling a customer to choose the level of credit required. This workflow may include the following steps:

1. New customers sign up for a credit card. The initial process involves checking the user's details. Several parallel Lambda functions can be invoked to perform the required verification. For example, in the UK, the customer's name and address could be verified against the electoral roll.

2. If the automatic verification process is successful, then a review is not required, and the next step is invoked. If a review is required, then human intervention may be required to perform additional verification tasks. If the human verification is successful, the next step can be invoked; otherwise, the application can be rejected.

3. Next, the customer is offered an auto-approval credit amount. The customer is also given the choice to select a higher credit value, which will be subject to additional reviews. If the customer chooses a credit value that is within the auto-approval credit amount, then the application is automatically approved. If the customer chooses a credit value higher than the auto-approval credit amount, then another human intervention step is required. Here, the credit card company may request additional information such as salary slips to check the customer's credit-worthiness.

4. If the human intervention is successful, the application is approved; if not, the application will be rejected. There may be additional steps in the workflow for appealing against the rejection decision.

Step Functions enables us to build distributed application solutions and design microservices interactions to provide a complete end-to-end solution. Next, we will look at the different types of workflows you can set up for Step Functions.

Workflow types

With Amazon Step Functions, you can configure two types of workflows, as follows:

- **Standard workflow**: These have an exactly once workflow execution and can run for up to 1 year. Standard workflows are ideal if you require human interaction and approval processes as part of your workflow. You are charged on a per-state transition basis, which is each time a step in your execution is completed. Standard workflows also provide access to execution history and visual debugging.

- **Express workflow**: These have at least once workflow execution but can only run for up to 5 minutes. Express workflows are ideal for automated tasks and high event rate workloads, such as streaming data processing and IoT data ingestion. You are charged based on the number and duration of executions. Express workflows also offer unlimited state transition rates. Finally, all execution history is sent to Amazon CloudWatch.

Here is an example of a simple workflow that creates a task timer. In this example, a Step Functions state machine is being configured that implements a **Wait** state and uses a Lambda function to send out an Amazon SNS notification after the waiting period is over. The message that's sent out by the task is a simple *Hello World* message. The following screenshot shows the workflow and associated JSON:

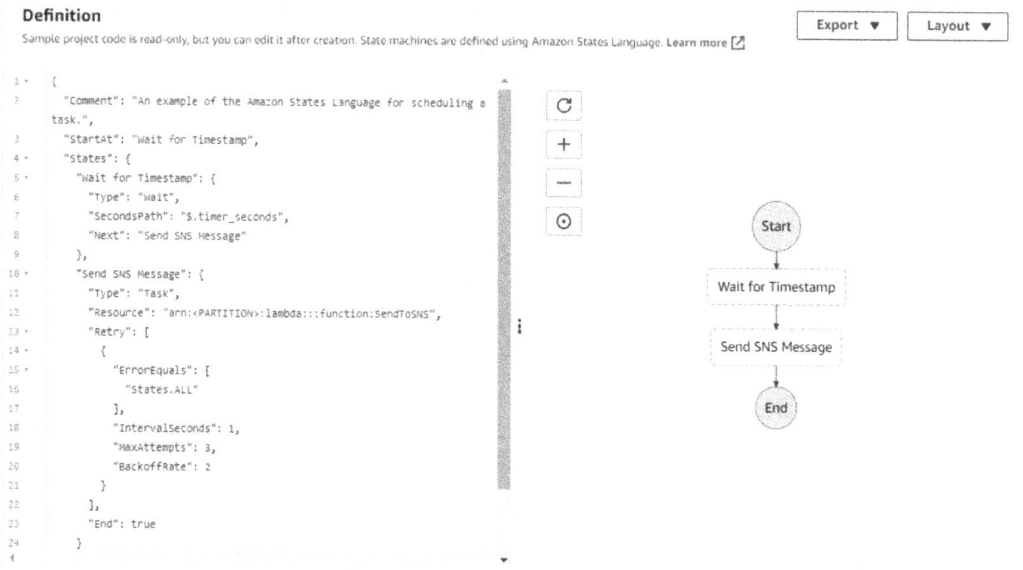

Figure 10.9 – AWS Step Functions task timer example

In this section, we looked at the two types of Step Functions execution types and identified the core difference between the two. Step Functions workflows can run for up to 1 year (when using the standard workflow type), so they are particularly useful for long-running application models that may also require human interaction. An example is a health insurance claim application process, which may require human intervention to verify the hospital bills and treatment that's dispensed to the claimant. In the next section, we will examine another task-oriented application integration service known as Amazon SWF.

Amazon Simple Workflow Service (SWF)

Amazon SWF is another task-oriented application integration service that allows you to coordinate work across distributed components of your application. Such coordination of tasks may involve processes such as managing dependencies, scheduling tasks, and handling the retries and timeouts of tasks to complete the logical workflow for your application.

Amazon SWF has a concept where you implement "workers" to complete your tasks. Workers can run either in the cloud across AWS EC2 instances or on the compute services available in your on-premises locations. As part of the logical workflow, Amazon SWF also allows you to incorporate human interaction within the logical distribution of tasks like Amazon Step Functions, as discussed in the previous section. With Amazon SWF, you can store tasks, assign them to workers, track progress, and maintain states.

While all of this sounds fairly like Amazon Step Functions, a key difference is that with Amazon SWF, you must write decider programs in any language that gets the latest state of each task from Amazon SWF and uses it to initiate subsequent tasks.

Amazon Step Functions, on the other hand, offers a fully managed service that has a more productive and agile approach to coordinating application components using visual workflows. If you are building new applications on AWS, you should consider using Amazon Step Functions. However, if you require external services to interact with your processes or you need to launch nested processes where a child process needs to return results to a parent process, you should consider using Amazon SWF.

In this section, we provided a brief introduction to Amazon SWF. Amazon SWF enables you to coordinate tasks across your distribution application components while offering capabilities such as maintaining their execution state durably and reliably.

Exercise 10.1 – Amazon S3 event notification using Amazon SNS

In the previous chapter, you designed, architected, and deployed a complete web application using several AWS services. One such service was the Amazon S3 service, where you created a bucket to host your application source code repository. The source code was comprised of multiple files that helped you build your web application.

Maintaining this source code is of paramount importance and any changes that are made to the code need to be monitored. There are several best practice strategies you can use to manage your source code, including using DevOps principles. In this exercise, your senior administrator, **Alice**, would like to know whenever a new file (object) gets uploaded to this source code repository, which is stored in the Amazon S3 bucket.

Amazon S3 comes with a feature known as event notifications. This feature enables you to receive notifications when certain events occur in your S3 bucket, such as an object being created or deleted. The service can be configured to send out such notifications to an Amazon SNS topic, which an administrator can subscribe to using an email as the endpoint. Let's configure an Amazon S3 notification to send email alerts to **Alice** whenever a new file is uploaded (that is, created) to the S3 bucket that hosts the source code repository.

This exercise is divided into four main steps, as described in the following sub-sections.

Step 1 – creating an SNS topic and subscribing to the topic

The first step is to create an SNS topic that will be used as the logical access point that Alice will subscribe to. Messages sent to this topic will then be emailed to Alice:

1. From the AWS Management Console, search for SNS in the top search box and select the service to be taken to the Amazon SNS dashboard.

2. If you have never created an SNS topic before, you should see the Amazon SNS splash screen.

3. Click on the far left-hand menu icon, denoted by the three lines, to expand the sidebar.

4. Next, click on the **Topics** link from the menu.

5. Click the **Create topic** button in the right-hand pane of the screen.

6. On the **Create topic** page, in the **Details** section, select the **Standard** type under **Type**.

7. Enter a name for the topic; for example, source-code-changes. Next, enter a display name for the topic; for example, Source Code Changes Alert.

8. Leave all the remaining settings as their default values and click the **Create topic** button.

9. Once the topic has been created, you will be redirected to the topic page. Make a note of the topic's ARN, as per the following screenshot:

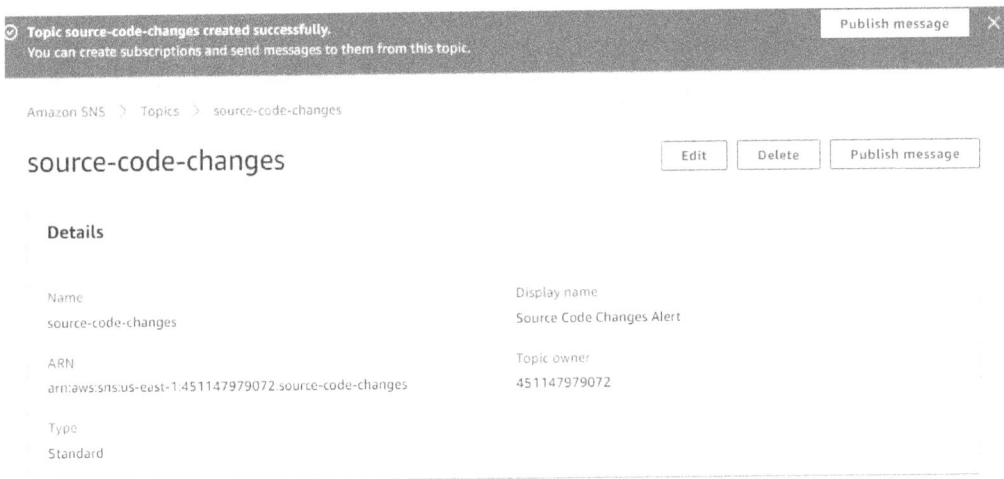

Figure 10.10 – Amazon SNS topic

10. Now that you have created a topic, you can create a subscription for it. We will be using email as the endpoint for notifications, and you can use your email address to receive the notifications.

11. In the bottom pane of the topics page, as per the previous screenshot, you will find a section to create subscriptions. Click on the **Create subscription** button.

12. On the **Create subscription** page, you will note that the topic ARN is already selected. If not, ensure that you paste the topic ARN that you made a note of earlier.

13. Next, under **Protocol**, select **Email** from the drop-down list.

14. In the text box under **Endpoint**, provide your email address.

15. Click the **Create subscription** button at the bottom of the page.

16. You will get a confirmation statement to say that your subscription has been created. However, its status will be set to **Pending confirmation**. AWS will have sent you a confirmation request to your email account. You will need to log into your email account and confirm the subscription to activate it. I have just logged into my email account to do the same, as per the following screenshot of my Gmail account:

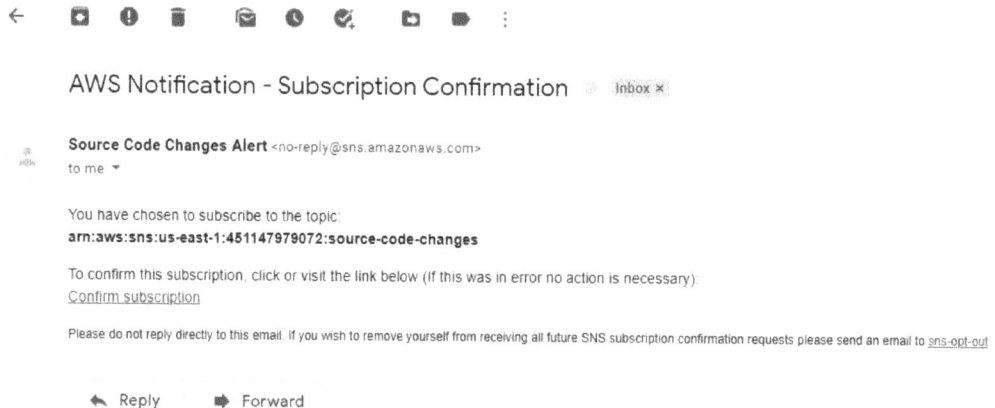

Figure 10.11 – Email subscription request for topic subscription

17. Once you confirm your subscription, return to the Amazon SNS dashboard and click on the **Topics** link from the left-hand menu.

Now that you have confirmed your subscription to the topic, you can configure an access policy that will grant the Amazon S3 service the permissions to send notifications to the topic.

Step 2 – configuring your SNS topic policy

For Amazon S3 to send notifications to the SNS topic you just created, you will need to configure an **access policy**. An access policy defines who or what can access your topic and publish messages to it. We have provided a sample policy document in the GitHub repository for this study guide that you will need to amend `https://github.com/PacktPublishing/AWS-Certified-Cloud-Practitioner-Exam-Guide`. You will need to have the following information before editing the policy:

- The ARN of the SNS topic, which you made a note of earlier.

- The Amazon ARN of the S3 bucket, which you created in the previous chapter. You can find the bucket ARN by clicking on the **Properties** tab on the bucket details page within your Amazon S3 dashboard.

- The AWS account ID (which you can obtain by clicking on your account name in the top right-hand corner of the screen and making a note of the 12-digit number next to **My Account**).

Open the sample access policy document in Notepad or a text editor of your choice, as per the following screenshot:

```
{
 "Version": "2012-10-17",
 "Id": "example-ID",
 "Statement": [
  {
   "Sid": "example-statement-ID", ⬅
   "Effect": "Allow",
   "Principal": {
    "Service": "s3.amazonaws.com"
   },
   "Action": [
    "SNS:Publish"
   ],
   "Resource": "SNS-topic-ARN", ⬅
   "Condition": {
      "ArnLike": { "aws:SourceArn": "arn:aws:s3:*:*:bucket-name" }, ⬅
      "StringEquals": { "aws:SourceAccount": "bucket-owner-account-id" } ⬅
   }
  }
 ]
}
```

Figure 10.12 – Sample access policy

Replace the values in the policy, as highlighted by the arrows in the preceding screenshot, with the following:

- For `Sid`, change `example-statement-ID` to any relevant information you would like to use; for example, `source-code-change-policy`.

- For `Resource`, change `SNS-topic-ARN` to the ARN of your topic, making sure to place the ARN in double quotes.

- For `ArnLike`, change `arn:aws:s3:*:*:bucket-name` to the ARN of your bucket name.

- For `StringEquals`, change `bucket-owner-account-id` to your AWS account ID.

Save the file and keep it handy for the next step of steps:

1. Navigate back to the Amazon SNS dashboard and from the left-hand menu, click ron **Topics**.

2. Click on your SNS topic in the middle pane, which will redirect you to the topic's details page, as per the following screenshot:

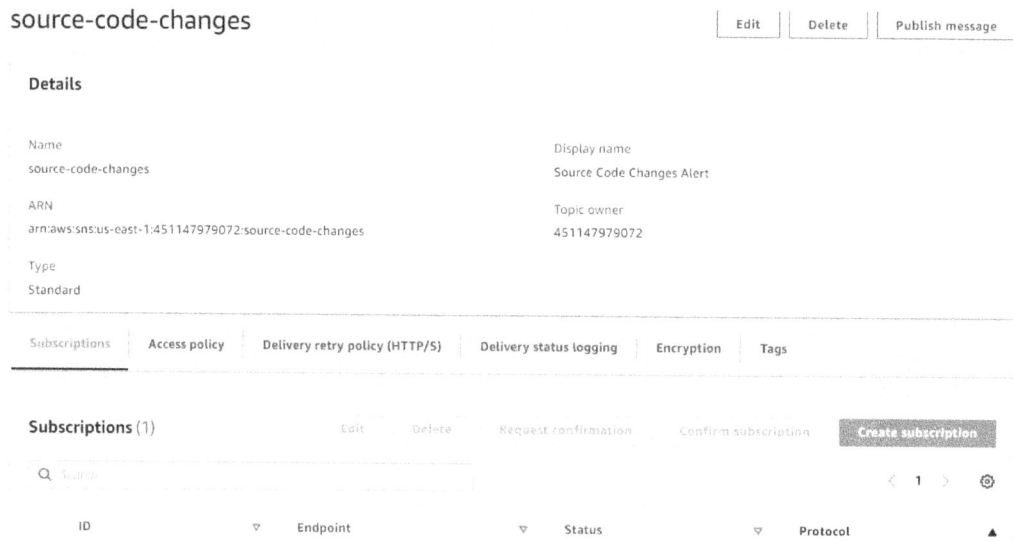

Figure 10.13 – SNS topic details page

3. Click on the **Access policy** tab in the bottom section of the pane.

4. You will find a default access policy that allows only the topic owner to publish to the topic.

5. In the top half of the pane, click on the **Edit** button.

6. Next, expand the **Access policy – optional** section.

7. Next, highlight and delete the existing policy that is in the JSON editor, and paste in a copy of your amended access policy instead.

8. Finally, click the **Save changes** button at the bottom of the page.

Now that you have set up the SNS topic and an appropriate access policy, it is time to set the Amazon S3 event notification service.

Step 3 – setting up the Amazon S3 event notification service

In this step, you will configure the event notification service on your Amazon S3 source code bucket, which hosts your application repository to send out alerts every time a new file is uploaded to the bucket:

1. Navigate to the Amazon S3 dashboard and click on the **Buckets** link from the left-hand menu.

2. From the right-hand pane, click on the Amazon S3 bucket that you created in the previous chapter to host your source code files.

3. Next, click on the **Properties** tab and scroll down until you reach the **Event Notifications** section.

4. Click the **Create event notification** button.

5. Enter a name for your event, such as New files added alert.

6. In the **Event types** section, tick the box that states **All object create events**.

7. Scroll further down until you reach the **Destination** section.

8. Select **SNS topic** from the **Destination** options.

9. Under **Specify SNS topic**, select the SNS topic that you created earlier in *Step 1* from the **SNS topic** drop-down list.

10. Finally, click the **Save changes** button.

Now that you have configured S3 to send event notifications to your SNS topic, it is time to test the configuration.

Step 4 – testing the configuration

In this step, we will test out the configuration of our Amazon S3 event notification service:

1. In the Amazon S3 dashboard, from the left-hand menu, select **Buckets**.

2. From the right-hand pane, select your Amazon S3 bucket, which will contain the source code files.

3. Next, click on the **Upload** button.

4. Go ahead and upload any random file you have access to. Alternatively, you can create a text file, save it, and then upload that text file instead. You can either use the **Add files** button to browse for a file on your computer or simply drag and drop a file from another file explorer window into the upload area.

5. Upload your file to the Amazon S3 bucket by clicking on the **Upload** button at the bottom of the page.

6. Once the upload has succeeded, click the **Close** button. Your object should be visible in the list of objects in the bucket.

7. Access your email account once again and check whether you have received a notification from AWS, alerting you to the fact that a new object has been uploaded. Refer to the following screenshot as an example:

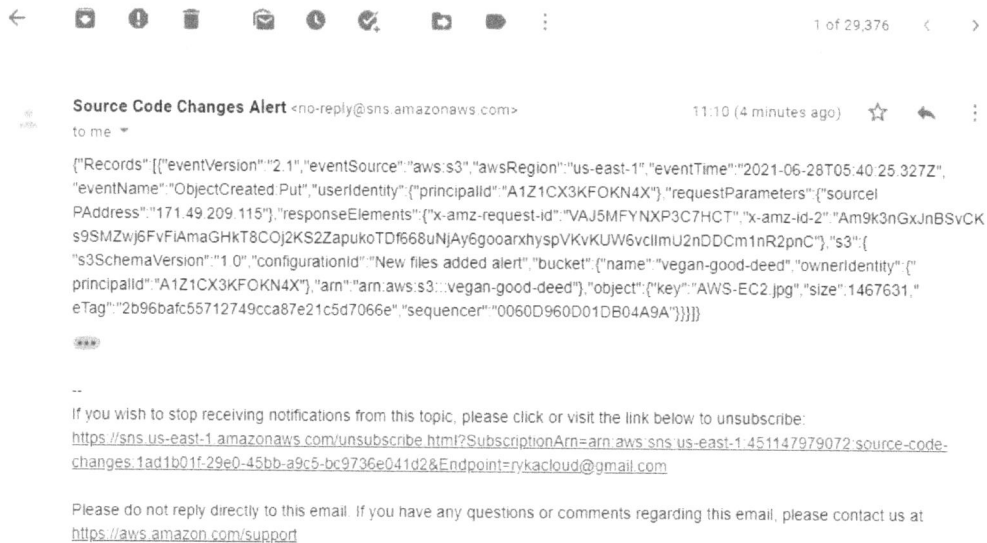

{"Records":[{"eventVersion":"2.1","eventSource":"aws:s3","awsRegion":"us-east-1","eventTime":"2021-06-28T05:40:25.327Z", "eventName":"ObjectCreated:Put","userIdentity":{"principalId":"A1Z1CX3KFOKN4X"},"requestParameters":{"sourceI PAddress":"171.49.209.115"},"responseElements":{"x-amz-request-id":"VAJ5MFYNXP3C7HCT","x-amz-id-2":"Am9k3nGxJnBSvCK s9SMZwj6FvFiAmaGHkT8COj2KS2ZapukoTDf668uNjAy6gooarxhyspVKvKUW6vcIImU2nDDCm1nR2pnC"},"s3":{ "s3SchemaVersion":"1.0","configurationId":"New files added alert","bucket":{"name":"vegan-good-deed","ownerIdentity":{" principalId":"A1Z1CX3KFOKN4X"},"arn":"arn:aws:s3:::vegan-good-deed"},"object":{"key":"AWS-EC2.jpg","size":1467631," eTag":"2b96bafc55712749cca87e21c5d7066e","sequencer":"0060D960D01DB04A9A"}}}]}

Figure 10.14 – Amazon S3 event notification alert email

As you can see, AWS has sent me an email, alerting me to the fact that an object was uploaded (created) in my Amazon S3 bucket. The email contains lots of information, including the time of the event, the alert's name, the bucket in question, the name of the object that was uploaded, as well as its size. As you can appreciate, this can be very useful for auditing purposes.

Amazon S3 event notifications can use other destinations too, such as an SQS queue or a Lambda function. In this exercise, you learned how Amazon SNS can be used to push out notification messages to an email address.

In the next exercise, you will perform a cleanup to remove any resources that are no longer required from our AWS account.

Exercise 10.2 – cleaning up

In this exercise, you will delete the resources that you created in the previous exercise as part of the cleanup process:

1. Navigate back to the Amazon SNS console.

2. From the left-hand menu, select **Topics**.

3. Next, from the right-hand pane, select the **source-code-changes** topic. Click the **Delete** button.

4. You will be prompted to confirm the delete request with a dialog box. Type `delete me` in the confirmation text box and then click the **Delete** button. The topic will be deleted.

Now that you have deleted the Amazon SNS topic, you can also delete the Amazon S3 bucket as we no longer require it:

1. Navigate to the Amazon S3 console.

2. From the left-hand menu, click on **Buckets**.

3. From the right-hand pane, select the bucket that you uploaded your source code to earlier.

4. You can only delete buckets if they are empty. This means that you have to delete the objects in your bucket first. With the bucket selected, click the **Empty** button.

5. Next, you will be prompted to confirm that you wish to delete the objects by typing `permanently delete` in the confirmation text box. Then, you can click on the **Delete** button to empty the bucket.

6. Now that the bucket has been emptied, you can delete it.

7. Click the **Exit** button to go back to the list of buckets. With the bucket still selected, click the **Delete** button. Next, in the confirmation text box, type in the name of the bucket to confirm that you wish to delete it and click the **Delete bucket** button.

Your Amazon S3 bucket will be successfully deleted.

Next, we will provide a summary of this chapter and the key concepts that you learned.

Summary

In this chapter, we examined the key application integration services that allow you to build highly robust and distributed application solutions. The array of services offered by AWS for application integration enables communication between the decoupled components of your applications, allowing you to move away from a monolithic architecture to one that can be built using microservices. The application integration tools available from AWS also help you design serverless solutions more easily, allowing you to further benefit from cost savings associated with server-based solutions.

The various services you learned about in this chapter included Amazon SNS, Amazon SQS, and Amazon MQ, which are message-oriented application integration services. These enable communication between application components, which allows you to build loosely coupled application architectures.

Amazon Step Functions and Amazon SWF are task-oriented application integration services that offer workflows that run for up to 1 year and can incorporate human intervention as part of the workflow process. Amazon Step Functions also helps you coordinate application components using visual workflows.

Finally, we looked at Amazon EventBridge, which is a serverless event bus service that makes it easier to build event-driven applications. EventBridge can ingest and process events that are generated by your applications, partner **Software-as-a-Service (SaaS)** applications, and other AWS services.

In the next chapter, we will look at a wide range of analytical services that are on offer from AWS that allow you to stream data from a wide range of sources, perform complex queries on ingested data, build data lakes, and build visualization dashboards and reporting.

Questions

Answer the following questions to test your knowledge of this chapter:

1. Which AWS services does Amazon CloudWatch use to send out email alerts to administrators when alarms are triggered and enter the `Alarm` state?

    A. Amazon SNS

    B. Amazon SES

    C. Amazon CloudTrail

    D. Amazon Email

2. Which feature of Amazon CloudWatch enables you to create a visualization of metrics by resource type and service?

   A. CloudWatch Events

   B. CloudWatch Logs

   C. CloudWatch Alarms

   D. CloudWatch dashboards

3. Which AWS application integration service can be configured to offer A2P communication using mobile SMS to send out text alerts?

   A. Amazon SQS

   B. Amazon SNS

   C. Amazon Amplify

   D. Amazon Workspaces

4. You need to configure your Amazon SNS topic to push out messages of newly uploaded videos to an Amazon S3 bucket, across three different SQS queues. Each queue is designed to encode the raw video into a different resolution. Which feature of Amazon SNS enables you to push out such notifications in parallel?

   A. Amazon SNS standard topic

   B. Amazon SNS FIFO topic

   C. Fanout scenario

   D. Amazon EventBridge

5. Which Amazon SQS queue type offers maximum throughput, best-effort ordering, and at least one delivery?

   A. SQS standard queue

   B. SQS power queue

   C. SQS FIFO queue

   D. SQS LIFO queue

6. Which AWS service is designed to help you build a decoupled application architecture where incoming web requests can be held in a queue until a backend application can retrieve and process the request?

   A. Amazon SQS

   B. Amazon SWF

   C. Amazon SNS

   D. Amazon Step Functions

7. You are required to configure an SQS queue for your application where the order of messages needs to be preserved for the application to function correctly. Which type of queue do you need to configure?

   A. SQS standard queue

   B. SQS power queue

   C. SQS FIFO queue

   D. SQS LIFO queue

8. To reduce costs, you have been asked to automate the shutdown of a fleet of UAT test servers every weekday at 7 P.M. and then restart them the following weekday at 8 A.M. The servers should remain in the shutdown state at weekends.

   Which AWS service can help you achieve the preceding requirements?

   A. Amazon SQS

   B. Amazon Athena

   C. Amazon EventBridge

   D. Amazon SNS

9. Which AWS service enables you to manage application workflows as state machines by breaking them into multiple steps, adding flow logic, and tracking the inputs and outputs between the steps?

   A. Amazon Step Functions

   B. Amazon SQS

   C. Amazon SNS

   D. Amazon SWF

10. Which AWS service offers an orchestration service to coordinate work across application components that make use of decider programs to determine the latest state of each task and use it to initiate subsequent tasks?

A. Amazon SNS

B. Amazon EventBridge

C. Amazon SQS

D. Amazon SWF

11

Analytics on AWS

In this age of information, understanding your data has become extremely important. With current cutting-edge technologies, extensive amounts of data are generated every second – data that needs to be stored and analyzed. Companies perform data analytics to explain, predict, and ultimately gain a competitive advantage in business. Traditional analytics would include retail analytics, supply chain analytics, or stock rotation analytics. With **machine learning (ML)** and **artificial intelligence** taking a firm hold on the economy, new evolutions of analytics have come into play, such as cognitive analytics, fraud analytics, and speech analytics. The list is almost endless but suffice it to say that understanding your raw data has required considerable effort and a whole business unit dedicated to **data analytics** alone.

AWS offers a vast array of analytics tools that you can use to ingest, store, and effectively understand the data that's generated by your business. In this chapter, we will we look at some of those services in detail.

In this chapter, we will cover the following topics:

- Learning about data streaming with Amazon Kinesis
- Learning how to query data stored in Amazon S3 with Amazon Athena
- Introduction to Amazon Elasticsearch
- Overview of Amazon Glue and QuickSight
- Additional analytics services

Technical requirements

To complete the exercises in this chapter, you will need access to your AWS account and be logged in as the IAM user **Alice**.

Learning about data streaming with Amazon Kinesis

To analyze your business data, you need to ingest that data into a service that can perform the required analysis on it. Businesses generate tons of data from a wide range of sources, including logs generated by applications, content such as videos, images, and documents, clickstream data from websites, IoT data, and more. Ingesting this data is the first step toward understanding it.

However, rather than ingesting all the data first and then figuring out how you would go about understanding that data, **Amazon Kinesis** lets you process and analyze data as it arrives and respond to it instantly. Amazon Kinesis is a fully managed service that enables you to process streaming data at any scale in a cost-effective manner. Furthermore, it is **serverless**, meaning that you do not need to set up and manage expensive infrastructure to process your data. Amazon Kinesis is comprised of the following four key services:

- Amazon Kinesis Data Firehose

- Amazon Kinesis Data Streams

- Amazon Kinesis Data Analytics

- Amazon Kinesis Video Streams

Let's look at each of these services in detail.

Amazon Kinesis Data Firehose

Modern business approaches and strategies to keep customers loyal and engaged have resulted in an insurmountable amount of data to collect, process, and make sense of. Whether you are trying to analyze what products your customers click on your website, make recommendations based on their product searches, or alert your security team about potentially fraudulent transactions, you need to collect and process data as it is being generated. Traditionally, you would have had to build the infrastructure to provide this kind of backend ingestion and processing of data, which can be cost-prohibitive for many businesses – not to mention the management overhead associated with maintaining hundreds of servers, storage, and network components.

Amazon Kinesis Firehose is a fully managed service that can ingest and deliver streaming data to AWS data stores such as *Amazon S3*, *Redshift*, and *Amazon Elasticsearch* for near real-time analytics with existing **business intelligence** (**BI**) tools. The service workflow can be illustrated with the following diagram:

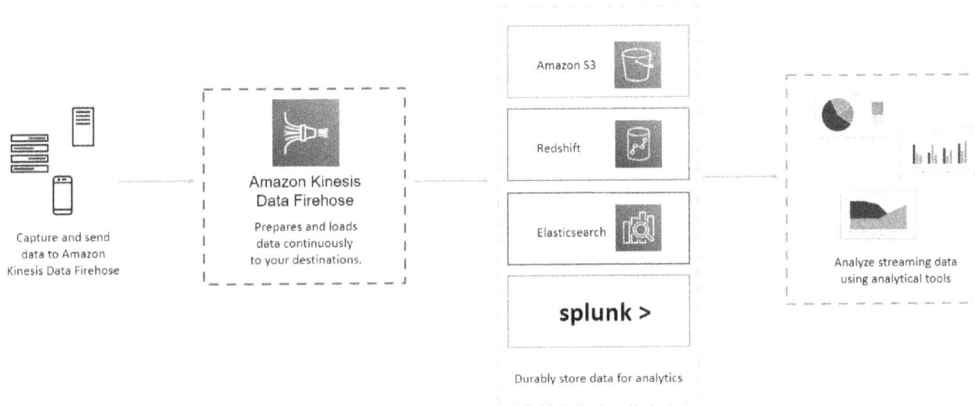

Figure 11.1 – Kinesis Firehose

Amazon Kinesis Firehose can also deliver data to third-party services such as *Datadog*, *New Relic*, *MongoDB*, and *Splunk*. Amazon Kinesis Firehose will also allow you to batch process, compress, transform, and even encrypt data before loading it into a service, which means you reduce overall storage costs and enhance security.

In addition, incoming data streams can be automatically converted into open standard formats such as **Apache Parquet** and **Apache ORC**. Finally, with Amazon Kinesis Firehose, there are no infrastructure setup costs to worry about. You simply pay for the data you transfer through the service, any data conversion costs, delivery to Amazon VPCs, and data transfers.

Next, we will look at the Kinesis Data Streams service.

Amazon Kinesis Data Streams

Whereas Kinesis Firehose is designed to load massive amounts of data into **data stores** such as Amazon S3 or Redshift for *near real-time*, **Amazon Kinesis Data Streams** is a fully managed *real-time* continuous data streaming service that allows you to capture gigabytes of data per second and stream it into custom applications for processing and analysis.

Amazon Kinesis Data Streams can make streaming data available to several analytical applications, such as Amazon S3 and AWS Lambda, within 70 milliseconds of the data being collected. It offers high levels of durability by replicating your streaming data across three Availability Zones.

Kinesis Firehose does not offer any data storage capabilities. However, Amazon Kinesis Data Streams will store and make your data accessible for up to 24 hours by default, but this can be raised to 7 days by enabling the extended data retention feature, or even up to 365 days by enabling the long-term data retention feature.

You ingest and store streaming data for processing to build real-time applications services such as **real-time dashboards**, **real-time anomaly detection**, **dynamic pricing**, and so on. Like Kinesis Firehose, you are charged on a pay-as-you-go basis with no upfront cost nor minimum fees. However, there is a fundamental difference in that Kinesis Data Streams uses the concept of *shards*, which uniquely identify the data records in a stream. A stream can be comprised of multiple shards that determine the overall capacity. Specifically, each shard represents up to five transactions per second for reads with a maximum data read rate of 2 MB per second. For writes, you can have up to 1,000 records per second and a total data write rate of 1 MB per second (including partition keys). The total capacity of the stream is the sum of the capacities of its shards. The important concept to appreciate here is that you are charged for each shard that's provisioned per hour, regardless of whether you use it or not.

In *Chapter 10*, *Application Integration Services*, we discussed a service called **Amazon SQS**. Now, it may seem that Kinesis and SQS do the same thing, but they are very different. Amazon SQS is a message queueing service that helps store messages while they travel between the different components of your application. Amazon SQS helps you decouple your application stack so that individual messages can be tracked and managed independently, and so that the different components of your application can work independently.

Next, we will look at the Kinesis Data Analytics service.

Amazon Kinesis Data Analytics

Kinesis Data Analytics lets you query and analyze stream data in real time. Data can be streamed into the Kinesis Data Analytics application from various sources, including Amazon **Managed Streaming for Kafka** (**MSK**) and **Amazon Kinesis Data Streams** (discussed earlier). With Kinesis Data Analytics, you do not need to build complex streaming integrated applications with other AWS services. Instead, you can use standard programming and database query languages such as Java, Python, and SQL to query streaming data or build streaming applications. The following diagram illustrates these key features:

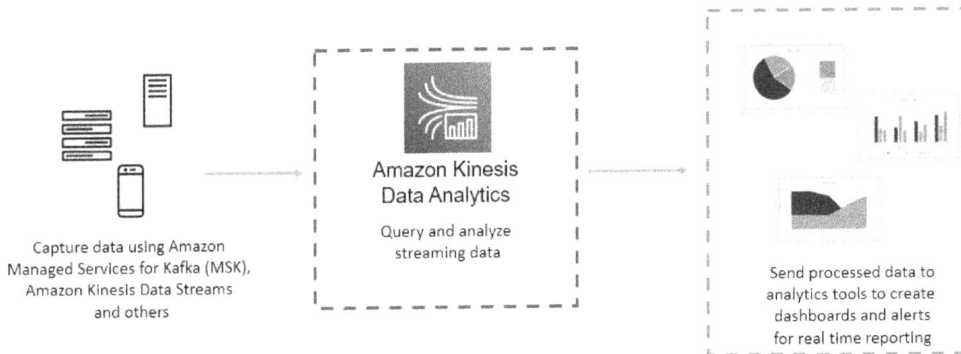

Figure 11.2 – Amazon Kinesis Data Analytics

Amazon Kinesis Data Analytics also enables you to analyze streaming data in real time and build streaming applications using open source libraries and connectors for **Apache Flink**. Apache Flink is a fully open source, unified stream processing and batch processing framework developed by the **Apache Software Foundation**.

Finally, you have access to the **Kinesis Data Analytics Studio**, which allows you to build sophisticated stream processing applications using SQL, Java, Python, and Scala.

In terms of pricing, you are only charged for the resources that you use to run your streaming applications and there are no upfront commitments.

Next, we will look at the Amazon Kinesis Video Streams service.

Amazon Kinesis Video Streams

If you are looking to stream video devices to AWS for analytics, ML, playback, and other processing services, then the Amazon Kinesis Video Streams service is going to be the tool you use. Amazon Kinesis Video Streams can also ingest data from edge devices, smartphones, security cameras, and more.

Amazon Kinesis Video Streams makes use of Amazon S3 as the underlying storage repository from your streaming videos, which, as you already know, offers high levels of **data durability**. In addition, you can search for and retrieve video fragments based on devices and timestamps. Your videos can be encrypted and indexed as well.

With Amazon Kinesis Video Streams, you can play back videos for live or on-demand viewing. In addition, you can use Kinesis Video Streams to help you build applications that make use of computer vision video analytics technologies on AWS such as **Amazon Rekognition**. Incidentally, Amazon Rekognition is a fully managed image and video analysis service that can be used to identify objects, people, text, scenes, and activities in images and videos. Amazon Rekognition can also be used to detect any inappropriate content. The following diagram illustrates the Amazon Kinesis Video Streams service:

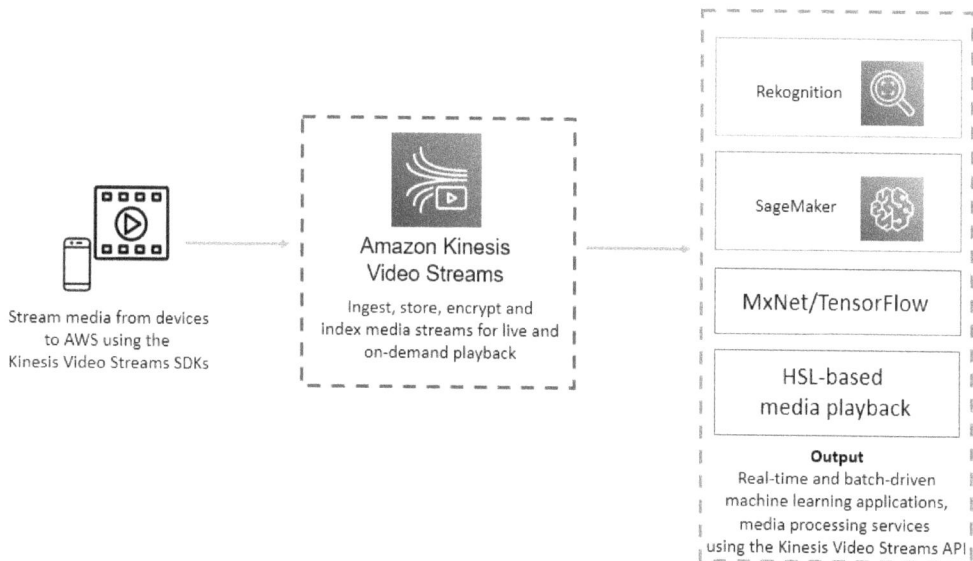

Figure 11.3 – Amazon Kinesis Video Streams

Amazon Kinesis Video Streams enables you to design applications for a wide range of use cases. One such use case includes the ability to stream video and audio for smart home devices such as doorbells. Amazon Kinesis Video Streams will ingest, index, and store the media streams and your application can use HTTP live streaming to play the stream to a smartphone app, allowing you to monitor and communicate with the person *knocking* on the door.

In this section, we looked at Amazon Kinesis and discussed its four key offerings. In the next section, we will introduce you to another AWS analytics service known as Amazon Athena, which is a fully managed, serverless interactive query service that enables you to analyze data in Amazon S3 using standard SQL.

Learning how to query data stored in Amazon S3 with Amazon Athena

Businesses store vast amounts of data in repositories such as Amazon S3. A lot of this data is not necessarily being hosted on regular Amazon RDS or NoSQL databases. In many cases, this is because the dataset is not being regularly updated and queried. Previously, even if you wanted to perform ad hoc queries or analysis against some of that data, you would need to ingest it into a database and then run your queries against the database.

Amazon Athena is a fully managed serverless solution that allows you to interactively query and analyze data directly in **Amazon S3** using standard SQL. There is no infrastructure to provision, and you only pay for the queries you run.

Amazon Athena uses **Presto**, which is an open source SQL query engine that's designed to allow you to perform ad hoc analysis. You can use standard ANSI SQL, which provides full support for large joins, window functions, and arrays.

Data can be presented to Amazon Athena in a variety of formats, such as CSV, JSON, ORC, Avro, or Parquet. Furthermore, you can use Athena's JDBC driver to connect to a wide range of BI tools.

Amazon Athena allows you to present unstructured, semi-structured, and structured data to it, which you can then use to run queries to analyze that data. This process involves creating a database within the Athena service and one or more tables for each specific dataset that you want to query and analyze. These tables allow you to define metadata that tells Athena where the data is held in S3 and the structure of that data; for example, the column names and data types.

Tables need to be registered in Athena to perform queries and have the results returned. These tables can be created automatically or manually. Once your tables have been registered, you can use SQL SELECT statements to query them. Your query results can also be stored in Amazon S3 in a location you specify.

In an upcoming exercise, *Analyzing your sales report with Amazon Athena and AWS Glue*, we will upload some data to Amazon S3 and use Amazon Athena to query it. In this section, we introduced you to the Amazon Athena service and provided an overview of how you can use Athena to query data stored directly in Amazon S3.

In the next section, we will look at another AWS analytics tool known as Amazon Elasticsearch.

Introduction to Amazon Elasticsearch

Elasticsearch is an open source text search and analytics engine that's capable of storing, analyzing, and performing search functions against big volumes of data in near real time. You can use Elasticsearch to analyze all types of data such as textual, numerical, geospatial, structured, and unstructured data.

Amazon's offering of Elasticsearch as a service comes as a fully managed service with no need to set up and manage any infrastructure, allowing you to focus on your applications and their functionalities. Following the same pay-as-you-consume model, there are also no upfront costs, although you can reserve instances for a 1- or 3-year term for a significant discount over the on-demand pricing model.

Amazon Elasticsearch is designed to be highly scalable and can index all types of content to help you deliver applications for use cases such as the following:

- Website search
- Application search
- Logging and log analytics
- Infrastructure metrics and monitoring
- Security analytics

Raw data such as log files, messages, metrics, documents, and lists are ingested, normalized, and then indexed in Elasticsearch. You can then run complex queries against this data and use aggregations to review data summaries.

Amazon Elasticsearch also offers integration with **Kibana**, a data visualization tool that's used to analyze large datasets to help you produce visual representations of that data in the form of graphs, pie charts, heat maps, and much more.

Another service that Amazon Elasticsearch integrates with is **Logstash**, which is an open source, server-side data processing pipeline that allows you to ingest data from a wide range of sources and transform it and send it to a **stash** such as Elasticsearch. Elasticsearch, Kibana, and Logstash are often referred to by the acronym **ELK**.

Finally, Amazon Elasticsearch supports querying your cluster using standard SQL, making it easy for your developers to start using the service. You can also connect to your existing SQL-based BI and ETL tools using a JDBC driver.

In this section, we introduced you to the Amazon Elasticsearch service, which allows you to create highly scalable, secure, and available Elasticsearch clusters and offers full integration with Kibana and Logstash for a complete managed ELK solution.

In the next section, we will look at Amazon Glue and QuickSight.

Overview of Amazon Glue and QuickSight

Business data can often be stored in a wide range of services – databases, storage buckets, spreadsheets, and more. Being able to bring all the relevant data together for analysis can sometimes be a big project. Later, you may wish to extract and present that data in a manner that is easy to digest and understand using BI tools or seamlessly integrate insights from that data into your applications, dashboards, and reporting. Two services offered by AWS that can help with these types of requirements are **Amazon Glue** and **QuickSight**. We'll take a quick look at each of these services next.

Overview of Amazon Glue

Amazon Glue is a serverless **Extract, Transform, and Load** (**ETL**) service. With Amazon Glue, you can discover, prepare, enrich, clean, and transform your data from various sources. You can then load the data into databases, data warehouses, and data lakes. Data from streaming sources can also be loaded for regular reporting and analysis. This data can then be used for analytics, as per your business requirements, and help with decision-making.

Amazon Glue comes with a **Data Catalog**, which is a central metadata repository that stores information about your data, such as table definitions. You use a **crawler** service to scan various repositories, classify data, and *infer* schema information such as its format and data types. The metadata is then stored as tables in the Data Catalog and used to generate ETL scripts to transform, flatten, and enrich your data. The data is then populated into your chosen data warehousing solution or data lakes, for example.

Amazon Glue also comes with the **AWS Glue console** service to help you define and orchestrate your ETL workflow. It lets you do the following:

- Define Glue objects such as jobs, crawlers, tables, and so on.
- Schedule **crawler** run operations.
- Define events or schedules for job triggers.
- Search for and filter lists of objects in Glue.
- Edit transformation scripts directly or by using the visual tools provided.

Amazon Glue is a fully managed service and scales resources as needed to run your jobs. It handles errors and retries automatically. With Amazon Glue, you are charged an hourly rate, which is billing by the second. This pricing is based on running crawlers (for discovering data) and performing ETL jobs (for processing and loading data). In addition, you pay a monthly fee to store and access the metadata in the AWS Glue Catalog.

Next, we will look at the Amazon QuickSight service.

Overview of Amazon QuickSight

Amazon QuickSight is a serverless and fully managed BI service in the cloud that can be used to create and publish interactive BI dashboards for your business data. This provides you with access to meaningful information for your business so that you can make decisions.

Amazon QuickSight can connect to your data wherever it is stored – whether it's stored in **AWS services**, **on-premises databases**, **spreadsheets**, **SaaS data**, or **B2B data**. This data can then be transformed into rich dashboards and reporting tools that can help your business understand operations, sales figures, profits, successes, and where there may be room for improvement. Amazon QuickSight can publish dashboards securely to enable collaborative efforts from your organization's workforce via mobile phones, email, or web applications.

Amazon QuickSight also integrates with ML services, which allows it to build and deliver deeper insights from your data. With **ML Insights**, you can discover hidden insights across your datasets, such as any anomalies and variations, enabling you to quickly act to changes that occur. You can also schedule automatic anomaly detection jobs. With ML Insights, you can perform better forecasting, which can be used to perform accurate *what-if* analysis. Finally, you can summarize your data into easy-to-consume natural language narratives, which can help you deliver better contextual information in your dashboards and reporting services.

Amazon QuickSight's pricing model is offered as a pay-as-you-use service and is determined by who is using the service; for example, admins, authors, and readers. Therefore, the pricing is based on the number of users, similar to a user-based license. Additional charges are incurred for services such as alerting and anomaly detection too. You can view the pricing overview at `https://aws.amazon.com/quicksight/pricing/`.

In this section, we reviewed two additional AWS services that fall within the analytics category. We introduced you to the Amazon Glue service, which is a fully managed and serverless ETL solution. We also provided an overview of the cloud-native BI tool, which uses ML to offer greater business insights from your data.

In the next section, we will cover a couple of additional tools as part of the overall analytics offering.

Additional analytics services

In this section, we will take a very quick look at some other AWS analytics services that you need to be aware of. Specifically, we will look at the **Elastic Map Reduce** (**EMR**) service, **CloudSearch**, and **Data Pipeline**:

- **AWS EMR**: This provides a managed **Hadoop framework** to enable you to process vast amounts of big data. You can use open source tools such as Apache Spark, Apache Hive, Apache HBase, Apache Flink, Apache Hudi, and Presto. Amazon EMR comes with an **integrated development environment** (**IDE**) called **EMR Studio** to help you develop, visualize, and debug data engineering and data science applications written in R, Python, Scala, and PySpark. You can run your EMR workloads on EC2 Instances, Amazon **Elastic Kubernetes Service** (**EKS**) clusters, and on-premises using the AWS Outpost service. In terms of pricing, you are charged at a per-instance rate for every second used, with a 1-minute minimum charge.

- **AWS Data Pipeline**: This is a web service that allows you to schedule and automate how your data is moved and transformed from various sources, including **on-premises servers**, into services such as Amazon S3, RDS, DynamoDB, and EMR. With AWS Data Pipeline, you can create workflows that transfer and transform data at scheduled intervals to ensure alignment with the application processes. For example, you can archive your web server logs to an Amazon S3 bucket daily, and then run weekly Amazon EMR jobs to analyze those logs and generate traffic reports that can be consumed by your application.

- **AWS CloudSearch**: This is a fully managed service that enables you to deploy, manage, and scale a search solution for your web applications. Amazon CloudSearch supports 34 languages and adds rich search capabilities to your website, including free text, Boolean, and faceted search. It also offers features such as automated suggestions, highlighting, and more.

In this section, we looked at some additional services offered by AWS that fall within the analytics category. In the next section, we will move on to this chapter's exercises.

Exercise 11.1 – analyzing your sales report with Amazon Athena and AWS Glue

In this exercise, you will need to download a sample CSV file, which is available in the Packt GitHub repository for this chapter `https://github.com/ PacktPublishing/AWS-Certified-Cloud-Practitioner-Exam-Guide`. This is a simple CSV file that contains some sales data for the Vegan Studio, the fictitious company that you have been carrying out a series of exercises for in the previous chapters.

You will need to store this CSV file in an Amazon S3 bucket and then use Amazon Athena to run queries against the data. Ensure that you have downloaded the CSV file and stored it on your computer before you start this exercise.

Step 1 – Amazon S3

1. Log into your AWS account using the IAM user ID of our senior administrator, **Alice**.

2. Navigate to the Amazon S3 dashboard.

3. Create two new buckets with appropriate names. For example, I have named my buckets `vegan-sales-report` (to store the CSV file) and `vegan-query-results` (to store the Athena query results). Since I have taken these names, you will not be able to use them since bucket names must be unique in the AWS ecosystem. Ensure that the buckets are created in the `us-east-1` (N.Virginia) Region.

4. Next, upload your CSV file to the bucket that will be used to host the data. Recall the steps required from the previous chapters to complete the upload. (Tip: try to do this from memory as this will help you build your confidence.)

Step 2 – Amazon Athena and Amazon Glue

5. Navigate to the Amazon Athena dashboard. You can search for the Athena service from the top search bar in your AWS Management Console.

6. If this is the first time you are accessing Amazon Athena, you should see a splash screen. Click **Get Started**. If you do not see the **Get Started** option, this is because you are using the new user interface. AWS is notorious for making changes to the UI. If you do see the new console, then you will need to click on the **Explore the query editor** button. For this lab, we suggest that you use the old console for now. To access the old console, click on the ellipsis (three dashes) in the top far left of the console and switch the toggle to disable the **New Athena experience** option. This will take you back to the old console interface:

Amazon Athena   ✕

New Athena experience
Let us know what you think

Query editor

Workgroups

Data sources

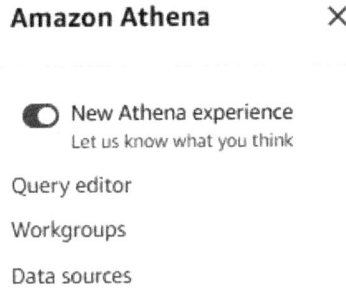

Figure 11.4 – Disabling the New Athena experience toggle switch

7.   From the top right-hand corner, click on the **Settings** link. You will be presented with the following dialog box. You will need to provide the S3 bucket details to store your query results. The format should be `s3://bucket-name`. You can also store your queries in a sub-folder and choose to encrypt your query results:

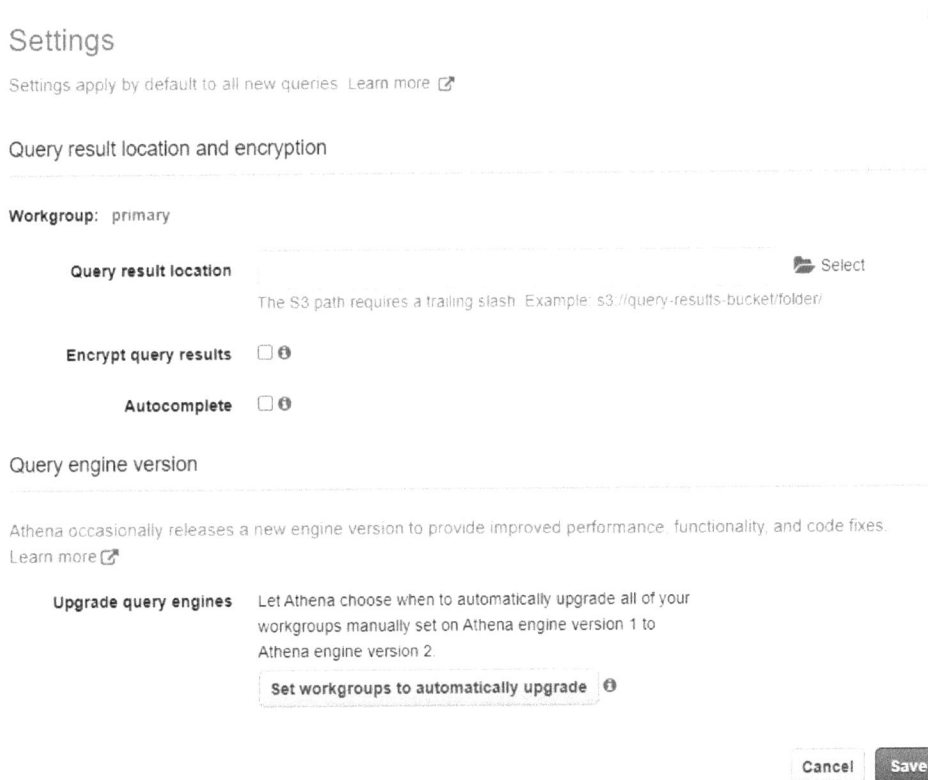

Settings

Settings apply by default to all new queries Learn more ⬚

Query result location and encryption

Workgroup: primary

Query result location   📂 Select

The S3 path requires a trailing slash Example s3://query-results-bucket/folder/

Encrypt query results   ☐ ❶

Autocomplete   ☐ ❶

Query engine version

Athena occasionally releases a new engine version to provide improved performance, functionality, and code fixes.
Learn more ⬚

Upgrade query engines   Let Athena choose when to automatically upgrade all of your
workgroups manually set on Athena engine version 1 to
Athena engine version 2.

Set workgroups to automatically upgrade ❶

Cancel   **Save**

Figure 11.5 – Amazon Athena – Settings

8.   Next, click the **Save** button.

9. Next, from the left-hand menu, click on **Connect data source**, as per the following screenshot:

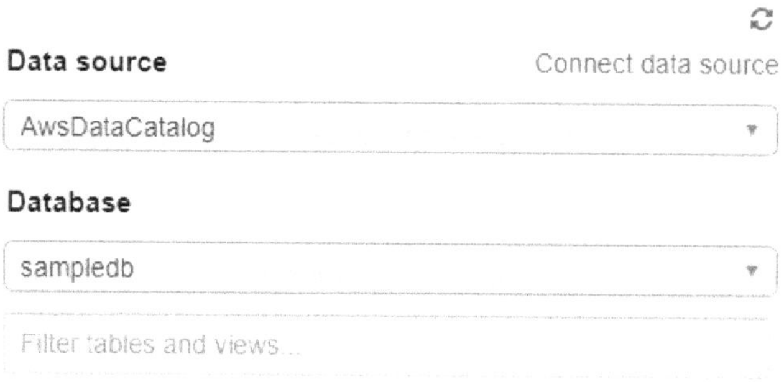

Figure 11.6 – Amazon Athena – Connect data source

10. Under **Choose where your data is located**, ensure that **Query data in Amazon S3** is selected.

11. Under **Choose a metadata catalog**, ensure that **AWS Glue Data Catalog** is selected. For this exercise, you will use AWS Glue to crawl your data and create a schema. There is a slight charge to this, but it is very minimum, and you will only need to do this once.

12. Click the **Next** button on the right-hand pane of the page.

13. Under **Connection details**, ensure that **AWS Glue Data Catalog in this account** is selected. Then, under **Choose a way to create a table**, select **Create a crawler in AWS Glue**.

14. Click the **Connect to AWS Glue** button. This will launch **AWS Glue** in a new browser tab. Switch over to this tab to configure Amazon Glue. Do not close the Amazon Athena browser tab as we will return to this later.

15. If you see the splash screen, click the **Get started** button.

16. From the left-hand menu, click **Crawlers**.

17. From the right-hand pane, click the **Add crawler** button. You will be presented with the **Add crawler** wizard:

    A. Provide a name for the crawler. Click **Next**.

    B. On the **Specify crawler source type** page, ensure that the **Data stores** and **Craw all folders** options are selected.

    C. Click **Next**.

    D. On the **Choose a data store** page, ensure that **S3** is selected as your chosen data store.

    E. Next, under **Craw data in**, select the **Specified path in my account** option. For **Include path**, provide the S3 bucket URL of your S3 bucket that hosts the sales report. In this case, the URL is in the format of `s3://bucket-name`. Note that at the end of the path defined for your S3 bucket, ensure that you add another slash(`/`). **For example, my bucket path reads** `s3://vegan-sales-report/`.

    F. Click **Next**.

    G. On the **Add another data store** page, click **No** and then **Next**.

    H. Next, on the **Choose an IAM Role** page, select the **Create an IAM role** option and type a name in the text box next to **AWSGlueServiceRole-**. This will form the path of your IAM **role** name. For example, I have typed in `VeganSalesRole` so that it is self-explanatory. Click **Next**.

    I. On the **Create a schedule for this crawler** page, set **Frequency** to **Run on demand**. Click **Next**.

    J. On the **Configure the crawler's output** page, click the **Add database** button.

    K. In the **Add database** dialog box that appears, provide a name for your database, such as `vegansalesdb`. Click **Create**.

    L. You will be taken back to the **Configure the crawler's output** page, with your newly created database name shown in the **Database** text box. Click **Next**.

    M. On the **Review all steps** page, click **Finish**.

18. You will be redirected to the **Crawlers** page. Here, you will be able to see that your crawler has been created. Click on the checkbox next to your crawler and click the **Run crawler** button, as per the following screenshot:

Crawlers A crawler connects to a data store, progresses through a prioritized list of classifiers to determine

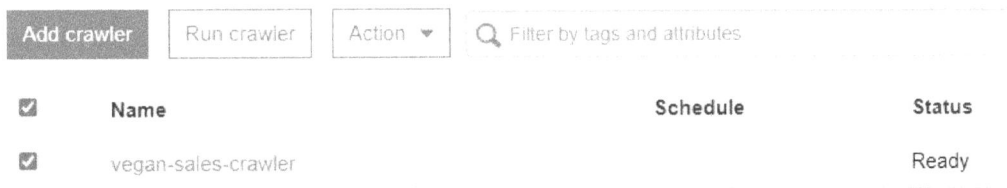

Add crawler	Run crawler	Action ▾	Q Filter by tags and attributes

☑	Name	Schedule	Status
☑	vegan-sales-crawler		Ready

Figure 11.7 – Amazon Glue – Run crawler

19. After a minute or two, you should find that its **Status** is set to **Ready** and that the crawler has successfully run. You will see that a table has been added, as per the following screenshot:

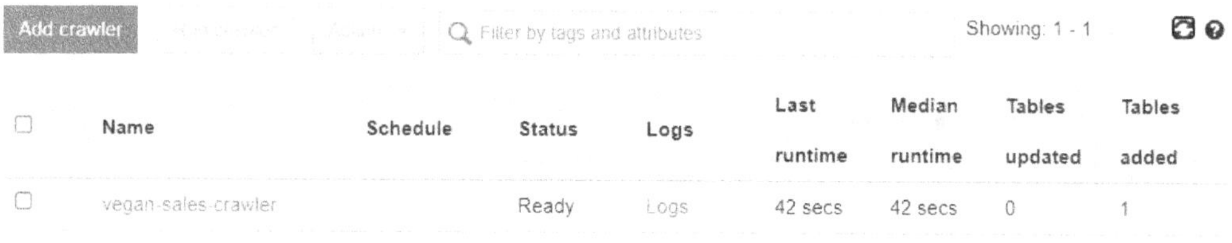
Add crawler
Filter by tags and attributes
Showing: 1 - 1
Name
Schedule
Status
Logs
Last runtime
Median runtime
Tables updated
Tables added
vegan-sales-crawler
Ready
Logs
42 secs
42 secs
0
1

Figure 11.8 – Amazon Glue – Crawl complete

20. Now that the crawl is complete, we can return to the Amazon Athena browser tab.

Step 3 – Amazon Athena

21. Back in the Athena browser tab, go ahead and click on the **cancel** button at the bottom right-hand corner of the page. This will take you back to the main **Amazon Athena Data sources** page, where you will see that your recently created Glue Catalog is listed under **Data sources**, as per the following screenshot:

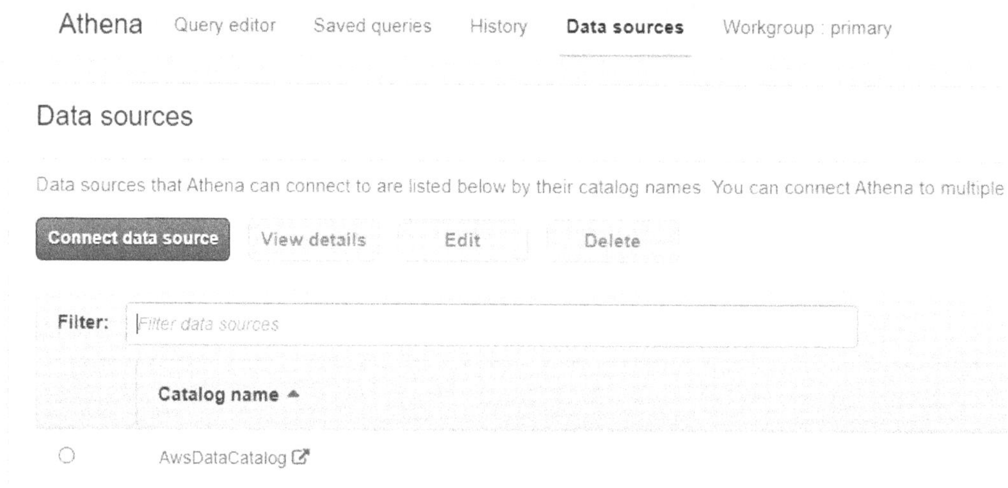
Athena
Query editor
Saved queries
History
Data sources
Workgroup : primary
Data sources
Data sources that Athena can connect to are listed below by their catalog names. You can connect Athena to multiple
Connect data source
View details
Edit
Delete
Filter:
Filter data sources
Catalog name
AwsDataCatalog

Figure 11.9 – Amazon Athena – Data sources

22. Next, click on the **Query editor** tab.
23. From the left-hand menu, select the database you created earlier in *Step 2*, which will also reveal the table that you created within the database, as per the following screenshot:

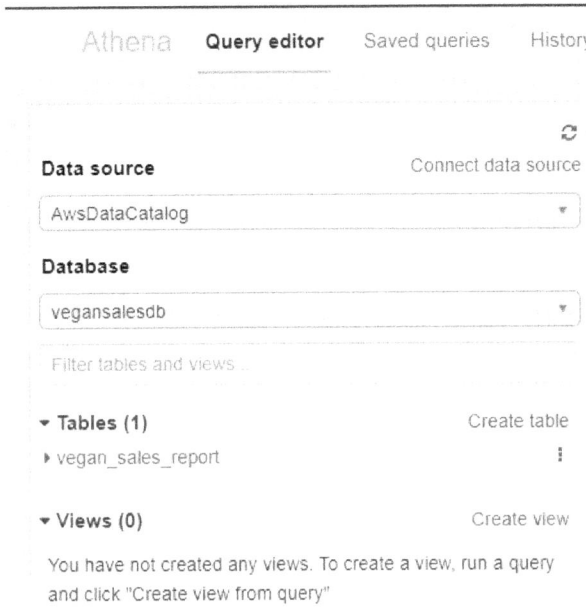

Figure 11.10 – Amazon Athena – Query editor

24. Now, you can easily preview the data held in Amazon S3 by clicking on the ellipsis (the three dots next to your table name) and then clicking on **Preview table** from the context menu that appears. This will run a SQL query and retrieve the sample data from your table, as per the following screenshot:

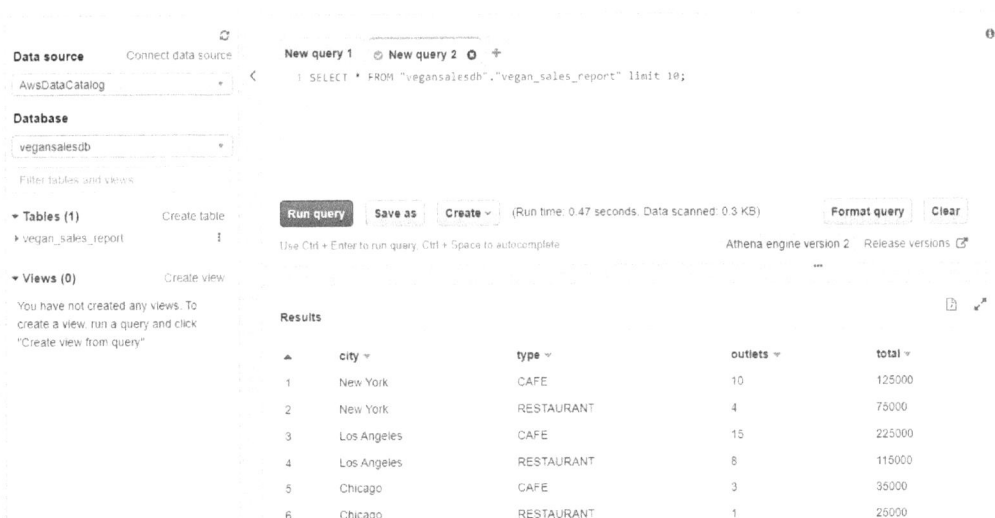

Figure 11.11 – Amazon Athena – Sample query

25. You can run additional queries. For example, you can replace the SQL statement in the top half of the pane with the following:

```
SELECT * FROM "vegansalesdb"."vegan_sales_report" WHERE
total >=100000;
```

The preceding statement will showcase all the cities where the sales that were achieved were equal to or above $100,000, as per the following screenshot:

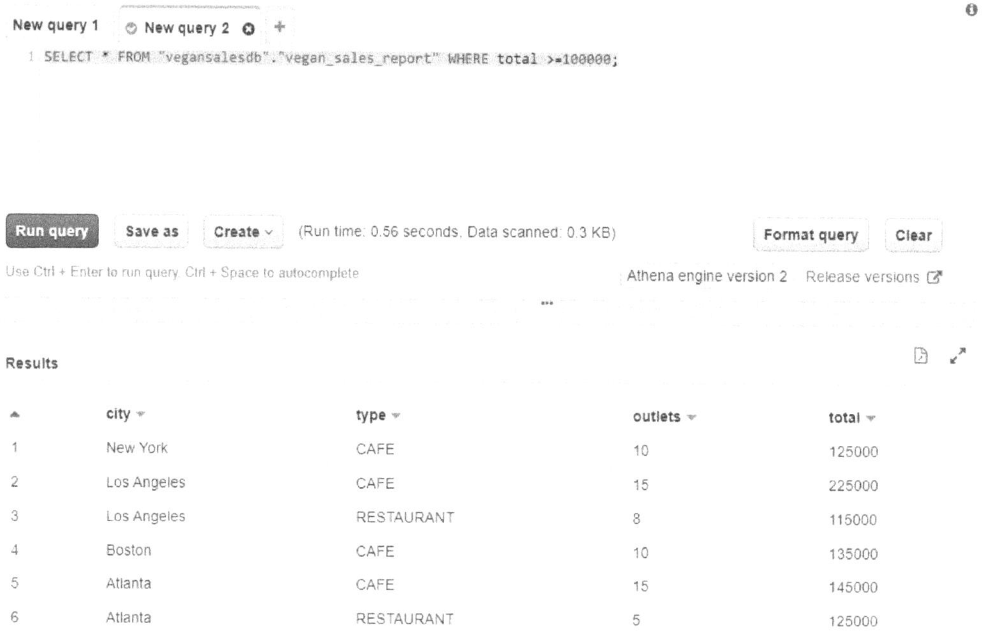

Figure 11.12 – Amazon Athena – Query to identify those cities where
sales were greater than or equal to $100,000

As you can see, Amazon Athena is extremely powerful in being able to access and query your raw data in Amazon S3. You do not need to set up and deploy servers or run expensive databases for such ad hoc analysis of your data.

Now, we will perform a cleanup exercise to remove unwanted resources from our AWS account.

Exercise 11.2 – cleaning up

In this exercise, you will delete the resources you created in the previous exercise to ensure that there are no unwanted costs:

1. Navigate to the Amazon Glue console.

2. From the left-hand menu, click the **Crawlers** link. In the right-hand pane, select **vegan-sales-crawler**. From the **Actions** drop-down list, click the **Delete Crawler** option and then confirm the delete operation.

3. Next, from the left-hand menu, click **Databases**. In the right-hand pane, select the **vegansalesdb** database. Then, from the **Actions** drop-down list, click the **Delete database** option.

4. Click the **Delete** button in the **Delete Database** confirmation dialog box that appears.

Next, you will need to delete the Amazon S3 buckets as they are no longer required:

1. Navigate to the Amazon S3 console. From the left-hand menu, click on **Buckets**.

2. In the right-hand pane, select the **vegan-query-results** bucket and then click the **Empty** button. Confirm that you want to empty the bucket by typing `permanently delete` in the confirmation text box. Next, click the **Empty** button. You will get a confirmation message, stating that the bucket has been successfully emptied. Click the **Exit** button.

3. Next, with the **vegan-query-results** bucket still highlighted, click the **Delete** button.

4. Confirm the delete operation by typing the bucket's name in the confirmation text box and then clicking on the **Delete bucket** button.

5. Repeat *Steps 1* to *4* for the **vegan-sales-report** bucket.

Now that you have completed the cleanup exercise, we will summarize this chapter.

Summary

In this chapter, we discussed several services from AWS that fall within the analytics category. Businesses today possess a vast array of data and being able to analyze and make sense of that data is extremely important. Information that's obtained from this data can help businesses respond to their customers' needs and demands, address potential issues, and even predict future growth. Ultimately, businesses can gain an advantage over competitors.

In this chapter, you learned about services such as Amazon Kinesis, which allows customers to stream and respond in real time and near real time to data. You also learned about services that can be used to quickly query your data, such as Amazon Athena, as well as services to help you present that data using BI tools. Most of these analytical services are also offered as fully managed services on a pay-as-you-consume pricing model, making them very affordable for almost any business.

In the next chapter, you will learn about various deployment and orchestration tools on AWS that can help you provision and deploy your applications in the cloud without extensive manual configuration. We will look at **Infrastructure as Code** (**IaC**), which has taken the IT world by storm as you can design and deploy end-to-end infrastructure solutions in a matter of minutes using predefined templates. We will also look at how to automate common IT tasks using serverless compute services such as AWS Lambda.

Questions

Answer the following questions to test your knowledge of this chapter:

1. Which AWS service can help you ingest and deliver massive amounts of streaming data into Amazon Redshift for near real-time analytics?

   A. Amazon Athena

   B. Amazon Kinesis Firehose

   C. Amazon Kinesis Video Streams

   D. Amazon RDS

2. Which AWS service can help you query streaming data using standard SQL queries in real time?

   A. Amazon Kinesis Data Streams

   B. Amazon Kinesis Data Analytics

   C. Amazon Glue

   D. Amazon QuickSight

3. You are planning on building an application that will capture video streams from speed cameras on country roads for analysis. You need to be able to capture all the vehicles that break the speed limit and identify the offending drivers via the vehicles' license plates. Which two services on AWS can help you achieve these requirements? (Choose 2 answers.)

   A. Amazon Athena

   B. Amazon Kinesis Data Analytics

   C. Amazon Kinesis Video Streams

   D. Amazon Elasticsearch

   E. Amazon Rekognition

4. Which AWS service enables you to index all types of content, offers integration with **Kibana**, and helps you build data visualization tools to analyze large datasets?

   A. Amazon Elasticsearch

   B. Amazon Glue

   C. Amazon Athena

   D. Amazon Kinesis Firehose

5. You store several network log files (in CSV format) in an Amazon S3 bucket. You have been asked to analyze the contents of a specific file for possible malicious attacks. Which AWS service can help you analyze raw data in Amazon S3 and perform the necessary ad hoc analysis?

   A. Amazon Glue

   B. Amazon QuickSight

   C. Amazon Athena

   D. Amazon Data Pipeline

6. Which AWS service can be used to perform serverless ETL functions to discover, prepare, enrich, clean, and transform your data from various sources for analysis?

    A. AWS Glue

    B. AWS Athena

    C. AWS QuickSight

    D. AWS Rekognition

7. Which AWS service enables you to create and publish interactive BI dashboards for your business data to provide access to meaningful information for your business to make decisions?

    A. AWS Kinesis Data Analytics

    B. AWS Glue

    C. AWS QuickSight

    D. AWS Kinesis Firehose

12
Automation and Deployment on AWS

So far, you have learned how to configure and deploy various services on AWS. However, most of your configuration has been manual, with very little automation. For example, in *Chapter 9, High Availability and Elasticity on AWS*, you deployed an application in a multi-tier design. This consisted of application servers configured in an auto-scaling group, an application load balancer, a database to store application data, and an S3 bucket to host your source code. All of this was deployed in a private network in the form of **Amazon VPC**.

While building your application stack, you had to manually configure the various services on AWS with all the necessary resources to deploy your application. For example, with the VPC, you had to configure subnets, IP address ranges, security groups, NAT gateways, and much more. Now, imagine having to perform this sort of manual labor every time you need to create a new environment to host your applications. It would, without a doubt, be very time consuming.

AWS offers various deployment and automation tools to help you architect, provision, and build infrastructure components to host your applications. Some of these tools fall within the PaaS model, where you simply need to focus on the application build and deployment strategy, and AWS will provision the necessary infrastructure to support that application. Other services enable you to define a template for your infrastructure that you may need to provision repeatedly. Using **Infrastructure as Code (IaC)**, you can draft a template that describes how your services and resources need to be configured. AWS will then build the infrastructure, along with all its components, which can then be used to host your application. Automating your infrastructure deployments offers several benefits, including being able to scale your infrastructure globally by sharing your *templates* across your organization to enforce corporate standards and security best practices. Furthermore, by avoiding manual configuration efforts, you reduce the human error element.

In this chapter, we will look at some of the core automation and deployment tools to help you get your solutions to market faster.

In this chapter, we will cover the following topics:

- Understanding application deployment with Amazon Elastic Beanstalk
- Understanding the benefits of IaC using Amazon CloudFormation
- Introduction to the orchestration of Chef and Puppet solutions using AWS OpsWorks
- IT automation with Lambda

Technical requirements

To complete the exercises in this chapter, you will need to log into your AWS account as the IAM user **Alice**, which you had set up in *Chapter 4, Identity and Access Management*.

Understanding application deployment with Amazon Elastic Beanstalk

Amazon Elastic Beanstalk is a service that enables you to deploy your application without having to manually configure the underlying infrastructure that will support the application. Amazon Elastic Beanstalk does all the heavy lifting, which involves provisioning the necessary infrastructure to host and manage your application. This includes capacity provisioning, scaling, load balancing, and health monitoring.

If your application is developed in one of the supported languages, which includes Go, Java, .NET, Node.js, PHP, Python, and a few others, Amazon Elastic Beanstalk will build out the platform using one or more AWS resources, such as EC2 instances, to run your application. All you need to do is follow a prescribed process to deploy your application.

Another option to consider is to use Elastic Beanstalk to deploy Docker containers. This gives you a lot more flexibility because, with Docker containers, you can configure your runtime environments and choose a programming language, supported platform, and any dependencies as required. Thus, you are not restricted to any limitations that are usually associated with the other platforms.

Amazon Elastic Beanstalk does not restrict you to one specific underlying infrastructure design either. You can modify how the underlying infrastructure components will be deployed – for example, you can specify the EC2 instance type and size that is deployed or enforce that a set minimum number of EC2 instances are deployed as part of an auto-scaling group. Ultimately, you retain control of how the infrastructure is configured but without the complex manual configuration, which Amazon Elastic Beanstalk takes care of. Once your application has been deployed, you can manage your environment and deploy application updates or versions later.

Now, let's take a look at the core components of the Amazon Elastic Beanstalk service.

Core components of Amazon Elastic Beanstalk

The infrastructure components that AWS Elastic Beanstalk creates to host your application are called an *environment*. Each environment can only run a single application version, but you can create many environments for different application versions simultaneously. You can modify your environment if you need to and deploy upgrades to your application when you release new updates.

As part of launching an Elastic Beanstalk environment, you need to select the *environment tier*. The environment tier you choose is determined by the type of application you are deploying and ultimately determines the resources Elastic Beanstalk will configure to support your application. We have two environment tiers that can be configured with Elastic Beanstalk: **web server environment tier** and **worker environment tier**.

Web server environment tier

This environment is designed for the frontend tier of your application stack, such as the frontend **user interface** (**UI**) of your e-commerce application. AWS Elastic Beanstalk will provision an environment that is designed to accept inbound traffic to your web application via an elastic load balancer, offer scalability in the form of auto-scaling group configurations, and one or more EC2 instances.

The following diagram illustrates the typical architecture that's deployed:

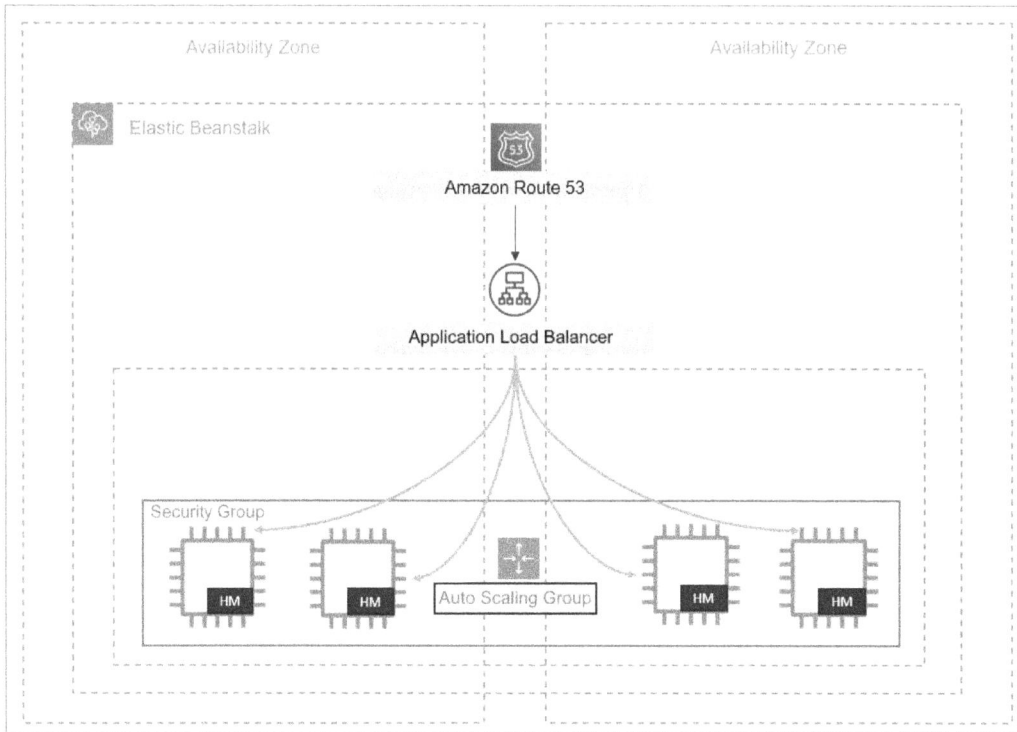

Figure 12.1 – AWS Elastic Beanstalk – web server environment tier

As part of the *web server environment tier*, AWS Elastic Beanstalk will deploy EC2 instances across multiple AZs. Your application is then deployed onto those instances. Furthermore, Amazon Auto Scaling will be configured as part of the environment to scale out as demand for your application increases or scale back in when demand drops. You will have full control over the scaling policies that are used and their parameters. AWS Elastic Beanstalk will also deploy a software stack, depending on the *container type*, which is the infrastructure topology and software stack to be used for that environment. For example, if you were deploying a .NET application on Windows, the environment that would be configured by Elastic Beanstalk would be comprised of the Microsoft Windows operating system, Windows EC2 instances, and a version of **Internet Information Services+ (IIS)**.

AWS Elastic Beanstalk also configures an environment URL that points to the load balancer in the format of `app-name.region.elasticbeanstalk.com`. This URL is aliased with the elastic load balancer's URL. You can also alias this URL with a CNAME record of your choice, such as your company domain name, using Amazon Route 53. This will allow you to use more friendly names such as `myapp.mycompany.com`.

In addition, each EC2 instance will be configured with a software component called the **host manager** (**HM**). The HM is responsible for the following:

- Deploying the application

- Aggregating events and metrics for the servers

- Generating instance-level events

- Monitoring the application log files for critical errors

- Monitoring the application server

- Patching your EC2 instances

- Rotating your application's log files and publishing them to Amazon S3

Next, we'll look at the worker environment tier.

Worker environment tier

This environment is intended for backend operations. In a multiple-tier application architecture, this would represent the middleware, application, and backend database layers of your application stack. AWS Elastic Beanstalk provisions an environment that's appropriate for such backend operations and consists of an Auto Scaling group, one or more Amazon EC2 instances, and an IAM role. In addition, AWS Elastic Beanstalk will also provision an Amazon SQS queue if you have not configured your own yet.

AWS Elastic Beanstalk will also install a *daemon* on each EC2 instance in the fleet. This daemon will read messages from the Amazon SQS queue and send this data to the application servers running in the worker environment tier for processing. This process is illustrated in the following diagram:

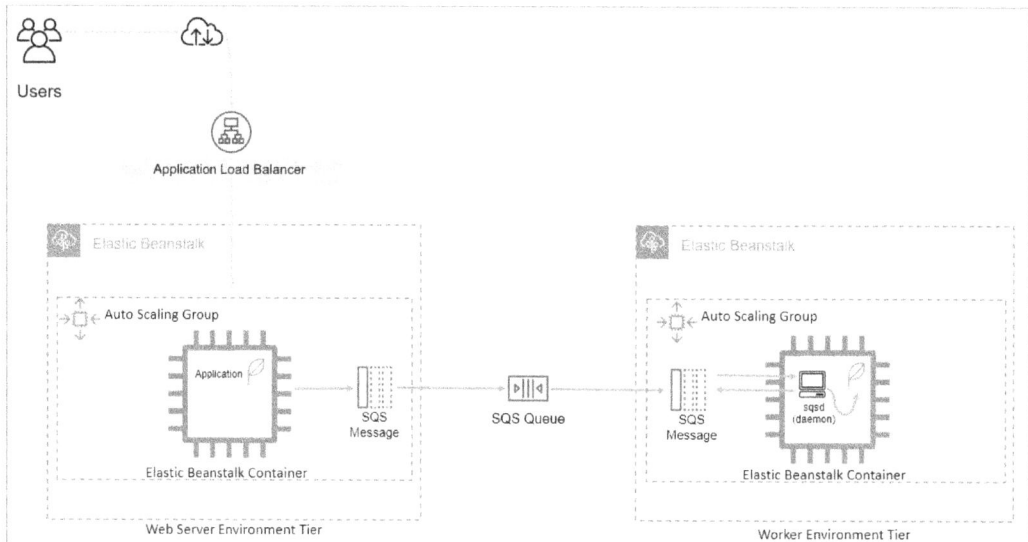

Figure 12.2 – AWS Elastic Beanstalk – worker environment tier

AWS Elastic Beanstalk will also monitor the health of your EC2 instances using a service called **Amazon CloudWatch**. We will look at Amazon CloudWatch in *Chapter 13, Management and Governance on AWS*.

Apart from this, AWS Elastic Beanstalk can help you deploy and manage your application stack using various deployment types, including *all at once, rolling, rolling with additional batch, immutable*, and *traffic splitting*. These deployment types enable you to design your application deployments and any ongoing updates that suit your specific business requirements.

In this section, we learned about the AWS Elastic Beanstalk service, which is an application deployment and management service on AWS. AWS Elastic Beanstalk allows developers to focus on their application code rather than having to minutely configure infrastructure components to support the application. AWS Elastic Beanstalk will provision a highly available and scalable infrastructure architecture and gives you control over how those infrastructure components are configured to support your specific requirements.

In the next section, we will look at the Amazon CloudFormation service, which enables you to architect your underlying infrastructure resources using templates written in JSON or YAML code.

Understanding the benefits of IaC using Amazon CloudFormation

The approach of using code to describe and deploy your infrastructure components automatically is known as IaC. IaC is a fundamental component that enables you to automate infrastructure builds in the cloud.

Building your infrastructure using code greatly improves deployment processes because the code is executed by machines. This also means that any infrastructure that's deployed using code is less prone to human errors, which is inherent in manual deployments. Furthermore, you can create templates for repeat deployments and enable versioning for those templates. Often, you want to mimic the testing and production environments with each other so that you are assured that once your application has passed the testing phase, it can be easily deployed to production. Using templates that describe your infrastructure is the best way to avoid any discrepancies between those environments.

Amazon CloudFormation is a solution that can help you design, build, and deploy your infrastructure using code. It helps you define the resources to be created and configured and how those resources interact with each other.

CloudFormation templates

Amazon CloudFormation uses *templates*, which are text files that are written in either JSON or YAML format. The code describes the resources that you wish to deploy, along with configuration information related to those resources. For example, if you need to deploy a **security group** as part of your VPC template, you can specify its name and the inbound and outbound rules that are permitted.

The best part about CloudFormation templates is that they contain descriptions of the resources that you want CloudFormation to deploy. This can double up as a technical document of your entire infrastructure. In addition, templates can be used repeatably to deploy infrastructure for different environments. This ensures that you have the same architecture for both your development and testing environments. While the configuration in each environment will be the same, they will be separate environments with different resource identifiers.

AWS CloudFormation templates can also be used to accept input parameter values. This can further help expedite the process of building different environments for experimentation and testing. For example, if you were testing a new application and you wanted to compare performance levels based on the underlying EC2 instance type and size that the application was hosted on, you could deploy multiple test environments with different EC2 instance types and sizes. The input parameters, at the time of deploying the template, will prompt you to specify which instance type and size to use for a given environment. This also makes it easier to use the same template for different situations.

CloudFormation templates are used to deploy CloudFormation stacks, which are a container that organizes the resources described in the template. We'll look at CloudFormation stacks next.

CloudFormation stacks

With CloudFormation, you can create *stacks*, which are a container to collectively manage all related resources as a single unit. So, for example, a stack could consist of the VPC, an Auto Scaling group, an elastic load balancer, and an Amazon RDS database instance. You can have multiple stacks that deploy a collection of resources that may also need to communicate with each other.

You can also delete a stack when you no longer need the resources contained within it. AWS will then delete every resource that was deployed when the stack was created using your template. This ensures that you do not have any resources lingering, which often happens when you are manually trying to delete resources from an environment. This also ensures that you can quickly get rid of test or development environments when they're no longer needed.

Sometimes, you may need to modify your stack – for example, if you need to upgrade the EC2 instance type or change the database endpoint. We will look at how to implement changes to your stack next.

Change sets

If you need to make changes to the resources that are deployed in your stack, you can create a **change set**. AWS CloudFormation creates this change set based on a summary of changes you make to the template and gives you the details of the proposed changes. You can then decide whether to roll out those changes or make further modifications first.

Change sets are particularly useful at highlighting changes that may have catastrophic results, ensuring that you can perform all precautionary tasks first. For example, if you were to simply change the name of your Amazon RDS database in a template, AWS CloudFormation would create a new database and delete the old one. If you did not make a backup of the RDS database before performing this action, you would end up losing your data. Change sets can help you identify where such irreversible changes will occur, and you can make sure that you first perform a manual snapshot of your RDS database before rolling out the change.

Drift detection

The recommended approach, when making any changes to your CloudFormation stack, is to use change sets. However, it is possible to make changes to the stack resources outside of CloudFormation. Sometimes, these may be accidental but at other times, you may wish to make these changes urgently, although any change should follow a proper change management process.

Another issue is that if changes are made outside of CloudFormation, then any stack update or deletion requests may fail. AWS offers a feature of the CloudFormation service known as *drift detection*. You can use drift detection to identify stack resources that configuration changes have been made to outside of CloudFormation management.

In this section, you learned about the AWS CloudFormation service, which uses an IaC approach to build and deploy infrastructure resources that can be used to support your application. CloudFormation enables you to design templates using JSON- and YAML-formatted text files that describe your resources and how they can be configured.

Then, CloudFormation does the heavy lifting of provisioning those resources just the way you want them. CloudFormation templates are used to build CloudFormation stacks, which collectively manage all the related resources as a single unit.

In the next section, we will look at another AWS service that can help you orchestrate and deploy applications that have been designed for the Chef and Puppet automation tools.

Introduction to the orchestration of Chef and Puppet solutions using AWS OpsWorks

In this section, we will provide a quick introduction to another configuration management and orchestration service known as **AWS OpsWorks**. AWS offers three different services that fall under the AWS OpsWorks offering. There are AWS OpsWorks stacks and AWS OpsWorks for Chef Automate, which let you use Chef cookbooks for configuration management. The third service is OpsWorks for Puppet Enterprise, which lets you configure a Puppet Enterprise master server in AWS.

OpsWorks for Puppet Enterprise provides a *fully managed Puppet master server* that is used to communicate with, configure, deploy, and manage your nodes – which could be EC2 instances or even on-premises servers. The service also takes care of handling tasks such as software and operating system configurations, package installations, database setups, and more.

AWS OpsWorks for Chef Automate lets you create *AWS-managed Chef servers*, including Chef Automate software solutions. AWS OpsWorks creates a Chef server that is used to manage all your nodes and acts as a central repository for your Chef cookbooks. Chef cookbooks contain recipes that are authored using the Ruby programming language and enable you to define a collection of resources and their attributes that need to be configured.

AWS OpsWorks for Puppet Enterprise and OpsWorks for Chef Automate require that you know how to operate the Chef and Puppet software. If you prefer a more IaC approach to deploying your application and are not familiar with Chef or Puppet, you can also use AWS OpsWorks stacks. Next, we will look at the AWS OpsWorks stacks service.

AWS OpsWorks stacks

By now, you are aware that deploying a single application can be comprised of several resources running behind the scenes, including the host EC2 instances that the application runs on, backend databases such as Amazon RDS to store the application data, load balancers to distribute traffic, and so on. This group of related resources is what we call a stack.

A stack is a container for your AWS resources such as your EC2 instances, RDS databases, and all the related components, which are logically managed together. Each stack will contain at least one layer but can have more.

Let's consider a typical example that consists of the following three layers to support an application:

- A backend database layer, which can either be a self-hosted database solution or a managed service such as Amazon RDS

- An application layer, which is comprised of EC2 instances or containers

- A load balancing layer, which hosts your application load balancer to distribute traffic to your application layer

The following diagram illustrates this configuration:

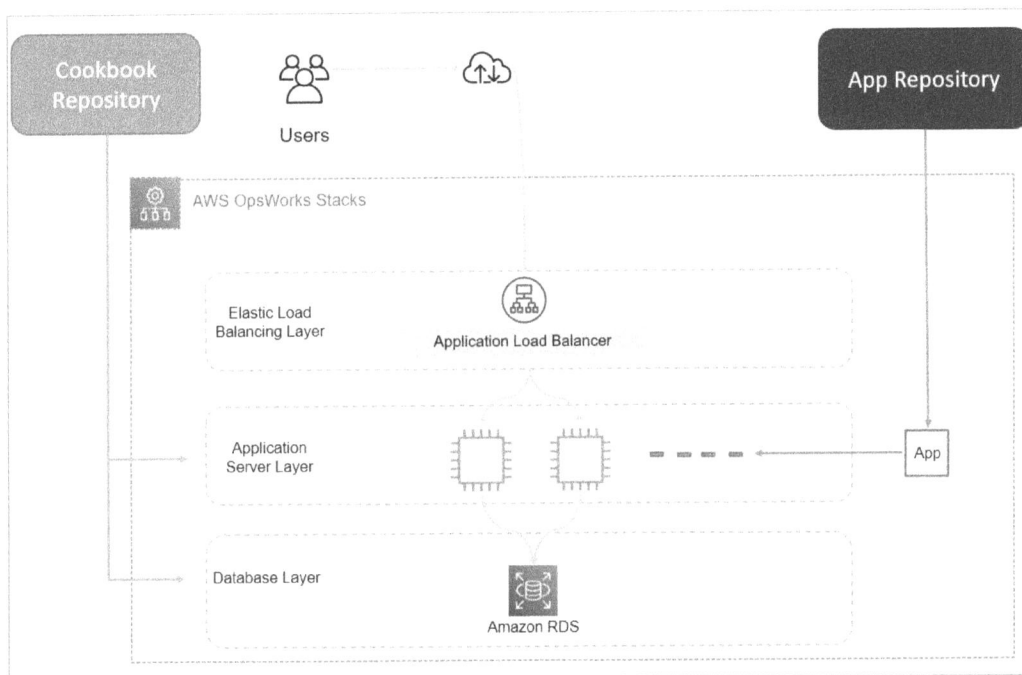

Figure 12.3 – AWS OpsWorks stacks

Amazon OpsWorks stacks will not only help you build out your stack but also monitor the stack's performance and health, security permissions, and more. With AWS OpsWorks stacks, you do not need to create or manage Chef servers as OpsWorks stacks performs some of the work for you.

There are two types of layers that AWS OpsWorks stacks uses, both of which we will discuss next.

AWS OpsWorks layers

An OpsWorks layer represents a particular service or task as part of your overall application stack. For example, your EC2 instances could represent the application layer, which is used to host the application. Next, your application load balancer will represent the load balancing layer, distributing traffic across the EC2 instances. Each layer is comprised of a template that allows you to specify the components of the resources that are being provisioned. For example, for the application layer, this would include your EC2 instance's security group or IP addressing details. You can provision any supported operating system and even add on-premises servers to the stack.

AWS OpsWorks stacks does not provision a Chef server or a Puppet Enterprise Master server; instead, all the management tasks are performed by AWS OpsWorks via the embedded Chef solo client that is installed on Amazon EC2 instances.

Service layers

AWS OpsWorks can also extend your stack to include service layers, which can be used to build a complete multi-tier application solution. These service layers can include the following:

- **Amazon Relational Database Service (RDS)**: This layer will enable you to integrate your application with an existing RDS database deployment.

- **Elastic Load Balancing**: This layer will enable you to distribute traffic to your fleet of EC2 instances across multiple Availability Zones while offering high-availability features.

- **Amazon Elastic Container Service (ECS) cluster**: This layer will enable you to create an ECS cluster layer, allowing you to connect your OpsWorks stacks to an ECS cluster running your Docker-based applications.

In this section, we looked at the AWS OpsWorks suite of services, which enables you to orchestrate application deployment solutions using Chef and Puppet. You also learned about the OpsWorks stacks service, which allows you to create and manage stacks and applications.

In the next section, we will look at the AWS Lambda service and how you can use it to automate administrative tasks on AWS.

IT automation with Lambda

In *Chapter 7, AWS Compute Services*, we learned that **AWS Lambda** is a serverless compute service that allows you to run code without having to provision or manage servers in the cloud. With AWS Lambda, you simply upload your code and have it executed based on a specific trigger. AWS Lambda will provision all the underlying infrastructure needed to run your code, be it compute power, memory, or temporary storage.

Your code can be automatically triggered from various AWS services and SaaS applications or even be called directly from any web or mobile application. You could use AWS Lambda in conjunction with other serverless offerings such as API Gateway, DynamoDB, and the Amazon S3 static website hosting service to build the ultimate serverless application for your business or clients.

In addition to this, Lambda can also be used to help automate a vast array of day-to-day administrative tasks. This can include any repetitive tasks that are triggered by a specific event or even on a specific schedule. For example, you can automate how firmware updates are installed on hardware devices, start and stop EC2 instances, schedule security group updates if you need to make changes, and much more.

In the upcoming exercise in this chapter, we will look at one such use case for using AWS Lambda to help you perform the typical administrative task of stopping and restarting EC2 instances at regular intervals.

Exercise 12.1 – stopping and starting EC2 instances at regular intervals using AWS Lambda

Imagine a scenario where you need to run a fleet of on-demand EC2 test servers that your **user acceptance testing** (**UAT**) team needs to perform multiple functional and technical tests on for an upcoming application that you are developing. Your UAT team only works from Monday to Friday, 9 A.M. to 5 P.M. The UAT team only needs access to the fleet of test servers during this time. Rather than have a technician manually start up all the servers in the morning and shut them down again at the end of the business day, you could automate the process using AWS Lambda. You would not want to have your on-demand EC2 instances running when they are not needed because you are charged for every hour that those servers are running.

In this exercise, we will look at how to configure AWS Lambda to automatically stop and then start your EC2 instances at defined schedules.

To complete the exercises in this chapter, you will need to download the sample IAM policy from this book's Packt GitHub repository at: `https://github.com/PacktPublishing/AWS-Certified-Cloud-Practitioner-Exam-Guide`

Step 1 - Launching an EC2 instance

To complete this excercise, you will need to deploy an EC2 Instance running the Linux 2 AMI:

1. Log into the AWS Management Console as the IAM user called **Alice**.

2. Navigate to the EC2 dashboard. Ensure that you are in the **us-east-1 (N. Virginia)** Region.

3. Click on **Instances** from the left-hand menu and then click on the **Launch instances** button.

4. For **Step 1: Choose an Amazon Machine Image (AMI)**, select the **Amazon Linux 2** AMI.

5. For **Step 2: Choose an Instance Type**, select the **t2.micro** instance type and click the **Next: Configure Instance Details** button.

6. For **Step 3: Configure Instance Details**, provide the following key details:

   - For **Network**, select the default VPC.

   - For **Subnet**, select the subnet ID that represents the default subnet in **us-east-1a**.

   - In the text box next to **Auto-assign Public IP**, ensure that **Enable** is selected.

   - Leave all the remaining options as their default values and click the **Next: Add Storage** button at the bottom of the page.

7. For **Step 4: Add Storage**, leave all the options as their default values and click the **Next: Add Tags** button at the bottom of the page.

8. For **Step 5: Add Tags**, click the **Add Tag** button and provide a key-value name for your EC2 instance. For example, for **Key**, type in Name, and for **Value**, type in UAT-Server-01.

9. Next, click on the **Next: Configure Security Group** button at the bottom of the page.

10. For **Step 6: Configure Security Group**, ensure that the **Create a new security group** option is selected and set **Security group name** to **UAT-SG**. This will represent the security group of our UAT test server(s).

11. Next, you need to configure the inbound rules. You should already have a pre-configured inbound rule defined that allows SSH traffic inbound from the internet, as per the following screenshot:

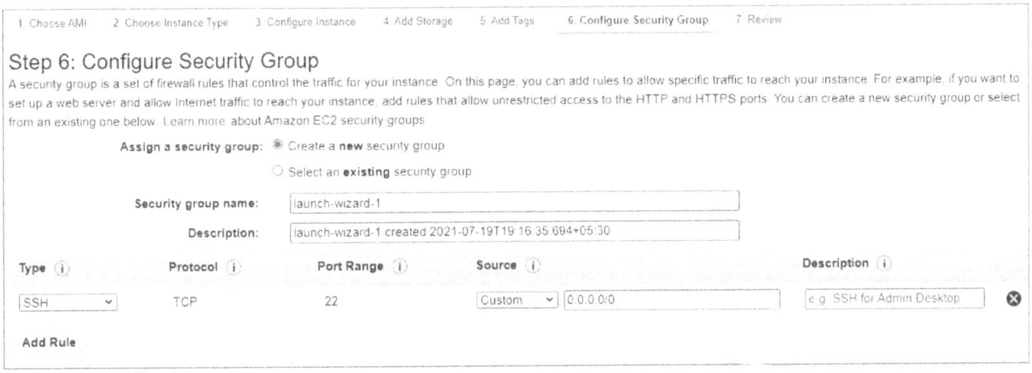

Figure 12.4 – Configure Security Group

12. Click the **Review and Launch** button at the bottom of the page.

13. On the next page, review your configuration and then click the **Launch** button at the bottom of the page.

14. You will be prompted to choose an existing key pair or create a new one. You can use the existing key pair you created for the previous exercise. You will need to click on the checkbox to acknowledge that you have access to this key pair and then click on the **Launch instances** button.

15. Finally, click on the **View Instances** button to be taken back to the EC2 dashboard, where you can see the instance you just launched.

Because we will be stopping and starting these EC2 instances using a Lambda function, you will need to make a note of the EC2 instance ID, as per the following screenshot. Note that your EC2 instance ID will be different from the one shown in this screenshot:

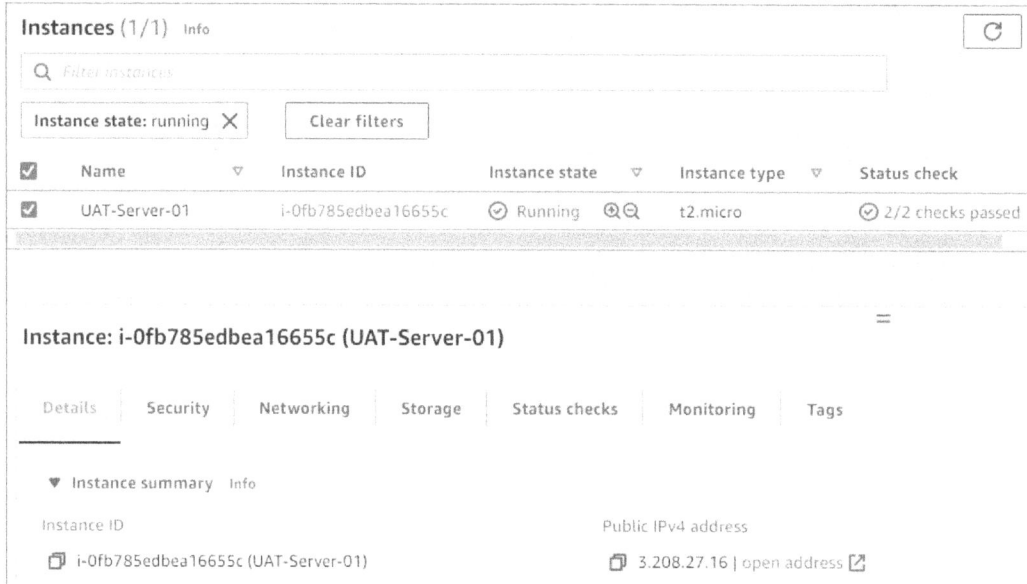

Figure 12.5 – EC2 instance ID

Once your instance is up and running, we can start creating the necessary IAM policy and IAM role to enable our Lambda functions to start and stop our EC2 instance.

Step 2 - Creating an IAM policy and execution role for your Lambda function

For your Lambda function to be able to start and stop your EC2 instances, it needs to have the necessary permissions. In this step, you will configure an IAM policy and an IAM role to enable your Lambda function to perform the start and stop operations on your EC2 instances:

1. Navigate to the IAM dashboard.

2. From the left-hand menu, click on **Policies**.

3. Click the **Create Policy** button from the right-hand pane.

4. Select the **JSON** tab and delete the default JSON text in the editor tool.

5. Next, copy and paste the following JSON policy document into the policy editor (you can also download this policy from the Packt GitHub repository for this book https://github.com/PacktPublishing/AWS-Certified-Cloud-Practitioner-Exam-Guide):

```
{
    "Version": "2012-10-17",
    "Statement": [
        {
            "Effect": "Allow",
            "Action": [
                "logs:CreateLogGroup",
                "logs:CreateLogStream",
                "logs:PutLogEvents"
            ],
            "Resource": "arn:aws:logs:*:*:*"
        },
        {
            "Effect": "Allow",
            "Action": [
                "ec2:Start*",
                "ec2:Stop*"
            ],
            "Resource": "*"
        }
    ]
}
```

6. Click the **Next: Tags** button at the bottom of the page.

7. On the **Add tags (Optional)** page, click the **Add tag** button to provide a tag name for your policy. For example, for **Key**, type in Name, and for **Value**, type in Lambda-EC2-Access-Policy.

8. Click **Next: Review**.

9. On the **Review Policy** page, type in a name for your policy; for example, Lambda-EC2-Access-Policy.

10. Click the **Create policy** button at the bottom of the page. You will receive a notification that your policy has been created.

11. Next, you will need to create an IAM role for Lambda. From the left-hand menu, click on **Roles**.

12. Click **Create role** from the right-hand pane.

13. On the **Create role** page, ensure that the **AWS service** option is selected under **Select type of trusted entity**.

14. Next, under **Choose a use case**, select **Lambda**.

15. Click the **Next: Permissions** button at the bottom of the page.

16. On the **Attach permissions policies** page, search for the policy you just created in the search box next to **Filter policies**. In the search box, start by typing in Lambda-EC2, which should filter the list down to the policy you just created. Select this policy and click on the **Next: Tags** button at the bottom of the page.

17. On the **Add tags (Optional)** page, click the **Add tag** button and provide a tag name for your policy. For example, for **Key**, type in Name, and for **Value**, type in Lambda-EC2-Start-Stop-IAM-Role.

18. Click the **Next: Review** button and set **Role name** to Lambda-EC2-Start-Stop-IAM-Role.

19. Click the **Create role** button at the bottom of the page.

20. You will get a notification, stating that the IAM role has been created.

Now that your role has been created, you can create Lambda functions to stop and start your EC2 instances.

Step 3 - Creating Lambda functions that stop and start your EC2 instances

Now, let's create a Lambda function to stop and start your EC2 instance at a predefined schedule:

1. From the **Services** drop-down list at the top of the AWS Management Console page, select **Lambda**, which is located under the **Compute** category.

2. From the left-hand menu, click on **Functions**.

3. Click the **Create function** button in the right-hand pane.

4. Select the **Author from scratch** option.

5.   Under **Basic information**, provide the following details:

  ▪   For **Function name**, provide a name that represents the purpose of the function. For example, if you are creating the Stop EC2 instance function first, in the name field, type in StopEC2Instances.

  ▪   For **Runtime**, select **Python 3.8**

  ▪   Next, under **Permissions**, expand **Change default execution role**.

  ▪   Under **Execution role**, select **Use an existing role**. Next, from the **Existing role** drop-down list, select the IAM role you created earlier.

6.   Click the **Create function** button.

7.   You will be redirected to the **StopEC2Instances** function page.

8.   Next, in the **Code source** section, in the **Code** tab, you will be able to add your function code. From the left-hand **Environment** menu, expand the StopEC2Instance folder and double-click on the lambda_function.py file. In the right-hand pane, you will notice some sample code.

9.   Delete the sample code and replace it with the following code:

```python
import boto3
region = 'us-west-1'
instances = ['i-12345cb6de4f78g9h',
'i-08ce9b2d7eccf6d26']
ec2 = boto3.client('ec2', region_name=region)

def lambda_handler(event, context):
    ec2.stop_instances(InstanceIds=instances)
    print('stopped your instances: ' + str(instances))
```

> **Important Note**
>
> You will need to amend the code as follows.
>
> For the Region, replace the us-west-1 Region with the Region your EC2 instance is in. In our example, the Region would be us-east-1.

10. Next, you will notice that the sample code refers to two EC2 instances. Replace this with the instance ID of the server you deployed in *Step 1 - Launching an EC2 instance* of this exercise, ensuring that the instance ID is placed within single quotes.

11. Next, click on the **Configuration** tab.

12. Click the **Edit** button from the **General configuration** pane.

13. Next, set the timeout value to 10 seconds.

14. Click the **Save** button. This will take you back to the **Function** page. Click the **Code** tab.

15. Recheck your Lambda code and ensure that you have made all the preceding changes. Next, click on the **Deploy** button.

16. If necessary, click on the ellipsis icon in the far left-hand pane to bring up the main menu and click on the **Function** link.

17. You should now see a list of your functions, which will include the StopEC2Instances function you just created.

18. Repeat *Steps 1* to *15* to create another function. This time, you will be creating a function to start your EC2 instances. For *Step 5*, enter a different function name than the one you used before; for example, **StartEC2Instances**.

19. For *Step 9*, copy and paste the following code into the editor pane, remembering to delete the sample code that is already there:

```python
import boto3
region = 'us-west-1'
instances = ['i-12345cb6de4f78g9h',
'i-08ce9b2d7eccf6d26']
ec2 = boto3.client('ec2', region_name=region)


def lambda_handler(event, context):
    ec2.start_instances(InstanceIds=instances)
    print('started your instances: ' + str(instances))
```

20. Remember to also change the Region to us-east-1 and amend the instance ID to the ID of your EC2 instance.

21. Once you have deployed your function, go back to the list of functions by clicking on the link from the left-hand menu (if necessary, by first clicking on the ellipsis icon).

You should now have two functions that will be used to stop and start your EC2 instances, as per the following screenshot:

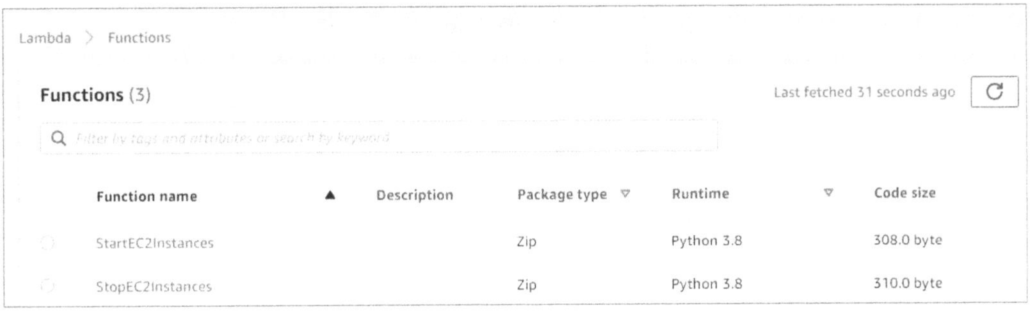
Lambda > Functions
Functions (3)
Last fetched 31 seconds ago
Function name
Description
Package type
Runtime
Code size
StartEC2Instances
Zip
Python 3.8
308.0 byte
StopEC2Instances
Zip
Python 3.8
310.0 byte

Figure 12.6 – Functions

Next, you will create CloudWatch event rules to help you execute the Lambda functions at scheduled times. We discussed CloudWatch events briefly in *Chapter 13, Management and Governance on AWS*. We will look at CloudWatch in more detail in the next chapter.

Step 4 - Creating CloudWatch event rules to trigger your Lambda functions

In this step, you will learn how to create CloudWatch event rules to trigger your Lambda functions at a given schedule:

1. Navigate to the CloudWatch dashboard. You can either search for CloudWatch from the top search bar of the AWS Management Console page or click the **CloudWatch** link from the **Management & Governance** category in the list of services.
2. From the left-hand menu, click on **Rules**, under **Events**.
3. Click the **Create rule** button.
4. For **Step 1: Create rule**, select **Schedule** under **Event Pattern**.
5. Next, select **Cron expression**. A CRON expression is a string comprised of six fields separated by white spaces that represent a set of times. These can be used as a schedule to execute a particular task regularly.
6. In the text field next to **Cron expression**, you will need to type in an expression that tells Lambda when to stop your instances. To learn more about how to define your expressions, visit `https://docs.aws.amazon.com/AmazonCloudWatch/latest/events/ScheduledEvents.html`.
7. For this exercise, we wish to stop our EC2 instances at 6 P.M. Our UAT testers normally leave work at 5 P.M., but just in case anyone decides to work a bit late, we can execute the Lambda Stop function at 6 P.M.

8. The cron expression that you can use to execute the Lambda Stop function is **0 18 ? * MON-FRI ***. This will also display the next 10 triggers and their time of execution, as per the following screenshot:

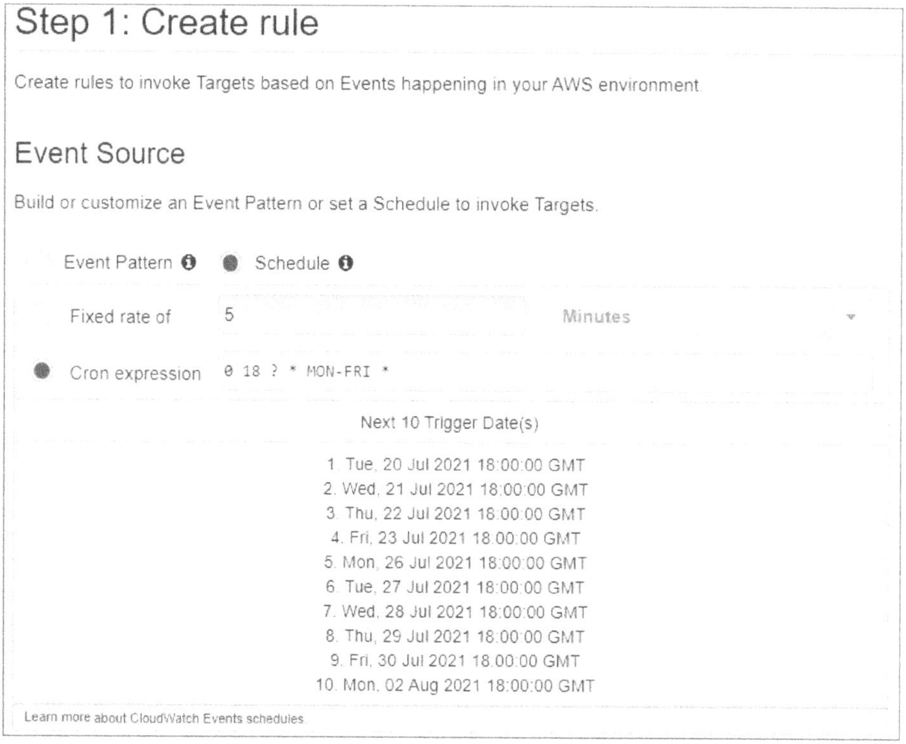

Figure 12.7 – AWS CloudWatch event rule cron expression

9. Next, click on the **Add target** button under **Targets**.

10. From the drop-down list that appears, select **Lambda function**.

11. In the drop-down list next to **Function**, select the **StopEC2Instances** function. As you may recall, this is a Lambda function that will stop your EC2 instance.

12. Next, click on the **Configuration details** button.

13. For **Step 2: Configure rule details**, type an appropriate name in the **Name** field under **Rule definition**. For example, you can type in StopUATInstances.

14. In the **Description** field, type Stops UAT instances at 6 PM Monday to Friday.

15. Next, click the **Create rule** button.

16. Repeat *Steps 3* to *15* to start your EC2 instances. For this rule, ensure that, in *Step 6*, for **Cron expression**, you provide an expression that defines a start time of 8 A.M. Monday to Friday. Here, again, our UAT engineers will start work at 9 A.M. but some may come in a bit early. You can refer to `https://docs.aws.amazon.com/AmazonCloudWatch/latest/events/ScheduledEvents.html` to help you build the cron expression. For this particular exercise, the cron expression that's required is **0 8 ? * MON-FRI ***. Furthermore, in *Step 11*, for **Function**, select the **StartEC2Instances** function. Also, in *Step 13*, enter `StartUATInstances` for the name and for the description, `Starts UAT instances at 8 AM Monday to Friday`.

17. At this point, you have ensured that your scheduled events automatically trigger the relevant Lambda functions, as per the following screenshot:

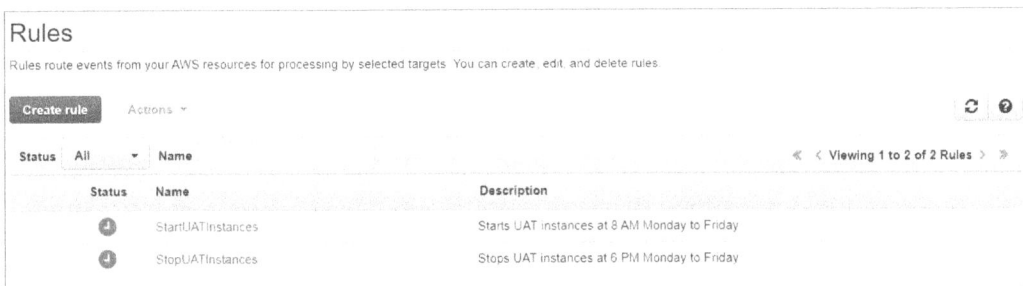

Figure 12.8 – CloudWatch event rules

You can wait for the designated times to check whether your EC2 instance has been stopped and then restarted. Alternatively, you can just test your Lambda function, which is what we will be doing next.

Step 5 - Testing your Lambda function

We can test our Lambda function rather than waiting for the scheduled times to see whether the functions work:

1. Navigate back to the Lambda dashboard.

2. From the left-hand menu, select **Functions**. Then, from the right-hand pane, select the **StopEC2Instances** function.

3. Next, from the **Actions** drop-down list, select **Test**.

4. For **StopEC2Instances**, click on the **Test** button in the **Test event** pane.

5. If the function has been configured correctly, you should see the **Execution result: succeeded** message and an option to expand the **Details** pane. This will provide details of the execution, as per the following screenshot:

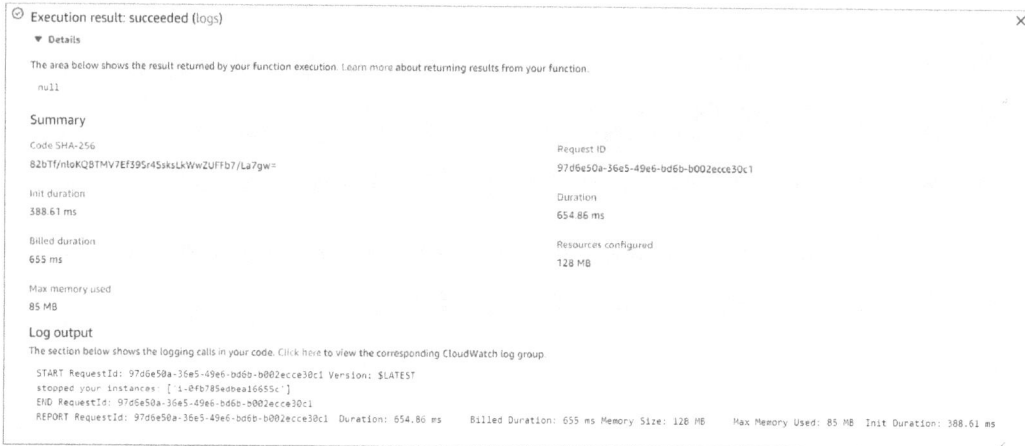

Figure 12.9 – The Lambda function's execution details

6. In another browser window, navigate to your EC2 dashboard. Here, you should find that your EC2 instance has been stopped.

7. Return to the AWS Lambda browser tab and click on **Functions** from the far left-hand menu.

8. Repeat *Steps 2* to *5*, this time selecting the **StartEC2Instances** function.

9. Within a few seconds, you should find that the function was executed successfully, and you can cross-reference this with the EC2 dashboard. You should find that your EC2 instance has started again.

In this exercise, you learned how to create Lambda functions that can be used to perform repetitive IT tasks for your organization and help automate various processes. You also learned how to schedule those repetitive tasks using AWS CloudWatch event rules and cron expressions.

Exercise 12.2 – cleaning up

In this exercise, you will terminate your EC2 instance and delete your Lambda functions to avoid any unnecessary charges to your AWS account:

1. Navigate to the EC2 dashboard and ensure that you are in the **us-east-1** (North Virginia) Region.

2. Click on **Instances** from the left-hand menu.

3. In the right-hand pane, under **Instances**, select the EC2 instance you launched earlier. Then, from the **Instance state** drop-down menu, click **Terminate instance**. Confirm that you wish to terminate the instance; AWS will terminate it.

4. Next, navigate to the Lambda dashboard.

5. Click on **Functions** from the left-hand menu. In the right-hand pane, for each function, select the function. Then, from the **Actions** drop-down list, click **Delete**. Click the **Delete** button in the pop-up dialog box to delete the function.

6. Next, navigate to the CloudWatch management console.

7. From the left-hand menu, click on **Rules**, under **Events**.

8. For each rule, select the rule. Then, from the **Actions** drop-down list, click **Delete**. Click the **Delete** button in the pop-up dialog box to delete the rule.

Your resources have now been removed from your AWS accounts. In the next section, we will provide a summary of this chapter.

Summary

AWS offers numerous services and tools to help you architect efficient application deployment strategies with a focus on the application rather than the underlying infrastructure that supports it. Amazon Elastic Beanstalk helps your developers focus on the application code rather than how to go about configuring every minute infrastructure component to support it.

AWS also offers you tools to build end-to-end infrastructure components using an IaC approach, which drastically improves speed and agility and reduces the human error element that manual configurations are prone to. AWS CloudFormation is an intelligent tool that accepts code in its declarative form to help an architect complete infrastructure deployments.

You also learned how to automate day-to-day administrative tasks using serverless **Function-as-a-Service** (**FaaS**) tools such as AWS Lambda.

In the next chapter, we will look at various management and governance tools that will help you efficiently manage your AWS resources and design for better performance, security, and operations.

Questions

Answer the following questions to test your knowledge of this chapter:

1. Which AWS service automatically provisions the necessary infrastructure (for example, load balancing, auto-scaling, and health monitoring) and enables developers to automatically deploy an application's built-in supported languages such as Node.js, PHP, and Python?

   A. AWS CloudFormation

   B. AWS Lambda

   C. AWS Elastic Beanstalk

   D. AWS Deployer

2. You work for a web application development company and have been asked to design an infrastructure solution that can be repeatedly created using scripted templates. This will allow you to create individual sandbox environments for your developers to use. Some infrastructure components will include the setup and configuration of a VPC, EC2 instances, S3 buckets, and more. Which AWS service enables you to design an infrastructure template that can be deployed to create repeatable infrastructure for your developers to use as a sandbox environment?

   A. AWS Systems Manager

   B. AWS CloudFormation

   C. AWS Config

   D. AWS FSx for Lustre

3. Which two file formats are used when creating CloudFormation templates? (Choose 2.)

   A. JSON

   B. YAML

   C. XML

   D. HTML

   E. Java

4. Which AWS service provides integration with Chef recipes to start new application instances, configure application server software, and deploy the application?

    A. Amazon CloudFormation

    B. Amazon Elastic Beanstalk

    C. Amazon OpsWorks

    D. Amazon Cookbook.

5. Which type of environment do you need to configure for an Elastic Beanstalk deployment to host backend application layer services?

    A. Web server environment tier

    B. Worker environment tier

    C. Backend environment tier

    D. Hybrid environment

6. Which feature of the Amazon CloudFormation service enables you to review any proposed changes you wish to make to an environment and identify how those changes will impact your environment?

    A. Drift detection

    B. Change sets

    C. Stack sets

    D. Change management

13
Management and Governance on AWS

Managing and monitoring your resources on AWS is a crucial part of ensuring that your applications perform as expected, are highly available and secure, and run in the most cost-efficient manner. You want to be able to monitor how your applications are being consumed, identify any technical issues that may affect performance and availability, and ensure that only authorized entities are granted access. Furthermore, you need to be able to audit your environment, with access to information such as access patterns, and identify any anomalies that may indicate potential performance or security issues. You also want to be able to enforce change management processes to ensure that any modifications or changes made to your AWS resources are accounted for and approved.

Finally, as part of your day-to-day administrative tasks in maintaining your workloads, you want to be able to effectively manage your resources, such as patching, performing updates, and automating tasks wherever possible.

In this chapter, we will look at a series of AWS services that allow you to effectively monitor, report, and audit your AWS resources, enabling you to incorporate change management processes and implement a centralized administration of day-to-day tasks. We will also look at tools that offer guidance on where you can improve the performance, fault-tolerance, security, and cost-effectiveness of your resources.

The following topics are dealt with in this chapter:

- The basics of Amazon CloudWatch

- Meeting compliance requirements with AWS CloudTrail

- Learning about change management with AWS Config

- Managing your AWS resources with AWS Systems Manager

- Learning how to use AWS Trusted Advisor

- Understanding the AWS Well-Architected Framework

Technical requirements

To complete the exercises in this chapter, you will need to log in to your AWS account as the **Identity and Access Management (IAM)** user **Alice**.

The basics of Amazon CloudWatch

Amazon CloudWatch enables you to monitor your AWS resources and applications, running on AWS as well as on-premises. With Amazon CloudWatch, you can see how your resources are performing in real time. Using CloudWatch, you can collect resource and application metrics, logs, and events, and have these recorded into CloudWatch for analysis and identifying trends. A metric represents a time-ordered set of data points that are published to CloudWatch.

Amazon CloudWatch can be used to configure alarms whereby if those metrics breach certain thresholds for a specified period, you can generate an alarm on which action can be taken to remediate.

With Amazon CloudWatch, you can track and collect metrics for your Amazon EC2 instances, Amazon DynamoDB tables, Amazon **Relational Database Service (RDS)** instances, and more. Every AWS service publishes metrics to Amazon CloudWatch. You get basic metrics, which are offered free of charge, and detailed metrics, for which you pay additional charges.

Some typical use cases of Amazon CloudWatch include the following:

- **Infrastructure monitoring and troubleshooting**: Monitor key metrics, logs and visualize trends over time to identify any potential issues and bottlenecks, enabling you to conduct root-cause analysis that helps to resolve both incidents and problems.

- **Proactive resource optimization**: Configure alarms that monitor metric values and are triggered if breaches occur. Define an automatic remediation action, such as configuring auto-scaling, to launch new instances and terminate failed instances. Send out notification alerts to administrators and system operators.

- **Log analytics**: Analyze log information from various sources to help address operational issues, potential security attacks, or application performance issues, and take effective actions to remediate.

Let's look at metrics in more detail.

CloudWatch metrics

CloudWatch metrics represent variables that you can monitor as a time-ordered set of data points. Any service that you consume will report metrics to CloudWatch. For example, you could monitor the CPU utilization of an EC2 instance over a period that will help you track the performance of that instance.

Each data point will consist of a timestamp. If the resource does not provide a timestamp as it publishes the metric to CloudWatch, then CloudWatch will create a timestamp based on the time that the data point was received. The following diagram illustrates the CPU utilization metric of Linux-based EC2 instances over the past three days:

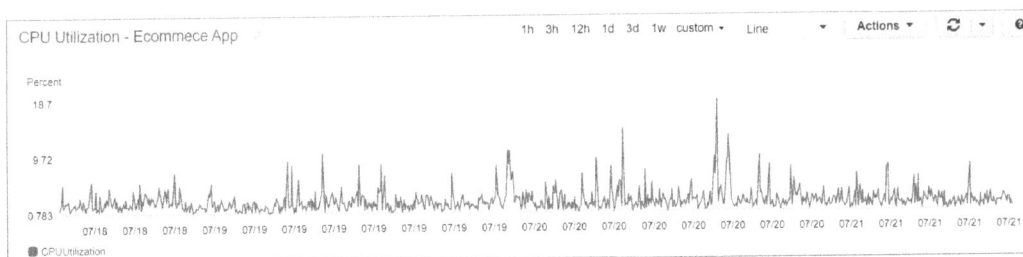

Figure 13.1 – CPU utilization metrics for an EC2 instance running an e-commerce application

Amazon CloudWatch offers a wide range of built-in metrics, but you can also create custom metrics based on your requirements:

- **Built-in metrics**: Amazon CloudWatch allows you to collect default metrics from the vast array of AWS services, such as Amazon EC2, Amazon S3, Amazon RDS, and more, out of the box. An example includes CPU utilization, disk read/write, and data-transfer metrics for your EC2 instance.

- **Custom metrics**: Amazon CloudWatch also allows you to collect custom metrics from your own applications to monitor performance and troubleshoot any bottlenecks. Examples include web application load times, request error rates, and so on. Another example of a custom metric is *operating system memory consumption*. This is because unlike CPU utilization, which monitors the underlying hardware CPU usage patterns, memory metrics are at the OS-level and cannot be monitored by default. To ingest custom metrics, you need to use the CloudWatch Agent or the `PutMetricData` API action to publish them to CloudWatch.

An important point to note is that metrics exist in the Region in which they are created. However, you can configure *cross-account cross-Region dashboards*, which will allow you to gain visibility of CloudWatch metrics, logs, and alarms across related accounts and understand the health and performance of your applications across Regions.

CloudWatch will store your metrics for up to 15 months. Data points older than 15 months are expired, as new data points come in on a rolling basis.

Next, let's look at how you can use AWS CloudWatch dashboards to get centralized visibility on your resources.

Dashboards

You can create one or more dashboards on Amazon CloudWatch that allow you to visualize and monitor your resources and the metrics that are important, all within a single view pane. As well as visualizing your metrics through a variety of charts and graphs, you can also publish your CloudWatch alarms (discussed next) on your dashboard for high visibility on any potential issues.

CloudWatch dashboards can be configured to provide insights on your resources' health across AWS accounts and Regions, using the cross-account functionality. This functionality is integrated with *AWS Organizations*, enabling you to efficiently build your cross-account dashboards. Here is a quick screenshot of a CloudWatch dashboard:

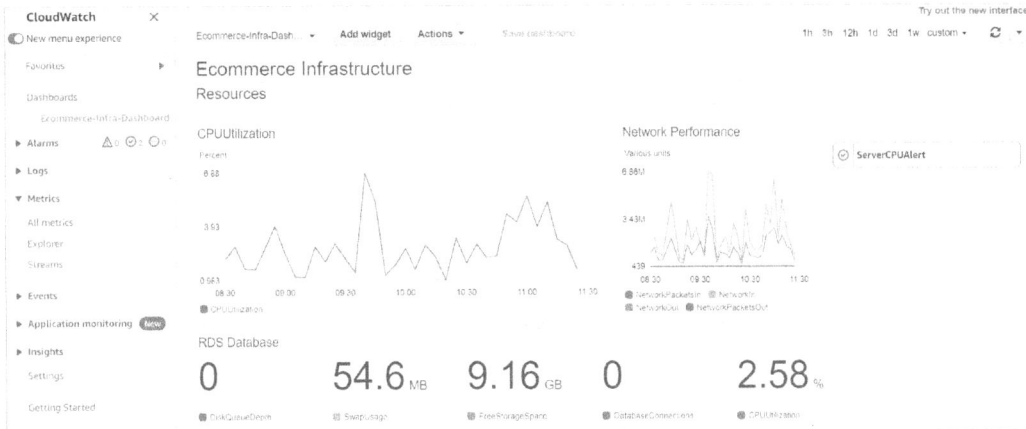

Figure 13.2 – A CloudWatch dashboard

Alarms

You can configure CloudWatch alarms to monitor a given resource metric, for example, the average CPU utilization of an EC2 instance. If the metric crosses a specific threshold for a specified period, then the alarm can be triggered to take a certain action. The alarm only triggers if the threshold has been breached for a specified period, and this is important. You do not want the alarm triggering just because there happens to be a momentary spike. So, for example, you would want the alarm to be triggered if the average CPU utilization on your EC2 instance goes above 80% for 15 minutes.

Alarms can be in one of three states:

- **OK**: Occurs when a metric is within the range defined as acceptable.

- **Alarm**: Occurs when a metric has breached a threshold for a period.

- **Insufficient data**: Occurs when the data needed to make the decision is missing or incomplete. This also generally happens when you first configure an alarm while it waits to receive and analyze data.

Once an alarm has been triggered, an automatic action can be taken to respond to it, which can include the following:

- **Simple Notification Service (SNS) notification**: You can send out automatic alerts to an administrator (**application-to-person** or **A2P**) or push a notification to an application to take some action (**application-to-application**, or **A2A**).

- **Auto Scaling action**: The EC2 Auto Scaling service can be triggered to add or remove an EC2 instance in response to an alarm state. Auto Scaling groups can be configured to launch additional instances if the average load across the servers is above a given threshold for a period of time, for example, if average CPU is above 80% for more than 10 minutes.

- **EC2 action**: You can have an alarm trigger an EC2 action, such as stopping an EC2 instance, terminating it, restarting it, or recovering it. Recovery of an instance simply means that the instances are migrated onto another host, something you would do if there was an underlying issue with the host hardware running the instance. Recovery action is only initiated based on the failure of *system status check* errors.

Finally, you can also publish your alarms in your CloudWatch dashboards, which can give you a quick visual of your alarm statuses.

Next, we look at Amazon CloudWatch Logs.

CloudWatch Logs

Amazon CloudWatch offers a feature to centrally collect and store logs from both AWS and non-AWS sources. These AWS sources could be EC2 instances, CloudTrail logs (discussed later in this chapter), Route 53 DNS queries, and VPC flow logs. You can also ingest logs from non-AWS sources, such as your web applications access logs, error logs, and operating system event logs.

CloudWatch gives you a central view of all your logs regardless of their source – log events generated as part of your CloudWatch logs are essentially a time-ordered series of events that you can then query, analyze, search, and filter for specific patterns or error codes, and so on. You can also visualize your log data using dashboards and ingest these logs into other applications for more complex querying.

In terms of retention, logs are kept indefinitely and never expire. However, you can adjust the retention policy by choosing a retention period between 1 day and 10 years. You can further archive your log files into Amazon S3 using one of the Glacier classes for long-term storage.

Let's look at some of the key components of CloudWatch Logs next:

- **Log events**: This represents some event or activity recorded by the application or resource being monitored. The log event will include the timestamp and the log message.

- **Log streams**: This consists of a sequence of log events that share the same source. CloudWatch Logs groups log events from the same source into a log stream. For example, a log stream may be associated with a web server access log on a specific host.

- **Log groups**: CloudWatch then organizes these log streams into a *log group*. A log group represents log streams that have the same retention, monitoring, and access control settings. For example, you can create a log group to collect and organize all related log streams that relate to web server access logs from multiple hosts in a fleet.

- **Metric filters**: These allow you to extract specific data from ingested events and then have those data points as custom metrics in CloudWatch. You can then use these filtered metrics to generate visualization representations and perform the necessary analysis. For example, you may wish to run a filter to identify how many times users trying to access your application were not able to connect with a `504 Gateway Timeout` error message, indicating some communication or network problem.

In this section, we learned about Amazon CloudWatch Logs, which can be used to ingest log information into CloudWatch from both AWS and non-AWS sources. You can use CloudWatch Logs to analyze access patterns, identify security and technical issues, and assist in triaging bottlenecks.

Next, we will look at Amazon CloudWatch Events, which is a near real-time stream of system events related to your Amazon resources.

Amazon CloudWatch Events

With Amazon CloudWatch Events, you create *rules* that continuously monitor your AWS resources and then respond with an action when a given event occurs. Amazon CloudWatch delivers a near real-time stream of system events, and the rules you define can trigger some action when those events occur.

An example of an event is when an EC2 instance enters the *stop state* because someone performed a shutdown operation on the instance. Another example is when an IAM user logs into the AWS Management Console. Every *write API operation* is an event that has occurred, and you can choose which events to monitor and what action to take, if necessary, when those events occur.

When you define a rule to monitor an event, you also specify an action by selecting an appropriate target. Targets can be a *Lambda* function, an *EC2 instance action*, an *SQS queue* or *SNS topic* to post a message to, an *ECS task*, and more. For example, let's say you want to trigger a Lambda function to process an image as soon as it is uploaded to an S3 bucket. You create an event so that when the upload is complete, the Lambda function is called to process the image in some way, such as creating multiple formats of the image or adding a watermark to it.

You can also configure Amazon CloudWatch Event rules to trigger an action at a given schedule, using *standard rate* and *cron expressions*. This can be particularly useful to perform day-to-day operation tasks. In *Chapter 12*, *Automation and Deployment on AWS*, you completed an exercise that involved automatically starting an EC2 instance at 8 A.M. and then shutting it down at 6 P.M., Monday to Friday.

In this section, we learned about Amazon CloudWatch Events, which allow you to use simple rules to perform some action when a given event takes place for a particular resource. CloudWatch Events can help you respond to operational changes to complete workflows and task, or take any corrective action if required. CloudWatch events can also be used to schedule automated actions that trigger at certain times to help repeatable day-to-day operations.

> **Important Note**
>
> In *Chapter 10*, *Application Integration Services*, you learned about Amazon EventBridge. Amazon CloudWatch Events and EventBridge use the same underlying service and API, and Amazon recommends EventBridge as the preferred way to manage your events as it offers more features.

Next up, you will learn about Amazon CloudTrail, which is a service you use to enforce governance, compliance, and operational and risk auditing of your AWS account.

Meeting compliance requirements with Amazon CloudTrail

AWS CloudTrail is a service that enables you to log every action taken in your AWS account, allowing you to track user activity and API usage. CloudTrail is enabled by default on your AWS account when you create it. It stores event history accessible within the CloudTrail dashboard for every activity that occurs in your AWS account. The following screenshot shows an example of the CloudWatch event history:

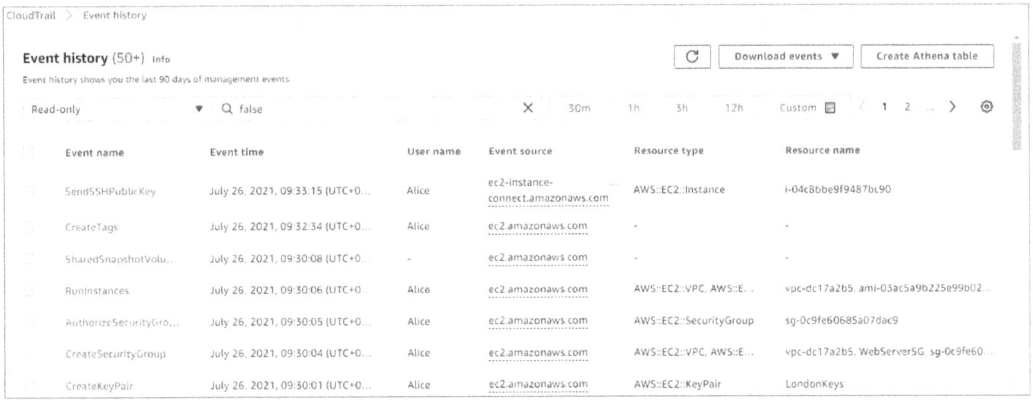

Figure 13.3 – AWS CloudTrail

You can use CloudTrail for enforcing and managing your overall compliance and governance requirements since it can provide you with a time-ordered series of events that have taken place in your account. You can also respond to events as they occur by ingesting them into Amazon CloudWatch and then configuring alarms or event rules accordingly to react to specific events. AWS CloudTrail events provide a history of both API and non-API activity. API activity includes actions such as launching a new EC2 instance or creating a new IAM user. Non-API activity refers to other types of actions, such as logging into the AWS Management Console.

AWS CloudTrail enables you to record three different types of events, as follows:

- **Management events**: Provides information about management operations you carry out in your AWS account. These are also known as *control plane operations* and include *write-only events*, such as the RunInstances API operation to launch an EC2 instance as shown in the preceding screenshot, which was performed by our IAM user, **Alice**. Write-only events refer to operations that may create/modify a resource. Management events can also include *read-only events*, such as the DescribeInstances API operation, which will return a list of EC2 instances but where no changes are made. Finally, management events can also include non-API activity, such as when a user logs into the AWS Management Console.

- **Data events**: Provides information about operations performed on or in a resource within your AWS account, for example, creating an object (such as uploading a file) in an Amazon S3 bucket using the `PutObject` API operation. Data events are also known as *data plane operations* and tend to be high-volume. Other examples of data events include Lambda function executions using the `Invoke` API and the `PutItem` API operation on a DynamoDB table, which results in new items being added to the table. Data events are not logged by default. To record data events, you need to create a trail and explicitly add supported resources or resource types for which you wish to collect events.

- **Insights events**: Provides information on unusual activities in your AWS account. Once enabled, CloudTrail will detect any unusual activity, which is determined if there are any API operations that significantly differ from the account's typical usage patterns. For example, typically your account logs no more than 10 Amazon S3 `deleteBucket` API calls per minute, but all of a sudden, it starts to log an average of 200 `deleteBucket` API calls per minute. Insight events contain related information about the event, such as API, incident time, and statistics, that help you understand and act on unusual activity.

CloudTrail will log all API management events automatically when you open a new AWS account. The event history service will store 90 days' worth of management events that can be viewed and downloaded. The event history will not contain any data events – for this, you will need to create a trail. Let's look at trails next.

Trails

If you wish to configure CloudTrail to store specific management events or data events and require more than 90 days' worth of event history, you can configure a *trail*. A trail is a configuration that enables CloudTrail to record specific types of events and have them delivered to an Amazon S3 bucket, CloudWatch Logs, and CloudWatch Events.

Trails can be configured to record events from a single Region or across Regions:

- **Single Region trail**: CloudTrail will records events that are specific to the Region you specify, and those log files are delivered to an Amazon S3 bucket you specify. Multiple trails can be delivered to the same S3 bucket or separate buckets. Single trails are viewable only in the AWS Regions where the logs are created.

- **All Regions trail**: CloudWatch will record events in each Region and deliver the event log files to the S3 bucket you specify. Note that if Amazon adds a new Region after you have created a trail that applies to *all Regions*, then the new Region is automatically included as well.

By default, an *all-Regions trail* is the default option when you create a trail in the CloudTrail console. Furthermore, CloudTrail will deliver log files from multiple Regions to a single Amazon S3 bucket and a CloudWatch Logs log group.

Most AWS services are Region-specific and so events are recorded in the Region in which the action occurred. However, for global services like **IAM**, events are recorded as occurring in the US East (N. Virginia) Region.

You can also create an *AWS organization trail*, which is a configuration option that allows you to record events in the management account and all member accounts in AWS Organizations. Using an organization trail will ensure that you record important events across all your AWS accounts.

In this section, you learned about AWS CloudTrail, which is a service that enables you to audit your AWS account and enable compliance, monitoring, and governance. AWS CloudTrail is not designed for performance and system health monitoring. Other monitoring services, such as Amazon CloudWatch, are designed to help you address performance or system health-related matters. Often, you are going to be using these tools together as part of the overall management of your AWS services and resources.

In the next section, we will look at the AWS Config service, which allows you to assess, audit, and evaluate the *configuration changes* of your AWS resources, and help in your overall change management process.

Learning about change management with AWS Config

AWS Config is a service that allows you to gain visibility into how your AWS resources are configured and deployed in your AWS account. With AWS Config, you can see how resources are related to each other, how they were configured in the past, and historical changes to those resources over time.

This can be particularly useful when you start running multiple environments, such as development and production, where within those environments are hosted countless resources across the vast array of services on AWS. For example, you would want to be aware of how your VPCs have been configured, what subnets and security groups are attached to them, what routes have been added to your route tables, and so on. AWS Config can help you maintain an accurate database of all this information as well as track changes as they occur in your AWS accounts.

You can use AWS Config to ensure that your resources have been configured in accordance with internal guidelines that fulfill compliance requirements. The service enables you to effectively implement security analysis, change management processes, and troubleshooting exercises.

Let's look at some of the core components of AWS Config.

Configuration items

A **Configuration Item** (**CI**) refers to a *point-in-time* snapshot of various attributes for a given AWS resource in your AWS account. A configuration item will include information such as metadata, attributes, relationships with other resources, and its current configuration. A new configuration item is created whenever a change is detected to a resource that is being recorded. You will be able to see what the configuration was previously and what changed. The configuration item is detailed in a JSON diff file, highlighting the fields which have changed, as shown in the following screenshot:

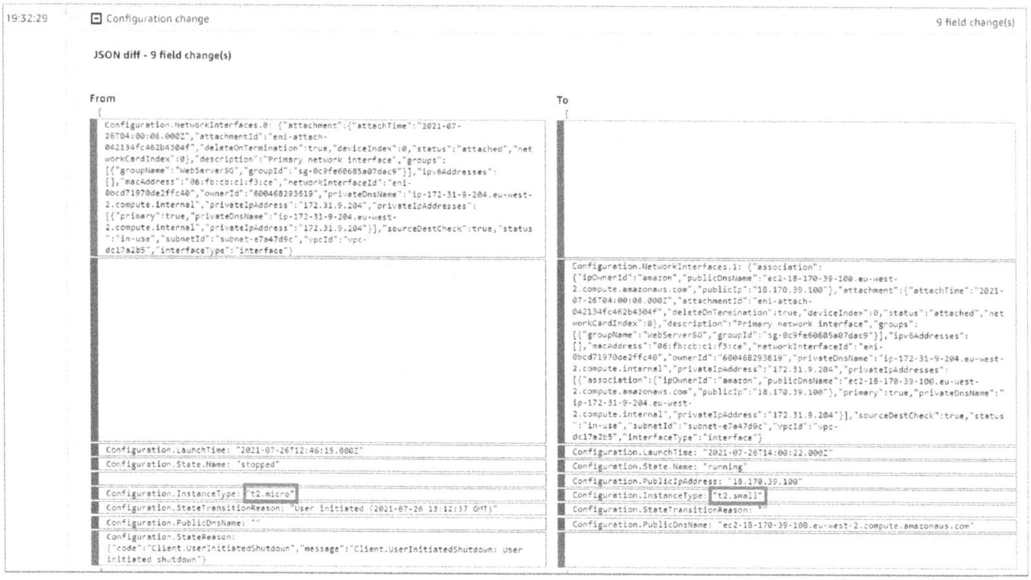

Figure 13.4 – AWS Config JSON showing a configuration change for an EC2 instance

As noted in the preceding screenshot, you can see that AWS Config has noted various changes that were made to our EC2 instance, such as the upgrade of the instance type from t2.micro to t2.small.

Configuration history

This is particularly useful when you want to review changes to your resources over time. The configuration history is a record of configuration items for a given resource over a period. Using the configuration history, you can answer questions such as when a resource was first created, what changes were made over the last week, and what configuration changes were introduced two days ago at 4 P.M. The configuration history file for each resource type is stored in an Amazon S3 bucket that you specify.

Configuration recorder

The configuration recorder must be enabled and started for AWS Config to start recording changes made to your resources and to create CIs. The configuration record can be configured to record all resources or only specific resources in each Region, as shown in the following screenshot:

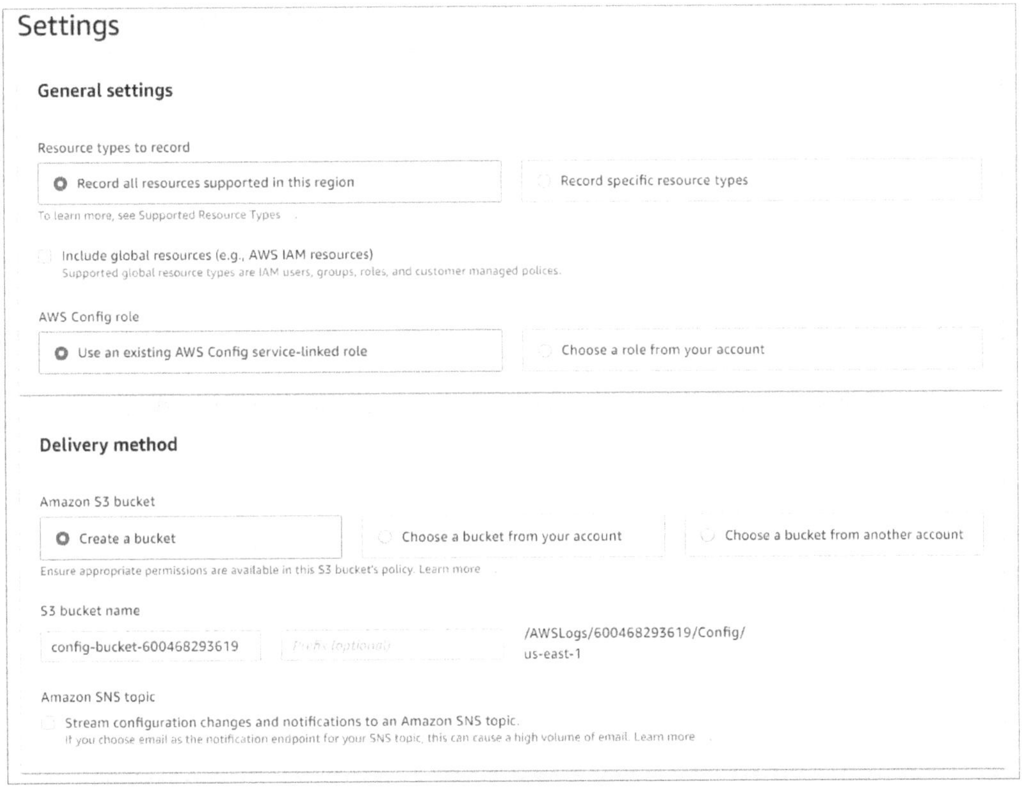

Figure 13.5 – AWS Config settings

You will also note that you can also choose to record global resources, such as IAM resources.

Configuration snapshot

This is a collection of configuration items for your resources represented as a point-in-time image in your AWS environment. The snapshot can be used to validate configurations and identify any that are incorrectly configured. The snapshots can be stored in a predefined Amazon S3 bucket and can be viewed in the AWS Config console.

Configuration stream

As you create, modify, or delete resources in your AWS account, new CIs are created, which are added to the configuration stream. The configuration stream uses an Amazon SNS topic to send out notifications every time a change to your resources occurs. This can be used to alert an administrator, for example, and potentially watch out for any technical or security issues.

In this section, you learned about the AWS Config service, which is a service to help you manage changes to your resources in your AWS account. You can use AWS Config to monitor how your resources relate to each other and how the configurations and relationships change over time.

In the next section, we look at AWS Systems Manager, which enables you to track and resolve operational issues across your AWS accounts, automate day-to-day operational tasks to manage your resources, and enforce security measures for your applications.

Managing your AWS resources with AWS Systems Manager

AWS Systems Manager is a service that enables you to centrally manage your AWS resources. With AWS Systems Manager, you can gain visibility of your resources across AWS services, perform configuration management, and automate day-to-day operational tasks. Previously known as **SSM**, AWS Systems Manager can help you can enforce compliance with desired configuration states and take corrective action on any policy violations where necessary.

AWS Systems Manager uses the concept of documents (written in JSON or YAML), which define the actions that Systems Manager performs on your managed resources. The documents are used by AWS Systems Manager to fulfill its various capabilities, such as operational management, change management, application management, and node management. AWS Systems Manager comes with a vast collection of predefined documents, and you can create your own. For example, the **AWS-CreateRdsSnapshot** document can be used to create an RDS snapshot for an RDS instance, as shown in the following screenshot:

AWS Systems Manager > Documents > AWS-CreateRdsSnapshot

AWS-CreateRdsSnapshot

Delete   Actions ▼   **Execute automation**

Description   Content   Versions   Details

Document version
1 (Default)

Document description

Platform	Created	Owner	Target type
Windows, Linux, MacOS	Thu, 14 Nov 2019 18:15:51 GMT	Amazon	/AWS::RDS::DBInstance

Status
⊘ Active

Creates an RDS Snapshot for an RDS instance. This automation does not support encrypted snapshots.

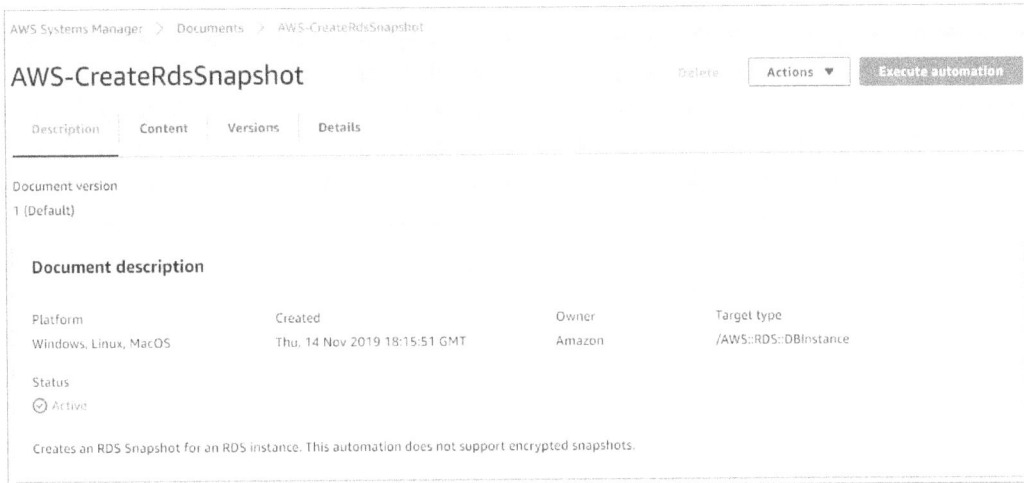

Figure 13.6 – AWS Systems Manager predefined documents

AWS Systems Manager offers a wide range of capabilities, including the following:

- **Run Command**: Part of the node management capability, the Run Command capability enables you to remotely run Linux shell scripts and Windows PowerShell commands on your fleet of EC2 instances. This can be used to perform configuration changes, and install and update applications.

- **State Manager**: This ensures that your managed instances are configured to a predefined state, enabling you to maintain consistency in your configurations across a fleet of instances, such as firewall configuration, antivirus configurations, and more.

- **Inventory**: This enables you to gather software configuration information about your EC2 instances. The Inventory capability will provide information on applications, files, components, and patches across your managed instances.

- **Maintenance Window**: Part of the change management capability, the Maintenance Window service enables you to schedule times when administrative tasks such as installing patches and updates can be performed so as not to disrupt your team during normal business hours.

- **Patch Manager**: This enables you to automate patching of your EC2 instances, which can comprise security and application updates. Note that updates for applications on Windows servers are limited to those released by Microsoft.

- **Automation**: This enables you to automate various maintenance tasks, such as updating AMIs, creating snapshots of **Elastic Block Store** (**EBS**) volumes, resetting passwords, and launching or terminating EC2 instances, among others.

- **Parameter Store**: This offers a means of securely storing configuration data and secret information. For example, in *Chapter 9, High Availability, and Elasticity on AWS*, you will recall configuring the database connection strings for the *good deed of the month* application. The database connection strings consisted of sensitive username and password information, which was saved in plain text in the Amazon S3 bucket repository. Furthermore, the database connections file when copied from the S3 bucket and stored in the HTML directory of your server is a security risk. Sensitive information should always be stored and managed more securely. AWS Systems Manager's Parameter Store enables you to store sensitive information such as passwords and database strings as parameter values. These values can be stored encrypted, and your application can be configured to securely retrieve these values as they are needed from the Parameter Store.

- **Distributor**: This enables you to create and deploy application packages to your managed instances. You can create your software packages as executables that operating systems recognize, enabling easy deployments. Distributor can also reinstall new package versions and perform in-place updates.

- **Session Manager**: In *Chapter 7, AWS Compute Services*, we discussed the importance of bastion hosts, which act as an entry point to administer other EC2 instances you deploy in a VPC, across your public and private subnets. While bastion hosts are intended to be highly secure because of how they are configured or accessed, you still need to manage and maintain the servers, including performing security updates, ensuring performance levels, and so on. AWS Systems Manager offers **Session Manager**, which allows remote access to your EC2 instances using a browser-based shell or the CLI. Session Manager provides secure and auditable instance management without the need to open inbound ports, maintain bastion hosts, or manage SSH keys.

- **Incident Manager**: This is another Systems Manager offering that enables you to manage and resolve incidents affecting AWS-hosted applications. The Incident Manager service offers a management console to track all your incidents and notify responders of impact, identify data that can help with troubleshooting, and help you get services back up and running.

These are just some of the capabilities on offer from AWS Systems Manager. Here is a quick screenshot of the Inventory console, showcasing information from a single EC2 instance that was deployed in my AWS account:

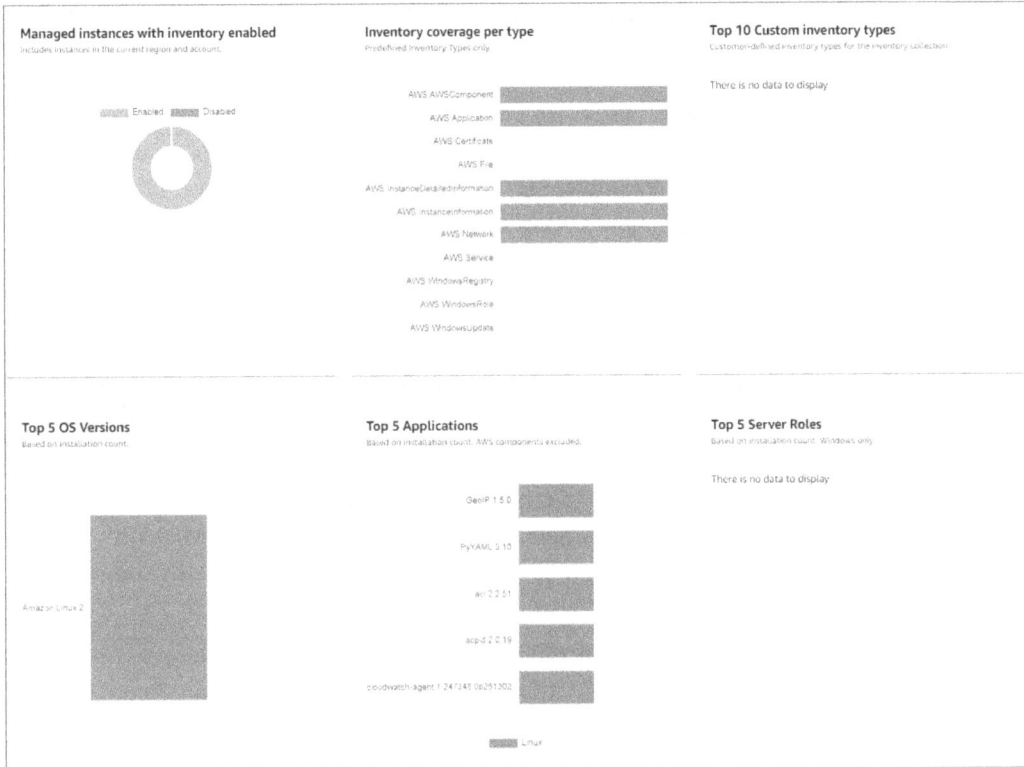

Figure 13.7 – AWS Systems Manager Inventory

So far, we have looked at the AWS Systems Manager service, which offers a suite of capabilities to centrally manage and automate day-to-day operations for your AWS resources.

In the next section, you will learn about the AWS Trusted Advisor service, which enables you to inspect your AWS environment and identify whether your resources have been configured in accordance with AWS best practices.

Learning how to use AWS Trusted Advisor

The AWS Trusted Advisor service analyzes your resources and how they have been configured. The service helps to measure the configuration of your resources against best practices and identify opportunities to save money, improve system availability and performance, or address security concerns.

Specifically, the Trusted Advisor service will report on its analysis against the following core categories:

- **Cost optimization**: Performs checks on your resources to identify which ones are underutilized. AWS Trusted Advisor will then offer recommendations on where you could reduce your costs. For example, Elastic IP addresses are only free if they are attached to a running EC2 instance. AWS charges you an hourly fee for provisioning Elastic IP addresses if they are not being consumed, that is, not attached to any instance, or attached to an instance that is in a stopped state.

- **Performance**: Offers recommendations on where you can improve the responsiveness of your applications. For example, if you are using a gp2 EBS volume type for an EC2 instance that seems to be heavily utilized, it can recommend you an upgrade to an io1 EBS volume, which will improve performance.

- **Security**: Reports on any resources that have not been configured in accordance with security best practices. For example, if you have not configured MFA on the root account, then AWS will highlight this as a potential security risk and recommend that you configure MFA.

- **Fault tolerance**: Identifies options for increasing the resiliency of your AWS solutions. For example, AWS will identify any RDS instance that has not been configured with multi-AZ as a risk factor.

- **Service limits**: Checks your AWS account to identify whether you are approaching any service limits or quotas. For example, when using the AWS Auto Scaling service, you have a default limit of configuring up to 200 launch configurations per Region. Should you start to exceed more than 80% of this limit, you will see an alert in Trusted Advisor.

AWS Trusted Advisor and Support plans

The AWS Trusted Advisor service offers different levels of checks based on the AWS Support plan that you have subscribed to. If you are only on the *Basic* Support plan, then you only have access to six checks in the security category and all checks in the service limits category.

To access the full range of checks across all categories, you must be subscribed to either the *Business* or *Enterprise* Support plans. With either of these plans, you can also use Amazon CloudWatch Events to monitor the status of Trusted Advisor checks.

In this section, we discussed the AWS Trusted Advisor service, which is a reporting tool that enables you to identify whether your resources have been configured in accordance with best practices and whether there are opportunities to save on costs.

In the next section, you will learn about the AWS Well-Architected Framework, which offers a series of recommendations to help you build secure, high-performing, resilient, and efficient application infrastructure.

Understanding the AWS Well-Architected Framework

The AWS Well-Architected Framework consists of a set of design principles and architectural best practices that you can follow when building solutions for the cloud. AWS offers the Well-Architected Tool, which can be used to review the state of your applications and resources, and compares them to the latest AWS architectural best practices.

The Well-Architected Framework comprises the following *five pillars*.

Reliability

Applications deployed in the cloud must be resilient to failures. The resources that your applications depend on (compute, storage, networks, and databases) must be available and reliable. Any technical issues on any of these resources will cause your application to become unreliable and potentially fail.

The reliability pillar also focuses on how quickly you can recover from failure based on your architectural design. This is because failures are bound to happen and your architecture must be able to recover from these failures swiftly. One key concept that you should also consider is the fact that replacing a failed component is often better than trying to figure out why the component failed and attempting to resolve the issue that caused the failure. This is because as you spend time trying to troubleshoot the failure, you risk increasing your **Recovery Time Objective** (**RTO**). For example, with EC2, you can deploy your application across multiple instances and multiple AZs. You can then configure Elastic Load Balancers and the Auto Scaling service to ensure that if an EC2 instance fails, traffic is routed to only those instances that are healthy, while the failed EC2 instance is replaced automatically in the background.

Performance efficiency

When architecting your cloud solutions, you want to offer the best performance while still ensuring that you are optimized for cost. This means you should always try to select the resource types and sizes based on your performance needs, while monitoring your resources consistently to ensure you maintain those levels of performance in accordance with demand. Performance should not suffer if demand increases. At the same time, you should only provide resources as they are required to avoid underutilization of those resources.

You'd often need to incorporate resources that fall across different AWS services, such as compute, storage, and networking, and architecting your solution requires careful planning in addition to configuring each of those resources. For example, if your application is hosted in the London Region, your users in London may experience good performance. However, if you have users in South America, you may experience poor performance due to network latency issues. You can consider using Amazon CloudFront to cache your application content at edge locations closer to your South American users, improving overall application performance.

Security

You should always keep security in mind whenever you are designing your cloud solutions. You want to ensure that your applications are accessed securely by only authorized users. You also want to ensure data integrity, privacy, and sovereignty. Assigning permissions to users must always be based on the principle of least privilege, ensuring that access is granted only where required to fulfill the job function and nothing more.

You should incorporate a backup and disaster recovery strategy for your application solutions, which would also comprise securing the underlying resources that power your application. For example, you must ensure that your databases are backed up or that you regularly create EBS snapshots. Another example is to configure Amazon S3 bucket replication and, if possible, to use cross-regional replication configurations. Finally, you must be able to audit every activity that takes place in your account, and you can use tools such as AWS CloudTrail to maintain an audit log.

Operational excellence

This pillar focuses on achieving operational excellence by making frequent, reversible, and continuous changes to your workloads. Your aim should be to achieve continuous improvements in your processes and procedures. Furthermore, automating operational tasks will strengthen the other pillars, and using **Infrastructure as Code** (**IaC**) with tools such as CloudFormation can help to avoid human error and enable consistency in how you respond to events.

The operational excellence pillar also suggests anticipating failure and to consider performing fail tests and recovery exercises, from which you can learn how to remove potential sources of failure and mitigate risks. Understanding how your applications fail will also help you design automatic recovery solutions, which will offer consistency and rapid recovery.

Cost optimization

This pillar focuses on ensuring that you architect and build solutions in a manner that avoids unnecessary costs. At the same time, you want to be able to ensure that your applications are highly performant, reliable, operationally efficient, and secure. To achieve cost optimization, you should first understand your spending patterns and analyze where the money is going. Using tools such as Cost Explorer and Cost and Usage Reports will help you with this.

Next, you must always try to adopt a consumption model. If you are running development and test servers that only going to be used for 8 hours a day, Monday to Friday, it makes sense to consider procuring those EC2 instances using the On-Demand pricing option. You should then ensure that those servers stopped outside of normal business hours for a potential cost saving of up to 75% (40 hours versus 168 hours a week). Remember that you can automate the startup and shutdown of your EC2 instances using Lambda functions and CloudWatch Events, as discussed in *Chapter 12, Automation and Deployment on AWS*.

Other areas where you can design for cost optimization include using managed service offerings instead of performing heavy-lifting, data center-style operations. For example, it is much more cost-effective to host your databases on Amazon RDS and have AWS perform all the management functions for you than spinning up EC2 instances on which you install your database software. The latter results in more management efforts to ensure your database servers are patched, backed up, and secure.

In this section, you learned about the AWS Well-Architected Framework, which is a set of guiding principles and best-practice recommendations to help you design and run your solutions in the cloud. In the next section, we move on to some exercises for this chapter.

Exercise 13.1 – Reviewing the Trusted Advisor reports in your AWS account

In this exercise, you will log into your AWS account and review the Trusted Advisor service:

1. Log in to your AWS Management Console as the IAM user **Alice**.

2. Navigate to the **Trusted Advisor** console, which is located under the **Management & Governance** category in your list of services.

3. You will be redirected to the Trusted Advisor dashboard.

4. Because you may only have subscribed to the Basic Support plan, you will note that only a few checks are visible. From the main dashboard, note the **Checks Summary** section on the main pane of the dashboard.

5. From the right-hand pane, click on the **Security** category.

6. In the right-hand pane, you will note various checks, such as the one for MFA being enabled on your root account.

7. Expand the check labeled **Security Groups - Specific Ports Unrestricted**. This check analyzes how your security groups have been configured, highlighting specific ports that should provide restricted access. For example, the number 22 port that enables **SSH** remote connections should not be opened to the internet. Ideally, you should restrict access to this port to a specific IP range, such as the IP block of your corporate on-premises network. In our previous exercises, we created inbound rules on this port from the entire internet and, therefore, AWS Trusted Advisor will highlight it as a potential security issue.

    In the following screenshot, you will note that we have three security groups that have been configured to allow inbound traffic on the number 22 port from the internet unrestricted:

Alert Criteria
Green: Access to port 80, 25, 443, or 465 is unrestricted.
Red: Access to port 20, 21, 1433, 1434, 3306, 3389, 4333, 5432, or 5500 is unrestricted.
Yellow: Access to any other port is unrestricted.

Recommended Action
Restrict access to only those IP addresses that require it. To restrict access to a specific IP address, set the suffix to /32 (for example, 192.0.2.10/32). Be sure to delete overly permissive rules after creating rules that are more restrictive.

Additional Resources
Amazon EC2 Security Groups
List of TCP and UDP port numbers (Wikipedia)
Classless Inter-Domain Routing (Wikipedia)

Security Groups - Specific Ports Unrestricted (3)
3 of 6 security group rules allow unrestricted access to a specific port.

Exclude & Refresh   Included items ▼

‹ 1 › ⚙

	Status ▽	Region ▽	Security Group Name ▽	Security Group ID ▽	Protocol ▽	From Port ▽	To Port ▽
☐	⚠	eu-west-2	MyWebServerSG	sg-00bcd2a8ef5831d76	tcp	22	22
☐	⚠	us-east-1	BastionHost-SG	sg-081e3d08d428cc717	tcp	22	22
☐	⚠	us-east-1	launch-wizard-1	sg-0345f63096fa63009	tcp	22	22

Figure 13.8 – AWS Trusted Advisor dashboard

From this exercise, you should've learned how to review the AWS Trusted Advisor dashboard. You were able to identify some security alerts based on your configuration of security groups from earlier exercises that did not adhere to best practices. You can then use this report to identify which security groups need to be amended to increase security.

Summary

In this chapter, we looked at several services on AWS that can help you manage and govern your applications in the cloud. We discussed various monitoring and logging tools, such as Amazon CloudWatch, AWS CloudTrail, and AWS Config, and how you can use these services to ensure performance, reliability, security, and effective change management. You also learned about the AWS Systems Manager service, which offers a suite of capabilities to centrally track and resolve operational issues across your AWS resources. We used AWS Systems Manager to automate day-to-day administrative tasks, ensure compliance of your resource configurations, offer incident and change management services, and improve visibility and control.

The AWS Trusted Advisor service offers a wide range of reports that allow you to cross-reference your resource configurations with various best-practice design principles. Finally, we discussed how you must follow the design principles and recommendations offered by the AWS Well-Architected Framework when building solutions for the cloud.

In the next chapter, we will cover AWS security concepts and look at various security tools you should use when architecting and managing cloud applications.

Questions

1. Which AWS service enables you to track all API activity in your AWS account, regardless of whether the activity was performed using the AWS Management Console of the CLI?

    A. AWS CloudTrail

    B. AWS Config

    C. AWS Trusted Advisor

    D. Application load balancer logs

2. As part of implementing change management, which AWS service can be used to assess, audit, and evaluate change configurations of your AWS resources, enabling you to identify whether a change was the cause of an incident?

    A. AWS Config

    B. AWS CloudTrail

    C. Amazon CloudWatch

    D. AWS Outposts

3. Which AWS service can be used to monitor your company's fleet of EC2 instances, which can be used to identify performance issues related to CPU utilization or memory consumptions?

    A. Amazon CloudWatch

    B. AWS Cloud Monitor

    C. AWS EC2 Monitor

    D. AWS CloudTrail

4. Which AWS service helps you identify potential unused resources, such as Elastic IP addresses, that are not attached to a running instance and thus highlight opportunities to save on costs?

    A. AWS Cost Explorer

    B. AWS Trusted Advisor

    C. AWS Resource Manager

    D. AWS Budgets

5. Which capability of the AWS Systems Manager service enables you to remotely connect to your Linux EC2 instances without having to use bastion hosts in your VPC?

    A. Session Manager

    B. Parameter Store

    C. Run Command

    D. Incident Manager

Section 3: AWS Security

In this section, we look at how to implement security on AWS. We learn about the Shared Responsibility Model and examine AWS security and compliance concepts. We will also look at a number of core AWS security services, such as Amazon WAF, Shield, and Inspector.

This part of the book comprises the following chapters:

- *Chapter 14, Implementing Security on AWS*

14
Implementing Security in AWS

Architecting and implementing security solutions in your cloud journey is going to be of paramount importance if you are to convince businesses to migrate their on-premises workloads to the cloud. There are always going to be some businesses who just feel that managing all things IT within the confines of their data center is most secure. However, this is far from the truth, given that companies couldn't possibly afford to spend the kind of money that cloud providers such as AWS do to offer highly secure environments for their clients to work in.

AWS enables businesses to design and run their applications in the cloud, with stringent security services and controls on offer. Businesses are encouraged to use the vast array of security tools from AWS and follow security guidelines and principles when architecting their applications for the cloud. The onus on designing and implementing security measures falls on both the customer and AWS in what we call the **Shared Responsibility Model**, which we will discuss in this chapter.

AWS offers several security tools and services that can help clients protect data, enforce authentication and authorization protocols, secure network and application access, monitor and detect threats, and enforce compliance and privacy measures.

In this chapter, we will cover the following topics:

- Understanding the Shared Responsibility Model
- Introduction to the AWS compliance programs and AWS Artifact
- AWS vulnerability scanning
- Overview of data encryption services on AWS
- Protecting cloud resources and applications with AWS WAF and AWS Shield
- Assessing and securing your EC2 instances with AWS Inspector
- Other AWS security services

Let's start by understanding the principles behind the Shared Responsibility Model.

Understanding the Shared Responsibility Model

AWS offers public cloud services that allow customers to build isolated environments within the cloud platform. AWS manages and secures all the underlying infrastructure such as network, storage, and compute services, as well as host hypervisor software, among others. AWS also takes care of physical security, which includes access to its data centers, where your workloads are hosted. AWS gives the customer access to those infrastructure components to build their cloud applications and solutions. One customer's approach to designing an application architecture will be different from another and to facilitate different requirements, AWS shares security and compliance responsibilities with the customer.

While the customer can rest assured that all the underlying physical infrastructure, access to data center buildings, and host hypervisor systems are secured and stringently managed by AWS, how their applications are designed, built, and deployed on AWS is the customer's responsibility.

This is known as the **AWS Shared Responsibility Model**. You must understand your responsibilities as a customer to ensure that your applications are highly secure and meet any compliance or regulatory requirements. The Shared Responsibility Model is based on the concept of *security of the cloud*, which is AWS's responsibility, and *security in the cloud*, which is the customer's responsibility. The following diagram illustrates a basic understanding of the demarcation of responsibility between AWS and the customer:

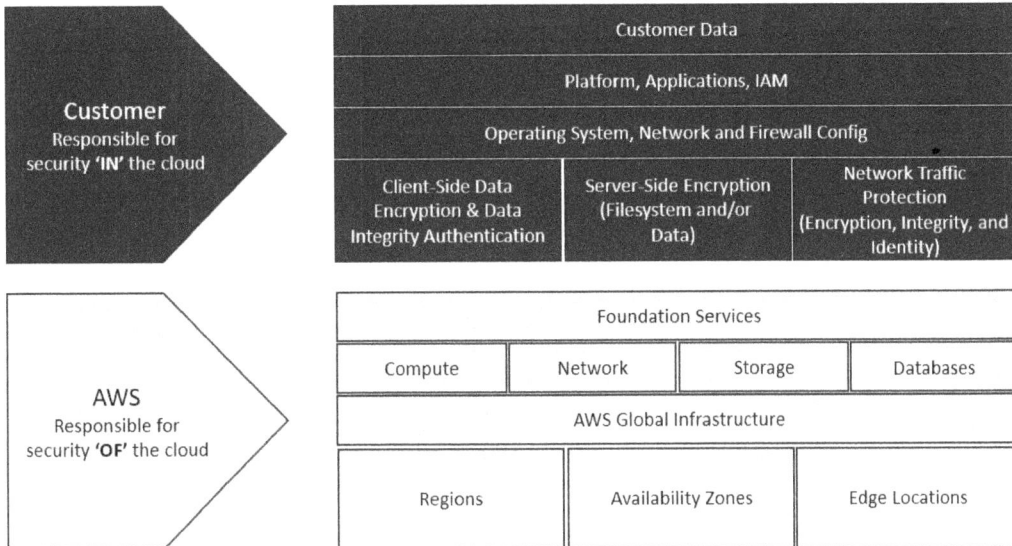

Figure 14.1 – AWS Shared Responsibility Model

Next, let's understand the responsibilities of AWS and the customer in more detail.

Security of the cloud

AWS is responsible for the *cloud* itself. It will enforce strict security protocols and measures for the underlying global infrastructure – its data centers, which are located in Availability Zones within Regions, and its points of presence infrastructure, which are its **edge** locations. AWS is also responsible for the underlying compute, storage, network, and database services, among others. This includes the **hypervisor** software that's used to enable customers to launch their EC2 instances.

Security in the cloud

The customer is responsible for security *in* the cloud. At a simple level, this refers to the fact that the customer is responsible for how their applications are configured, as well as how their data is managed and accessed. The customer's responsibility varies based on the type of service that the customer chooses. Because different services fall within different cloud computing models (IaaS, PaaS, and SaaS), the level of responsibility on the customer will differ. For example, Amazon EC2 falls within the IaaS computing model. Here, the customer is responsible for every aspect of security from the EC2 instance's guest operating system and above. The customer will need to ensure that the EC2 instance is patched, has antivirus software installed, and has properly configured security groups to ensure only the required network traffic is allowed into the instance. Any applications that are installed on the EC2 instance also need to be maintained and secured, which is the customer's responsibility.

This is in contrast to a managed service such as Amazon RDS, where the underlying database instance and attached storage are abstracted from the customer. The maintenance and how the database software is patched is also AWS's responsibility. The customer can only configure specific service-level components, including creating security groups to enable connections to the database on specific ports and from specific sources, enabling multi-AZ, and defining backup retention periods. However, the customer does not need to patch or install antivirus software on the database instance itself or determine where to store the backups.

The following diagram illustrates the level of responsibility that a customer might have in enforcing security measures, depending on the cloud computing model used:

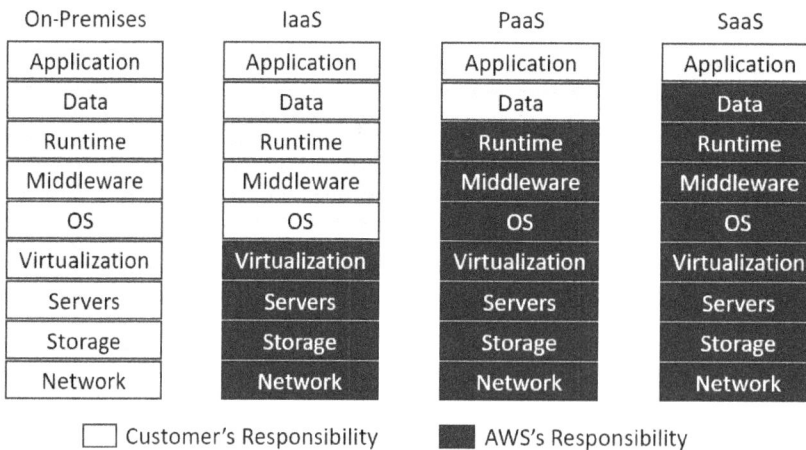

On-Premises	IaaS	PaaS	SaaS
Application	Application	Application	Application
Data	Data	Data	Data
Runtime	Runtime	Runtime	Runtime
Middleware	Middleware	Middleware	Middleware
OS	OS	OS	OS
Virtualization	Virtualization	Virtualization	Virtualization
Servers	Servers	Servers	Servers
Storage	Storage	Storage	Storage
Network	Network	Network	Network

☐ Customer's Responsibility   ■ AWS's Responsibility

Figure 14.2 – The customer's responsibility varies, depending on the type of cloud computing model

Where possible, AWS always recommends considering opting for managed versions of their products and services rather than adopting a data center mindset management approach. For example, if you require a MySQL database solution, it makes sense to opt for Amazon RDS rather than to provision EC2 instances and deploy the MySQL database software on those instances. By hosting your databases on EC2 instances, you become solely responsible for all the management and maintenance tasks associated with the upkeep of the database solution.

Ultimately, you need to take full responsibility for the applications and data you put in the cloud to ensure that your solutions are secure and reliable.

In this section, we discussed the importance of the AWS Shared Responsibility Model and how you can share security responsibility with AWS for the applications and workloads that you deploy in the cloud. In the next section, we will quickly review some key compliance programs that AWS adheres to, which can help you fulfill your regulatory requirements.

Introduction to the AWS compliance programs and AWS Artifact

Depending on the nature of your business and the applications you plan to host on AWS, you need to ensure that you meet any compliance or regulatory requirements. For example, if you plan to process and store credit card information, you must ensure that your application meets the **Payment Card Industry Data Security Standard (PCI DSS)**. This allows you to store, process, or transmit **cardholder data (CHD)** or **sensitive authentication data (SAD)**.

Similarly, if you are in the health care industry and based in the US, you are subject to the US **Health Insurance Portability and Accountability Act (HIPAA)** of 1996. This means that in addition to various business processes, any applications that are hosted on AWS must also be aligned with HIPAA compliance with regards to how **personal health information (PHI)** is processed and maintained.

In this example, AWS needs to ensure that its services are aligned with HIPAA compliance if its healthcare customers are to consume those services for their applications. Let's take the example of Amazon Athena, which, as discussed in *Chapter 11*, *Analytics on AWS*, is an interactive query service that enables you to analyze data held directly in Amazon S3 using standard SQL. AWS confirms that Amazon Athena can be used to process data containing PHI. Features such as encrypting data in transit between Amazon Athena and Amazon S3 are provided by default. Furthermore, Amazon Athena uses AWS CloudTrail to log all API calls so that an audit trail is maintained. The customer is required to ensure that they meet HIPAA compliance by enabling encryption at rest while the data resides in Amazon S3. The customer can also use server-side encryption to encrypt query results from and within Amazon Athena. Again, this feature is offered by AWS for the customer to configure.

Ultimately, compliance is a shared responsibility between AWS and the customer. As another example, AWS manages and controls the components of the host operating system and virtualization layer down to the physical infrastructure for Amazon EC2. The customer is responsible for the guest operating system (which will include patching, performing security updates, and more), as well as any applications that have been installed and their security group configurations.

Next, let's learn where customers can obtain information that confirms that AWS meets various compliance requirements.

About AWS Artifact

Customers have access to various compliance reports to confirm whether the services offered by AWS meet their specific and regulatory requirements. These reports are available via a portal on AWS known as **AWS Artifact**. These reports include AWS **System and Organization Controls** (**SOC**) reports, **Payment Card Industry** (**PCI**) reports, and certifications from accreditation bodies across different Regions.

Customers can download these reports and submit security and compliance documents to their auditors. These can be used to demonstrate how various AWS services and their underlying infrastructure fulfill the security and compliance requirements that the customer is consuming.

These **artifacts** can also be used by the customers to evaluate their application architecture and cloud configurations and assess the effectiveness of their business's internal controls.

Finally, customers can review, accept, and track the status of various AWS agreements such as the Business Associate Addendum, where the customer is subject to HIPAA guidelines to ensure that PHI information is properly safeguarded. If you have multiple AWS accounts, you can accept agreements on behalf of those accounts and manage them with your AWS organizations.

In this section, we discussed AWS compliant and the Artifact service, which allows you to gain confirmation from AWS on how they meet compliance requirements. AWS Artifact is your single source of truth and gives you access to various compliance documents such as AWS ISO certifications, PCI and SOC reports, and more.

In the next section, we will discuss how AWS allows you to proactively test your cloud deployments for vulnerabilities and protects your applications in the cloud.

AWS vulnerability scanning

As an AWS customer, you are going to be consuming various services to build and deploy your cloud applications. You want to make sure that your cloud solutions are highly secure and protected. To that end, you will follow key guidelines and industry best practices while implementing security controls and procedures at the different levels of your cloud ecosystem.

But how you can confirm that the level of protection you have implemented is sufficient and whether the controls you have put in place work?

AWS allows its customers to conduct **penetration testing** on their workloads in the AWS cloud. Also known as *pen testing*, this is a simulated cyber-attack against your computer systems to check for vulnerabilities. This is usually conducted by your internal or appointed security team.

As a customer, you need to follow the service policy for penetration testing, which includes permitted services and prohibited activities. For example, you are prohibited from using any tools to perform **denial of service** (**DoS**) attacks or simulations, even if it is against your AWS assets. **Distributed denial of service** (**DDoS**) simulation testing can only be carried out by an **AWS Partner Network** (**APN**) partner that has been pre-approved by AWS to conduct DDoS simulation tests.

The following table gives you a quick view of some of the permitted services you can run tests against, as well as their prohibited activities:

Permitted Services	Prohibited Activities
Amazon EC2 instances, NAT gateways, and elastic load balancers	DNS zone walking via Amazon Route 53 hosted zones
Amazon RDS	DoS, DDoS, simulated DoS, and simulated DDoS
Amazon CloudFront	Port flooding
Amazon Aurora	Protocol flooding
Amazon API Gateway	Request flooding
AWS Lambda and Lambda@Edge functions	
Amazon Lightsail resources	
Amazon Elastic Beanstalk environments	

Table 14.1 – AWS permitted services and prohibitive activities for penetration testing

You can also perform **network stress testing** and other simulated events where you will need to request authorization from AWS by filling in a simulated events form.

In this section, we provided a quick overview of how customers can perform penetration testing against their workloads on AWS. In the next section, we will provide a quick overview of encryption services on AWS.

Overview of data encryption services on AWS

Encrypting your data is a critical step in ensuring that you protect its integrity and avoid data being readable by unauthorized parties. AWS enables you to encrypt your data both in transit (while it is being transmitted from a source to a destination) and at rest (while it resides on a disk).

To protect data in transit, you must transmit the data using **Secure Socket Layer/ Transport Layer Security** (**SSL/TLS**) or some form of client-side encryption. SSL/TLS requires you to make use of certificates, which are used to encrypt and decrypt the data.

To protect data at rest, you must create and use encryption keys to encrypt and decrypt your data. Encryption keys are data files containing a long series of numbers or letters that is used by a cryptographic algorithm to encode and decode data. Examples of algorithms you may have heard of include **Triple DES** or **Advanced Encryption Standard** (**AES-256**) bit encryption.

An encryption key can either be **symmetric** (where you use the same key to encrypt and decrypt the data) or **asymmetric**, which is a public/private key combination. Here, you use the public key to encrypt the data, and only users or applications that have access to the private key can decrypt the data.

AWS takes away a lot of the overhead of creating and managing encryption keys with its **Key Management Service** (**KMS**). You can use KMS to add encryption services for your data that's held in various AWS services, including Amazon EBS volumes, Amazon RDS, Amazon S3, and others.

With AWS KMS, you create and manage **customer master keys** (**CMKs**), which are stored within the KMS service and used to encrypt and decrypt your data. AWS KMS will create the *material* that's used to generate these CMKs. A key material is a secret string of bits that's used in a cryptographic algorithm. With KMS, this material cannot be extracted, exported, or viewed in any way. You can also configure multi-Region CMKs, which let you encrypt data in one AWS Region and decrypt it in a different AWS Region.

There are two types of CMKs, as follows:

- **Customer-managed CMKs**: These are CMKs that you create, own, and manage. You can rotate their cryptographic material, add tags, specify key policies and create aliases.

- **AWS-managed CMKs**: These are CMKs that are created, managed, and used on your behalf by an AWS service that is integrated with AWS KMS. You cannot manage these CMKs, rotate them, or change their key policies.

AWS KMS also comes with additional features, including a built-in auditing feature since it integrates with Amazon CloudTrail. Using AWS KMS, you can record and track all API requests, including key management actions and the usage of your keys.

With Amazon S3, additional encryption options are offered. We will look at this next.

Amazon S3 encryption

To protect data at rest in Amazon S3, you have the following types of encryptions:

- **Server-side encryption**: Amazon S3 will encrypt your data before saving it to disk and decrypt it when you attempt to access your data.

- **Client-side encryption**: You encrypt the data on the client side before uploading it to Amazon S3. With client-side encryption, you manage the encryption process, keys, and tools.

Concerning server-side encryption, Amazon S3 will encrypt your data at the object level as it writes it to disks, and then decrypt it when you access those objects. This process is transparent to the user as AWS does all the heavy lifting for you. There are three options available with server-side encryption:

- **Server-Side Encryption with Amazon S3-Managed Keys** (**SSE-S3**): Amazon S3 encrypts each object with a unique key and then encrypts the key with a *master key*. It uses AES-256 to encrypt your data. Key rotation is managed by AWS as a backend process.

- **Server-Side Encryption with CMKs Stored in AWS KMS** (**SSE-KMS**): This is similar to SSE-S3 but with additional features such as separate permission requirements for using the CMK and the ability to provision an audit trail that shows when your CMK was used and by whom. As a customer, you can create and manage these CMKs or use AWS-managed CMKs that are unique to your account, service, and Region.

- **Server-Side Encryption with Customer-Provided Keys** (**SSE-C**): You manage the encryption keys and AWS will manage the process of encrypting and decrypting objects using your keys.

As well as protecting data at rest, you can protect your data in transit using SSL/TLS, as we discussed previously.

AWS KMS is backed by the **Federation Information Processing Standard** (**FIPS**) 140-2 **hardware security modules** (**HSMs**), which KMS manages. FIPS 140-2 is a security accreditation program for validating cryptographic modules. AWS KMS uses HSMs, which are multi-tenanted. If you must host your own isolated HSMs for additional regulatory requirements, then you should consider using **AWS CloudHSM**, which we will discuss briefly next.

AWS CloudHSM

AWS CloudHSM is a dedicated **hardware security module** (**HSM**) that allows you to generate and manage encryption keys in the cloud. You are provided with dedicated FIPS 140-2 Level 3-validated HSM devices. These are placed in your VPC and are fully managed for you by AWS. This is particularly useful if you must manage your own cryptographic keys. You are responsible for generating, storing, importing, exporting, and managing your cryptographic keys as AWS does not have access to your keys. You can use CloudHSM for both symmetric and asymmetric key pairs.

AWS CloudHSM also lets you integrate with your applications using industry-standard APIs such as the PKCS#11, **Java Cryptography Extension (JCE)**, and **Microsoft CryptoNG (CNG)** libraries.

In this section, you were provided with an overview of the encryption services in AWS. We discussed AWS KMS, which allows you to also create and manage your CMKs to encrypt and decrypt data. AWS KMS also offers auditing capabilities, which allow you to enforce compliance requirements.

We also looked at the different encryption options available for Amazon S3. To implement data at rest with Amazon S3, you need to decide on either **server-side encryption** or **client-side encryption**. With **server-side encryption**, you can either implement **SS3-S3**, which uses Amazon S3-managed keys, SS3-KMS, which enables you to create CMKs and offers an audit trail of when the keys are used and by whom, or SS3-C, which is where you manage the encryption keys as a customer.

Finally, you learned about the AWS CloudHSM service, which offers dedicated HSMs to enable you to generate and manage keys that are isolated from other AWS customers. CloudHSM devices are FIPS 140-2 Level 3 compliant and useful if you need to fulfill extra regulatory requirements.

In the next section, we will look at a couple of AWS security firewall solutions, namely **AWS Web Application Firewall (AWS WAF)** and AWS Shield.

Protecting cloud resources and applications with AWS WAF and AWS Shield

AWS offers various security tools and services to help protect your cloud workloads from attacks. We will look at two of these services in this section – AWS WAF and AWS Shield.

Protecting applications with AWS WAF

AWS WAF is a web application firewall designed to protect any applications that are made available via Amazon CloudFront, the Amazon API Gateway REST API, application load balancers, or the AWS AppSync GraphQL API.

AWS WAF can help protect applications at layer 7 of the **Open Systems Interconnection (OSI)** model, which helps you monitor and protect traffic over HTTP and HTTPS. This allows you to protect your content from common web exploits, such as **SQL injection** and **cross-site scripting**.

You use AWS WAF to control access to your content by specifying **web access control lists (web ACLs)**. You define rules that specify an inspection criterion with an action to take if a web request meets the criteria.

When using AWS WAF, you are charged based on the web ACLs and rule groups you create and for the number of HTTP(S) requests that AWS WAF inspects.

Next, we will look at the AWS Shield service.

Protecting network attacks with AWS Shield

AWS Shield is a fully managed service that offers protection against **DDoS** attacks. This is a type of attack where the perpetrator attempts to overwhelm your network and servers using a flood of internet traffic, preventing authorized users from accessing your service.

AWS offers two tiers of the Shield service, as follows:

- **AWS Shield Standard**: This is available at no additional cost for AWS customers and protects against common network and transport layer DDoS attacks on your websites and applications. AWS Shield Standard can be used with Amazon CloudFront and Route 53 to provide layered protection against all known infrastructure attacks.

- **AWS Shield Advanced**: This provides additional protection against attacks on your EC2 instances, elastic load balancers, CloudFront, Global Accelerator, and Route 53 resources. The service also offers detection and real-time visibility into attacks, allowing you to protect against large and sophisticated DDoS attacks. In addition, AWS Shield Advanced gives you access to the AWS **Shield Response Team** (**SRT**) 24/7 to assist you in handling such attacks. AWS Shield is a chargeable service at the rate of $3,000 per **AWS organization** and on a per-GB fee on data transfer out, depending on the service being used, such as elastic load balancers or CloudFront, Route 53, Global Accelerator, and elastic IPs.

> **Important Note**
> The cost of AWS Shield Advanced includes the charges of AWS WAF and some other services.

In this section, we discussed two key firewall tools offered by AWS. AWS WAF protects your applications from common web exploits at layer 7 of the OSI model, whereas AWS Shield offers protection against DDoS attacks.

In the next section, we will look at the AWS Inspector service, which can be used to protect your EC2 instances.

Assessing and securing your EC2 instances with AWS Inspector

AWS Inspector is a security tool that's used to assess the network accessibility and application security state of your EC2 instances. You can use AWS Inspector to detect open network ports, any root logins being enabled, and vulnerable software versions on your EC2 instances. You can then take appropriate action, as per the best practices.

The AWS Inspector service uses *rules packages*, which are a collection of security checks that are used in your *assessment runs*. There are two types of rules packages, as follows:

- **Network reachability rules package**: Designed to check for access to your EC2 instances from the internet and identify open ports.

- **Host assessment rules packages**: Designed to check for software vulnerabilities and insecure configurations, including **Common Vulnerabilities and Exposures (CVE)**, **Center for Internet Security (CIS)**, as well as other security best practices.

If you need to perform a host assessment to get a detailed analysis of your host, application configurations, and filesystem access, you will need to install the **AWS Inspector agent**. The AWS Inspector agent will collect telemetry data formatted as JSON documents from its assessment of your EC2 instances and store it in a managed S3 bucket for up to 30 days.

These findings can then be viewed in the AWS Inspector dashboard, as per the following screenshot:

Figure 14.3 – AWS Inspector dashboard

You can then drill down into these findings and see specific recommendations for your EC2 instances. The following screenshot shows findings stating that automatic Windows updates should be enabled:

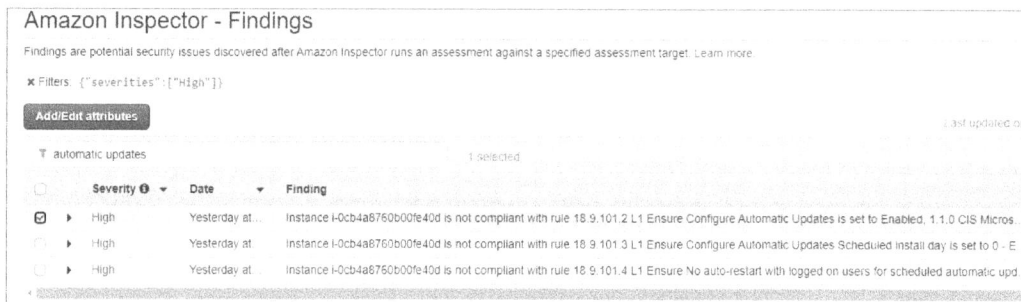

Figure 14.4 – AWS Inspector – Findings

You can install the Amazon Inspector agent using the **Systems Manager Run Command** on multiple instances, manually install it on each EC2 instance, or configure an AMI that's prebaked with the agent installed.

> **Important Note**
> If you are just using AWS Inspector to check for network ports that can be reached from outside the VPC, you do not need to install the AWS Inspector agent.

In this section, you learned about the AWS Inspector service, which allows you to perform an automatic security assessment of your EC2 instances to identify exposure, vulnerabilities, and deviations from best practices.

In the next section, we will discuss additional security services offered by AWS.

Other AWS security services

In this section, we will examine a few additional security tools and services on AWS. We will start by looking at Amazon Macie, which recognizes sensitive data such as **personally identifiable information (PII)**.

Amazon Macie

Amazon Macie uses machine learning and pattern matching techniques to detect and alert you to any sensitive data, such as PII, stored in Amazon S3. You can also use Macie to send alerts on S3 buckets that are unencrypted, publicly accessible, and shared with other AWS accounts outside of your AWS organizations.

AWS Macie will monitor how your data is accessed in Amazon S3, identify any anomalies, and generate alerts if it detects unauthorized access. AWS Macie also provides you with a dashboard that provides a summary of all its findings, as per the following screenshot:

Figure 14.5 – Amazon Macie – Summary

Next, we will look at AWS GuardDuty, which is designed to detect malicious activity and unauthorized behavior against your AWS accounts.

AWS GuardDuty

AWS GuardDuty is a threat detection service that can analyze and detect malicious activity against your AWS accounts and application workloads. The service uses various intelligence feeds to be aware of malicious IP addresses and domains. It also uses machine learning to help detect anomalies by analyzing data from your CloudTrail event logs, Amazon VPC Flow Logs, and DNS logs.

AWS GuardDuty can detect the use of exposed credentials, any communication with malicious IP addresses and domains, as well as irregular activities carried out in your AWS account. This includes EC2 instances being launched in Regions you do not normally work in.

Some of the threats that AWS GuardDuty can detect include EC2 instance compromises, such as those associated with cryptocurrency mining, or S3 bucket compromises, such as unusual S3 API activity from a remote host or unauthorized access from known malicious IP addresses.

Pricing for Amazon GuardDuty is based on the number of CloudTrail events and the volume of VPC Flow Logs and DNS log data to be analyzed. There is a free 30-day trial available for the service.

Next, we will look at another service, Amazon Detective, that can help identify the root cause of potential security issues or suspicious activities.

Amazon Detective

Several different AWS services can be used to identify potential security issues, such as the previously discussed Amazon Macie and Amazon GuardDuty. However, if you are trying to determine the root cause of those security findings, you would generally need to process vast amounts of logs and use an **Extract, Transform, and Load** (**ETL**) solution to perform the necessary deep-dive analysis required.

Amazon Detective can extract time-based events such as logins, network traffic from AWS CloudTrail and Amazon VPC Flow Logs, as well as ingesting your GuardDuty findings. Amazon Detective then helps by producing visualizations using the information that's been ingested, which can help you identify resource behaviors over time and interactions between those resources, ultimately assisting in identifying the root cause of your security issues.

This pricing is based on the volume of data that's ingested from AWS CloudTrail logs, Amazon VPC Flow Logs, and Amazon GuardDuty findings. Amazon Detective is available as a 30-day free trial.

Next, we will look at the **AWS Certificate Manager** (**ACM**) service, which enables you to create, store, and renew SSL/TLS certificates.

AWS Certificate Manager

To protect your AWS resources with secure encryption and verify the identity of websites over the internet, you need to provision and manage **SSL/TSL** certificates. Certificates are issued by a **certificate authority** (**CA**), which is a trustworthy third party that will authenticate both ends of the transaction.

ACM enables you to create, manage, and deploy your SSL/TLS certificates for use with AWS resources and is integrated with several AWS services that need to use SSL/TLS connectivity for secure communications. You can use ACM to issue and manage certificates for your elastic load balancers, Amazon CloudFront distributions, and APIs on an API gateway. AWS ACM can also manage all your certificate renewals.

For ACM-integrated services, there is no charge for provisioning certificates for your elastic load balancers and API gateways. If you need private certificates, **ACM Private CA** lets you pay monthly for the services and certificates you create.

Next, we will look at the AWS Secrets Manager service, which protects the secrets you need to access applications, databases, and other AWS services.

AWS Secrets Manager

In *Chapter 13*, *Management and Governance on AWS*, we looked at the **AWS Systems Manager** (**ASM**) Parameter Store service. This service can be used to securely store parameter values such as your database username and passwords, rather than you having to hardcode this information in your application. Storing such sensitive information in your application code is not a best practice and carries an administrative overhead. With the SSM Parameter Store, your applications can retrieve the necessary parameters to make calls to the database.

However, the service does have some limitations, and this is where AWS Secrets Manager comes to the rescue. In addition to allowing you to store *secrets information*, the service also makes it easy to manage and rotate those database credentials. AWS Secrets Manager offers native support for rotating credentials for databases hosted on Amazon RDS, Amazon DocumentDB, and clusters hosted on Amazon Redshift. This rotation can be automated via a schedule, which adds another layer of security. Furthermore, you can use AWS Secrets Manager to rotate other secrets by modifying sample Lambda functions. This can be used, for example, to rotate OAuth refresh tokens, which are used to authorize applications or passwords used for MySQL databases that are hosted on-premises. Your applications then retrieve secrets by replacing hardcoded secrets with a call to Secrets Manager APIs.

AWS Secrets Manager also encrypts your *secrets* using encryption keys, which you create and manage with Amazon KMS. The service also enables you to replicate your secrets across AWS Regions for multi-Region application support and disaster recovery purposes.

In terms of costs, you are charged based on the number of secrets that are managed in Secrets Manager and the number of Secrets Manager API calls that are made.

Next, we will look at another identity service that enables you to use external public identity providers to grant access to your AWS resources, namely Amazon Cognito.

Amazon Cognito

To authenticate users and grant them access to your AWS resources, you need to create and manage their identities and offer a mechanism for authentication and authorization. AWS IAM is an identity and access management solution you are already familiar with but is designed for users who typically work with your AWS services to build solutions, such as internal staff members. If you are building applications that need to authenticate against backend resources, you could use IAM user accounts, but there are several limitations, including the maximum 5,000 IAM user accounts that you can create per AWS account.

Amazon Cognito lets you set up identity and access control solutions for your web and mobile applications using standards such as OAuth 2.0, SAML 2.0, and OpenID Connect. With Amazon Cognito, you can create **user pools** and **identity pools**:

- **Amazon Cognito user pools**: Used for verifying your users' identities, you can create user pools to host millions of users or federate their access via a third-party **identity provider** (**IdP**), such as Apple, Facebook, Google, or Amazon, and Microsoft Active Directory. In the case of the latter, the user pool will manage tokens issued by your third-party IdP. You use user pools to set up the *sign-up* and *sign-in* features for your web pages or app. You can also use user pools to track the user devices, locations, and IP addresses of your users.

- **Amazon Cognito identity pools**: These allow you to define authorization by creating identities for users who need access to specific AWS services, such as an Amazon S3 bucket or a DynamoDB table. You can also use identity pools to generate temporary AWS credentials for unauthenticated users.

Next, we will look at AWS Directory Service, which is particularly useful if you wish to use the Microsoft Active Directory IAM service.

AWS Directory Service

AWS Directory Service for Microsoft Active Directory (also known as **AWS Managed Microsoft AD**) is a fully managed Active Directory service on the AWS platform. The service enables you to use standard tools such as Active Directory Users and Computers and take advantage of features such as trusts, **Distributed File System** (**DFS**), Group Policy, and **single sign-on** (**SSO**).

You can deploy Microsoft Active Directory-aware applications on your EC2 instance, as well as joining your Windows EC2 instances to the Active Directory domain. You can also join and manage **Amazon RDS for SQL Server** instances to a domain that provides seamless authentication services to your SQL databases running on Amazon RDS.

In addition, you can join **Amazon WorkSpaces** to the domain. Amazon WorkSpaces is an end user computing service that enables you to deploy virtual Linux and Windows desktops in the cloud. AWS manages the virtual desktops, including security patching and managing the operating system. With Amazon WorkSpaces, you can consider migrating away from your on-premises desktop infrastructure to a **virtual desktop infrastructure (VDI)** solution. This enables you to manage your **total cost of ownership** (**TCO**) and opt for an OPEX mode of investment in IT.

Another service that integrates with AWS Managed Microsoft AD is **Amazon WorkDocs**. Amazon WorkDocs is a content creation, storage, and collaboration service. Amazon WorkDocs enables you to store Microsoft Word, PowerPoint, Excel, and various other document types. You and your colleagues can use Amazon WorkDocs for collaboration and to share work documents. Like other collaboration tools such as Microsoft SharePoint, the service also enables you to access APIs so that you can develop content-rich applications.

In this section, we looked at several AWS security tools and services to help you design and architect security measures that will help protect your applications and workloads on AWS. You must examine the vast array of security services on offer and align their usage with your specific business objectives, all while considering any compliance or regulatory requirements.

Next, we will conduct an exercise to demonstrate one of the security tools we covered in this section. Specifically, you will learn how to use Amazon Macie for a simple real-world example.

Exercise 14.1 – preventing data leaks with Amazon Macie

In this exercise, you will use Amazon Macie to monitor a single Amazon S3 bucket and identify whether any PII was stored in the bucket. Imagine a scenario where a user in your organization has uploaded a sensitive file to the wrong Amazon S3 bucket. In our example, we have a *product details* bucket, which would contain product information that can be accessed by the marketing team. However, because of poorly configured access policies, a member of the HR team has uploaded sensitive employee information into this bucket.

This could result in data leaks. While you want to ensure that users are restricted to which buckets they can access, sometimes, accidents do happen. Amazon Macie can detect content that's uploaded to S3 buckets and identify specific types of sensitive data. You can then take the appropriate action.

Step 1 – creating a new Amazon S3 bucket

1. Navigate to Amazon S3 and click on the **Buckets** link from the left-hand menu.

2. In the right-hand pane, click the **Create bucket** button.

3. Provide a bucket name and select the **us-east-1** Region to create your bucket in. I have named my bucket **justdessertsproducts**, so you will need to select a different name that is unique to your bucket. We are going to assume that someone has accidentally uploaded sensitive PII to this bucket.

4. Click the **Create bucket** button at the bottom of the page.

5. Next, create a simple CSV file using Excel or Google Sheets and upload it to the S3 bucket. Ensure that the CSV file contains some data that you can classify as PII. For example, in the following screenshot, you will note that I have added some dummy data that contains sensitive employee information. In this example, **EmployeeID** can be used to identify a specific employee and their salary information:

	A	B	C	D
1	EmployeeID	Name	Department	Salary
2	JD-8976	Richard	IT	$110,000.00
3	JD-9899	Mark	Sales	$120,000.00
4	JD-1212	Fernando	Sales	$120,000.00
5	JD-2343	Chekov	Catering	$90,000.00
6				

Figure 14.6 – Salary information of employees

When using Amazon Macie to identify information that may be PII-related, you would need to be able to identify the format of the data so that Amazon Macie knows what to look for. Note that the format of the EmployeeID data, as per the preceding screenshot, is *two characters (capital letters)*, followed by a *hyphen (-)* and then *four numbers*.

Step 2 – configuring Amazon Macie to identify sensitive employee data

1. From the AWS Management Console, search for Amazon Macie in the top search bar and click on the service from the filtered search results.

2. If this is the first time you have navigated to the Amazon Macie console, you will see a splash screen, as shown in the following screenshot:

Security, Identity, & Compliance
Amazon Macie
Discover and protect your sensitive data at scale
Amazon Macie is a data security and data privacy service that uses machine learning to help you identify and protect your sensitive data in AWS.
Get started with Macie
Automatically discover sensitive data across all of your organization's S3 buckets.
Review detailed findings to take remediation action.
Get started

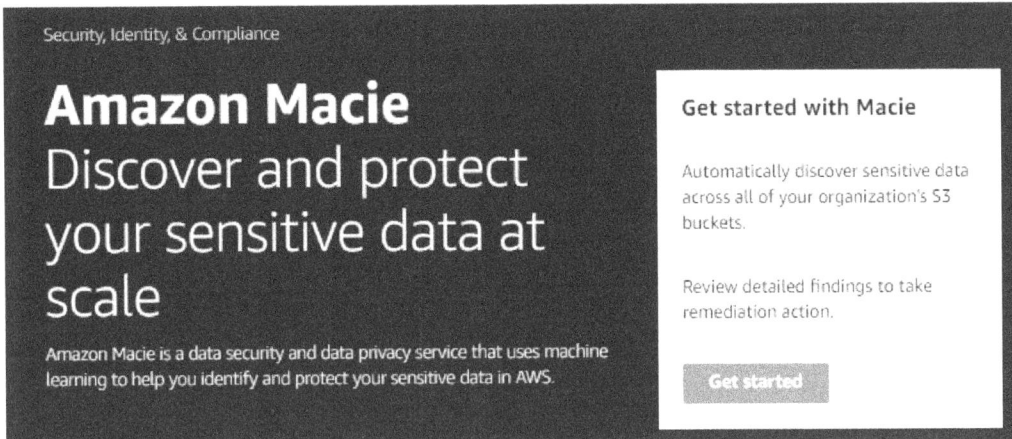

Figure 14.7 – Amazon Macie

3. Click the **Get started** button.
4. Amazon Macie requires permission to access information about the data that you store in Amazon S3. You will be provided with the option to enable Amazon Macie with the necessary access, which will create an IAM role.
5. Go ahead and click the **Enable Macie** button, as per the following screenshot:

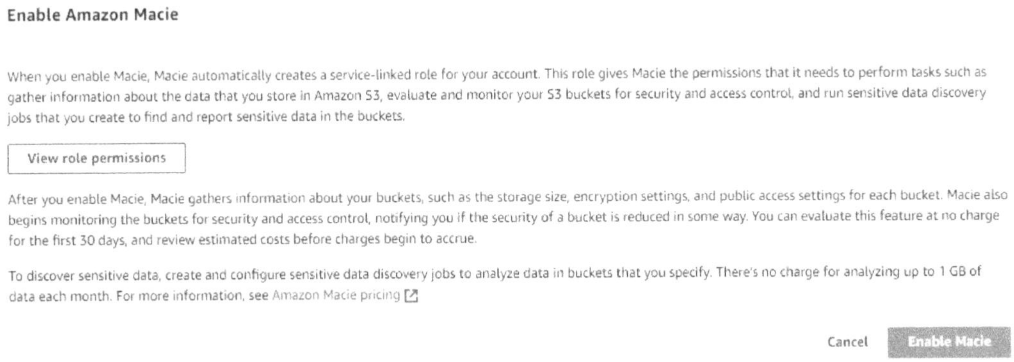

Enable Amazon Macie
When you enable Macie, Macie automatically creates a service-linked role for your account. This role gives Macie the permissions that it needs to perform tasks such as gather information about the data that you store in Amazon S3, evaluate and monitor your S3 buckets for security and access control, and run sensitive data discovery jobs that you create to find and report sensitive data in the buckets.
View role permissions
After you enable Macie, Macie gathers information about your buckets, such as the storage size, encryption settings, and public access settings for each bucket. Macie also begins monitoring the buckets for security and access control, notifying you if the security of a bucket is reduced in some way. You can evaluate this feature at no charge for the first 30 days, and review estimated costs before charges begin to accrue.
To discover sensitive data, create and configure sensitive data discovery jobs to analyze data in buckets that you specify. There's no charge for analyzing up to 1 GB of data each month. For more information, see Amazon Macie pricing
Cancel
Enable Macie

Figure 14.8 – Enable Macie

6. Once enabled, Amazon Macie will analyze your environment and, after a short time, provide a dashboard, as shown in the following screenshot:

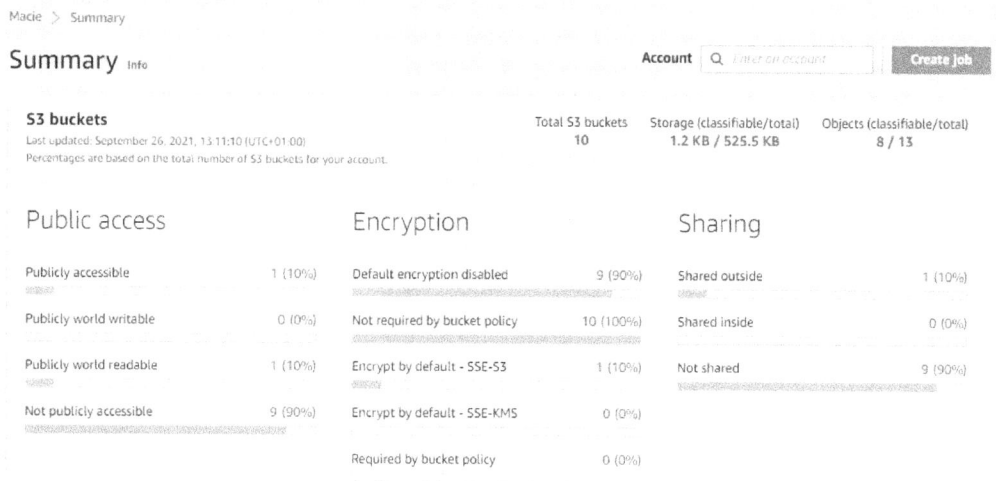

Macie > Summary

Summary Info

Account [Enter an account]   **Create job**

S3 buckets
Last updated: September 26, 2021, 13:11:10 (UTC+01:00)
Percentages are based on the total number of S3 buckets for your account.

	Total S3 buckets	Storage (classifiable/total)	Objects (classifiable/total)
	10	1.2 KB / 525.5 KB	8 / 13

Public access

Publicly accessible	1 (10%)
Publicly world writable	0 (0%)
Publicly world readable	1 (10%)
Not publicly accessible	9 (90%)

Encryption

Default encryption disabled	9 (90%)
Not required by bucket policy	10 (100%)
Encrypt by default - SSE-S3	1 (10%)
Encrypt by default - SSE-KMS	0 (0%)
Required by bucket policy	0 (0%)

Sharing

Shared outside	1 (10%)
Shared inside	0 (0%)
Not shared	9 (90%)

Figure 14.9 – Amazon Macie – Summary

7. While several items from the preceding summary need to be examined, for this exercise, we are primarily interested in ensuring that PII is not uploaded to the specific bucket we configured earlier. For this, we need to create an **Amazon Macie job**.

8. From the left-hand menu, select the **Jobs** link.

9. From the right-hand pane, click the **Create job** button. This will launch the job creation wizard, which you will need to complete as follows:

   ▪ For **Step 1**, select the bucket you created to upload the CSV file to. In my case, it is the **justdessertsproducts** bucket.

   ▪ Click **Next**.

   ▪ For **Step 2**, confirm that the correct bucket has been selected and click **Next**.

   ▪ For **Step 3, Refine Scope**, select the **One-time job** option and then click **Next**. Normally, you would set a schedule to perform the discovery daily, but for this exercise, we are only going to perform the discovery once.

   ▪ Under **Additional settings**, select **File name extensions** under **Object criteria**.

   ▪ In the text box below, type in csv and then click the **Include** button.

   ▪ Click **Next**.

- For **Step 4, Select managed data identifiers**, select **None** and then click **Next**.

- For **Step 5, Select custom data identifiers**, click on the **Manage custom identifiers** link to open the **Custom data identifiers** web page in another browser tab, as per the following screenshot:

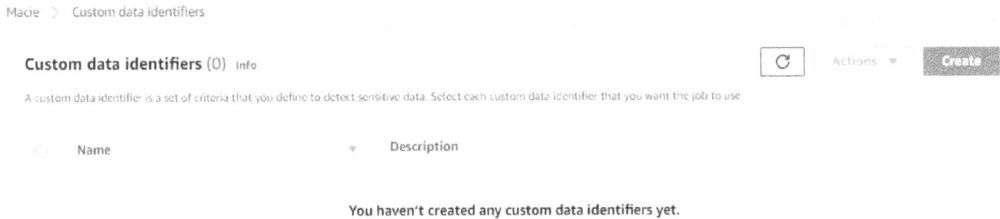

Macie > Custom data identifiers

Custom data identifiers (0) Info   C   Actions ▾   **Create**

A custom data identifier is a set of criteria that you define to detect sensitive data. Select each custom data identifier that you want the job to use

☐   Name   ▾   Description

You haven't created any custom data identifiers yet.

Figure 14.10 – Creating custom data identifiers

- Click the **Create** button.

- Give the identifier a name and an optional description.

- Next, you will need to provide a **regex** pattern that matches your EmployeeID format. Regex is short for *regular expression*, which is a sequence of characters that specifies a search pattern. Often, you can use regex patterns to search for specific types of data or to validate data. For this exercise, your regex pattern will be *[A-Z] {2}-[0-9]{4}*. Input this pattern into the **Regular expression** text box.

- Next, click the **Submit** button, which will create the custom data identifier for this exercise.

- Navigate back to the **Job creation** wizard, which should still be visible in the previous browser tab. You should still be in **Step 5**. Click on the refresh icon to access the **EmployeeID** custom data identifier, as per the following screenshot:

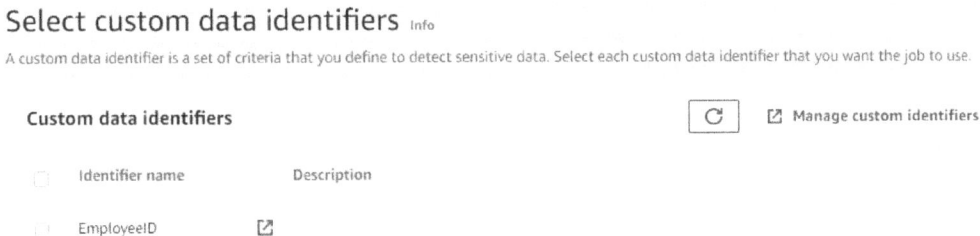

Select custom data identifiers Info

A custom data identifier is a set of criteria that you define to detect sensitive data. Select each custom data identifier that you want the job to use.

Custom data identifiers   C   ↗ Manage custom identifiers

☐   Identifier name   Description

☐   EmployeeID   ↗

Figure 14.11 – Select custom data identifiers

- Select the **EmployeeID** custom data identifier and click the **Next** button.

- For **Step 6**, provide a name for the job and then click on the **Next** button.

- For **Step 7**, review the configuration settings and then click the **Submit** button at the bottom of the page.

10. Your job has now been created and will be in the **Active (Running)** state, as per the following screenshot:

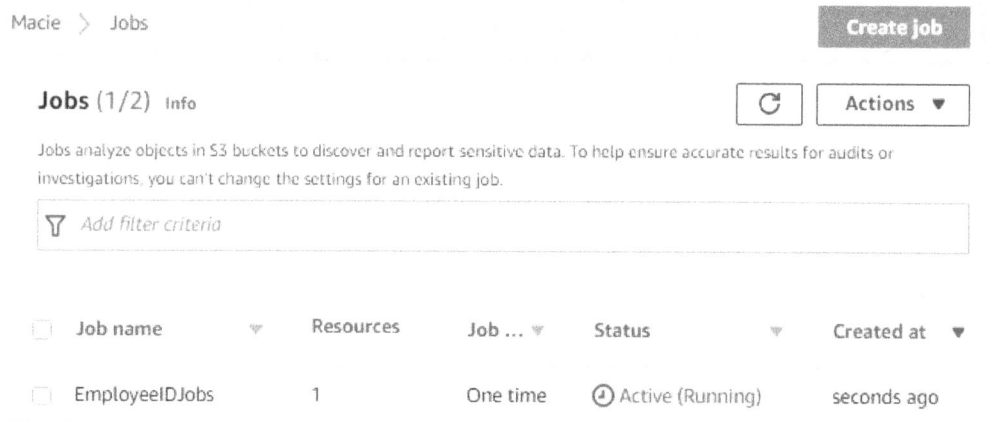

Figure 14.12 – Amazon Macie S3 job created

11. The job will take a few minutes to complete. Once the job completes, click on the **Show results** drop-down arrow and select **Show findings**. You will see that it has found the CSV file we uploaded earlier, as per the following screenshot:

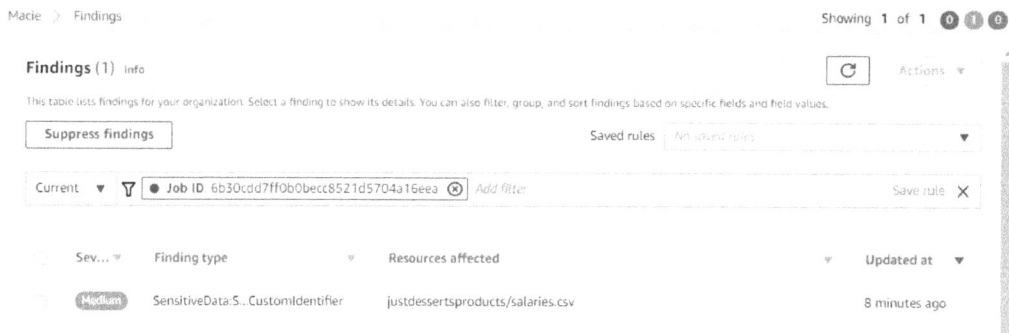

Figure 14.13 – Macie findings

12. Select the findings. Then, from the **Actions** drop-down list, select **Export (JSON)**.

13. You will be able to review the JSON file, which clearly shows that it identified four counts of PII data, as per the following screenshot:

Read-only ⓘ

```
1 ▾ [
2 ▾   {
3         "accountId": "451147979072",
4         "archived": false,
5         "category": "CLASSIFICATION",
6 ▾       "classificationDetails": {
7           "detailedResultsLocation": "s3://[export-config-not-set]/AWSLogs/451147979072/Macie/us-east-1
              /6b30cdd7ff0b0becc8521d5704a16eea/451147979072/b29381d3-fae2-31e4-b8ee-a673d2422eb9.jsonl
              .gz",
8           "jobArn": "arn:aws:macie2:us-east-1:451147979072:classification-job
              /6b30cdd7ff0b0becc8521d5704a16eea",
9           "jobId": "6b30cdd7ff0b0becc8521d5704a16eea",
10 ▾       "result": {
11           "additionalOccurrences": false,
12 ▾         "customDataIdentifiers": {
13 ▾           "detections": [
14 ▾             {
15                 "arn": "c674bf71-5f9d-40d7-828f-49a4a4e21c9f",
16                 "count": 4,
17                 "name": "employeeid",
18 ▾               "occurrences": {
19 ▾                 "cells": [
```

Figure 14.14 – Amazon Macie job findings

14. The JSON will also highlight the columns and rows where the PII data was identified.

As you can see, Amazon Macie is an extremely powerful tool that can help you understand your data access patterns and alert you of potential data breaches.

In the next exercise, you will clean up the resources you created.

Exercise 14.2 – cleaning up

In this exercise, you will clean up the resources you created in this chapter:

1. Navigate to the Amazon Macie console.

2. From the left-hand menu, select **Settings**.

3. Next, from the right-hand pane, click on the **Disable Macie** button. Amazon Macie will permanently delete all your existing findings, classification jobs, and other Macie resources.

4. In the dialog box that appears, type **Disable** in the text box and click the **Disable** button.

5. Next, from the Amazon S3 console, you will need to empty the S3 bucket you created and then delete the bucket, as you did previously.

Next, we will provide a summary of this chapter.

Summary

Security in the cloud is a shared responsibility between the customer and AWS. AWS will take responsibility for the security *of* the cloud, while the customer is responsible for security *in* the cloud. This distinction must be understood. You have access to a vast array of security tools and services that can help you build a highly robust and secure environment, within which you can host your application workloads.

This chapter also discussed concepts related to compliance and how AWS services are aligned with various regulatory bodies. As a customer, you are responsible for ensuring that your applications are built while following any necessary protocols. So, for example, if you are in the healthcare industry and based in the US, then your application processes and workflows must adhere to HIPAA compliance.

We also discussed encryption and how you must implement encryption both at rest and in transit. AWS offers KMS services to help you manage your encryption needs. If you have specific requirements to create, manage, and rotate your encryption keys using dedicated FIPS 140-2 Level 3 hardware security modules, then you can also opt for the AWS CloudHSM service.

AWS also offers several tools to protect your network and applications with firewall solutions such as AWS Shield and AWS WAF. AWS Shield offers protection against DDoS attacks, whereas AWS WAF is an application tier firewall design that can protect you from common web exploits such as SQL injection and cross-site scripting.

Finally, we examined various specific AWS security tools and services, including AWS Inspector, Amazon Macie, and AWS Secrets Manager. These tools can help you build a complete end-to-end secure environment to protect your data, grant access to authorized users, and prevent attacks.

In the next chapter, we will discuss the principles of billing and pricing on AWS. You will learn about cloud economics and how to architect solutions that are cost-effective for your business. We will discuss cost management strategies, as well as looking at various tools that can help you monitor costs and compare pricing options.

Questions

Answer the following questions to test your knowledge of the topics covered in this chapter:

1. Which of the following is part of the customers' responsibility regarding the Shared Security Model? (Choose 2.)

   A. Patch Windows EC2 instances with the latest security patches.

   B. Configure NACL to only allow inbound ports `80` and `443` to Linux web servers from the internet.

   C. Update the network cabling in the `us-east-1` data centers.

   D. Upgrade the underlying infrastructure support for the Lambda service.

   E. Upgrade the biometric readers in the London Region.

2. Which service in AWS protects your virtual network and resources from common DDoS attacks?

   A. AWS WAF

   B. AWS Shield

   C. AWS Detective

   D. Amazon Macie

3. Which of the following AWS Security tools can protect your web applications or APIs against common web exploits that may affect availability, compromise security, or consume excessive resources?

   A. AWS WAF

   B. AWS GuardDuty

   C. AWS Shield

   D. AWS NACL

4. Which AWS service uses machine learning to classify sensitive information stored in your Amazon S3 buckets and monitor access patterns for anomalies that indicate risks or suspicious behavior, such as large quantities of source code being downloaded?

    A. Amazon Macie

    B. Amazon X-Ray

    C. AWS Shield

    D. AWS WAF

5. Which AWS service enables companies looking to migrate to the AWS cloud to obtain copies of various compliance documents such as ISO certifications, **PCI, and SOC** reports?

    A. AWS Artifact

    B. AWS Config

    C. AWS CloudWatch

    D. AWS security reports

6. To fulfill strict compliance requirements, you need to create and manage your encryption keys using FIPS 140-2 Level 3-validated HSM devices. Which type of encryption service would you recommend?

    A. AWS KMS

    B. AWS CloudHSM

    C. Certificate Manager

    D. BitLocker

Section 4: Billing and Pricing

Moving to the cloud has to be cost effective if you want customers to migrate from on-premises environments. AWS offers competitive pricing models to help you design, build, and deploy applications that reduce your costs. Understanding AWS pricing strategies and best practices will help your organization become cost efficient.

This part of the book comprises the following chapters:

- *Chapter 15, Billing and Pricing*
- *Chapter 16, Mock Tests*

15
Billing and Pricing

Many companies looking to move to the cloud have heard of the cost benefits associated with shifting from a CAPEX model of investment in IT to an OPEX model. CAPEX refers to capital expenditure for the procurement of long-term assets such as infrastructure equipment, vehicles, and buildings. A company that purchases such assets will own those assets, but this tends to tie up capital that could otherwise have been spent on other resources that directly benefit the business. Consider a company having access to spare capital to invest in research and development of their products and services, as opposed to simply purchasing IT infrastructure equipment.

OPEX refers to operating expenditure and is an ongoing cost for running a product, business, or system. In the context of cloud computing, OPEX is the ongoing cost of leasing resources from the cloud provider, such as running a fleet of web servers, where the business is charged on a pay-as-you-go model. The idea that you do not need to tie up vast amounts of capital in expensive hardware equipment (which depreciates rapidly), or expensive software licenses for your on-premises infrastructure, is a critical factor for your company's financies.

However, simply shifting to the cloud is not necessarily going to save you money. The solutions you design to host your applications in the cloud need to be cost-optimized and architected to ensure you get the best value for your dollar. The good news is that with a cloud provider such as AWS, the solutions you design can be highly cost-effective. For one thing, you can ensure that you only pay for services when you consume them. In addition, each service offering has various options to minimize costs that you can configure, based on your access patterns and core business requirements.

One thing to be careful about is that you should not consider cost optimization in isolation. Minimizing cost is important for the business, but so is ensuring application availability, reliability, and scalability.

In this chapter, we look at the fundamentals of cost optimization and the key components of AWS billing and pricing. We discuss key principles and look at various tools to help you gain insight and manage your costs effectively.

The topics covered in this chapter are as follows:

- An overview of billing and pricing on AWS
- Understanding AWS cost optimization
- Learning about AWS Billing and Cost Management tools
- Learning how to use the AWS pricing and TCO calculators – pricing calculator and migration calculator

Technical requirements

To complete the exercises in this chapter, you will need to log in to your AWS account as the IAM user **Alice**.

An overview of billing and pricing on AWS

AWS charges you based on three core components, which are the amount of *computing*, *storage*, and *outbound data transfers* you perform. There are some deviations to this, but these are the primary cost drivers. With compute, this could represent the number of hours that your **On-Demand Instance** is in the running state or the number of Lambda executions you perform. With storage, you are charged per GB. In both cases, you are charged for any data transferred out of the Region or AWS. The data transfer out rate is dependent on the service and the Region from which the data originates.

A couple of key points to note here are as follows:

- For data transfer within a Region:

  - If an **internet gateway** is used to send data to an AWS public service (Amazon S3, DynamoDB, and so on) in the same Region, then there are no data transfer charges.

  - If a **NAT gateway** is used to access those public services in the same Region, then there is a data processing charge, calculated at a per-GB rate.

- Data transfers between workloads within the same **Availability Zone** are free. For certain services, such as EC2, data transfers across Availability Zones will incur a data transfer charge.

- For data transfers across Regions, there is always a data transfer charge.

You are not charged for data transferred into AWS, for example, from your on-premises data center.

Let's examine some key principles of pricing on AWS and the strategies you should adopt when moving your workloads to the cloud.

Understanding AWS cost optimization

Depending on your workloads and specific business requirements, you have access to a wide range of service options and tools to help you optimize for costs, ensuring that you monitor and control your spending. Choosing the right combination of services and deploying the correct resource configurations will help you run a tight ship. Here are some key concepts to bear in mind when architecting solutions for the cloud:

- **Choose the right pricing model**: You can use On-Demand Instances when you know that you only need the compute resources for a limited amount of time and for a specific short-term project. From a reliability point of view, you need to ensure that there are no interruptions to the service, but you are not going to be using the resources 24/7, 365 days a year. In this case, an on-demand pricing model works well. Alternatively, you can invest in **reserved capacity**, which can save up to 72% over equivalent on-demand capacity price, but only if you need the resources 24/7, and for a much longer duration, for example, a year. You can also consider using **Spot EC2 instances** for applications that are fault-tolerant, scalable, or flexible, and where your application can tolerate interruptions. Spot instances can save you up to a 90% discount off on-demand prices and there is no upfront commitment. Finally, consider opting for the **compute Savings Plans** to reduce EC2, Fargate, and Lambda costs. EC2 Savings Plans can deliver savings of up to 72% on your AWS compute usage. Here, you make a commitment to a certain amount of compute usage, and you are automatically charged at the discounted savings plan prices.

- **Match capacity with demand**: AWS offers several tools to help you identify usage patterns. AWS Cost Explorer resource optimization (discussed later in this chapter) can help you identify instances that are idle or have low utilization. You can save money by either reducing the capacity provisioned or if possible, stopping the instances during periods when they are not in use. You can use **AWS Instance Scheduler**, which is a CloudFormation template offered by AWS that can help reduce costs by configuring start and stop schedules for your Amazon EC2 and Amazon RDS instances. This is like the exercise you completed in *Chapter 12, Automation and Deployment on AWS*, where you configured Amazon CloudWatch Events and Lambda to automatically start and stop EC2 instances at scheduled times. Another service that can be stopped when not in use is Amazon RDS. In the case of Amazon Redshift, you can pause the service.

- **Implement processes to identify resource waste**: Carefully monitoring your AWS resources will help you identify ones that are sitting idle and not being used. For example, if you have EBS volumes where utilization is less than 1 IOPS per day for a period of 7 days, then it is likely that volume is not being used. You can save on costs by taking a snapshot of those volumes and deleting them. For Amazon S3, you can use the **Amazon S3 analytics** service to understand storage access patterns that will help you decide when to transition less frequently used objects to infrequent storage classes. You can also use the results from the tool to help build your Amazon S3 life cycle rules. Another area where you can save costs is on the networking front. Amazon elastic load balancers are chargeable and if you happen to have any idle ELBs, then it makes sense to terminate them. You can analyze metrics from your ELBs and watch out for ones that have a `RequestCount` value of less than 100 over the past 7 days, which could be ideal candidates to terminate.

> **Important Note**
> Amazon S3 analytics will not provide recommendations for transitions to the One Zone-IA or S3 Glacier storage classes.

In this section, we examined some core concepts to ensure that the workloads you deploy in the cloud are cost-optimized. In the next section, we will look at various tools and services that will help you monitor, control, and manage your costs effectively.

Learning about AWS billing and cost management tools

In this section, we'll look at several AWS billing and cost management tools, enabling you to visualize your usage and spend across your AWS accounts as well as implementing better cost control strategies. We'll start by looking at the Cost Explorer service.

AWS Cost Explorer

AWS Cost Explorer enables you to monitor and visualize your costs via dashboards and reports. You can access 12 months of usage and spending data as well as forecasting what your future costs will be for the next 12 months. You need to enable the Cost Explorer service and, once enabled, the current month's reporting data is accessible after 24 hours. Additional data from previous months will start to appear after a few more days.

The AWS Cost Explorer home page displays a snapshot of your spending. Here, you will find **Current month costs** as well as **Forecasted month end costs**, as per the following screenshot:

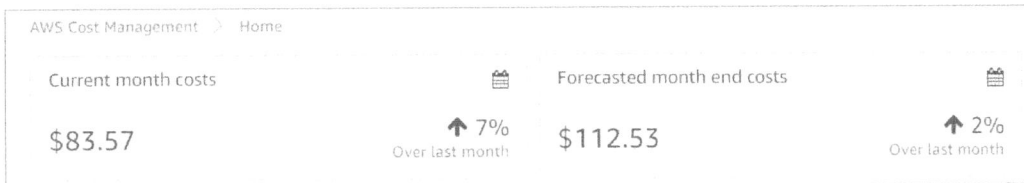

Figure 15.1 – Cost Explorer home page

AWS Cost Explorer offers several types of reports. For example, in the following screenshot, you will see a *monthly cost by service* report for one of my AWS accounts:

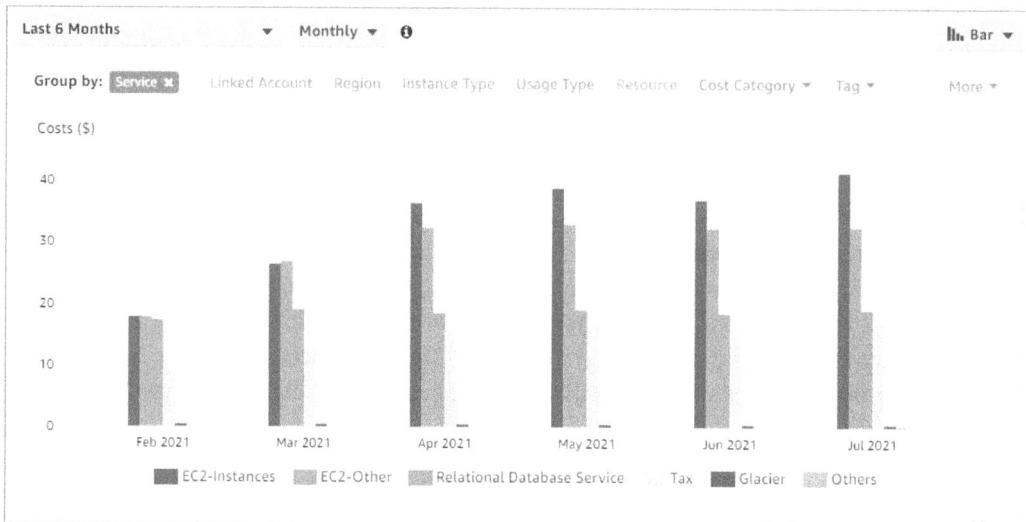

Figure 15.2 – Cost Explorer – monthly cost by service report

The report shows the last 6 months of spending broken down by service, comprising EC2 instances, Amazon RDS, and storage archives in Amazon Glacier.

Additional features of Cost Explorer include granularly filtering your data to understand daily and hourly cost patterns and by resource types.

AWS Cost Explorer also provides various recommendations where you can save some money. For example, from the left-hand menu of the AWS Cost Explorer console, you can select **Recommendations**, under the **Reservations** category, to see where Cost Explorer suggests purchasing EC2 **Reserved Instances** (**RI**) for your existing on-demand workloads as per the following screenshot:

Figure 15.3 – Cost Explorer RI recommendations

As you can see, purchasing RI for the two on-demand instances I have running would potentially save me 33% off my final costs.

AWS Cost Explorer can also be used to send you EC2 *rightsizing recommendations* to help you examine your historical usage patterns and identify resources that are being underutilized. AWS Cost Explorer will be able to identify idle instances and allow you to take action to terminate or resize your underutilized resources.

Another report you can view within Cost Explorer is the **RI utilization report**. The RI utilization report enables you to identify the savings made from using RI over on-demand options for Amazon EC2, Amazon Redshift, Amazon RDS, and so on. The RI report also provides details of how much you may have overspent because you are underutilizing your RI.

Finally, AWS Cost Explorer comes with a security feature to help you detect anomalies in your spending patterns, and the root causes, and can be configured to send you alerts using Amazon SNS.

In this section, you have learned about the AWS Cost Explorer tool, which enables you to visualize and manage your costs and allows you to create custom reports to identify usage patterns and trends.

In the next section, we discuss how cost allocation tags can help you identify which services and resources you are consuming and spending on.

Cost allocation tags

Previously we have discussed how important tagging your resources is. It allows you to not only identify your resources, but you can also automate and run management tasks against your resources using these tags as a form of identification. Additionally, you can also use tagging to help you identify which resources are costing you money and by using cost allocation tags, you can track usage by different projects, departments, business units, and so on.

Tags consist of a *key* and a *value*, and for every resource, you can have up to 50 key-value tags. In addition, the tag key must be unique, and each key can only have one value. You need to first activate cost allocation tags, which AWS will then use to help you analyze your costs using those tags. There are two types of cost allocation tags:

- **AWS-generated tags**: These are created and defined by AWS and the AWS Marketplace.
- **User-defined tags**: These can be created and defined by you based on your business cost management and naming conventions.

Activating the tags can be done from the billing management console and each *tag key* will need to be activated individually. Once activated, AWS will generate **comma-separated values (CSV)** file reports with your usage and costs grouped by your active tags.

Next, let's look at the cost and usage report.

Cost and usage report

AWS offers a cost and usage report tool that provides a very detailed breakdown of the resources you use and the associated costs. The report includes usage amounts and units consumed, rates, costs, and product attributes. Reports are published to an Amazon S3 bucket of your choice, and the reports can be configured to provide a breakdown of your costs by hour, day, or month. These reports are made available in your S3 bucket in CSV format, which you can then import into Microsoft Excel or other spreadsheet tools.

In addition, you can also integrate the reports for use with Amazon Athena, Redshift, and QuickSight, as well as setting the compression type to GZIP, ZIP, or Parquet.

Next, let's examine the AWS Budgets service.

AWS Budgets

AWS Budgets enables you to understand how money is being spent on the workloads you deploy in the cloud. You can use AWS Budgets to track your cost and usage across different types of projects you may be running in your organization. You may have different departments requiring different cloud services and, based on their requirements, specify monthly budgets for their spending.

If your budgets exceed the threshold (whether actual or forecasted), you can define specific actions to take on your AWS account. At present, you can define three types of actions:

- An **Identity and Access Management (IAM)** policy
- **Service Control Policies (SCPs)**
- Target running instances (EC2 or RDS)

For example, you could specify an IAM policy that denies the ability to launch new EC2 instances if the monthly budget for EC2 has been exceeded. You can then apply this policy to the IAM user, group, or role used to launch instances.

AWS Budgets actions can be executed automatically when the threshold has been exceeded or can be configured for manual approval if required.

With AWS Budgets, you can also set up alerts using Amazon SNS to be notified (or perform some event-driven action) if the actual or forecast spending is likely to cross the budgeted threshold values. You can also configure variable target amounts, so that the budgeted amount increases by a certain percentage over a period because consumption is likely to increase.

AWS Budgets can also be used to ensure that you stay within specified quotas and limits, including staying under a specific AWS Free Tier offering.

There are several types of budgets you can create, including the following:

- **Cost budgets**: Define specific amounts you are willing to spend.

- **Usage budgets**: Define how much you want to use a specific service.

- **RI utilization budgets**: Specify a utilization threshold and be alerted if you use less than the threshold value. This enables you to identify those RI resources that may be underutilized and reconsider whether those reservations are required.

- **RI coverage budgets**: Identify how much of your instance usage is covered by the reservation.

- **Saving plan utilization budgets**: Identify when your usage of savings plans falls below utilization threshold values, enabling you to identify under-utilized savings plans.

- **Saving plan coverage budgets**: Identify how much of your instance usage is covered by saving plans.

In this section, we looked at AWS Budgets, which can be used to ensure you do not overspend on the resources you deploy on AWS.

In the next section, we move on to look at some AWS pricing and cost calculators.

Learning how to use the AWS pricing and TCO calculators

AWS offers calculators to help you estimate costs associated with setting up and deploying resources on AWS. Two key calculators are the **AWS Pricing Calculator** and **Migration Evaluator** (previously known as the TCO Calculator).

The AWS Pricing Calculator

The AWS Pricing Calculator is a tool used to calculate the estimated monthly costs of deploying and configuring AWS services and resources. You can use the AWS Pricing Calculator to work out estimated monthly charges for your projects and applications that you wish to deploy on AWS. You can view prices per service or per group of services to analyze your proposed architecture costs.

You can then send the estimate's unique link to other team members, finance, and your clients to help analyze the potential cost of the project.

To see how the AWS Pricing Calculator works, let's take an example of a simple application solution comprising elastic load balancers, EC2 instances, and Amazon RDS databases. Refer to the following architectural diagram for the proposed design:

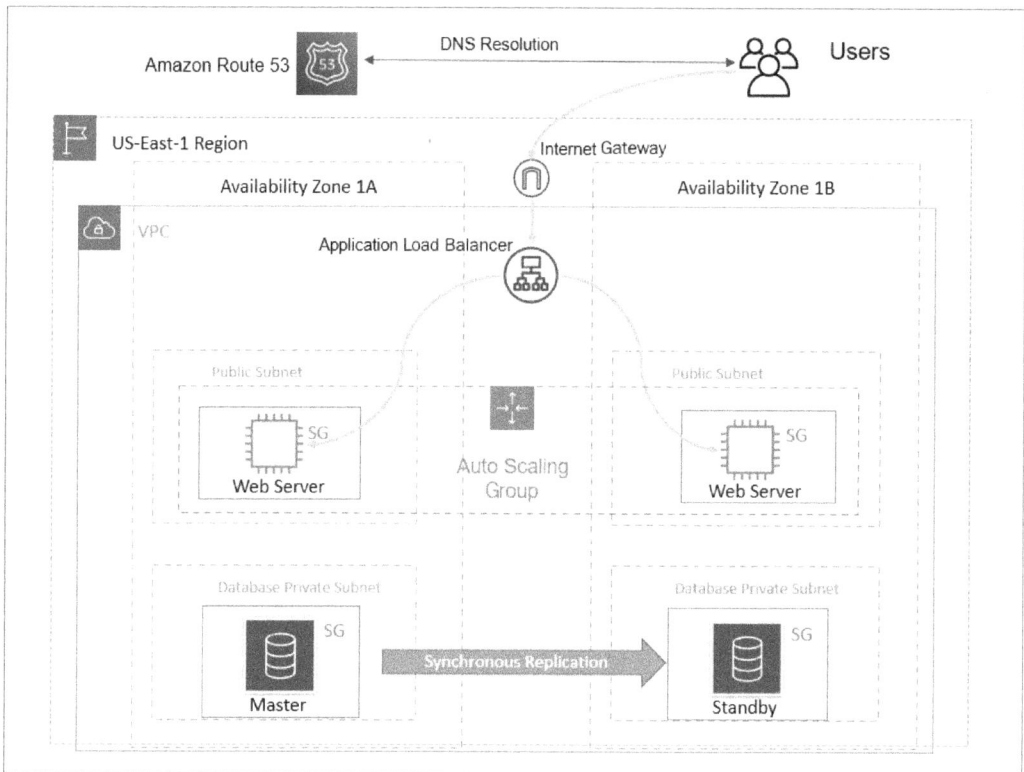

Figure 15.4 – AWS Pricing Calculator example – application architecture

In the preceding architecture, the core resources deployed are as follows:

- A VPC in the **US-East-1** Region, with six subnets, spread across two Availability Zones.

- An internet gateway to allow traffic into and out of the VPC.

- An **Application Load Balancer** (**ALB**) to allow users on the internet to access the web application running on the **app servers**.

- Two EC2 instances, one in each Availability Zone, within private subnets. An **Auto Scaling group** designed to scale out/in as required. For this example, we will keep the instance fleet size to two EC2 instances, replacing any failed EC2 instances if required.

- An Amazon RDS database running the MySQL engine, configured with **Multi-AZ** to offer high availability and resilience.

Determining the estimated costs for provisioning these resources also requires you to be aware of the type and amount of traffic that will be transferred in and out. These details are going to be derived from setting a test environment and monitoring your workloads or perhaps from previous deployments.

To simplify this example, let's assume that your architecture is provisioned as per the following table:

Characteristic	Estimated Usage	Description
Utilization	100%	All infrastructure components run 24 hours per day, 7 days per week
Instance	t2.medium	4 GB memory, 2 vCPUs. On-demand pricing option
Storage	Amazon EBS SSD gp2	1 EBS volume per instance with 30 GB of storage per volume
Data backup	Daily EBS snapshots	1 EBS volume per instance with 30 GB of storage per volume, 50 MB change per snapshot
Data transfer	Data in 1 TB/month Data out: 1 TB/month	10% incremental change per day
Instance scale	2	There are 2 EC2 instances running in the fleet and for Auto Scaling, we will maintain the fleet size to 2
Load balancing	10 GB/Hour	Elastic load balancing is used 24 hours per day, 7 days per week. It processes a total of 10 GB/Hour (data in + data out)
Database	MySQL, db.m5.large instance with 8 GB memory, 2 vCPUs, 200 GB storage	Multi-AZ deployment with synchronous standby replica in separate Availability Zone

Table 15.1 – AWS Pricing Calculator example architecture

To determine the estimated monthly costs for the set of resources as detailed in the preceding table, you can use the AWS Pricing Calculator available at `https://calculator.aws/`.

Within the AWS Pricing Calculator, you will need to select the various services and provide resource configuration information as per the example shown in the screenshot:

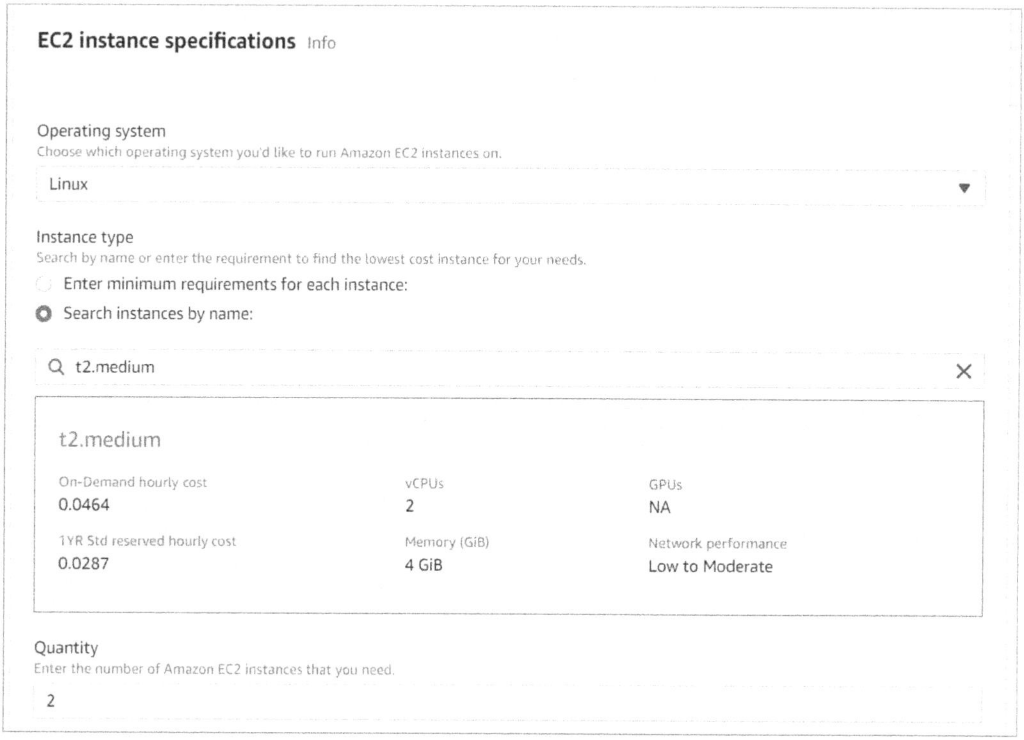

EC2 instance specifications Info

Operating system
Choose which operating system you'd like to run Amazon EC2 instances on.

| Linux | ▼ |

Instance type
Search by name or enter the requirement to find the lowest cost instance for your needs.

○ Enter minimum requirements for each instance:
● Search instances by name:

| Q t2.medium | ✕ |

t2.medium

On-Demand hourly cost	vCPUs	GPUs
0.0464	2	NA

1YR Std reserved hourly cost	Memory (GiB)	Network performance
0.0287	4 GiB	Low to Moderate

Quantity
Enter the number of Amazon EC2 instances that you need.

| 2 |

Figure 15.5 – AWS Pricing Calculator – EC2 instance specifications

Once you have provided all the various configuration items as well as an indicative quantity of data transferred in and out, you will be provided with an estimate table, like the following:

Service	Monthly	Annual	Configuration summary
Amazon Virtual Private Cloud (VPC)	$92.07	$1,104.84	DT Inbound: Internet (1 TB per month), DT Outbound: Internet (1 TB per month), DT Intra-Region: (0 TB per month), Data transfer cost (92.07)
Application Load Balancer	$74.83	$897.96	Number of Application Load Balancers (1)

Service	Monthly	Annual	Configuration summary
Amazon EC2	$47.90	$574.80	Operating system (Linux), Quantity (2), Pricing strategy (EC2 Instance Savings Plans 1 Year No Upfront), Storage amount (30 GB), Instance type (t2.medium)
Amazon Elastic Block Store (EBS)	$9.37	$112.44	Number of instances (2), Average duration each instance runs (730 hours per month), Storage amount (30 GB), Snapshot Frequency (Daily), Amount charged per snapshot (250 MB)
Amazon RDS for MySQL	$295.66	$3,547.92	Storage for each RDS instance (General Purpose SSD (gp2)), Storage amount (200 GB), Quantity (1), Instance type (db.m5.large), Deployment option (Multi-AZ), Pricing strategy (OnDemand)
Amazon Route 53	$0.50	$6.00	Hosted Zones (1), Basic Checks Within AWS (1)
Total Project Costs	$520.33	$6,243.96	

Table 15.2 – AWS Pricing Calculator – estimated costs

Note that the aforementioned costs are just estimates and to keep the example simple, don't cover minute configuration details for your resources. For example, no backup storage is included with Amazon RDS and no routing policies are defined for Route 53. In addition, the prices exclude taxes.

Next, let's look at a brief overview of Migration Evaluator.

AWS Migration Evaluator

Many companies looking to move to the cloud will need to understand the cost implication of migration projects required to complete the shift from their on-premises environments. As well as designing the technical steps required to perform the migration, companies need to perform estimated cost analysis and the potential savings likely to be gained by moving to the cloud.

AWS Migration Evaluator removes the guesswork in planning any migration project to AWS. Your organization can work with the Migration Evaluator team to help capture millions of real-time data points related to your IT environment and review recommendations for the right sizing and right costing of workloads on AWS. With the Migration Evaluator service, you can use **AWS Application Discovery Service**, the **TSO Logic agentless collector**, or third-party tools to discover and gain insights into your current compute, storage, and total cost of ownership. The agentless collector tool can analyze your on-premises resources requiring just read-only access to your VMware, Hyper-V, Windows, Linux, Active Directory, and SQL Server infrastructure.

Migration Evaluator then provides extensive analysis and matches your existing workloads with appropriate AWS resources, such as the correct EC2 instance types, storage platform and database services, and so on, that can be provisioned in the cloud for your applications' workloads. The service can also help perform *what-if* analysis such as identifying cost savings by selecting different purchase options for your EC2 instances as well as incorporating elasticity features to eliminate waste by provisioning resources only when there is demand.

Ultimately, Migration Evaluator helps companies build business case reports and illustrate the cost benefits of moving to the cloud.

In this section, we looked at two types of AWS calculators to help identify costs associated with hosting workloads and applications in the cloud as well as performing total cost of ownership analysis with a view to helping customers make the right migration decisions.

In the upcoming exercise for this chapter, we look at how to work with AWS Budgets.

Exercise 15.1 – setting up cost budgets on AWS

In this exercise, you will create a total monthly cost budget for your workloads. This will allow you to monitor your spending and avoid unintentional cost overruns:

1. Log in to the AWS Management Console using the credentials of the IAM user *Alice*.

2. Search for AWS Budgets from the top search bar on the console screen, and navigate to the AWS Budgets console page.

3. From the right-hand pane, click on the **Create a budget** button.

4. Next, from the list of **Budget types**, select **Cost budget – Recommended**.

5. Click the **Next** button in the bottom-right corner of the page.

6. In the **Set budget amount** pane, configure the following:

    A. Set the period to **Monthly**.

    B. Select **Recurring budget**.

    C. Set the start month of your choice.

    D. Select the **Fixed** option under **Choose how to budget**.

    E. Next, enter your budget amount. I have set mine to $20 as per the following screenshot:

Set budget amount

Period
Daily budgets do not support enabling forecasted alerts, daily budget planning, or attaching actions.

Monthly   ▼

Budget effective date

● Recurring budget
   Recurring budgets renew on the first day of every monthly billing period.

○ Expiring budget
   Expiring monthly budgets stop renewing at the end of the selected expiration month.

Start month

Aug ▼   2021 ▼

Choose how to budget

| ● Fixed | ○ Monthly budget planning |
| Create a budget that tracks against a single monthly budgeted amount. | Specify your budgeted amount for each budget period. |

Enter your budgeted amount ($)
Last month's cost: $0.60

20.00

Figure 15.6 – AWS Budgets – setting up monthly cost budgets

F. Scroll further down to review your **Budget scoping** pane. Here, you will see
    whether your monthly cost budget is reasonable based on historic costs. As you
    will note from the following screenshot, for my AWS account, the budget of $20
    is acceptable because, since February 2021, I have been spending less than $15,
    which is within my budget:

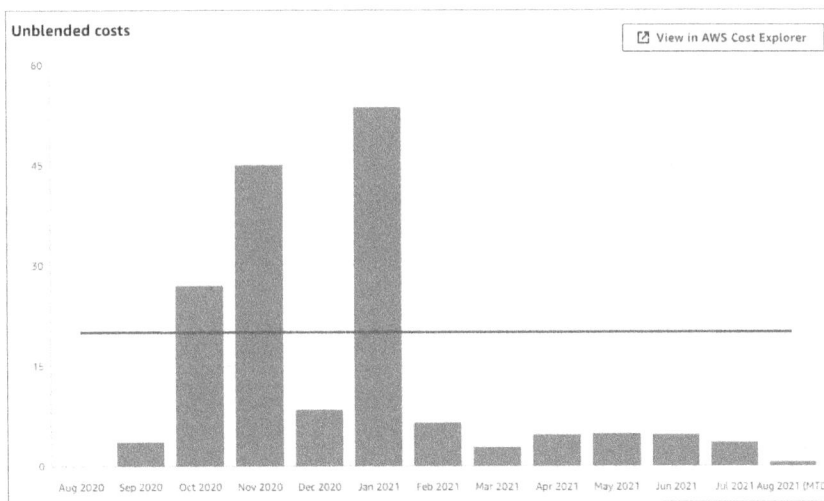

Figure 15.7 – AWS Budgets – budget scoping

G. Finally, provide a name for your budget and click **Next**.

H. Next, you can configure an alerts threshold, where you can be notified if you cross the threshold. Click on the **Add an alert threshold** button.

I. Under **Set alert threshold**, specify a threshold of 80% of the budgeted amount and set the trigger to **Actual** as per the following screenshot:

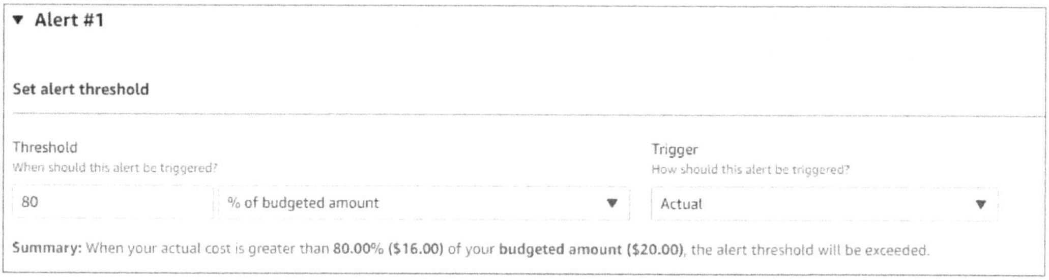

Figure 15.8 – Setting alert threshold

J. Next, under **Email recipients**, specify an email address to send the alerts to.

K. You can also set Amazon SNS alerts and Amazon Chatbot alerts.

L. Click on the **Next** button in the bottom-right corner of the page.

M. You can also define **Actions** to take when the threshold is crossed, for which you will need to configure appropriate IAM roles. For now, just click on the **Next** button.

N. Review your settings and click the **Create budget** button in the bottom right-hand corner of the page.

O. Your cost budget will be created, and you will be able to view it in the **Overview** console as per the following screenshot:

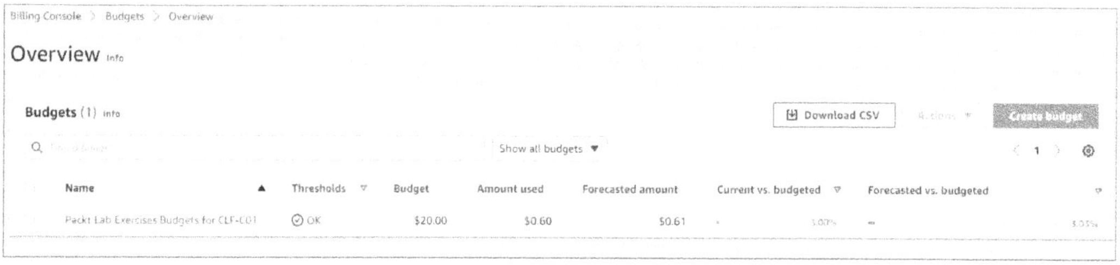

Figure 15.9 – AWS Budgets – Overview console

Note that you can set up five additional alerts for your budgets, such as the following:

- When current monthly costs exceed the budgeted amount
- When current monthly costs exceed 80% of the budgeted amount
- When forecasted monthly costs exceed the budgeted amount

In this section exercise, you learned how to create monthly cost budgets for your AWS account. Setting custom budgets can alert you if you are crossing specific threshold values or even forecasted costs in the future.

Next, we provide a summary of this chapter.

Summary

Businesses across the globe of all sizes can benefit from moving their applications and workloads to the cloud and gain a competitive advantage from the vast array of services on offer by AWS. AWS enables customers to build highly available, fault-tolerant, scalable, and resilient cloud solutions. AWS offers some of the latest and greatest cutting-edge technologies to help you design and build robust application solutions. Without AWS, the cost of provisioning the vast pool of services and resources alone would make it impossible for many companies, including start-ups, to innovate.

One of the prime drivers of cloud adoption is the fact that companies can save on capital expenses and opt for a variable expense model for all their IT investments. This allows businesses to channel more of the previous capital into improving their products and services. However, careful planning and monitoring are required. It is all too easy to spin the instances and other resources and incur unnecessary costs, especially where you may have overprovisioned the infrastructure required to support your applications. This clearly defeats one of the primary reasons for moving to the cloud.

In this chapter, you learned about the principles of cost optimization, and the approaches you can take to ensure you consume the right services at the right price. You learned how to provision resources when you need them most and not have idle resources simply sitting around gathering dust. Different pricing options for different workload types enable you to make sensible decisions about your investment. In addition, AWS offers a number of different management and monitoring tools to help you stay well informed about your spending. Using Cost Explorer, cost and usage reports, and AWS Budgets, you can run a tight ship if you implement best practices and plan your solutions carefully.

Finally, we also reviewed a couple of cost calculators that help you estimate the investment required for implementing solutions on AWS and how migrating your workloads to the cloud could save you money using Migration Evaluator.

In the next chapter, we offer two complete exam-style practice questions to help you gauge your understanding of the syllabus for the AWS Certified Cloud Practitioner exam and help you prepare for the certification.

Passing the AWS Certified Cloud Practitioner exam will help you progress your career in cloud computing. This exam guide has been designed to help you not only achieve exam success, but also gain valuable hands-on experience in architecting, building, and managing real-world cloud solutions. With a vast array of exercises in this study guide, we hope you are now feeling much more confident in tackling real-world business use cases.

I would like to take this opportunity to thank you for your time in going through this material and I hope you have gained valuable knowledge and some hands-on experience in designing cloud solutions.

Questions

1. Which AWS service enables you to specify a monthly cost amount for your AWS account and send out an alert if the actual or forecast spending is likely to cross the budgeted threshold values?

    A. AWS Budgets

    B. AWS Cost Explorer

    C. AWS Config

    D. AWS Pricing Calculator

2. Which AWS service enables you to access 12 months of usage and spending data as well as forecasting what your future costs will be for the next 12 months?

    A. AWS Cost Explorer

    B. AWS Billing alarms

    C. AWS Config

    D. AWS annual report

3. The finance department would like to get a report on the total monthly spending for all AWS resources broken down by business unit/project name. The purpose of which is to understand how much each business unit/project costs. Which feature of the AWS Billing and Cost Management services enables you to achieve this requirement?

   A. Cost allocation tags

   B. Cost and usage report

   C. AWS Budgets

   D. AWS CloudWatch

4. Which AWS service can analyze your monthly compute usage and offer recommendations for purchasing EC2 **RI** for your existing on-demand workloads?

   A. AWS CloudTrail

   B. AWS Budgets

   C. AWS Cost Explorer

   D. AWS Migration Evaluator

5. You are planning on architecting and building a three-tier application solution, comprising EC2 instances, load balancers, the Auto Scaling service, and a backend database. Which AWS service can you use to produce an estimated monthly cost for your proposal?

   A. AWS Cost Explorer

   B. AWS Pricing Calculator

   C. AWS Cost Optimization Analyzer

   D. AWS SNS

16
Mock Tests

The AWS Certified Cloud Practitioner exam is a 90-minute, multi-choice, and multi-answer-style exam. AWS charges USD 100 to enroll for the exam, which can be booked via their website at `https://www.aws.training/certification`. The exams are delivered at test centers located across several cities in many countries or can be taken online at home.

In this chapter, we will offer you two complete exam-style mock tests that you can use to gauge your readiness for the official AWS Certified Cloud Practitioner exam.

These tests comprise 65 questions each with full answer explanations available at the end of this chapter. To effectively use these tests, ensure you follow these steps:

- Set aside 90 minutes for each test, where you will not be disturbed. You can use your phone to set an alarm.

- Grab some paper and a pen and for each question, jot down the question's number and the Correct answer.

- Ensure that if you have time left over when you complete the test, you review your answers once more.

- At the end of the 90 minutes, review your answers and compare them with the answer explanation section at the end of this chapter.

You want to be able to answer all the questions correctly before you enter the exam room. You should use these tests to identify any questions that you did not answer correctly and then go back and review the chapter that covers the concepts examined in the question. This way, you will fill any gaps in your knowledge and be in a better position to get a high score on the official exam.

Mock test 1

1. Which type of cloud service model is most like an on-premises environment, where you configure virtual infrastructure components such as compute, network, and storage services that you can host your applications on?

    A. **Software as a Service (SaaS)**

    B. **Platform as a Service (PaaS)**

    C. **Infrastructure as a Service (IaaS)**

    D. **Function as a Service (FaaS)**

2. Your company is looking to move all its applications and services to the cloud but would like to migrate workloads in stages. This would require you to ensure that there is connectivity between the on-premises infrastructure and the applications you deploy on AWS for a while. What cloud deployment model would you need to establish?

    A. Private cloud

    B. Public cloud

    C. Hybrid cloud

    D. Multi-cloud

3. Which of the following statements are valid reasons for choosing a specific AWS Region to deploy your applications in? (Choose two)

    A. Your organization would choose a specific AWS Region that enables you to ensure that your applications are closer to your end users, thereby reducing any latency.

    B. If your organization has specific compliance or data residency laws to follow, then your choice of an AWS Region will be dictated by this requirement.

    C. Your organization would choose a Region closer to its location since your IT staff will need to visit the AWS data centers to set up servers and networking equipment.

D. Your organization would choose a Region-based location where your business has an established legal presence. This is because you cannot access other Regions unless you have a legal establishment in that Region.

E. Your organization would select an AWS Region that offered higher variable costs but lower upfront costs.

4. Which component of the AWS Global Infrastructure enables you to cache content (videos, images, and documents) and offer low-latency access when your users try to download them?

   A. AWS Regions

   B. Availability Zones

   C. Edge locations

   D. Local Zones

5. Which of the following AWS services can help you design a hybrid cloud architecture and enable your on-premises applications to get access to Amazon S3 cloud storage?

   A. Amazon Snowball Edge

   B. AWS Storage Gateway

   C. Amazon Elastic Block Store

   D. Amazon CloudFront

6. You are planning on using AWS services to host an application that is still under development, and you need to decide which AWS support plan you should subscribe to. You do not need production-level support currently and are happy with a 12-hour response time for any system-impaired issues. Which is the most cost-effective support plan you should subscribe to?

   A. Basic Support plan

   B. Developer Support plan

   C. Business Support plan

   D. Enterprise Support plan

7. Which of the following are regarded as global services on AWS? (Choose two)

   A. AWS IAM

   B. Amazon Route53

   C. Amazon EC2

   D. Amazon EFS

   E. Amazon RDS

8. Which of the following statements closely relates to the advantage of cloud computing that discusses the ability to *go global in minutes*?

   A. The ability to trade capital expenses for variable expenses and thus avoid huge CAPEX.

   B. The ability to provision resources just in time for when you need them using tools such as Auto Scaling.

   C. The ability to deploy your applications across multiple Regions with just a few mouse clicks.

   D. The ability to focus on experimentation and the development of your applications rather than infrastructure builds, management, and maintenance.

9. Which AWS service can you configure to send out an alert to an email address if your total expenditure crosses a predefined monthly cost?

   A. Set up a billing alarm in Amazon CloudWatch

   B. Set up a billing alarm in Amazon CloudTrail

   C. Set up a billing alarm in Amazon Config

   D. Set up a billing blarm in Amazon Trusted Advisor

10. Which of the following resource types is tied to the Availability Zone that it was launched in?

   A. **Elastic Block Store (EBS)**

   B. **Elastic File Store (EFS)**

   C. Amazon Route53 Hosted Zones

   D. Amazon DynamoDB

11. As part of enhancing the security of your AWS account, you need to ensure that all IAM users use complex passwords comprising of at least one capital letter, a number, a symbol, and a minimum of 9 characters. Which AWS IAM feature can you use to configure these requirements?

    A. Password policies

    B. Permission boundaries

    C. Service Control Policies (SCPs)

    D. Resource policies

12. As a recommended best practice, what additional authentication security measure can you implement for your root user and IAM users?

    A. Implement MFA.

    B. Implement LastPass.

    C. Implement AWS WAF.

    D. Implement AWS Shield.

13. What is the easiest way to assign permissions to many IAM users who share a common job function?

    A. Create a customer-managed IAM policy and attach the same policies to all IAM users who share a common job function.

    B. Create an IAM Group, add IAM users who share the common job function to that group, and apply an IAM policy to the group with the necessary permissions.

    C. Create an SCP to restrict users who share a common job function for specific permissions.

    D. Create an IAM role with the necessary permissions and assign the role to all IAM users who share the common job function.

14. You have outsourced the development of your application to a third-party provider. This provider will require temporary access to your AWS account to set up the necessary infrastructure and deploy the application. What type of identity should you configure for the provider to use to gain access?

    A. IAM User

    B. IAM Group

    C. IAM role

    D. Root user

15. Which tool on AWS can be used to estimate your monthly costs?

    A. AWS Pricing Calculator

    B. AWS TCO Calculator

    C. AWS Free Tier Calculator

    D. AWS Monthly Calculator

16. You need to differentiate the cost of running different workloads in your AWS account by business unit and department. How you can identify your resources, as well as their owners, in the billing reports generated by AWS?

    A. Designate specific tags as cost allocation tags in the AWS Billing and Cost Management Console.

    B. Set up an SNS alert for each department.

    C. Create a billing alarm.

    D. Configure consolidated billing in AWS Organizations.

17. Which AWS tool enables you to view your **Reserved Instance** (**RI**) utilization?

    A. AWS Cost Explorer

    B. AWS Config

    C. AWS CloudTrail

    D. AWS Personal Health Dashboard

18. Which set of credentials do you need to configure for IAM users who need to access your AWS account via the **command-line interface (CLI)**?

    A. IAM username and password

    B. IAM access key ID and secret access key

    C. IAM MFA

    D. IAM key pairs

19. An application is to be deployed on EC2 instances that will need to access an Amazon S3 bucket to upload any artifacts that are created. Which security option is considered a best practice to grant the application running on the EC2 instances the necessary permissions to upload files to the Amazon S3 bucket?

    A. Create an IAM user account with a set of access keys and assign the required level of permissions using an IAM policy. Hardcode the application with the access keys.

    B. Create an IAM user account with a username and password and assign the required level of permissions using an IAM policy. Hardcode the application with the username and password.

    C. Create an IAM role with the required level of permissions using an IAM policy. Attach the role to the application running on the EC2 instance.

    D. Create an IAM role with the required level of permissions using an IAM policy. Attach the role to the EC2 instances that will host the application.

20. Which AWS service enables you to troubleshoot IAM policies by identifying which set of permissions are allowed and which are denied?

    A. AWS Policy Simulator

    B. AWS Policy Manager

    C. AWS CloudTrail

    D. AWS **SCPs**

21. As part of your regular compliance processes, you are required to regularly audit the list of your IAM users and review information such as if they have been configured with passwords and access keys, as well as if MFA has been enabled on those accounts. Which AWS IAM service enables you to produce regular reports containing the preceding information?

    A. IAM Credentials Report

    B. IAM MFA Report

    C. AWS CloudWatch

    D. AWS Config

22. Which type of AWS policy enables you to define boundaries against what an IAM user or IAM role can be permitted to do in your AWS account?

    A. IAM policies

    B. Resource-based policies

    C. **SCPs**

    D. Permission boundaries

23. Which type of AWS policy enables you to control the maximum set of permissions that can be defined for AWS member accounts of an organization?

    A. IAM policies

    B. Resource-based policies

    C. **SCPs**

    D. Permission boundaries

24. Which of the following Amazon S3 storage classes can help you reduce the cost of storage for objects that are infrequently accessed, and yet still give you instant access when you need it?

    A. Amazon S3 Standard-IA

    B. Amazon S3 Glacier

    C. Amazon S3 Glacier Deep Archive

    D. Amazon S3 Standard

25. You are hosting an Amazon S3 bucket that contains important documents, and you want to enhance security whereby IAM users who try to access the objects can only do so from within the corporate office network. How would you configure your S3 bucket to fulfill this requirement?

    A. Create a resource policy granting the necessary level of access with a condition statement that defines and specifies the corporate office IP block.

    B. Create a resource policy granting the necessary level of access with a condition statement that specifies your corporate IAM users' accounts.

    C. Create an SCP granting access with a condition statement that specifies the corporate office IP block.

    D. Create an Amazon S3 **Access Control List** (ACL) with a condition statement that specifies your corporate IAM users' accounts.

26. Which type of Amazon S3 Storage class is cost-effective where you are unsure of your access patterns for the data contained within the S3 bucket?

    A. Amazon S3 Standard storage class

    B. Amazon S3 Standard-IA storage class

    C. Amazon S3 One-Zone IA

    D. Amazon S3 Intelligent Tiering

27. Your junior colleague accidentally deleted some financial data that was stored in an Amazon S3 bucket. How can you prevent such accidental deletions of data in Amazon S3?

    A. Do not give junior administrators access to Amazon S3.

    B. Set up Amazon S3 Versioning on your S3 bucket.

    C. Set up Amazon S3 Lifecycle Management.

    D. Set up Amazon S3 Termination Protection.

28. Which feature of Amazon S3 enables you to create a secondary copy of your objects in a given S3 bucket that will be stored in a different Region for compliance purposes?

    A. Amazon S3 **Cross-Region Replication** (CRR)

    B. Amazon S3 Same Region Replication

    C. Amazon S3 Versioning

    D. Amazon S3 Multi-Copy

29. Company policy dictates that objects stored in Amazon S3 must be encrypted at rest. It is also mandated that your choice of encryption should offer an auditing feature that shows when your **Customer Master Key (CMK)** was used and by whom. Which type of Amazon S3 encryption option will you need to configure to fulfill the requirements?

   A. **Server-Side Encryption with Amazon S3-Managed Keys (SSE-S3)**

   B. Client-Side Encryption

   C. **Server-Side Encryption with KMS keys stored in AWS Key Management Service (SSE-KMS)** Bitlocker

30. You need to retrieve a small subset of some archive data urgently to resolve a pending investigation. The data is stored in the Amazon S3 Glacier storage class. Which retrieval option can use to access the data urgently?

   A. Standard retrieval option

   B. Expedited retrieval option

   C. Bulk retrieval option

   D. Power retrieval option

31. You have a team of remote workers who need to upload research documents and videos to your Amazon S3 bucket hosted in the us-east-1 Region. You would like to ensure that your remote staff can upload research material with low latency access. What can you do to reduce speed variability for uploads, which are often experienced due to the architecture of the public internet?

   A. Enable Amazon **S3 Transfer Acceleration (S3TA)** for your bucket.

   B. Configure an IPSec site-to-site VPN connection between your remote workers and the VPC in the us-east-1 Region.

   C. Use the Amazon Storage Gateway service.

   D. Set up Amazon Express Route.

32. You need to transfer large amounts of data from your on-premises network to the Amazon S3 platform. The total data capacity is around 400 TB. You have decided to opt for the Amazon Snowball Edge service to complete the transfer. No data compute or processing is required. Which *flavor* of the Amazon Snowball Edge service would you recommend?

    A. Snowball Edge Compute Optimized

    B. Snowball Edge Storage Optimized

    C. Snowball Edge Data Optimized

    D. Snowball Edge Function Optimized

33. You host several Microsoft Windows applications on-premises that need low latency access to large amounts of storage. You would like to use the Amazon Storage Gateway service to host all application-level data. Which gateway option would you recommend?

    A. Amazon S3 File Gateway

    B. Amazon FSx File Gateway

    C. Volume Gateway Cached Mode

    D. Tape Gateway

34. Following best practices, you have deployed your application servers within the private subnets of a VPC. However, these servers require internet access to download updates and security patches. Which type of resource can enable you to grant internet access to EC2 instances in private subnets without having to assign public IP addresses to those instances?

    A. Internet gateway

    B. NAT gateway

    C. Subnet

    D. Route table

35. Which of the following statements is true about security groups?

    A. Security groups are stateful and you need to configure both inbound and the corresponding outbound rules for traffic to flow bidirectionally.

    B. Security groups are stateless and you do not need to configure both inbound and the corresponding outbound rules for traffic to flow bidirectionally.

    C. Security groups can be used to explicitly deny inbound traffic from a specific IP address range.

    D. Security groups are used to limit what actions IAM users that are members of the group can perform.

36. Which feature of the AWS VPC service enables you to connect multiple VPCs so that traffic between those VPCs can be sent using private IP address space?

    A. VPC peering

    B. VPC Flow Logs

    C. Subnets

    D. VPC endpoints

37. Which service enables you to reduce the complexity associated with establishing multiple VPC peering connections?

    A. AWS Transit Gateway

    B. AWS VPC Manager

    C. AWS Direct Connect

    D. IPSec VPN Tunnel

38. Which AWS service enables you to connect your on-premises network to your AWS account using a dedicated private connection that bypasses the internet altogether?

    A. IPSec VPN

    B. Express Route

    C. Direct Connect

    D. Snowball

39. Which AWS feature can help you establish connectivity between your on-premises network and your AWS VPC using an IPSec tunnel?

    A. Direct Connect

    B. **Virtual Private Network (VPN)**

    C. AWS Outposts

    D. Amazon SNS

40. You are about to publish your web application using an **Application Load Balancer (ALB)** and would like to use a friendly domain name to advertise the site to your users rather than the ALB's DNS name. Which AWS service can you use to configure the alias's name so that when users type in the friendly domain name into the browser, they are directed to the ALB's DNS URL?

    A. Amazon Route53

    B. Amazon CloudFront

    C. Amazon S3

    D. Amazon Direct Connect

41. Which AWS service enables you to purchase and register new domain names that can be used to publish your website on the internet?

    A. Route53

    B. VPC

    C. RDS

    D. Elastic Beanstalk

42. You have developed a web application that you want to offer redundancy and resilience for. Which feature of the Amazon Route53 service can help you design your web application with a primary site where all users' traffic is directed to, by default, and if the primary site is offline, then users are redirected to a secondary site located in a different Region.

    A. Simple routing policy

    B. Weighted routing policy

    C. Failover routing policy

    D. Geolocation routing policy

43. You plan to host a new Amazon S3 static website through which you will offer free recipe guides. The site is going to be accessed by users across the globe. The site contains lots of videos and images about the recipes you offer. Which AWS service can help you cache your digital assets locally to where users are located and thus reduce latency when your users access content on your website?

    A. Amazon Route53

    B. Amazon VPC

    C. Amazon CloudFront

    D. Amazon Cloud9

44. You have created an EC2 AMI that contains the base operating system and all necessary corporate settings/configurations. Your colleagues in another Region are trying to launch new EC2 instances but they are unable to access your AMI. What do you need to do so that your colleagues can use the new image?

    A. Copy the AMI to other Regions.

    B. Set up a VPC endpoint between the Regions to allow your colleagues to download the AMI.

    C. Copy the AMI to an S3 bucket.

    D. Use the Amazon Snowball service to send a copy of the AMI to your colleagues.

45. Which EC2 instance type is designed for floating-point number calculations, graphics processing, or data pattern matching?

    A. General Purpose

    B. Memory-Optimized

    C. Compute Optimized

    D. Accelerated Computing

46. You need to deploy a certain third-party application on an EC2 instance where the licensing term is based on a per-CPU core/socket basis. Which EC2 pricing option do you need to use for this requirement?

    A. On-Demand

    B. Reserved Instance

    C. Spot Instance

    D. Dedicated Host

47. You are currently running a test phase for a new application that is being developed in-house. Your UAT testers will need to access test servers for 3 hours a day, three times a week. The test phase is supposed to last 5 weeks. You cannot afford any interruptions to the application while the tests are being run. Which EC2 pricing option will be the most cost-effective?

    A. On-Demand

    B. Reserved

    C. Spot

    D. Dedicated Host

48. Which EBS volume type is designed for critical, I/O-intensive databases and application workloads?

    A. gp2

    B. st1

    C. sc1

    D. io1

49. Which of the following payment options will help you achieve the maximum discount for your RIs?

    A. A 1-year commitment with payment made using the **Partial Upfront** option.

    B. A 1-year commitment with payment made using the **All Upfront** option.

    C. A 1-year commitment with payment made using the **No Upfront** option.

    D. A 3-year commitment with payment made using the **All Upfront** option.

50. Which AWS service enables you to quickly deploy a **Virtual Private Server** (**VPS**) that comes preconfigured with common application stacks, SSD storage, and fixed IP addresses for a fixed monthly fee based on the configuration of the server?

    A. Amazon EC2

    B. Amazon Lightsail

    C. Amazon ECS

    D. Amazon ECR

51. You are planning on deploying a Docker application on AWS. You wish to deploy your Docker image without having to manage EC2 instances such as provisioning and scaling clusters, or patching and updating virtual servers yourself. Which service enables you to fulfill this requirement?

    A. Amazon ECS deployed using the EC2 Launch Type

    B. Amazon ECS deployed using the Fargate Launch Type

    C. Amazon ECS deployed using ECR

    D. Amazon ECS deployed with Lambda functions to manage your servers

52. Which of the following services is part of the AWS serverless offering that allows you to run code in response to a trigger or event?

    A. Amazon ECS

    B. AWS Lambda

    C. Amazon EC2

    D. AWS CloudFront

53. Which AWS storage option is designed to offer file sharing capabilities for Windows-aware applications and offers options for integration with Microsoft Active Directory?

    A. AWS FSx for Lustre

    B. Amazon FSx for Windows File Server

    C. AWS Elastic File Syste

    D. AWS instance store volumes

54. You are planning on deploying 10 EC2 instances across two Availability Zones that will host the new line of business applications. All the servers will need to share common files and will run the Amazon Linux 2 operating system. Which storage architecture would you recommend to host the shared files for your application servers?

    A. Amazon **Elastic File System** (**EFS**)

    B. Amazon FSx Lustre

    C. Amazon S3

    D. Amazon EBS

55. You have just launched a Windows EC2 instance. How can you obtain the Windows local administrator password?

    A. Raise a support request with Amazon to obtain the password.

    B. The password is sent to you automatically via email.

    C. The password is sent to you via an SMS text message to your registered mobile.

    D. Use the key pair to decrypt the password.

56. Which AWS service enables you to configure a hybrid solution by extending AWS Infrastructure so that EC2 and EBS services can be hosted in your on-premise data center?

    A. AWS RDS

    B. AWS Direct Connect

    C. AWS Outposts

    D. AWS Route53

57. Your company provides spread betting services. You wish to run an end of day analysis against the day's transaction costs and carry out the necessary market analysis. Which AWS service dynamically provisions the necessary compute services that will scale based on the volume and resource requirements of your submitted jobs?

    A. AWS Batch

    B. AWS CloudFront

    C. AWS Lambda

    D. AWS Blockchain

58. Which AWS service can help you deploy, manage, and scale containerized applications using Kubernetes on AWS?

    A. Amazon ECS

    B. Amazon EKS

    C. Amazon MFA

    D. Amazon EC2

59. Which of the following statements is an example of an advantage of using Amazon RDS over databases installed on EC2 instances?

   A. Amazon RDS is a fully managed database where AWS manages the underlying compute and storage architecture, as well as patching and updates.

   B. Amazon RDS grants you access to the operating system, allowing you to fine-tune the database for the operating system it is running.

   C. Amazon RDS is faster than running the Microsoft SQL Server database on EC2 instances.

   D. Amazon RDS automatically enables encryption of the data in Amazon RDS.

60. Which feature of Amazon RDS enables you to create a standby copy of the database and offer failover capabilities if the master copy fails?

   A. Read Replicas

   B. Multi-AZ

   C. Failover policy

   D. Snapshots

61. Your company is planning to migrate its on-premises MySQL database to Amazon RDS. Which service will enable you to perform the migration?

   A. Amazon **Server Migration Service (SMS)**

   B. Amazon **Database Migration Service (DMS)**

   C. Amazon VM Import Export

   D. Amazon Redshift Migration Utility

62. Which feature of AWS Redshift allows you to perform SQL queries against data stored directly on Amazon S3 buckets?

   A. Redshift leader node

   B. Redshift Spectrum

   C. Redshift Copy

   D. Redshift Streams

63. Which Amazon RDS engine offers high resilience with copies of the database placed across a minimum of three Availability Zones?

    A. MySQL

    B. PostgreSQL

    C. Microsoft SQL Server

    D. Amazon Aurora

64. Which AWS-managed database service enables you to store data using complex structures with options for nested attributes, such as a JSON-style document?

    A. Amazon RDS

    B. Amazon Redshift

    C. Amazon DynamoDB

    D. Amazon Aurora

65. Which AWS database service is designed to store sensitive data that is immutable and where the transactional logs are cryptographically verifiable?

    A. AWS QLDB

    B. Amazon Neptune

    C. Amazon Aurora

    D. Amazon RDS

Mock test 2

1. You are currently performing a manual snapshot of your single instance MySQL Amazon RDS database every 4 hours. Some users have complained that the application that connects to the database experiences a brief outage when the backup process initializes. What can you do to resolve this issue?

    A. Configure your Amazon RDS database with Read Replicas.

    B. Configure your Amazon RDS database with Multi-AZ.

    C. Configure an AWS backup to perform the RDS database backups.

    D. Use the DMS service to migrate the MySQL database to Microsoft SQL Server.

2. Your organization is in a healthcare industry based in New York. You are planning on using an in-memory caching engine to alleviate the load on your Amazon RDS database for frequently used queries. Which AWS in-memory caching engine offers Multi-AZ capabilities, encryption of data, and compliance with the **Health Insurance Portability and Accountability Act (HIPAA)**?

   A. Amazon Elasticache for Redis

   B. Amazon Elasticache for Memcached

   C. Amazon CloudFront

   D. Amazon DynamoDB DAX

3. Which AWS service offers a fully managed data warehousing capability and enables you to analyze large datasets using standard SQL and **Business Intelligence (BI)** tools?

   A. Amazon RDS

   B. Amazon QLDB

   C. Amazon Redshift

   D. Amazon Aurora

4. Which of the following services further increase your EC2 instances' costs? (Choose two)

   A. Detailed monitoring

   B. Use of Elastic Load Balancers

   C. S3 buckets that you connect to

   D. DynamoDB tables that you query

   E. Setting up multiple key pairs

5. Your developer team needs to deploy an Elastic Load Balancer that will direct traffic to your web servers based on the URL path and over the HTTPS protocol. Which Elastic Load Balancer would you recommend?

   A. Network Load Balancer

   B. ALB

   C. Gateway Load Balancer

   D. Classic Load Balancer

6. Which feature of the Elastic Load Balancer service is suitable for **Transmission Control Protocol (TCP)**, **User Datagram Protocol (UDP)**, and **Transport Layer Security (TLS)** type traffic and operates at layer 4 of the **Open Systems Interconnection (OSI)** model?

   A. Network Load Balancer

   B. ALB

   C. Gateway Load Balancer

   D. Classic Load Balancer

7. Which of the following statements is true about Elastic Load Balancers?

   A. Elastic Load Balancers act as firewalls to protect the application running on your EC2 instances.

   B. Elastic Load Balancers enable you to achieve high availability across multiple Regions by distributing incoming web traffic to targets located in multiple Regions.

   C. Elastic Load Balancers enable you to achieve high availability within a single Region by distributing incoming web traffic to targets located in multiple Availability Zones.

   D. Elastic Load Balancers enable you to scale horizontally by provisioning or terminating EC2 instances based on the demand of your resources.

8. Which component of an Elastic Load Balancer do you need to configure to ensure you accept traffic on a designated port and forward that traffic on a specific port to your EC2 instances behind the load balancer?

   A. Port forwarder

   B. NAT Gateway

   C. Listener

   D. Echo

9. You are building a multi-tier architecture with web servers placed in the public subnet and application servers placed in the private subnet of your VPC. Which type of load balancer would you choose to distribute traffic to your application servers?

   A. Internet-facing

   B. Internal load balancers

   C. Dynamic load balancers

   D. Static load balancers

10. Which configuration feature of the AWS Auto Scaling service enables you to define a maximum number of EC2 instances that can be launched in your fleet?

   A. Auto Scaling group

   B. Auto Scaling Launch Configuration

   C. Auto Scaling MaxFleet Size

   D. Auto Scaling policy

11. Which AWS service can help you provision only the necessary number of EC2 instances required to meet application demand, thus saving on costs usually associated with overprovisioning resources?

   A. Elastic Load Balancer

   B. Auto Scaling

   C. Cost Explorer

   D. EC2 Launcher

12. You have recently launched a new *free coupon* web application across a fleet of EC2 instances configured in an Auto Scaling group. Traffic has increased dramatically before the Black Friday sale and you have noticed that your Auto Scaling service is not launching any more EC2 instances, even though the threshold metrics have been crossed in CloudWatch. Your colleague tells you that you may have crossed a quota or limit on the number of EC2 instances you can launch. Which AWS service can offer you a quick look to determine this is the case?

   A. Personal Health Dashboard

   B. AWS Systems Manager

   C. AWS Config

   D. AWS Trusted Advisor

13. Which firewall protection service does the ALB offer to help protect against common web exploits such as cross-site scripting and SQL injection?

    A. AWS WAF

    B. AWS Shield

    C. Amazon Guard Duty

    D. **Network Access Control Lists (NACLs)**

14. Which dynamic scaling policy offered by the Amazon Auto Scaling service can help you launch or terminate EC2 instances in the fleet based on the target value of a specific metric?

    A. Target tracking scaling policy

    B. Step scaling policy

    C. Simple scaling policy

    D. Predictable scaling policy

15. You plan to use Amazon CloudWatch to send out alerts whenever the CPU utilization on your production EC2 instances is more than 80% for 15 minutes. Which AWS service can you use to send out this alert notification?

    A. Amazon SES

    B. Amazon SNS

    C. Amazon SQS

    D. Amazon MQ

16. Which feature of the Amazon SNS service enables you to push notification messages to multiple endpoints in parallel?

    A. You can use the SNS Fanout scenario to help you push notifications to multiple endpoints.

    B. You can use SNS FIFO topics to help you push notifications to multiple endpoints.

    C. You can change the timeout period to ensure that notifications are sent to multiple endpoints.

    D. To send out notifications to multiple endpoints, you will need to configure Amazon SQS to integrate with Amazon SNS.

17. Which AWS service enables you to design your application architecture by decoupling its components into distributed systems and facilitating the design and architecture of microservices?

    A. Amazon SNS

    B. Amazon Simple Queue Service (SQS)

    C. Amazon MQ

    D. Amazon Redshift

18. You plan to use Amazon SQS to help decouple your application components. Which queue type will help you ensure that the message order from one component to another is preserved?

    A. Configure Amazon SQS with a standard queue.

    B. Configure Amazon SQS with a FIFO queue.

    C. Configure Amazon SQS with a LIFO queue.

    D. Configure Amazon SQS with a DLQ.

19. You are planning on migrating an application to the cloud. Which message brokering service will enable you to continue to use Apache ActiveMQ and facilitate communications between application components?

    A. Amazon SQS

    B. Amazon MQ

    C. Amazon SNS

    D. Amazon SES

20. Which AWS service can help you trigger a Lambda function based on an event such as an object being deleted from an Amazon S3 bucket?

    A. AWS ECS

    B. AWS Batch

    C. AWS EventBridge

    D. Amazon CloudTrail

21. Your application architecture for an insurance claim solution has a workflow process that can take up to 30 days to complete and requires human intervention in the form of manual approval processes to follow. Which AWS service would you recommend for architecting the workflow process?

    A. Amazon SQS

    B. Amazon Step Functions

    C. AWS CloudFormation

    D. AWS Lambda

22. You plan to configure a Lambda function that will be used to automatically start and stop EC2 instances at the start and close of the business day, respectively. How can you automate the start and stop of EC2 instances according to a specified schedule?

    A. Configure Amazon SNS to send out an alert trigger to the Lambda function.

    B. Configure Amazon CloudTrail to trigger the Lambda function at the designated schedule.

    C. Configure Amazon CloudWatch Events with a rule to trigger the Lambda function at the designated schedule.

    D. Configure the Amazon Scheduler service.

23. You need to run certain SQL queries to analyze data from a streaming source and conduct analysis. Which of the following services can you use to analyze stream data in real time?

    A. Amazon SQS

    B. Amazon Kinesis Data Streams

    C. Amazon Kinesis Analytics

    D. Amazon Athena

24. You are required to run ad hoc test queries against weekly reports that are stored in Amazon S3. Which AWS service can you use to query raw data in Amazon S3 using standard SQL?

    A. Amazon Athena

    B. Amazon Kinesis

    C. Amazon RDS

    D. Amazon Redshift

25. Which AWS service can be used to load a massive amount of streaming data into your Redshift data warehousing solution in near real time?

   A. Amazon Kinesis Data Streams

   B. Amazon Kinesis Firehose

   C. Amazon Kinesis Video Streams

   D. Amazon Athena

26. Which AWS service can be used to create and publish interactive BI dashboards that can be embedded into your applications, websites, and portals using Amazon-provided APIs and SDKs?

   A. Amazon Athena

   B. Amazon QuickSight

   C. Amazon Config

   D. Amazon Glue

27. Which AWS service offers a serverless **Extract, Transform, and Load (ETL)** solution that's used to discover and extract data from various sources and perform any cleaning or normalization on data warehouses and data lakes, before loading them into databases?

   A. AWS QuickSight

   B. Amazon Athena

   C. Amazon Glue

   D. Amazon CloudTrail

28. As part of your migration to the cloud, you need to re-host an application that uses Apache Spark to process vast amounts of data for a big data project. Which service on AWS can you use to help with data transformation and perform ETL jobs such as sort, aggregate, and join on large datasets?

   A. AWS QuickSight

   B. Amazon EFS

   C. Amazon EMR

   D. Amazon S3

29. You need to regularly build test environments for new applications currently under development. Which AWS service can you use to automate the infrastructure build of your test environment and thus reduce the time taken to provision the infrastructure required?

    A. Amazon Elastic Beanstalk

    B. Amazon CloudFormation

    C. AWS OpsWorks

    D. AWS Systems Manager

30. Which service can be used to orchestrate and configure environments to deploy applications using the Chef and Puppet enterprise tools?

    A. Amazon CloudFormation

    B. AWS OpsWorks

    C. Amazon Elastic Beanstalk

    D. Amazon Cloud9

31. Which service enables developers to upload code to AWS and have the necessary infrastructure provisioned and managed to support that application?

    A. Amazon Elastic Beanstalk

    B. Amazon CloudFormation

    C. Amazon Cloud9

    D. AWS OpsWorks

32. Which of the following environment tiers within the Elastic Beanstalk architecture is designed to support backend operations?

    A. Web services tier

    B. Worker tier

    C. Backend tier

    D. Database tier

33. Which of the following formats are CloudFormation templates written in? (Choose two)

    A. YAML

    B. XML

    C. CSV

    D. JSON

    E. JAVA

34. Which of the following is an example of a custom CloudWatch metric?

    A. CPU utilization

    B. Disk read in

    C. Network bytes in

    D. Memory

35. Which feature of CloudWatch can help send you notification alerts via Amazon SNS whenever a particular threshold is breached for a specified period?

    A. Dashboards

    B. Alarms

    C. Logs

    D. Events

36. You plan to use CloudWatch Logs to monitor network traffic that enters the AWS environment that's been specifically destined for an EC2 instance. You would like to record all inbound network traffic on port 80 that was accepted. What service can you configure to help you achieve this requirement?

    A. ALB access logs

    B. VPC Flow Logs

    C. CloudTrail Logs

    D. Config logs

37. Which AWS service enables you to track user activity and API usage in your AWS account for auditing purposes?

    A. AWS Config

    B. AWS CloudWatch

    C. AWS CloudTrail

    D. AWS Trusted Advisor

38. Which AWS service can be used to see how resources are interrelated to each other, how they were configured in the past, and view historical changes to those resources over time?

    A. AWS Trusted Advisor

    B. AWS Systems Manager

    C. AWS Config

    D. AWS IAM

39. Which feature of the AWS System Manager service enables you to roll out security patches across EC2 instances and on-premises servers?

    A. Patch Manager

    B. Microsoft WSUS

    C. AWS Config

    D. SCCM

40. You are planning on deploying a three-tier application architecture that is comprised of a database backend. Your application has been hardcoded with the database connection strings and secrets such as username and password. The company's security policy dictates that this approach is unacceptable and they would like you to manage the secrets information more securely. What would you recommend?

    A. Store the configuration information in the SSM Parameter Store and reference the parameter name from your code to dynamically retrieve the connection information.

    B. Store the configuration information in Amazon Redshift and reference the connection details from your code to dynamically retrieve the connection information.

    C. Store the configuration information in Amazon S3 and reference the connection details from your code dynamically.

    D. Store the configuration information on an EBS volume and reference the connection details from your code dynamically.

41. Which AWS service can be used to manage and resolve incidents that affect their AWS-hosted applications?

    A. AWS Systems Manager Incident Manager

    B. AWS Systems Manager Event Manager

    C. Amazon EventBridge

    D. AWS **Personal Health Dashboard** (**PHD**)

42. Which AWS service can be used to identify resources that have not been configured by following security best practices?

    A. AWS CloudWatch

    B. AWS Trusted Advisor

    C. AWS IAM

    D. AWS CloudTrail

43. You are trying to review the AWS Trusted Advisor service to analyze potential cost savings opportunities for various workloads you have deployed on AWS. However, you have noticed that the Cost Optimization category is grayed out and there are no reports on current configuration states. What could be preventing you from viewing the Cost Optimization report?

    A. You do not have enough permissions to access the Cost Optimization category on AWS Trusted Advisor.

    B. You have not subscribed to either the business or enterprise support plans.

    C. You have logged in with an IAM account and only the root user can access pricing and cost information.

    D. The AWS account does not have an active debit/credit card associated with it.

44. Which Well-Architected Framework pillar suggests that replacing failed resources is often better than trying to figure out why the failure occurred? Identifying the reason for failure can be done later, but focusing on replacing the failed resource will help you get up and running quickly.

    A. Cost Optimization

    B. Fault Tolerance

    C. Reliability

    D. Performance

45. Which of the following services can help fulfill the guidelines provided in the performance pillar concerning ensuring low latency access to video content hosted in a single S3 bucket globally?

    A. Use AWS CloudFront to cache the video content closer to end users.

    B. Use AWS DynamoDB DAX to cache the video content closer to end users.

    C. Use Amazon Elasticache to cache the video content closer to end users.

    D. Use Amazon Kinesis to cache the video content closer to end users.

46. Which pillar of the Well-Architected Framework refers to selecting the appropriate pricing options that allow you to adopt a consumption model for provisioning various resources?

    A. Performance pillar

    B. Reliability pillar

    C. Fault Tolerance pillar

    D. Cost Optimization pillar

47. Regarding the AWS Shared Responsibility Model, who is responsible for patching Amazon RDS database instances?

    A. AWS

    B. Customer

    C. Database engine vendor

    D. Both the customer and AWS

48. Which AWS service gives customers access to various compliance reports that confirm if the services offered by AWS meet specific requirements and regulatory requirements?

    A. AWS CloudTrail

    B. AWS **Acceptable Usage Policy** (**AUP**)

    C. AWS Artifacts

    D. AWS Compliance Programs

49. AWS allows customers to run vulnerability scans and perform penetration testing. However, certain types of testing are not permitted. Which of the following actions is the customer prohibited from performing?

    A. Brute-force attacks by trying to guess your Amazon RDS database passwords.

    B. Running malware detection programs on your EC2 instances.

    C. Attempting to perform cross-site scripting or SQL injection tests via your ALB.

    D. Performing simulated **Distributed Denial of Service** (**DDoS**) attacks.

50. Which AWS service enables you to encrypt data stored in your Amazon S3 buckets with a **CMK** and offers auditing capabilities?

    A. **Server-Side Encryption with Amazon S3-Managed Keys (SSE-S3)**

    B. **Server-Side Encryption with CMKs Stored in AWS KMS (SSE-KMS)**

    C. **Server-Side Encryption with Customer-Provided Keys (SSE-C)**

    D. Client-Side Encryption with Amazon-Managed Keys

51. To meet strict compliance and regulatory requirements, you are required to encrypt the application data stored on your EC2 instances using dedicated FIPS 140-2 Level 3 validated devices. Which AWS service can you use to fulfill this requirement?

    A. AWS KMS

    B. AWS CloudHSM

    C. AWS TPM Hardware Modules

    D. AWS Certificate Manager

52. Which AWS security solution offers protection against **DDoS** attacks and features an AWS **Shield Response Team** (**SRT**) 24/7 to assist you in handling such attacks?

    A. AWS WAF

    B. AWS X-Ray

    C. AWS Detective

    D. AWS Shield Advanced

53. Which type of firewall solution integrates with Amazon CloudFront and ALBs to offer protection against common web exploits such as cross-site scripting and SQL injection?

    A. AWS WAF

    B. AWS Shield

    C. AWS X-Ray

    D. AWS Firewall Manager

54. You are planning lots of data on Amazon S3 and you would like to monitor how your data is accessed, particularly highlighting any sensitive information such as **personally identifiable information** (**PII**). Which AWS service can help you meet this requirement?

   A. Amazon Macie

   B. AWS GuardDuty

   C. AWS Detective

   D. AWS X-Ray

55. You are building a mobile application that will be publicly accessible and you would like to integrate a third-party identity provider for authentication purposes, such as Facebook or Google. Which AWS service can be used to set up identity and access control solutions for your web and mobile applications?

   A. AWS Cognito

   B. AWS IAM

   C. Active Directory

   D. AWS Certificate Manager

56. Which AWS service can help detect malicious activities by analyzing data from your CloudTrail event logs, Amazon VPC Flow Logs, and DNS logs?

   A. AWS Shield

   B. AWS Detective

   C. AWS GuardDuty

   D. Amazon Macie

57. Which AWS service can help you determine the root cause of security issues by extracting time-based events such as logins, network traffic from Amazon VPC Flow Logs, and data ingested from GuardDuty findings?

   A. AWS Shield

   B. AWS WAF

   C. AWS Detective

   D. Amazon Macie

58. You are planning on migrating your on-premises workloads and applications to the cloud. Which AWS service enables you to capture millions of real-time data points related to your IT environment and review recommendations for right sizing and appropriately costing workloads on AWS?

    A. AWS Pricing Calculator

    B. AWS Migration Evaluator

    C. AWS Hybrid Calculator

    D. AWS Cost Explorer

59. Which EC2 instance pricing model can offer up to a 90% discount off the on-demand price and be used in scenarios where interruptions to your instances will not impact the application workflow?

    A. Reserved Instances

    B. Spot Instances

    C. Dedicated Instances

    D. Dedicated Hosts

60. Which Amazon S3 storage class enables you to host 48 TB or 96 TB as part of the S3 storage capacity and provides the option to create a maximum of 100 S3 buckets on-premises?

    A. Standard storage class

    B. Standard One-Zone (IA)

    C. Glacier

    D. Amazon S3 on Outposts

61. Which type of policy can you create to grant anonymous access to the objects stored in an S3 bucket that can be used to host website assets?

    A. IAM policy

    B. IAM permission boundaries

    C. Resource policy

    D. SNS policy

62. Which AWS service enables you to register new domain names for your corporate business requirements?

   A. AWS DNS

   B. AWS Route53

   C. AWS VPC

   D. Amazon Macie

63. Which AWS service offers image and video analysis that can be used to identify objects, people, text, scenes, and other activities?

   A. Amazon Rekognition

   B. Amazon Kinesis Video Streams

   C. Amazon Prime

   D. Amazon Athena

64. Which AWS service offers text search and analytics capabilities that can store, analyze, and perform search functions against big data volumes in near real time?

   A. Amazon Redshift

   B. Amazon ElastiCache

   C. Amazon Elastisearch

   D. Amazon Search

65. You plan to migrate your entire on-premises network to the cloud and have also decided to move away from physical desktops and workstations to a complete VDI solution. Which service on AWS enables you to provision virtual desktops in the cloud, accessible via a web browser?

   A. Amazon EC2

   B. Amazon Lightsail

   C. Amazon WorkSpaces

   D. Amazon EKS

Answers

Chapter 1

1. D
2. A
3. A
4. A and B
5. A
6. C
7. C
8. C

Chapter 2

1. C and D
2. C
3. D
4. B
5. B
6. A
7. C
8. B
9. A and E
10. A and B
11. C
12. C

Chapter 3

1. A
2. A
3. A and B
4. A
5. A
6. B
7. B

Chapter 4

1. A and B
2. B
3. A
4. B
5. A
6. A
7. A
8. A
9. C
10. B

Chapter 5

1. A and D
2. B
3. C
4. D
5. B
6. C
7. A

8. A
9. B

Chapter 6

1. B
2. A
3. A
4. A
5. A
6. B
7. D
8. C
9. C
10. B

Chapter 7

1. D
2. A
3. C and E
4. A
5. A
6. C
7. A
8. A and B
9. A
10. B

Chapter 8

1. B
2. A
3. A, B, and C
4. C
5. B
6. B
7. C
8. C
9. B

Chapter 9

1. C
2. B
3. B
4. A
5. C
6. A

Chapter 10

1. A
2. D
3. B
4. C
5. A
6. A
7. C
8. C
9. A
10. D

Chapter 11

1. B
2. B
3. C and E
4. A
5. C
6. A
7. C

Chapter 12

1. C
2. B
3. A and B
4. C
5. B
6. B

Chapter 13

1. A
2. A
3. A
4. B
5. A

Chapter 14

1. A and B
2. B
3. A
4. A
5. A
6. B

Chapter 15

1. A
2. A
3. A
4. C
5. B

Chapter 16

Mock Test 1

1. C

   Infrastructure as a Service (IaaS) is a cloud service model that gives you access to virtualized infrastructure components comprising computing, network, and storage services. This is very similar to hosting your own VMware or Hyper-V virtualized platforms, where you deploy servers, attach storage volumes, and configure network connectivity services. However, the primary difference is that you do not have access to the underlying hypervisor platform with cloud-hosted IaaS solutions. IaaS offerings give the greatest amount of control over how the virtual components of your infrastructure are configured and also require you to take responsibility for managing, maintaining, and enforcing security measures for those components.

2. C

Companies who wish to move their entire suite of applications to the cloud would normally carry out a series of migration projects over time. During this migration phase, connectivity between the on-premises environment and the AWS cloud would be required to facilitate the migration. Many companies may also require a more permanent hybrid design architecture. This could be because certain types of applications need to be in much closer proximity to the on-premises infrastructure where users are based, such as your corporate office network, ultimately ensuring low-latency access, data residency requirements, or even unique local data processing requirements. AWS offers several services to build hybrid clouds from VPN technologies and Direct Connect services to offering on-premises services such as AWS Storage Gateway and AWS Outposts.

3. A and B

There are several reasons why you would choose a specific AWS Region to deploy your applications in, including the need for closer proximity to your end users and thus to reduce latency, data residence laws, regulatory requirements, the choice of services available, and costs.

4. C

Edge locations are AWS infrastructure facilities located across the globe that help cache content for Amazon CloudFront. Any content that is accessed from the origin is cached locally at one or more edge locations closer to users to access that content for a particular **time to live** (**TTL**). This way, repeated access to the same content is delivered over a low latency connection.

5. B

AWS Storage Gateway enables you to build hybrid cloud solutions by giving access to the Amazon S3 and Glacier environments from your on-premises network. You install and configure the gateway appliance at your on-premises location and can use any one of four gateway types to access unlimited storage in the cloud. These gateway types are Amazon S3 File Gateway, Amazon FSx File Gateway, Tape Gateway, and Volume Gateway.

6. B

    The Developer Support plan gives you access to technical support via email and chats only. This support plan is cheaper than the Business and Enterprise Support plans and is recommended for experimenting with or testing applications on AWS.

7. A and B

    Several services are configured from a global perspective on AWS. For example, with IAM, every IAM user in your AWS account is unique across the entire AWS Global Infrastructure. The same applies to Amazon Route53, where domain names configured in host zones have a global presence across all Regions on AWS. Another global service on AWS is Amazon CloudFront.

8. C

    One of the six advantages of cloud computing is the ability to **go global in minutes**. This is made possible because, as an AWS customer, you have access to all Regions and Availability Zones where you can provision the resources required to host your application in a matter of minutes.

9. A

    Amazon CloudWatch enables you to set up alarms that can be triggered when a particular threshold is crossed. You can set up an alarm for billing alerts that monitors when your total spend crosses a specified dollar amount. This alarm can be configured to send out an alert to an email address that uses the Amazon SNS service.

10. A

    Amazon EBS is like the virtual hard disks (volumes) that you attach to EC2 instances. They need to be provisioned in the same availability where the EC2 instance has been launched to be attached to that EC2 instance. Furthermore, you cannot attach an EBS volume to an EC2 instance in another Availability Zone. You could take a snapshot of the EBS volume and launch a new volume in another Availability Zone from that snapshot if required.

11. A

    AWS IAM Password Policies enable you to define custom policies on your AWS account to specify complexity requirements and mandatory rotation periods for your IAM users' passwords. This ensures that IAM user accounts are created with complex, hard-to-crack passwords.

12. A

Multi-Factor Authentication (**MFA**) is a recommended best practice authentication security measure. It requires you to authenticate with a username and password, as well as a security token generated by a physical or virtual device you own. Used together, these factors provide increased security for your AWS account settings and resources.

13. B

An IAM Group can be used to club users who share common job functions. You can then assign the necessary permissions to the IAM Group, which will filter down to the IAM users that are members of the group.

Where possible, entities that need access to services and resources on AWS ought to be granted access via IAM roles using temporary security credentials. For example, users who need to access resources in another AWS account can be granted permission to assume a role rather than having to create IAM accounts for them in the other account. This is considered best practice and AWS highly recommends this approach.

14. C

An IAM role enables you to grant external users access to resources in your AWS account using temporary credentials that are managed by the AWS **Security Token Service** (**STS**). An IAM role will also have an IAM policy attached to it that specifies the exact set of permissions the role will grant to the external user(s).

15. A

AWS Pricing Calculator can help you estimate the monthly costs of the resources you wish to provision on AWS. In addition to specifying the expected usage for the month, to calculate the cost of the resources, you can provide additional details such as the amount of data transferred in and out of the AWS Region and cross-Region to get a complete estimate.

16. A

You can use cost allocation tags to identify your resource costs on your cost allocation report and track your AWS costs by resources and who owns them. AWS provides two types of cost allocation tags: AWS-generated tags and user-defined tags.

17. A

AWS Cost Explorer enables you to view 12 months of usage and spending data, as well as a forecast of what your future costs will be for the next 12 months. Cost Explorer can also provide information on how much of your RI you have utilized, as well as the savings that have been made from using RIs over on-demand options for EC2 Amazon Redshift, Amazon RDS, and more.

18. B

To access an AWS account as an IAM user, you need to create a set of access keys, which is a combination of an access key ID and a secret access key. Note these keys are considered long-term access keys. One set of keys is associated with a specific IAM user.

19. D

You can create an IAM role with the necessary permissions to upload objects to a specific Amazon S3 bucket. You should then deploy your EC2 instances with this role attached (this involves creating an instance profile associated with the role). The IAM role will enable the application running on the EC2 instance to access the Amazon S3 bucket. Using an IAM role, the EC2 instance will obtain temporary credentials from the **Security Token Service (STS)**.

20. A

AWS Policy Simulator can help you identify which set of policies are allowed and which are denied against specific identities, groups of identities, and IAM roles. You can obtain granular visibility into specific permissions that have been allowed and denied, helping you troubleshoot access issues.

21. A

The IAM Credentials Report enables you to review IAM user accounts created in your AWS account. You can identify if an IAM user has been configured with a username and password, as well as access keys. You can also identify IAM users that may not have accessed resources in your AWS accounts recently, which may indicate that those accounts may not be required anymore. Regularly deleting unwanted IAM user accounts is part of the security best practice.

22. D

Permission boundaries enable you to define the maximum set of permissions that can be granted by an identity-based policy for an IAM user or IAM role.

23. C

SCPs is a feature of the AWS Organization service that enables you to set the maximum set of permissions that can be defined for member accounts. You can also set policies to prevent the root users of member accounts from removing the membership to an organization management account once the invitation to become a member account has been accepted.

24. A

Amazon S3 Standard-IA can be used to store objects that you are not going to frequently access, but at the same time, you have instant access to the data when you need it.

25. A

You can create a resource policy with a condition statement that allows you to restrict the application of the policy based on a predefined condition, such as the corporate office network IP block.

26. D

Amazon S3 Intelligent Tiering is ideal if you are unsure of what your object access patterns might be. Objects are automatically transitioned across four different tiers, two of which are latency access tiers, which are designed to move objects between frequently accessed and infrequently accessed tiers, while the other two are optional archive access tiers. For the infrequent access tier, if you do not access your objects for 30 days, then it transitions to the Amazon S3 Standard-IA storage class, which is cheaper. If you need to access the same objects again later, they are transitioned back to the Amazon S3 Standard storage class.

27. B

To protect against accidental deletions or overwriting, Amazon S3 Versioning can be enabled. This service ensures that if someone tries to perform a delete request on an object without specifying the version ID, it will not be deleted. Instead, a delete marker will be added and the object will be hidden from view. You can then delete this marker to reenable access to the object. You should also consider setting a bucket policy so that not all users can perform delete requests.

28. A

Amazon S3 **CRR** is used to asynchronously copy objects across AWS buckets in different AWS Regions. This feature can be used to fulfill compliance and regulatory requirements, which may require you to store copies of data thousands of kilometers away for **Disaster Recovery** (**DR**) purposes.

29. C

With SSE-KMS, you can encrypt your objects in Amazon S3. You can create and manage your **CMKs**, as well as benefit from the auditing feature, which shows when your CMK was used and by whom. This service integrates with Amazon CloudTrail to offer full auditing features.

30. B

If you need urgent access to just a subset of your archives, you can opt for the Expedited retrieval option. Expedited retrievals are made available within 1 to 5 minutes for archives of up to 250 **megabytes** (**MB**).

31. A

S3TA reduces this speed variability that is often experienced due to the architecture of the public internet. S3TA routes your uploads via Amazon CloudFront's globally distributed edge locations and AWS backbone networks. This, in turn, gives faster speeds and consistently low latency for your data transfers.

32. B

The Amazon Snowball Edge Storage Optimized device offers a larger storage capacity and is ideal for data migration tasks. With 80 TB of HDD and 1 TB of **serial advanced technology attachment** (**SATA**) SDD volumes, you can start moving large volumes of data to the cloud. The device also comes with 40 vCPUs and 80 GB of memory.

33. B

Amazon FSx File Gateway enables you to connect your on-premises Windows applications that need large amounts of storage to the cloud-hosted Amazon FSx service for Windows File Server with low latency connectivity. Amazon FSx File Gateway also supports integration with **Active Directory** (**AD**) and the ability to configure access controls using **ACLs**.

34. B

NAT gateways help relay outbound requests to the internet on behalf of EC2 instances configured to use them. The NAT gateway replaces the source IPv4 address of your EC2 instances with the private IP address of the NAT gateway, thus acting as a proxy. Response traffic is then redirected by the NAT gateway back to the private IP address of the EC2 instance that made the original request.

35. B

Security groups are stateful. This means that even if you have not configured any inbound rules, response traffic to any outbound requests will be permitted inbound by the Security Group. Similarly, if you configured any inbound rules, outbound response traffic to any inbound traffic is permitted, without you having to explicitly create those outbound rules.

36. A

A VPC peering connection is a private network connection between two VPCs. The service allows you to connect multiple VPCs so that instances in one VPC can access resources in another VPC over a private IP address space.

37. A

The problem with VPC peering, when you're configuring multiple VPC to connect, is that every VPC must establish a one-to-one connection with its peer. This can quickly create complex connections that are difficult to manage. Route tables for each VPC also need to be configured for every peering connection.

AWS Transit Gateway allows you to connect your VPCs via the gateway in a hub-and-spoke model, greatly reducing this complexity as each VPC only needs to connect to the Gateway to access other VPCs.

38. C

Direct Connect is a service that enables you to connect your corporate data center to your VPC and the public services offered by AWS, such as Amazon S3, via a dedicated private connection that bypasses the internet altogether. The service enables you to achieve bandwidth connectivity of up to 100 Gbps.

39. B

You can set up a **VPN** connection between your on-premises network and your VPC. This is a secure encrypted site-to-site tunnel that's established between two endpoints over the public internet. It offers AES 128 or 256-bit **IPsec** encryption, which means that you can transfer data between the two endpoints securely.

40. A

Amazon Route53 can help you create alias records so that when a user types in a corporate domain-friendly name into the browser, it will direct the traffic to an AWS service, such as an ALB, giving access to the web application.

41. A

Amazon Route 53 offers domain name registration. You can purchase and manage domain names such as `example.com` and Amazon Route 53 will automatically configure the DNS settings for your domains.

42. C

To offer high availability of your web application, you can host two copies of your resources ideally across different Regions. One set of resources will be designated as your primary resource and the other as a secondary resource. If the primary resource is offline, then users' requests are redirected to the secondary resource.

43. C

Amazon CloudFront is a **Content Delivery Network** (**CDN**) that helps you distribute your static and dynamic content globally over low latency connections. The service caches content at edge locations closer to where your users are accessing the website.

44. A

To launch an EC2 instance with a custom AMI that you have built in another Region, you need to ensure that you copy the AMI to that Region.

45. D

Accelerated Computing EC2 instance types are designed with hardware accelerators, or co-processors, to perform complex functions. They are best for processing complex graphics, number crunching, and machine learning.

46. D

A Dedicated Host is a physical host dedicated for your use alone and gives you additional control and management capability over how instances are placed on a physical server. In addition, dedicated hosts can help address certain third-party licensing terms that are based on a per-CPU core/socket basis.

47. A

On-Demand is ideal for users who need the flexibility to consume compute resources when required and without any long-term commitment. They are ideal for test/dev environments or for applications that have short spiky or unpredictable workloads.

48. D

Provisioned IOPS SSDs offering high-performance EBS storage is ideal for critical, I/O-intensive databases and application workloads.

49. D

Using an **All Upfront** payment option for your RIs means paying for the entire term of the RI upfront at the beginning of the contract. You do not get a monthly/hourly bill and you benefit from the maximum available discount. Furthermore, a 3-year commitment will offer a bigger discount than a 1-year one.

50. B

Amazon Lightsail is a VPS solution that comes pre-configured with common application stacks such as WordPress, Drupal, Plesk, LAMP, and your chosen operating system. You choose the size of the server and it comes preconfigured with SSD storage, an IP address, and more. The best part about Lightsail is that you have a fixed monthly fee based on the instance type and the associated operating system and applications that have been deployed.

51. B

The ECS Fargate Launch Type enables you to set up your ECS environment without having to spin up EC2 instances, provision and scale clusters, or patch and update virtual servers yourself. AWS will manage how the ECS tasks are placed on the cluster, scale them as required, and fully manage the entire environment for you.

52. B

Amazon Lambda is a serverless offering from AWS that allows you to run code and perform some tasks. Amazon Lambda is known as a **Function as a Service** (**FaaS**) solution that can be used to build an entirely serverless architecture comprised of storage, databases, and network capabilities where you do not manage any underlying servers.

53. B

Microsoft Windows-aware applications that need to share files can easily use FSx for Windows File Share, which offers support for the SMB protocol and Windows NTFS, **AD** integration, and **Distributed File System** (**DFS**).

54. A

Amazon EFS can be used by Linux-based EC2 instances as a centralized file storage solution. This is particularly useful when you have applications deployed across multiple EC2 instances that need to share common files. Amazon EFS can also be accessed from on-premises servers over a VPN or Direct Connect service.

55. D

When you launch a server (Windows or Linux), you must configure it to be associated with a key pair. This is an encrypted key where you will be able to use your private key to log in to Linux-based servers or decrypt the Windows administrator password using the AWS Management Console.

56. C

AWS Outposts is ideal when you want to run AWS resources with very low latency connections to your on-premises application or if you require local data processing due to any compliance and regulatory requirements. You can get AWS Outposts delivered to your local on-premises location as a 42U rack and can scale from 1 rack to 96 racks to create pools of compute and storage capacity.

57. A

AWS Batch can be used to run thousands of batch computing jobs on AWS for performing various types of analysis. AWS Batch will set up and provision the necessary compute resources to fulfill your batch requests. There is no need to deploy server clusters as AWS takes care of this for you.

58. B

Amazon EKS is designed to help you deploy, manage, and scale containerized applications using Kubernetes on AWS.

59. A

The primary advantage of using Amazon RDS over installing databases on EC2 is the fact that AWS manages all the compute and storage provisioning, as well as performing all management tasks on the database. This frees you up to focus on your application and the infrastructure components that host the database.

60. B

Amazon RDS offers a feature known as Multi-AZ where the primary (master) copy of your database is deployed in one Availability Zone and a secondary (standby) copy is deployed in another Availability Zone. Data is then synchronously replicated from the master copy to the standby copy continuously. If the master copy fails, AWS will promote the standby copy to become the new master and perform a failover.

61. B

Amazon offers a **Database Migration Service** (**DMS**) that can be used to migrate the data from one database to another. This migration can be performed from your on-premises network to the AWS cloud over a VPN connection or a Direct Connect connection. AWS DMS offers support for both homogeneous migrations, such as from MySQL to MySQL or Oracle to Oracle, as well as heterogenous migrations between engines such as Oracle to Microsoft SQL Server or Amazon Aurora.

62. B

Redshift Spectrum allows you to perform SQL queries against data stored directly on Amazon S3 Buckets. This is particularly useful if, for instance, you store frequently accessed data in Redshift and some infrequently accessed data in Amazon S3.

63. D

Amazon Aurora is an AWS proprietary database that maintains copies of the database placed across a minimum of three Availability Zones. Amazon Aurora is also five times faster than standard MySQL databases and three times faster than standard PostgreSQL databases. It also offers self-healing storage capabilities that can scale up to 128 TB per database instance.

64. C

DynamoDB supports both key-value and document data models such as JSON. DynamoDB is a NoSQL database solution that offers a flexible schema and offers single digit millisecond performance at any scale.

65. A

Amazon QLDB is a fully managed ledger database that provides a transparent, immutable, and cryptographically verifiable transaction log owned by a central trusted authority. Amazon QLDB can maintain a history of all data changes.

Mock test 2

1. B

   In scenarios where you have a single RDS database instance deployed, your users are likely to experience a brief I/O suspension when your backup process initializes. By configuring your Amazon RDS database with Multi-AZ, the backup is taken from the standby copy of the database instead of the master. This will ensure that users do not experience any brief outages when trying to access the database.

2. A

   Amazon Elasticache for Redis is designed for complex data types, offers Multi-AZ capabilities, encryption of data, and compliance with FedRAMP, HIPAA, PCI-DSS, as well as high availability and automatic failover options.

3. C

   Amazon Redshift is AWS's data warehousing solution that is designed for analytics and is optimized for scanning many rows of data for one or multiple columns. Instead of organizing data as rows, Redshift transparently organizes data by columns. You can use standard SQL to query the database and use it with your existing BI tools.

4. A and B

   Detailed monitoring and use of Elastic Load Balancers will increase your EC2 instances' costs. This is because Elastic Load Balancers are not part of the free tier and you are charged based on each hour or partial hour that a load balancer is running and the number of **Load Balancer Capacity Units** (**LCUs**) used per hour. Furthermore, whereas basic monitoring is offered free of charge, detailed monitoring is a chargeable service.

5. B

   ALBs are designed to distribute traffic at the application layer (using HTTP and HTTPS). Furthermore, with ALBs, you can have multiple target groups, allowing you to define complex routing rules based on the different application components. You can configure path-based routing, host-based routing, and much more. You can also configure Lambda functions as targets for your load balancer.

6. A

    Network Load Balancers are designed to operate at the fourth layer of the OSI model and can handle millions of requests per second. Network Load Balancers are designed for load balancing both TCP and UDP traffic and maintaining ultra-low latencies.

7. C

    Using a load balancer, you can direct incoming traffic to multiple registered EC2 instances across multiple Availability Zones within a given Region. This enables you to offer high availability in case any of the EC2 instances fails or even if an entire Availability Zone goes offline.

8. C

    You need to configure listeners to specific ports that you will accept incoming traffic on and the ports you will use to forward traffic to the EC2 instances.

9. B

    The nodes of an internal load balancer only have private IP addresses and allow communication between the web layer and the internal application layer. The DNS name of an internal load balancer can be publicly resolved to the private IP addresses of the nodes. Therefore, internal load balancers can only route requests from clients with access to the VPC for the load balancer.

10. A

    When configuring your Auto Scaling group, you can define the minimum and maximum size of your group. You can also choose to keep the size of the group to an initial size that does not expand the group size but ensures that you always have the exact number of EC2 instances running.

11. B

    Amazon Auto Scaling can help you provision new EC2 instances based on CloudWatch metrics that change according to the load on your instances. Similarly, when demand drops, you can configure Auto Scaling to terminate unwanted resources and thus save on costs.

12. D

   AWS Trusted Advisor is an online tool that can offer guidance on AWS infrastructure, improve security and performance, reduce costs, and monitor service quotas. The service quota (service limits) category of Trusted Advisor can notify you if you use more than 80% of a service quota for a specific service. You can then follow recommendations to delete resources or request a quota increase.

13. A

   The AWS ALB service provides integration with **Web Application Firewall** (**WAF**), which helps protect against common web exploits such as SQL injection and cross-site scripting.

14. A

   With the target tracking scaling policy, Auto Scaling will launch or terminate EC2 instances in the fleet based on the target value of a specific metric. The target tracking policy tracks the metric value and attempts to ensure the correct number of EC2 instances are running to meet that target value.

15. B

   Amazon **Simple Notification Service** (**SNS**) is a push-based messaging and notification system that can be used to allow one application component to send messages to another application component or directly to end users. So, Amazon SNS can be used to send out alerts when a particular CloudWatch metric crosses a threshold for a period and triggers an alarm.

16. A

   The Fanout scenario enables you to publish messages to an SNS topic for parallel asynchronous processing. You send the notifications to supported endpoints such as Kinesis Data Firehose delivery streams, Amazon SQS queues, HTTP(S) endpoints, and Lambda functions.

17. B

   Amazon **SQS** is a fully managed message queuing solution that enables you to decouple your application components into distributed systems and facilitates the design and architecture of microservices. A queueing system such as Amazon SQS can help different components of your application work independently. Queues can hold messages in the form of requests/tasks until capacity is available.

18. B

FIFO stands for **First-In-First-Out** and its queues are designed to preserve the order of your messages, as well as ensure only one-time delivery with no duplicates.

With FIFO queues, you can get a throughput at a rate of 300 transactions per second. If you use batching, you can get up to 3,000 transactions per second, per API method (`SendMessageBatch`, `ReceiveMessage`, or `DeleteMessageBatch`).

19. B

Amazon recommends using Amazon MQ for migrating applications from existing message brokers where compatibility with APIs such as JMS or protocols such as AMQP 0-9-1, AMQP 1.0, MQTT, OpenWire, and STOMP are required.

20. C

Amazon EventBridge is a serverless *event bus service* that allows you to stream real-time events from your applications to support targets such as Lambda functions, which can then be triggered to take some form of action. In the preceding example, EventBridge can trigger a Lambda function if someone tries to delete an object in your S3 bucket and some action can be taken in response to the event.

21. B

Amazon Step Functions enables you to define these workflows as a series of state machines that contain states that make up the workflow. These states make decisions based on input, perform some action, and produce an output to other states. Step Functions also allow you to integrate human interaction, particularly where manual intervention is required, and can run for up to 1 year.

22. C

You can configure CloudWatch Events with a rule to trigger a Lambda function at a defined schedule. You would need to create the Lambda function and then, in the rule settings for CloudWatch events, specify the Lambda function as a target to be triggered at the designated schedule.

23. C

Kinesis Data Analytics lets you query and analyze stream data in real time. You can use standard programming and database query languages such as Java, Python, and SQL to query streaming data as it is being ingested.

24. A

Amazon Athena is an interactive query service that can be used to analyze data in Amazon S3 using standard SQL. To set up the service, you need to specify the source S3 bucket and define a schema.

25. B

Amazon Kinesis Firehose is designed to capture, transform, and deliver streaming data to several AWS services, including Amazon S3 and Redshift in near real time. The service can also batch, compress, transform, and encrypt your data streams, thereby reducing storage usage and increasing security.

26. B

Amazon QuickSight is a serverless **Business Intelligence** (**BI**) service that can help you build interactive dashboards and embedded visualizations into your applications and web portals.

27. C

AWS Glue is a fully managed ETL service that makes it easy for customers to prepare and load their data for analytics.

28. C

Amazon EMR is a managed Hadoop framework that allows you to process vast amounts of big data. You can use open source tools such as Apache Spark, Apache Hive, Apache HBase, Apache Flink, Apache Hudi, and Presto.

29. B

Amazon CloudFormation is a solution that can help you design, build, and deploy your infrastructure using code. You can create templates that use a declarative approach to instruct CloudFormation to build a precise infrastructure repeatedly, as required, by your testing team. CloudFormation templates can also be configured to accept input parameters for environment-specific configurations and variations required in the build.

30. B

AWS OpsWorks is a configuration management and orchestration service that enables you to provision resources such as servers both in the cloud and on-premises using Chef and Puppet. With OpsWorks, you can define service layers for your application stack such as the database layer, load balancer layer, and more.

31. A

Amazon Elastic Beanstalk is a service that enables you to deploy your application without having to manually configure the underlying infrastructure that will support the application. You upload your code in a supported language and environment and AWS will provision the underlying infrastructure, such as the compute, storage, and network components, to support the application. Amazon ElasticBeanstalk will also enable you to specify how the underlying infrastructure components will be deployed – for example, you can specify the EC2 instance type and size that is deployed or enforce that a set minimum number of EC2 instances are deployed as part of an Auto Scaling group.

32. B

In a multi-tier application architecture, backend operations such as application, middleware, or database operations are performed by the worker tier of your Elastic Beanstalk configuration. AWS Elastic Beanstalk will also provision an Amazon SQS queue to facilitate communication between the web services tier and the worker tier.

33. A and D

CloudFormation templates can be written in both JSON and YAML format. These are declarative markup languages that can help CloudFormaton provision infrastructure in your AWS account.

34. D

Memory is a custom metric because memory metrics are at the OS level and cannot be monitored by default. To ingest custom metrics, you need to use the CloudWatch agent or the **PutMetricData** API action to publish them to CloudWatch.

35. B

You can configure CloudWatch alarms to monitor a given resource metric, such as the average CPU utilization of an EC2 instance. If the metric crosses a specific threshold for a specified period, then the alarm can be triggered to take a certain action. The alarm only triggers if the threshold has been breached for a specified period.

36. B

VPC Flow Logs can capture information about the IP traffic going to and from network interfaces in your VPC. You can configure VPC Flow Logs to capture all traffic to the VPC, a specific subnet, or a specific network interface of an EC2 instance.

37. C

AWS CloudTrail stores event history from within the CloudTrail dashboard for every activity that occurs in your AWS account. You can create trails to store specific management events or data events and if you require more than 90 days' worth of event history.

38. C

AWS Config is a service that allows you to gain visibility into how your AWS resources are configured and deployed in your AWS account. This includes configuration information, as well as changes to those configurations over time. You can also use AWS Config to enforce specific configurations rules and ensure that you follow internal guidelines that fulfill compliance requirements.

39. A

AWS Systems Manager's Patch Manager enables you to automatically patch your EC2 instances that are comprised of security and application updates. Note that updates for applications on Windows servers are limited to those released by Microsoft.

40. A

AWS Systems Manager's Parameter Store enables you to provide sensitive information such as passwords and database strings as parameter values. These values can be stored or encrypted, and your application can be configured to securely retrieve these values as they are needed from the Parameter Store.

41. A

The AWS Systems Manager Incident Manager service offers a management console to track all your incidents and notify responders of the impact, identify data that can help with troubleshooting, and help you get services back up and running.

42. B

AWS Trusted Advisor analyzes your resources and how they have been configured and compares those configurations against security practices to identify opportunities to save money, improve system availability and performance, or address security concerns.

43. B

The AWS Trusted Advisor service offers different levels of checks based on the AWS support plan that you have subscribed to. To access the full range of checks across all categories, you must be subscribed to either the Business or Enterprise Support plan. With either of these plans, you can also use Amazon CloudWatch Events to monitor the status of Trusted Advisor checks.

44. C

The Reliability pillar also focuses on how quickly you can recover from failure based on your architectural design. This is because failures are bound to happen and your architecture must be able to recover from these failures swiftly. One key concept that you should also consider is that replacing failed resources is often better than trying to figure out why the failure occurred and then attempting to resolve the issue that caused the failure.

45. A

When architecting solutions for the cloud, you must select the resource types and sizes based on your performance needs, while monitoring your resources consistently to ensure you maintain those levels of performance as per demand. Amazon CloudFront can help you improve performance by reducing the latency associated with accessing large amounts of content across the globe. It does this by caching content locally at edge locations as they are being accessed.

46. D

The Cost Optimization pillar focuses on ensuring that you architect and build solutions in a manner that avoids unnecessary costs. At the same time, you want to be able to ensure that your applications are highly performant, reliable, operationally efficient, and secure. To achieve cost optimization, you should understand your spending patterns and analyze where the money is going.

47. A

Amazon RDS is a managed database service. The customer can provision databases and select the instance type and size to power the database. However, the customer cannot manage the instance itself as this is taken care of by AWS, which includes patching and installing database updates. In contrast, patching EC2 instances is the customer's responsibility as EC2 is not a fully managed service.

48. C

Compliance reports and agreements are available via a portal on AWS known as **AWS Artifact**. These reports include AWS **Service Organization Control** (**SOC**) reports, **Payment Card Industry** (**PCI**) reports, and certifications from accreditation bodies across different Regions.

49. D

As a customer, you need to follow the service policy for penetration testing, which includes permitted services and prohibited activities. Such prohibited activities are **Denial of Service** (**DoS**), **DDoS**, simulated DoS, and simulated DDoS.

50. B

With SSE-KMS, you create and manage **CMKs**, and you use these keys to encrypt data keys and your data. SS3-KMS offers additional features such as auditing capabilities and integrates with CloudTrail.

51. B

AWS CloudHSM is a dedicated **Hardware Security Module** (**HSM**) that allows you to generate and manage your encryption keys in the cloud. You are provided with dedicated FIPS 140-2 Level 3 validated HSM devices, placed in your VPC, that are fully managed for you by AWS.

52. D

AWS Shield is a fully managed service offering protection against DDoS attacks. AWS Shield Advanced offers additional protection against attacks on your EC2 instances, ELBs, CloudFront, Global Accelerator, and Route53 resources. The service also offers a dedicated AWS **Shield Response Team** (**SRT**) 24/7 to assist you in handling such attacks.

53. A

AWS WAF can help protect applications at layer 7 of the **OSI** model, which helps you monitor and protect traffic over HTTP and HTTPS. This allows you to protect your content from common web exploits, such as SQL injection and cross-site scripting.

54. A

Amazon Macie uses machine learning and pattern matching techniques to detect and alert on any sensitive data, such as **PII**, stored in Amazon S3.

55. A

Amazon Cognito enables you to set up identity and access control solutions for your web and mobile applications using standards such as OAuth 2.0, SAML 2.0, and OpenID Connect. With Amazon Cognito, you can create user pools and identity pools.

56. C

AWS GuardDuty is a threat detection service that can analyze and detect malicious activity against your AWS accounts and application workloads. The service can detect the use of exposed credentials, any communication with malicious IP addresses and domains, as well as irregular activities carried out in your AWS account.

57. C

Amazon Detective can extract time-based events such as logins, network traffic from AWS CloudTrail and Amazon VPC Flow Logs, as well as ingest your GuardDuty findings to determine the root cause of those security findings.

58. B

With the Migration Evaluator service, you can use the AWS Application Discovery service, the TSO Logic agentless collector, or third-party tools to discover and gain insights into your current compute, storage, and total cost of ownership. The agentless collector tool can analyze any on-premises resources that just require read-only access to your VMware, Hyper-V, Windows, Linux, Active Directory, and SQL Server infrastructure.

59. B

Spot EC2 instances are ideal for applications that are fault-tolerant, scalable, or flexible, and where your application can tolerate interruptions. Spot Instances can save you up to 90% on On-Demand prices and there is no upfront commitment.

60. D

Amazon S3 on Outposts offers durability and redundancy by storing data across multiple devices and servers hosted on your outposts. It is ideal for low-latency access, while also enabling you to meet strict data residency requirements.

61. C

Resource policies are designed to enable access to resources such as objects in an Amazon S3 bucket. This policy enables you to identify a principal that you grant access to. With resource-based policies, you can configure the principal as a wildcard (*), which denotes anyone, and enables you to grant anonymous access.

62. B

Amazon Route53 offers complete domain name registration services. When you choose a name to register, you do so under a **top-level domain** (**TLD**) such as .com, .co.uk, .org, or .net. If the name of choice under a particular TLD is not available, you could try a different TLD.

63. A

Amazon Rekognition is a service that uses machine learning to identify objects, people, text, scenes, and activities in images and videos, as well as to detect any inappropriate content. Amazon Rekognition can be used for various application solutions such as identifying people, or sensitive data such as **PII** in images and videos.

64. C

Elasticsearch is an open source full-text search and analytics engine that can analyze all types of data such as textual, numerical, geospatial, structured, and unstructured data. Amazon Elasticsearch offers integration with Kibana, which is a data visualization tool, and Logstash, which is an open source, server-side data processing pipeline.

65. C

Amazon WorkSpaces is an end user computing service that enables you to deploy virtual Linux and Windows desktops in the cloud. AWS manages these virtual desktops, including security patching and managing the operating system. With Amazon WorkSpaces, you can consider migrating away from your on-premises desktop infrastructure to a **Virtual Desktop Infrastructure** (**VDI**) solution.

Packt>

Subscribe to our online digital library for full access to over 7,000 books and videos, as well as industry leading tools to help you plan your personal development and advance your career. For more information, please visit our website.

Why subscribe?

- Spend less time learning and more time coding with practical eBooks and Videos from over 4,000 industry professionals

- Improve your learning with Skill Plans built especially for you

- Get a free eBook or video every month

- Fully searchable for easy access to vital information

- Copy and paste, print, and bookmark content

Did you know that Packt offers eBook versions of every book published, with PDF and ePub files available? You can upgrade to the eBook version at packt.com and as a print book customer, you are entitled to a discount on the eBook copy. Get in touch with us at customercare@packtpub.com for more details.

At www.packt.com, you can also read a collection of free technical articles, sign up for a range of free newsletters, and receive exclusive discounts and offers on Packt books and eBooks.

Other Books You May Enjoy

If you enjoyed this book, you may be interested in these other books by Packt:

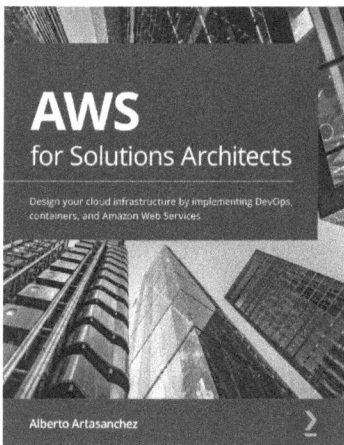

AWS for Solutions Architects

Alberto Artasanchez

ISBN: 9781789539233

- Rationalize the selection of AWS as the right cloud provider for your organization
- Choose the most appropriate service from AWS for a particular use case or project
- Implement change and operations management
- Find out the right resource type and size to balance performance and efficiency
- Discover how to mitigate risk and enforce security, authentication, and authorization

- Identify common business scenarios and select the right reference architectures for them

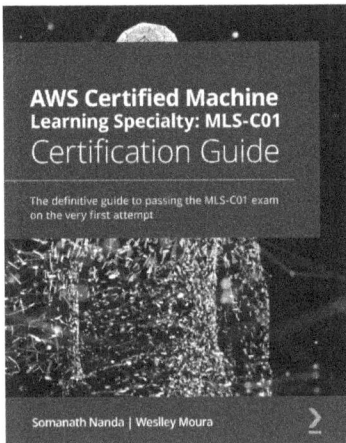

AWS Certified Machine Learning Specialty: MLS-C01 Certification Guide

Somanath Nanda, Weslley Moura

ISBN: 9781800569003

- Understand all four domains covered in the exam, along with types of questions, exam duration, and scoring

- Become well-versed with machine learning terminologies, methodologies, frameworks, and the different AWS services for machine learning

- Get to grips with data preparation and using AWS services for batch and real-time data processing

- Explore the built-in machine learning algorithms in AWS and build and deploy your own models

- Evaluate machine learning models and tune hyperparameters

- Deploy machine learning models with the AWS infrastructure

Packt is searching for authors like you

If you're interested in becoming an author for Packt, please visit `authors.packtpub.com` and apply today. We have worked with thousands of developers and tech professionals, just like you, to help them share their insight with the global tech community. You can make a general application, apply for a specific hot topic that we are recruiting an author for, or submit your own idea.

Share Your Thoughts

Now you've finished *AWS Certified Cloud Practitioner Exam Guide*, we'd love to hear your thoughts! Scan the QR code below to go straight to the Amazon review page for this book and share your feedback or leave a review on the site that you purchased it from.

`https://packt.link/r/180107593X`

Your review is important to us and the tech community and will help us make sure we're delivering excellent quality content.

Index

www.ingramcontent.com/pod-product-compliance
Lightning Source LLC
Chambersburg PA
CBHW060938210326

41598CB00031B/4662